Coastal Sedimentary Environments

Second Revised, Expanded Edition

With Contributions by

Paul B. Basan
Senior Geologist
ERCO Petroleum Services
Houston, Texas 77060
U.S.A.

Robert B. Biggs
Associate Professor
College of Marine Studies
University of Delaware
Newark, Delaware 19711
U.S.A.

Jon C. Boothroyd
Associate Professor
Department of Geology
University of Rhode Island
Kingston, Rhode Island 02881
U.S.A.

Michael J. Chrzastowski
Graduate Student
Department of Geology
University of Delaware
Newark, Delaware 19711
U.S.A.

Richard A. Davis, Jr.
Professor
Department of Geology
University of South Florida
Tampa, Florida 33620
U.S.A.

William T. Fox
Professor and Chairman
Department of Geology
Williams College
Williamstown, Massachusetts 01267
U.S.A.

Robert W. Frey
Professor
Department of Geology
University of Georgia
Athens, Georgia 30602
U.S.A.

Victor Goldsmith
Professor
Israel Oceanographic & Limnological
  Institute
Haifa, Israel and
Earth Resources Institute

University of South Carolina
Columbia, South Carolina 29208
U.S.A.

Thomas S. Hopkins
Physical Oceanographer
Brookhaven National Laboratory
Upton, Long Island, New York 11973
U.S.A.

George deVries Klein
Professor
Department of Geology
University of Illinois
Urbana, Illinois 61821
U.S.A.

John C. Kraft
Professor
Department of Geology
University of Delaware
Newark, Delaware 19711
U.S.A.

Maynard M. Nichols
Professor
Department of Geological
  Oceanography
Virginia Institute of Marine Science
Gloucester Point, Virginia 23062
U.S.A.

Alan W. Niedoroda
R. J. Brown and Associates
  of America, Inc.
2010 North Loop West—Suite 200
Houston, Texas 77018
U.S.A.

Donald J. P. Swift
Principal Research Geologist
ARCO Oil and Gas Company
Exploration and Production Research
Dallas, Texas 75221
U.S.A.

L. Donelson Wright
Professor and Head
Department of Geological
  Oceanography
Virginia Institute of Marine Science
Gloucester Point, Virginia 23062
U.S.A.

Coastal Sedimentary Environments

Second Revised, Expanded Edition

Edited by

Richard A. Davis, Jr.

With 376 Figures

Springer-Verlag
New York Berlin Heidelberg Tokyo

Richard A. Davis, Jr.
Department of Geology
University of South Florida
Tampa, Florida 33620
U.S.A.

Cover photo is of Egg Island, Alaska, taken in June, 1971, at low tide. It is courtesy of Miles O. Hayes, Research Planning Institute, Inc.

Library of Congress Cataloging in Publication Data
Main entry under title:
Coastal sedimentary environments.
    Includes bibliographies and index.
    1. Sediments (Geology) 2. Coasts. 3. Sedimentation
and deposition. I. Davis, Richard A. (Richard Albert),
1937–
QE471.2.C6   1985   551.4'57   84-23654

Typeset by Ampersand, Inc., Rutland, Vermont.
Printed and bound by Halliday Lithograph, West Hanover, Massachusetts.
Printed in the United States of America.

9 8 7 6 5 4 3 2 1

ISBN 0-387-96097-X Springer-Verlag New York Berlin Heidelberg Tokyo
ISBN 3-540-96097-X Springer-Verlag Berlin Heidelberg New York Tokyo

Preface

The zone where land and sea meet is composed of a variety of complex environments. The coastal areas of the world contain a large percentage of its population and are therefore of extreme economic importance. Industrial, residential, and recreational developments, as well as large urban complexes, occupy much of the coastal margin of most highly developed countries. Undoubtedly future expansion in many undeveloped maritime countries will also be concentrated on coastal areas. Accompanying our occupation of coasts in this age of technology is a dependence on coastal environments for transportation, food, water, defense, and recreation. In order to utilize the coastal zone to its capacity, and yet not plunder its resources, we must have extensive knowledge of the complex environments contained along the coasts.

The many environments within the coastal zone include bays, estuaries, deltas, marshes, dunes, and beaches. A tremendously broad range of conditions is represented by these environments. Salinity may range from essentially fresh water in estuaries, such as along the east coast of the United States, to extreme hypersaline lagoons, such as Laguna Madre in Texas. Coastal environments may be in excess of a hundred meters deep (fjords) or may extend several meters above sea level in the form of dunes. Some coastal environments are well protected and are not subjected to high physical energy except for occasional storms, whereas beaches and tidal inlets are continuously modified by waves and currents.

Because of their location near terrestrial sources, coastal environments contain large amounts of nutrients. The combination of this nutrient supply with generally shallow water gives rise to a diverse and large fauna and flora. Coastal areas also serve as the spawning and nursery grounds for many open-ocean organisms. Many species in coastal environments are of great commercial importance, such as clams, oysters, shrimp, and many varieties of fish. It is certain that we need better management of these resources. We also need to develop more and improved methods of cultivating these environments.

All of the above examples point out the need for a fundamental knowledge of modern coastal environments. An almost equally important aspect of coastal environments is their role in the geologic record. In order to properly interpret the relationships and depositional environments among various facies preserved in the rock record it is necessary to have a thorough understanding of modern environments (Law of Uniformitarianism). There is a great deal of economic significance to ancient coastal environments. Water, oil, and gas occur in large quantities in various reservoir sands of beach and dune complexes of ancient barrier island and inner shelf sands. The huge clastic wedge deposits of ancient deltaic environments are also a source of petroleum materials and coal.

All of the above examples indicate various reasons that a comprehensive knowledge of coastal environments is necessary for a geologist, engineer, oceanographer, or coastal manager, or for other persons involved in the coastal zone. This book is designed to provide such a background. Most of the important coastal environments are included. The discussions contain such descriptive aspects as morphology and sediment distribution but also emphasize physical processes and their interactions with the sediments and sediment body morphology. Primary consideration is given to the principles involved and to general considerations but numerous case history examples are included.

Our efforts here represent an unusual attempt to produce a text for student use in that each chapter is authored by a different individual or individuals. The subject matter makes this approach a practical one in that the discussion of each of the coastal environments is presented by a specialist who has considerable experience in the environment being discussed.

Each of the chapters that covers one of the coastal environments can stand alone. As a result, these chapters can be ordered to suit the reader or the instructor. The depth of treatment for each of these chapters shows some variation, in part because of the existing literature. For example, there are many volumes devoted to beaches, estuaries, and deltas in addition to the journal literature. Such coastal environments as dunes and marshes, in contrast, have not received the same coverage in the literature, especially by way of thorough summaries. These environments therefore have been treated in more detail than other environments. The chapters on dunes and marshes are probably the most comprehensive summaries available in the literature.

The last two chapters both involve applications of basic principles to the study of coastal environments or coastal systems. The chapter on sequences, especially Holocene stratigraphic sequences, provides an excellent framework for the reader to gain an understanding of Walther's Law as well as coastal systems. Application of computer techniques to coastal sedimentation has been an important and valuable tool. The discussion of this topic considers various approaches to modeling, including conceptual, simulation, and predictor types. No consideration is given to programming or computer language; only the principles of modeling and various examples are discussed.

Addendum

Much research has taken place in coastal geology since the publication of *Coastal Sedimentary Environments*. This additional research and the need to expand the coverage by adding chapters has stimulated this second edition. Each of the original authors has updated and modified his respective chapter. In two of the chapters, co-authors were taken on to help accomplish this task. Especially noticeable is the expansion of the chapters on coastal bays (estuaries) and on coastal sequences. Two new chapters have been added to make the text more comprehensive; one on intertidal flats and one on the shoreface. The need to include the former is obvious, but inclusion of the shoreface may not be. This portion of the inner shelf interfaces with various coastal environments and is integral to a complete understanding of process-response systems operating on the coast.

In preparing this revised edition, we have tried to incorporate suggestions made in the various published reviews of the first edition. Although the title implies comprehensive coverage of coasts, virtually all of the discussion is devoted to terrigenous sedimentary environments although basic principles apply to carbonate environments and sediments as well.

Authors

Each of the contributors is an active researcher in the coastal zone, with specialization in the environment for which he is responsible in the book. The authors represent a rather broad spectrum of backgrounds, with research experience being concentrated on the Atlantic and Gulf coasts of the United States, although most have some foreign experience as well.

Dr. Paul B. Basan is a sedimentologist and paleontologist currently involved in research applied to the energy industries. He conducted his doctoral studies on sedimentation in salt marshes under the supervision of his co-author, Dr. R. W. Frey, at the University of Georgia. In addition to his research on salt marshes, Dr. Basan has authored numerous articles on trace fossils.

Dr. Robert B. Biggs is a geological oceanographer with the College of Marine Studies and the Department of Geology at the University of Delaware. Dr. Biggs has studied extensively the trace element geochemistry of estuarine bottom and suspended sediments, quantification of the sources and sinks of suspended organics and organic matter in estuarine waters, environmental effects of dredging and dredge spoil disposal, and the Holocene and Pleistocene coastal features of the Delmarva Peninsula.

In addition to his research on coasts, Dr. Jon C. Boothroyd also is active in research on glaciofluvial environments and on the geomorphology and surface processes on Mars. Dr. Boothroyd's coastal research is now directed to the Alaskan and Rhode Island coasts. Most of his research has been supported by the Office of Naval Research, Coastal Engineering Research Center, Rhode Island Sea Grant Program, and the National Atmospheric and Space Administration.

Michael Chrzastowski is a doctoral candidate in geology at the University of Delaware under the supervision of his co-author, Dr. John C. Kraft. His current research concerns the sedimentology and stratigraphy of a coastal lagoon which has been influenced by human activity. Mr. Chrzastowski has had experience with the NOAA National Ocean Survey in Alaska, Hawaii, California, and Washington.

The editor, Dr. Richard A. Davis, Jr. is interested in coastal depositional systems with emphasis on barrier island complexes. He has published numerous research articles with W. T. Fox and has written *Principles of Oceanography* and *Depositional Systems* and has edited *Beach and Nearshore Sedimentation* and *Wave Dominated Coasts*. Major support for his coastal research has come from the Office of Naval Research, the Coastal Engineering Research Center and the Florida Sea Grant Program.

Dr. William T. Fox combines expertise in computer modeling with much experience in beach and nearshore dynamics. Much of his work has been with R. A. Davis and has been funded by the Office of Naval Research. Dr. Fox has recently authored *The Seas Edge*, a popular book on coastal environments.

Dr. Robert W. Frey is a paleoecologist who specializes in animal–sediment relationships and ichnology. He has 15 years experience on recent environments of the Georgia coast and shelf. In addition to his many research articles, Dr. Frey edited a comprehensive volume, *Trace Fossils*. Much of his research has been supported by the National Science Foundation.

After spending nearly 10 years on the Atlantic coast of the United States, Dr. Victor Goldsmith has relocated to the eastern Mediterranean coast. He has had much experience in dune sedimentation and beach and nearshore dynamics. Most of Dr. Goldsmith's research on dunes was funded by the Coastal Engineering Research Center.

Dr. Thomas S. Hopkins has studied continental shelf circulation along coasts of the United States, Peru, and the Mediterranean Sea. He specializes in observations and modeling of shelf processes, particularly, as related to geological and biological problems. At the Brookhaven National Laboratory, Dr. Hopkin's research has been supported by the Department of Energy and the National Oceanographic and Atmospheric Administration.

Dr. George deVries Klein's research interests are in coastal and deep-water marine sediment transport and deposition. He has done considerable work on intertidal sand bodies in the Bay of Fundy. Dr. Klein has participated on Legs 30 and 58 of the Deep Sea Drilling Project. His research has been supported primarily by the National Science Foundation and the Office of Naval Research. Dr. Klein has authored or edited *Clastic Tidal Facies, Sandstone Depositional Models and Exploration for Fossils Fuels, Holocene Tidal Sedimentation* and *Sedimentary Processes: Processes of Detrital Sedimentation.*

Holocene and Pleistocene coastal sequences have been studied by Dr. John C. Kraft for nearly two decades. Although most of his research has been along the North Atlantic coast of the United States, other areas have also been included with recent emphasis in the eastern Mediterranean Sea and Hawaii. Much of Dr. Kraft's research has been supported by the Office of Naval Research, the National Science Foundation and the Delaware Sea Grant Program.

Dr. Maynard M. Nichols is a geological oceanographer with specialization in estuarine sedimentation. His numerous publications on estuarine processes emphasize the dynamics of the turbidity maximum, responses to river flooding and estuarine–shelf interrelationships. Most of Dr. Nichols' research has been supported by the Army Research Office and the Environmental Protection Agency.

After completing his graduate work, Dr. Alan W. Niedoroda was on the faculty of the University of Massachusetts for several years conducting research on coastal dynamics. He is now a consulting oceanographer specializing in coastal and shelf, sediments and processes. Research from which this chapter was formulated was supported by the Office of Naval Research and the National Oceanographic and Atmospheric Administration.

Dr. Donald J. P. Swift spent many years in academics and in federal research agencies prior to assuming his present position. His research emphasis has been on shelf sedimentation. Dr. Swift has studied sediment dynamics on the Atlantic coast of North America, in the Bay of Fundy, on the Spanish Mediterranean coast and on the Argentine coast. He is co-author of *Shelf Sediment Dynamics, Marine Sediment Transport and Environmental Management* and of the reprint volume, *Coastal Sedimentation*. Dr. Swift's research has been supported by the National Science Foundation, the Department of Energy, the Corps of Engineers and a consortium of petroleum research companies.

Dr. L. Donelson Wright has investigated nearly every major delta in the world during his association with the Coastal Studies Institute of Louisiana

State University. He also investigated beach and nearshore dynamics throughout the coast of Australia and is now conducting similar studies on the mid-Atlantic coast of the United States. Most of Dr. Wright's research has been supported by the Office of Naval Research and the Australian Research Grants Committee.

Richard A. Davis, Jr.

Table of Contents

3.   Intertidal Flats and Intertidal Sand Bodies
George deVries Klein

4.   Coastal Salt Marshes
Robert W. Frey and Paul B. Basan

5.   Coastal Dunes
Victor Goldsmith

6.   Beach and Nearshore Zone
Richard A. Davis, Jr.

7. Tidal Inlets and Tidal Deltas
Jon C. Boothroyd

8. The Shoreface
Alan W. Niedoroda, Donald J. P. Swift, and Thomas S. Hopkins

9. Coastal Stratigraphic Sequences
John C. Kraft and Michael J. Chrzastowski

10. Modeling Coastal Environments
William T. Fox

Acknowledgments

A book of this type is not only the result of the efforts of its authors and editor, but also in large measure of various types of input from many other sources. These include many researchers whose results have been published elsewhere, numerous colleagues and students who are, or have been, associated with the authors, and the many agencies who have supported the authors' research efforts. In addition, several people have given freely of their time and expertise by way of reviewing chapters. Numerous people and publishers have given permission to reproduce illustrations from previous publications: their generosity is greatly appreciated. Specific acknowledgments are as follows:

Chapter 1. Most of the research on which this chapter is based was supported by the Office of Naval Research through the Coastal Studies Institute, Louisiana State University and by the Australian Research Grants Committee through the University of Sydney. Support of the preparation of this revised edition was provided by the Virginia Institute of Marine Science. The author is especially grateful to J. M. Coleman and B. G. Thom for cooperation in the field, in numerous discussions of ideas and concepts, and for collaboration on many of the research papers underlying this chapter.

Chapter 2. The authors' original research which contributed to this chapter was supported by the Army Research Office and the Environmental Protection Agency through the Virginia Institute of Marine Science and the University of Delaware Sea Grant Program (NA80AA-D-00106, Office of Sea Grant, NOAA, U. S. Dept. of Commerce). Additionally, this chapter synthesizes research results of many estuarine studies, notably those of G. P.

Allen, K. Kranck R. Gibbs, D. Haven, R. Krone, I. N. McCave, H. Postma, L. Clifton, and P. Scruton. Kay Stubblefield of VIMS drafted the figures.

Chapter 3. Financial support for the author's research was provided by the National Science Foundation through its grants GA-407, GA-1583, and GA-21141, and the Office of Naval Research through its grant Nonr-266(84). Numerous students and colleagues have provided assistance, suggestions and advice during the course of this work.

Chapter 4. Financial support for the authors' research has been provided by the National Science Foundation (GA-22710), the Georgia Sea Grant Program and a NSF-Institutional Grant from the University of Georgia. The critical assistance of several people is acknowledged; E. S. Belt, J. D. Howard, V. J. Hurst, K. B. MacDonald, R. Pestrong, P. R. Pinet, F. B. Phleger, L. R. Pomeroy, A. C. Redfield, and R. G. Wiegert. L. Gassert, and G. K. Maddock aided in the literature search.

Chapter 5. Most of the author's original research on Cape Cod was supported by the Coastal Engineering Research Center (Contract DACW 82-62-C-0004). The Virginia Institute of Marine Science and Virginia Sea Grant (Contract 04-5-158-49), have also provided some support. C. A. M. King and R. J. Byrne reviewed the first edition, and H. Tsoar reviewed and provided many useful suggestions for this revision.

Chapter 6. The author's research on beach and nearshore environments has been funded by the Office of Naval Research (Contracts 388-092 and 388-136) in cooperation with W. T. Fox and by the Coastal Engineering Research Center (Contracts DACW72-70-C-0037 and DACW72-73-C-0003). Numerous stimulating discussions with J. C. Boothroyd, M. O. Hayes, D. Nummedal, and H. E. Clifton have been very beneficial. A large number of graduate students has also provided significant input.

Chapter 7. Much of the discussion in this chapter has its roots in the association of the author with M. O. Hayes and Coastal Research groups at the universities of Massachusetts and South Carolina. Funds were provided through the Office of Naval Research (Contract N00014-67-A-0230-001), the Coastal Engineering Research Center (Contract DACW-72-70-C-0029) and the Rhode Island Sea Grant Program. Numerous colleagues and students have assisted in various aspects of this work.

Chapter 8. The author's shoreface research has been supported largely by the Office of Naval Research, Department of Energy, and NOAA. Field assistance was provided by H. Peg, D. Battisti, S. Wall, L. Wall, J. Collins, J. Leonard, G. Bowers, B. Brenninckmeyer, E. Divit, and D. Carlson. Various logistic support was provided by Southampton College, Boston

College, and Woods Hole Oceanographic Institution. ARCO Oil & Gas Company provided support for the preparation of the manuscript.

Chapter 9. Research on Holocene sequences has been supported by the Delaware Sea Grant College Program; State of Delaware, Dept. of Natural Resources and Environmental Control; Office of Naval Research (Contract N00014-69-A0407); the National Science Foundation (GP-5604); and the University of Delaware Research Foundation. Many graduate students at the University of Delaware have contributed both the data and the concepts incorporated in this chapter.

Chapter 10. The author is grateful to the late W. C. Krumbein who stimulated his interest in computer applications and provided his initial training at Northwestern University. Most of the author's research in this area was supported by the Office of Naval Research (Contracts N00014-69-C-0151 and N0014-77-C-0151). In addition to the assistance of numerous students, much input has been provided by colleagues, especially R. A. Davis, Jr. and P. D. Komar.

Finally, gratitude is expressed to the personnel of Springer-Verlag, Inc., New York for their patience, cooperation, and assistance in preparation and production of this revised edition. Appreciation is also extended to Wanda McClelland who retyped several of the revised chapters and to Iris Rose who typed much of the correspondence associated with the second edition.

1

River Deltas*

L. D. Wright

Introduction

Deltas result from interacting fluvial and marine forces. Since ancient times, river deltas have been of fundamental importance to civilization. Owing to their early significance as agricultural lands, deltas received attention from such scholars as Homer, Herodotus, Plato, and Aristotle. Today, deltaic accumulations play a role in accommodating the world's energy needs. Ancient deltaic sediments provide source beds and reservoirs for a large fraction of the known petroleum reserves.

The term *delta* was first applied by the Greek historian Herodotus, circa 450 B.C., to the triangular alluvial deposit at the mouth of the Nile River. For the purposes of this discussion, deltas are defined more broadly as coastal accumulations, both subaqueous and subaerial, of river-derived sediments adjacent to, or in close proximity to, the source stream, including the deposits that have been secondarily molded by waves, currents, or tides. By this definition, deltas include all delta plains, regardless of plan-view shape or of the suite of individual landforms present. Because the different processes that control delta development vary appreciably in relative intensity on a global scale, delta-plain landforms span nearly the entire spectrum of coastal features and include distributary channels, river-mouth bars, open and closed interdistributary bays, tidal flats, tidal ridges, beaches, beach ridges, dunes and dune fields, and swamps and marshes.

The range of deltaic environments is equally broad. Some deltas occur along coasts that experience negligible tides and minimal wave energy; others

*VIMS Contribution No. 1116.

have formed in the presence of extreme tide ranges or large waves. Deltas may accumulate in the humid tropics, where vegetation is abundant and biologic and chemical processes are of prime importance, or they may form in arid or arctic environments, where biologic activity is subdued. Despite these pronounced environmental contrasts, all actively forming deltas have, by definition, at least one common attribute: A river supplies clastic sediment to the coast and inner shelf more rapidly than it can be removed by marine processes.

Occurrence and Distribution of Deltas

Deltas may occur wherever a stream debouches into a receiving basin. This statement holds whether the receiving basin is an ocean, gulf, inland sea, bay, estuary, or lake. Consequently, deltas of various sizes can be found throughout the world. Table 1-1 gives the locations of some of the world's largest modern deltas. In addition to these major deltas, literally thousands of minor deltas are distributed over all the world's coasts.

Although major deltas occur on the shores of virtually all seas and at all latitudes except at the poles, there are certain requirements for their occurrence that are not met in many parts of the world. The first prerequisite for a significant deltaic accumulation is the existence of a major river system that carries substantial quantities of clastic sediment. For such a system to exist, there must be a large drainage basin within which precipitation accumulates, sediments are supplied by erosion, and individual tributaries coalesce to create a larger trunk stream. The sediment-water discharge from the drainage basin is then transported to the coast by way of the alluvial valley that confines the trunk stream. The general spatial relationships between the drainage basin, alluvial valley, and delta are illustrated in Figure 1-1.

Whether or not a river is sufficiently large and transports sufficient quantities of sediment to produce a major delta depends largely on the nature of the drainage basin. Drainage basin climate, geology, relief, and area are all critical determinants of river discharge. Unless the drainage basin experiences signifiant precipitation for at least part of the year, no river can arise. Lithologic resistance, combined with drainage basin pedology and vegetation, determines the rate at which sediments can be eroded from the basin to be supplied to the river. Through gravity, vertical relief provides the energy to erode the underlying rocks and transport the resulting sediment to the water accumulating in the streams and rivers of the basin.

Of all the requisites for the occurrence of a major river system, probably the most restrictive is the necessity for a large catchment area. The mean basin area of the rivers listed in Table 1-1 is only slightly less than 10^6 km^2. The drainage basin of the Mississippi, for example, covers 41% of the

Table 1-1. Some major deltas and their locations.

River	Land mass	Receiving body of water	Coordinates Lat.	Coordinates Long.
Amazon	South America	Atlantic Ocean	0	52°W
Burdekin	Australia	Coral Sea	19°S	147°E
Chao Phraya	Asia	Gulf of Siam	13°N	101°E
Colville	North America	Beaufort Sea	71°N	151°W
Danube	Europe	Black Sea	43°N	28°E
Dneiper	Asia	Black Sea	47°N	32°E
Ebro	Europe	Mediterranean Sea	41°N	02°E
Ganges–Brahmaputra	Asia	Bay of Bengal	32°N	90°E
Grijalva	North America	Gulf of Mexico	18°N	93°W
Huang Ho	Asia	Yellow Sea	37°N	118°E
Indus	Asia	Arabian Sea	24°N	67°E
Irrawaddy	Asia	Bay of Bengal	16°N	94°E
Klang	Asia	Straits of Malacca	3°N	101°E
Lena	Asia	Laptev Sea	73°N	125°E
Mackenzie	North America	Beaufort Sea	68°N	139°W
Magdalena	South America	Caribbean Sea	12°N	69°W
Mahakam	Borneo	Makassar Strait	1°S	117°E
Mekong	Asia	South China Sea	10°N	107°E
Mississippi	North America	Gulf of Mexico	30°N	90°W
Niger	Africa	Gulf of Guinea	4°N	7°E
Nile	Africa	Mediterranean Sea	32°N	31°E
Ord	Australia	Timor Sea	16°S	120°E
Orinoco	South America	Atlantic Ocean	8°N	62°W
Paraná	South America	Atlantic Ocean	33°S	58°W
Pearl	Asia	South China Sea	22°N	113°E
Pechora	Europe	Barents Sea	68°N	54°E
Po	Europe	Adriatic Sea	44°N	12°E
Purari	Papua New Guinea	Gulf of Papua	8°S	144°E
Red	Asia	Gulf of Tonkin	21°N	107°E
Sagavanirktok	North America	Beaufort Sea	70°N	148°W
São Francisco	South America	Atlantic Ocean	11°S	37°W
Senegal	Africa	Atlantic Ocean	17°N	16°W
Shatt-al-Arab	Asia	Persian Gulf	30°N	49°E
Shoalhaven	Australia	Pacific Ocean	35°S	151°E
Tana	Africa	Indian Ocean	2°S	42°E
Volga	Europe	Caspian Sea	47°N	48°E
Yangtze	Asia	East China Sea	32°N	122°E

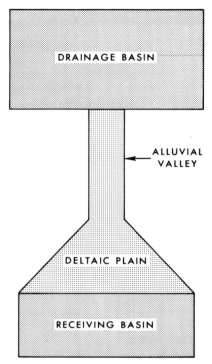

Figure 1-1. Major components of a river
system (from Coleman and Wright, 1971).

continental United States, or an area of 3.3×10^6 km², and the Amazon
(Brazil) arises from a basin 5.9×10^6 km² in area. Large, high-order
drainage basins of this type consist of numerous smaller, low-order basins
and result when tributaries join downstream to form a single large stream.
Because the process of tributary joining requires both relatively long periods
of time and moderately long distances, large river deltas normally are not
present along coasts that are highly active tectonically or that are in very
close proximity to drainage divides. Accordingly, the occurrence of major
river systems is also closely dependent on global tectonics. Inman and
Nordstrom (1971) have produced a macroscale classification of the world's
coasts in which they distinguish between the following tectonic classes and
subclasses (Inman and Nordstrom, 1971, p. 9):

1. Collision coasts
   a. *Continental collision coasts*, that is, collision coasts involving the
      margins of continents, where a thick plate collides with a thin plate
      (e.g., west coasts of the Americas)

    b. *Island arc collision coasts*, that is, collision coasts along island arcs, where a thin plate collides with another thin plate (e.g., the Philippines and the Indonesian and Aleutian island arcs)

2. Trailing-edge coasts
    a. *Neo-trailing-edge coasts*, that is, new trailing-edge coasts formed near beginning separation centers and rifts (e.g., the Red Sea and Gulf of California)
    b. *Afro-trailing-edge coasts*, that is, the opposite coast of the continent is also trailing, so that the potential for terrestrial erosion and deposition at the coast is low (e.g., Atlantic and Indian Ocean coasts of Africa)
    c. *Amero-trailing-edge coasts*, that is, the trailing-edge of a continent with a collision coast; and therefore "actively" modified by the depositional products and erosional effects from an extensive area of high interior mountains (e.g., east coasts of the Americas)

3. Marginal sea coasts, that is, coasts fronting on marginal seas and protected from the open ocean by island arcs (e.g., Vietnam, southern China, and Korea)

    In connection with their classification, Inman and Nordstrom (1971) assembled data on 58 major rivers having drainage areas in excess of 10^5 km². Of these, 46.6% were found to debouch along Amero-trailing-edge coasts (class 2 c), 34.5% along marginal seacoasts (class 3), 8.6% along Afro-trailing-edge coasts (class 2 b), and 1.7% along neo-trailing-edge coasts (class 2 a). Only 8.6% of the rivers examined [the Columbia (United States), Colorado (United States), Frazer (Canada), Ebro (Spain), and Po (Italy) Rivers] enter the sea along collision coasts, where tectonic activity is high and drainage divides are characteristically close to the sea.

    The absolute volumes of deltaic accumulations (subaqueous and subaerial) can be expected to vary with the magnitudes of sediment fluxes to the sea. Milliman (1981), and Milliman and Meade (1983) have synthesized data (mainly from Lisitzen, 1972, and Holeman, 1968) on the discharge of river-borne suspended load into the world ocean. Figure 1-2 (from Milliman, 1981) shows the geographic distribution of suspended sediment discharges. It is worth noting that by far the greatest sediment discharges are associated with Asian rivers.

General Characteristics of Deltas and Deltaic Environments

Deltas vary immensely in terms of morphologic suites, overall geometry, sediment properties, and dynamic environment. Consequently, with regard to these factors very few generalizations can be made until the causes and scope

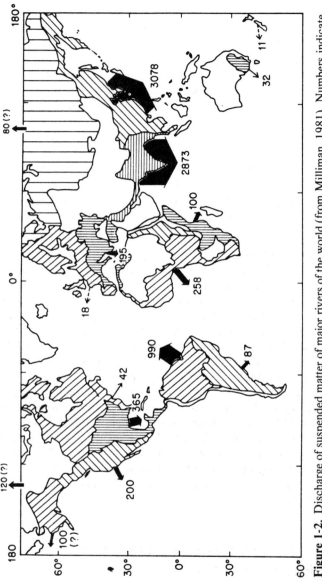

Figure 1-2. Discharge of suspended matter of major rivers of the world (from Milliman, 1981). Numbers indicate annual average input in tons $\times 10^6$.

of the variability are considered in the sections to follow. Nevertheless, there are a few general features that are almost universal among deltas and a few general processes that all deltas experience, although in varying intensities. These common denominators are the subjects of this section; in the following sections, the processes of deltaic development and the causes of deltaic variability are considered in more detail.

Delta Components

On a gross scale, delta plains can be subdivided into the basic physiographic zones illustrated in Figure 1-3. Generally, every delta consists of a subaqueous delta and a subaerial delta, even though the relative areas of these may vary considerably. The subaqueous delta lies below the low-tide water level; it is the foundation on which progradation of the subaerial delta must proceed. In general, the subaqueous delta is characterized by seaward fining of sediments, sand being deposited nearest the river mouths and fine silts and clays settling farther offshore from suspension in the water column. The seawardmost portion of the subaqueous delta is composed of the finest

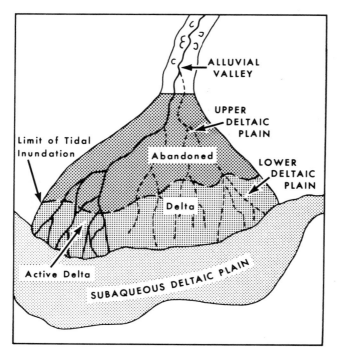

Figure 1-3. Components of a delta (from Coleman and Wright, 1971).

material deposited from suspension (primarily clays) and is referred to as the prodelta. The prodelta clays grade upward and landward into the coarser silts and sands of the delta front. The depositional features of the delta front are nearly as varied as deltas themselves and depend on the associated dynamic environments. In addition to a variety of types of distributary-mouth accumulations, the delta front may include such features as linear tidal ridges or shoreface beach deposits.

The subaerial delta is that portion of the delta plain above the low-tide limit. It often experiences a much broader range of processes than does the subaqueous delta and consequently displays a more complex assemblage of depositional forms and environments. However, in any one time frame, it is normally much thinner vertically than the subaqueous delta. In many cases it may be little more than a veneer capping a much thicker sequence of prodelta clays and delta-front sands. The subaerial delta typically consists of a lower delta plain and an upper delta plain. The lower delta plain lies within the realm of riverine–marine interaction and extends landward to the limit of tidal influence. The upper delta plain is the older portion of the subaerial delta, above significant tidal or marine influence. It is basically a seaward continuation of the alluvial valley and is dominated by riverine depositional processes. Salt- or brackish-water vegetation and fauna are not present within the upper delta plain. The landward extent of the lower delta plain relative to that of the upper delta plain depends largely on tidal range and the mean seaward slope of the delta surface. The lower delta plain is most extensive when tide range is large and the seaward gradients of the river channel and delta surface are low.

In many cases the subaerial and subaqueous deltas can also be subdivided into active and abandoned zones. The active delta plain is the accreting portion occupied by functioning distributary channels. The abandoned delta plain results when the river changes its lower course causing a shift in the locus of river-mouth sedimentation. The coastline of the abandoned depositional surface is then reworked by marine processes. The abandoned delta may prograde, provided that marine forces continue to introduce sediment. It may remain stable, or it may enter into a destructional phase, particularly if wave reworking is pronounced because of subsidence and rapid sediment compaction. In the case of the destructional phase, the delta shoreline will undergo a landward transgression, and the result will usually take the form of a coastal barrier, beach, or dune complex.

The Active Delta

The zones of active deltaic sedimentation and the associated processes are the primary concerns of this chapter. An active delta consists of one or more river-mouth systems to which prodelta and delta-front deposition are directly coupled, a distributary network, interdistributary and distributary-margin deposits, and a delta shoreline.

The river mouth is the most fundamental element of a delta system because it is the dynamic dissemination point for sediments that contribute to delta progradation. In the section that follows, river-mouth processes are examined in more detail; however, for the present the discussion is confined to the general descriptive aspects. The universal process characteristic of all river mouths was succinctly stated by G.K. Gilbert in 1884, p. 104: " . . . the process of delta formation depends almost wholly on the following law: the capacity and competence of a stream for the transportation of detritus are increased and diminished by the increase and diminution of the velocity." When a stream discharges into a receiving basin, its momentum is dispersed by its interaction with ambient sea water. The result is deceleration of the effluent, consequent loss of sediment-transporting ability, and deposition. The actual rate of deceleration and pattern of effluent dispersion are highly varied and depend on numerous factors, but, overall, there is a progressive seaward decrease in the concentration and grain size of sediments transported by the effluent. Figure 1-4, which is from Scruton's (1960) discussion of Mississippi Delta formation, illustrates the general process of river-mouth sedimentation and seaward fining of sediments. Except in cases where strong tidal currents dominate over effluent processes, the most rapid deposition and the deposition of the coarsest material take place a short distance from

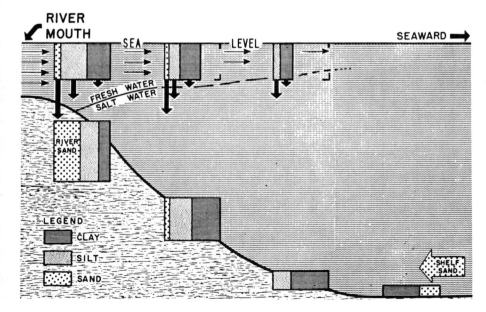

Figure 1-4. Deposition and sorting of sediments at a river mouth (from Scruton, 1960).

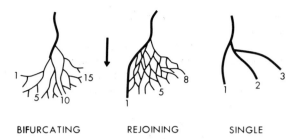

BIFURCATING   REJOINING   SINGLE

Figure 1-5. Three distributary network patterns (from Coleman and Wright, 1975).

the river mouth, on the upper parts of the delta front. At least a significant fraction of this deposition is from bed load. Toward the seaward margins of the delta front, the percentage of sand and the influence of bed shear diminish rapidly. Beyond the delta front, only fine silts and clays remain in suspension for deposition on the prodelta.

If the sediments deposited seaward of the river mouth accumulate faster than subsidence or removal of sediments by marine processes can take place, the prodelta, delta front, and eventually the river mouth itself prograde. As the system prograde, subaqueous natural levees form atop the uppermost portions of the delta front as incipient seaward extensions of the distributary banks; with continued accretion, subaqueous levees attain subaerial expression and the distributary channel is extended accordingly in length. If the channel is highly stable or confined, it may prograde seaward as a single digitate distributary; or it may bifurcate successively downstream; or it may alternately bifurcate and rejoin, to create a complex network. Three types of distributary network patterns are shown in Figure 1-5. Bifurcations tend to be most frequent when offshore slopes are flat or when sediment load consists of a high proportion of coarse material. In general, as the number of bifurcations increases, the width of the active subaerial delta and the width and continuity of the delta front increase, but the efficiency of the distributaries decreases. Quantitative descriptions of distributary networks have been offered by Smart and Moruzzi (1972).

The depositional surfaces marginal to and between the distributary channels account for most of the total area of the subaerial delta. The interdistributary and distributary-margin regions may be occupied by a multitude of coastal features. Immediately adjacent to the channels, natural levees, overbank splays, and associated minor channel systems are probably the most common depositional forms. Farther from the channel, in the interdistributary areas, the prevailing forms are determined by the dynamic environment. Coleman and Wright (1971) have described most of the more commonly occurring interdistributary landscapes. Examples of various interdistributary landform suites are illustrated in Figures 1-6 through 1-11.

Figure 1-6. ERTs satellite image of the Mississippi Delta (from Wright *et al.*, 1974).

Figure 1-7. Vertical aerial photo of the Colville Delta showing thaw lakes, patterned ground, and distributary channels (from Wright *et al.*, 1974).

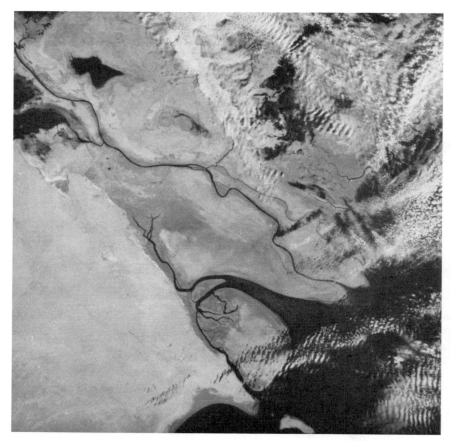

Figure 1-8. ERTs satellite image of the Shatt-al-Arab Delta, showing barren interdistributary flats (from Wright *et al.*, 1974).

Rapid distributary progradation, coupled with rapid subsidence and low nearshore wave energy, frequently results in the occupation of inter-distributary areas by shallow open or closed bays (Figure 1-6). If subsidence is slow, the bays may become completely infilled with marsh or mangrove swamp. In many arctic deltas, numerous freshwater lakes prevail (Figure 1-7). High tide range and arid climate often yield interdistributary evaporite or barren flats (Figure 1-8) separating river channels and tidal creeks and gullies. Intricate networks of tidal creeks, separated by densely vegetated mudflat surfaces (Figure 1-9), are common features of tide-influenced deltas in moister climates. Periodic high wave energy, alternating with rapid deltaic progradation, frequently produces interdistributary plains of broadly spaced, chenier-like beach ridges that are separated by wide tidal flat or marsh surfaces (Figure 1-10). Stronger and more persistent wave energy forms

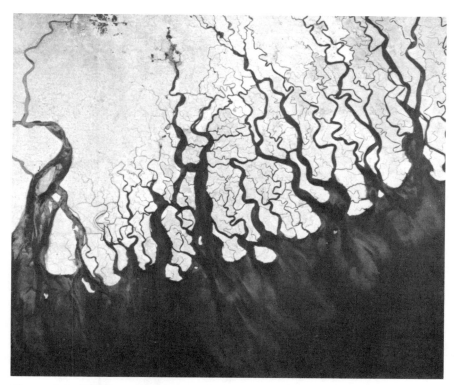

Figure 1-9. ERTs satellite image of the mouths of the Ganges–Brahmaputra, showing intricate tidal channel networks (from Wright *et al.*, 1974).

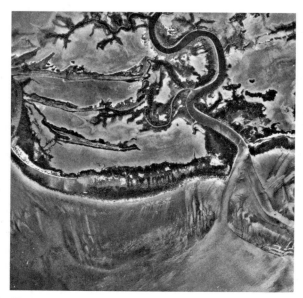

Figure 1-10. Vertical aerial photo showing intertidal flats, beach ridges, and barren supratidal flats in the Burdekin Delta (from Wright *et al.*, 1974).

Figure 1-11. Aerial photo of the São Francisco Delta.

continuous sandy plains of successive closely spaced beach ridges or dunes (Figure 1-11). Other types of interdistributary features include fields of eolian riverbank dunes and interdistributary highs of ancient rock.

The delta shoreline is the temporary land–sea interface and seaward boundary of the subaerial delta. Compared with the somewhat more stable shorelines of coastal barriers or beach systems, delta shorelines are ephemeral and unstable with respect to their position. In the case of rapidly prograding deltas experiencing minimal wave energy, the delta shoreline may be crenulate and highly irregular, consisting of mudflats or marsh (Figure 1-6). In higher energy situations, where powerful waves have slowed the advance of the delta and redistributed the deltaic sediments, the delta shoreline commonly tends to be straight or gently arcuate and to take the form of a sandy beach, such as prevails along the coast of the São Francisco Delta (Brazil)(Figure 1-11).

Delta-Forming Processes

Deltaic depositional and morphologic patterns are produced by a complexity of interacting dynamic factors, of which the most salient are river-mouth diffusion and deceleration processes; tidal transport processes; waves and their modification near shore; coastal currents; deformational processes; and a variety of geologic, biologic, and climatologic factors (Wright, 1977a). Before the significance of deltaic configurations and sedimentary sequences and their variability can be appreciated, the contribution of each of these dynamic factors must be understood.

River-Mouth Processes

Factors Affecting Effluent Type

A river-mouth system consists of an outlet, at which point the seaward-flowing river water leaves the confines of the channel banks and an associated river-mouth bar deposit seaward of the outlet. The river-mouth geometry and bar topography together make up a single unit that both influences and is influenced by effluent dynamics. The nature of this closely interacting morphodynamic system is determined by riverine flow characteristics, density contrasts between issuing and ambient water, water depths, bottom slope seaward of the mouth, tidal range and the degree to which tidal currents dominate within the lower river channel, and the ability of waves and other forces to obstruct the outlet. There have been several systematic and theoretical studies of river-mouth processes, and for in-depth study the interested reader is referred to the following: Bates (1953), Axelson (1967), Bonham-Carter and Sutherland (1968), Borichansky and Mikhailov (1966), Mikhailov (1966, 1971), Bondar (1970), Takano (1954a,b, 1955), Kashiwamura and Yoshida (1967, 1969, 1971), Jopling (1963), Waldrop and Farmer (1973), Scruton (1956, 1960), Wright (1970, 1971, 1977a,b), and Wright and Coleman (1971, 1974). River-mouth processes have been reviewed by Wright (1977a).

The various studies referred to above suggest that river-mouth effluent diffusion and sediment dispersion patterns depend on the relative roles of three primary forces: (1) the inertia of issuing river water and associated turbulent diffusion; (2) friction between the effluent and the bed immediately seaward of the mouth; and (3) buoyancy resulting from density contrasts between issuing and ambient fluids (Wright and Coleman, 1974; Wright, 1977a). Although all three forces are normally operative to some extent at all river mouths, the relative significance of each depends on outflow velocity, density stratification, and outlet geometry. When outflow velocities are high, depths immediately seaward of the mouth are relatively large, density contrasts are negligible, inertial forces dominate, and the effluent spreads and diffuses as a turbulent jet. In this effluent jet, the activity of turbulent eddies causes exchange of fluid and momentum between the effluent and the ambient water of the receiving basin. When water depths seaward of the mouth are shallow, turbulent diffusion becomes restricted to the horizontal, and bottom friction plays a major role in causing effluent deceleration and expansion. If sea water enters the mouth beneath seaward-flowing fresh water, vertical density stratification results, and the buoyancy of the river water becomes dominant; the effluent spreads and thins as a relatively discrete layer. Under these conditions the issuing river water is not in direct contact with the bottom and the effects of friction are minimized. Bates (1953) referred to outflows from river mouths having negligible density contrasts (for example, freshwater streams entering freshwater lakes) as *homopycnal* (i.e., having equal density), as opposed to *hypopycnal outflows*, which are characterized by buoyant river water issuing into denser basin water. In addition, he distinguished a third effluent type, *hyperpycnal outflows*, in which the issuing water is denser than, and plunges beneath, the basin water. Outflows from hypersaline lagoons or from rivers having extreme suspended-load concentrations (e.g., the Huang Ho River of China) fall into the last category. Unfortunately, however, very little is known at present about this relatively uncommon effluent type.

In the case of hypopycnal outflows, buoyant forces can be highly effective in suppressing the role of convective inertia and in inhibiting turbulence. The degree to which the effluent behaves as a turbulent or buoyant jet is in large part dependent on the densimetric Froude number, F', given by:

$$F' = U^2/(\gamma g h'), \qquad (1\text{-}1)$$

where U is the mean outflow velocity of the upper layer in the case of stratified flows, and:

$$\gamma = 1 - (\rho_f/\rho_s), \qquad (1\text{-}2)$$

where ρ_f and ρ_s are, respectively, the densities of fresh water and sea water; g is the acceleration of gravity; and h' is the depth of the density interface. Increasing F' values indicate increasing dominance of the inertial forces and are accompanied by an increase in turbulent diffusion; as F' decreases, turbulence decreases and the effects of buoyancy become more important. Turbulence tends to be suppressed when F' is near or less than 1.0 but generally increases in importance as F' increases beyond 1. Laboratory experiments by Hayashi and Shuto (1967, 1968) showed that fully turbulent effluent diffusion occurs when F' equals or exceeds 16.1. Because homopycnal outflows are, by definition, characterized by exceedingly low values of the density ratio, γ, it is expected that F' will exceed 16.1, whereas in hypopycnal outflows γ is relatively large and F' is reduced to the neighborhood of unity.

A readily applied engineering approach to estimating the likely degree of stratification or mixing, and hence the value of F' at the outlet, in terms of environmental conditions, is to use the *estuary number*, E_s (Harleman and Abraham, 1966; Turner, 1973), which indexes the ratio of tidal transport to freshwater (river) discharge and includes the effects of depth. The estuary number is given by:

$$E_s = \Pi_x \bar{U}_t^2/(ghQ_fT_t) \tag{1-3}$$

where Π_x is local tidal prism; $\bar{U}_t$ is the gross time and depth averaged tidal current velocity; Q_f is freshwater discharge; and T_t is tidal period. Harleman and Abraham (1966) estimated that estuaries can be expected to be stratified when $E_s < 0.03$ and well mixed when $E_s > 0.30$; the degree of mixing increases with increasing E_s. Buoyant effluents are most likely to occur when E_s is relatively small.

Probably the simplest river-mouth process model is that of a fully turbulent homopycnal effluent issuing into a deep basin where there is no interference from bottom topography. This situation is most common at the mouths of relatively steep-gradient streams entering deep, freshwater lakes or in cases where tidal mixing within the river channel is sufficient to destroy vertical density gradients. Gilbert's (1884) classic treatise on deltaic sedimentation was based on such a model. The details of the theory of turbulent jets can be found in Albertson *et al.* (1950), Schlichting (1955), Abramovich (1963), and Stolzenbach and Harleman (1971). However, the first attempt to apply the concept to deltaic sedimentation was offered by Bates (1953). Experiments by Kashiwamura and Yoshida (1967, 1969) suggest that effluent diffusion most nearly approximates Bates' turbulent jet model when outflow is of the homopycnal type. Townsend (1976) gives a detailed discussion of turbulent jets.

Two basic types of turbulent jet are recognized: (1) plane jets, in which a solid boundary (the bottom of the receiving basin) restricts expansion to the horizontal; and (2) axial jets, in which water depths beyond the outlet exceed the channel depth sufficiently to allow vertical as well as horizontal jet expansion. The type of jet depends on the ratio h_0/h_b, where h_b is the basin depth. In order for axial jet diffusion to occur, h_0/h_b must be appreciably less than 1.0. If it were possible to neglect the effects of bottom friction, plane and axial jets would be quite similar, except that expansion is three dimensional in axial jets and two dimensional in plane jets. In nature, however, plane jets are significantly affected by bottom friction, so that axial jets probably represent the simpler situation.

In turbulent jets, turbulent eddies are generated at the effluent boundaries in the fashion illustrated in Figure 1-12. These eddies are responsible for causing a two-way exchange of momentum between the effluent and ambient

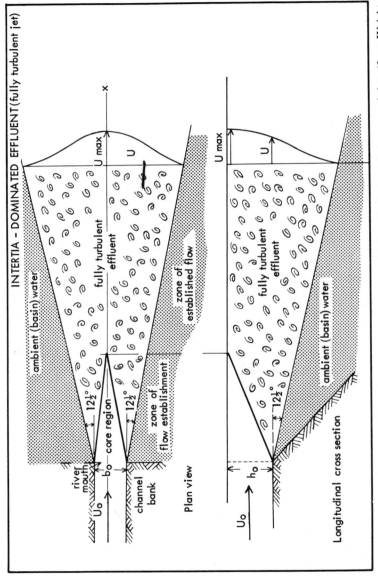

Figure 1-12. Spreading, diffusion, and deceleration pattern of fully turbulent axial river mouth jets (from Wright, 1977a).

fluids. Because of the progressive slacking of effluent momentum, outflow velocities and transporting capacities are diminished and sediment is deposited. Deceleration is accompanied by seaward growth of the turbulent region as the effluent expands laterally (and vertically, in the case of axial jets). From Figure 1-12 it can be seen that the jet consists of a zone of flow establishment and a zone of established flow. A core of constant velocity, extending from the outlet seaward to the point at which turbulent eddies penetrate to the effluent centerline, characterizes the zone of flow establishment. Albertson *et al.* (1950) observed the length, x_e, of the zone of flow establishment to equal 6.2 outlet diameters in the case of axial jets and 5.2 outlet widths in the case of plane jets. In natural situations, however, bottom friction contributes to the decrease of x_e in plane jets, so that x_e lengths of four channel widths or less are more common.

Seaward of x_e, within the zone of established flow, the entire effluent is fully turbulent, and even the maximum or centerline velocity, u_{max}, undergoes a progressive seaward decrease. Stolzenbach and Harleman (1971) found that within the zone of established flow of a fully turbulent jet, the ratio of the centerline velocity, u_{max}, to the outlet velocity, U_0, decreases at a linear rate with distance, x, from the outlet. For axial jets Stolzenbach and Harleman (1971) give:

$$u_{max}/U_0 = h_0(b_0/2)/(\varepsilon I x), \qquad (1\text{-}4)$$

and for plane jets:

$$u_{max}/U_0 = (b_0/2)/(\varepsilon I x), \qquad (1\text{-}5)$$

where, as before, h_0 and b_0 are, respectively, the outlet depth and width; x is the distance seaward of the outlet; I is a similarity integral, which is seen to have a constant value of 0.316 for the fully turbulent jet; and ε is the rate of jet expansion. According to Stolzenbach and Harleman (1971), ε is the same for both horizontal and vertical directions and has a constant value of 0.22 for fully turbulent jets or, in other words,

$$\varepsilon = \frac{d(b/2)}{dx} = \frac{dh}{dx} = 0.22. \qquad (1\text{-}6)$$

This value is equivalent to an angle of separation between the centerline and jet boundaries of 12°24′. Transverse to the effluent within the zone of established flow, outflow velocities decrease from u_{max} at the centerline to zero at the effluent boundaries. The resulting overall velocity profile follows a Gaussian distribution in the ideal case, shown in Figure 1-12 (Albertson *et al.*, 1950).

When flow velocities are diminished through the mechanism of jet diffusion, the ability of the flow to transport sediment is also diminished and is roughly in accordance with the pattern by which the outflow is dispersed and decelerated. Sediment begins to drop from the jet as soon as velocities fall below the competence and capacity levels required to transport the available sediment. In the ideal Gilbert-type situation, suspended sediment is deposited radially at the greatest distance from the core of constant velocity to form horizontally bedded bottomset deposits. As the river mouth progrades, coarser bed load is deposited at an angle just beyond the point at which effluent expansion is initiated to form basinward-dipping foreset beds. These deposits are finally capped by horizontal topset beds, which result from the advance of the channel deposits. The resulting depositional pattern is shown in Figure 1-13. Jopling (1960, 1963) describes laboratory experiments on the processes of deposition and bedding at homopycnal river mouths that have coarse-grained sediment loads and low values of the ratio of outlet depth, h_0, to basin depth, ($h_b < < 1.0$).

Plane Jets and the Effects of Bottom Friction

Continued discharge of sediment from the river mouth into the receiving basin eventually leads to shoaling in the region just beyond the mouth and consequently to an increase in the ratio h_0/h_b. Hence, under most natural conditions the basin depths within the zone of active jet diffusion are seldom significantly greater than the outlet depth and are frequently shoaler than at the outlet. Effluent diffusion and expansion are then restricted to the horizontal as a plane jet but, more importantly, shear between the outflow and the bottom introduces the added effects of friction. The result is more rapid deceleration and expansion than are predictable from the theory of turbulent jets alone. In this case, the bottom and the effluent behave as a closely coupled system wherein inertia and friction terms are largely responsible for the resulting process and form characteristics. This type of homopycnal outflow situation is not restricted to the mouths of streams entering freshwater lakes but is also common in nonstratified tidal inlets and at outlets where mixing by tides and other marine forces has broken down density contrasts.

In nature the theory of plane turbulent jets alone does not adequately explain the expansion and deceleration of effluents in contact with a shallow bottom. Borichansky and Mikhailov (1966) and Mikhailov (1971) concluded that under these conditions, effluent spreading and deceleration adjust so as to yield equilibrium between the forces of friction and inertia. Borichansky and Mikhailov (1966), Taylor and Dean (1974), and Bruun (1978) have analyzed the effects of bottom friction on effluent behavior. Bruun (1978) points out certain errors in the assumptions made by Borichansky and Mikhailov (1966) and follows Taylor and Dean (1974) who conclude that, considering friction effects alone, the effluent width, b, and maximum velocity, u_{max} relative to the corresponding quantities, b_0 and u_0, at the outlet are given by:

$$b/b_0 = e^{\mu_1 \, (x/b_0)} \tag{1-7}$$

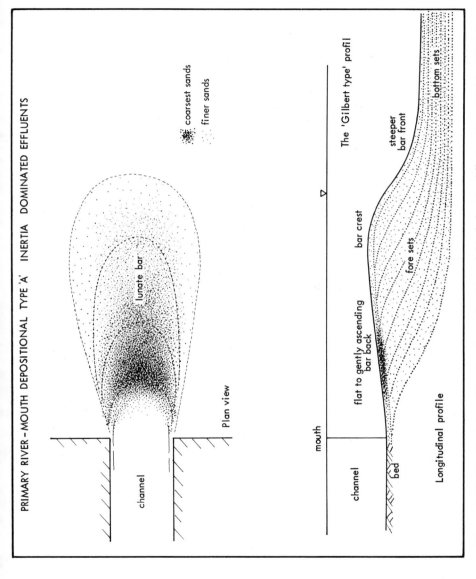

Figure 1-13. Idealized depositional pattern related to axial turbulent jet diffusion (from Wright, 1977a).

and:

$$\frac{u_{max}}{u_0} = e^{-\mu_1 \, (x/b_0)} \qquad (1\text{-}8)$$

where:

$$\mu_1 = C_D \, h_0/h \qquad (1\text{-}9)$$

is a friction parameter that incorporates the drag coefficient, C_D, and the relative depth, h_0/h. (Taylor and Dean, 1974, and Bruun, 1978, use the equivalent form $\mu_1 = f_D \, h_0/8h$ where f_D is the Darcy-Weisbach friction factor $= 8C_D$.) Equations 1-7 and 1-8 predict that expansion and "deceleration" rates will increase with increasing drag (and hence increasing roughness) and with increasing shoaling rates seaward of the outlet.

In actual fact, plane (as opposed to axial) turbulent jet diffusion acts together with bed friction to produce mixing and momentum exchange across the lateral boundaries of the effluent. As a consequence, expansion and deceleration rates of turbulent plane jets with bottom friction are greater than would be predicted from the equations for either mechanism acting alone. The combined effects have been considered by Ünlüata and Özsoy (1977) and Bruun (1978). The predicted effluent expansion rates are illustrated schematically in Figure 1-14. The most recent application of the plane jet model to deltaic deposition is offered by Wang (1984).

Because of the close interaction between the bottom and the issuing fluid, the dynamics of the effluent cannot be separated from the process of subaqueous bar development. Assuming that the receiving basin waters are tranquil, the zone of active effluent expansion can be considered an extension of the channel, lacking fixed lateral boundaries but subject to the same conditions for achieving and maintaining process-form equilibrium as alluvial channels. These conditions are (1) hydrodynamic continuity, (2) the tendency for the bottom and flow conditions to coadjust mutually to yield equal work per unit area of bed; and (3) satisfaction of conditions (1) and (2) in such a way as to result in minimum work for the system as a whole. Shoaling takes place seaward of the mouths of almost all rivers that debouch onto a shallow shelf or into a shallow lake, embayment, or estuary and results in the formation of a river-mouth or distributary-mouth bar. Bars are responses to effluent deceleration and can generally be regarded as striving to attain an equilibrium in which the conditions just stated are satisfied as nearly as possible.

However, sediment accumulates continuously at river mouths, so that the system may change its form faster than the equilibrium conditions can be attained. In addition, flow conditions at the mouths of natural rivers are subject to continuous change. Hence, the actual river-mouth and river-mouth–bar configurations are not dependent simply on the ideal equilibrium form toward which the system is approaching. The observed configurations will also reflect the rates at which sediment is supplied and at which the dynamic regime changes relative to the time required for equilibrium

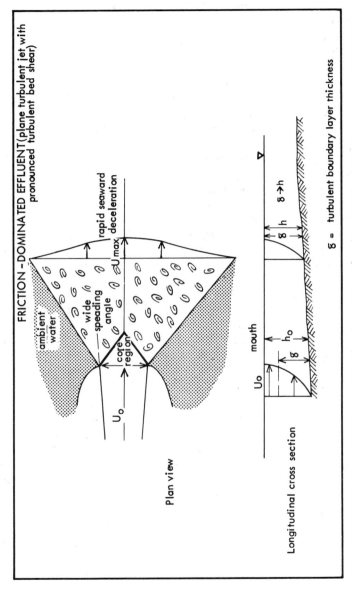

Figure 1-14. Spreading, diffusion, and deceleration pattern of a plane turbulent jet with pronounced bed friction (from Wright, 1977a).

adjustment to take place. Perhaps for this reason even the dimensionless forms of the river mouth and associated bar are rarely, if ever, completely static but are more often in a continuous state of evolving.

Despite the complexities, river-mouth morphologies are at least qualitatively predictable as functions of the dynamic environment. In the case of homopycnal, friction-dominated effluents issuing over a shallow basin bottom, there is a definite associated pattern of bar and subaqueous levee development. Initially, the rapid rate of effluent expansion that is characteristic of this type of river mouth produces a broad, arcuate radial bar. As the deposition continues, natural subaqueous levees develop beneath the lateral boundaries of the expanding effluent, where velocity gradients are steepest. These levees tend to inhibit any further increases in expansion rate, so that with continuing bar accretion continuity can no longer be satisfied simply by increasing effluent width. As the central portion of the bar grows upward, channelization takes place along the lines of maximum turbulence, which tend to follow the subaqueous levees. The result is the occurrence of a bifurcating channel that has a triangular middle-ground shoal separating the diverging channel arms (Figure 1-15). Observations show that flow tends to become reconcentrated into the divergent channels and to be tranquil over the middle-ground bar under normal discharge conditions (e.g., Mikhailov, 1966; Arndorfer, 1971; Welder, 1959). This situation should be more stable than in the case of the radial bar because the decreased bed width and greater depth in regions of strong flow permit the total work of the system to approach closer to the minimum while maintaining a uniform rate of work per unit area of bed. Accordingly, this type of bar pattern is most common where nonstratified outflows enter shallow basins. Classic examples of this type of bar can be found at the mouths of Mississippi River Delta crevasses (secondary channels cutting perpendicular to the main channel) that debouch into shallow interdistributary bays. Middle-ground shoals and bifurcating channels are also typical of many tidal deltas built into lagoons or bays by tidal inlets (e.g., Wright and Sonu, 1975). Patterson (1875, p. 46) reports that just following the formation of the Cubits Gap crevasse of the lower Mississippi Delta in the early 1860s, this crevasse was "a breach 2,700 feet wide, and with a maximum depth of 132 feet; but the water quickly shoals beyond the mouth of the crevasse, the depth in the course of a few hundred feet not being more than five feet at an average." The subsequent evolution of Cubits Gap has been traced from a series of hydrographic charts by Welder (1959), who showed that within a very few years after the initiation of the crevasse, the rapidly shoaling radial bar was replaced by a middle-ground bar flanked by diverging channels.

Coleman et al. (1964) analyzed the sedimentary structures and deposits associated with a middle-ground bar and bifurcating channel at the mouth of the Johnson Pass, a crevasse channel in the Mississippi Delta. This study showed the crest of the bar to be located at the extreme upstream end of the bar and to consist mainly of coarser bed-load deposits. The suspended load

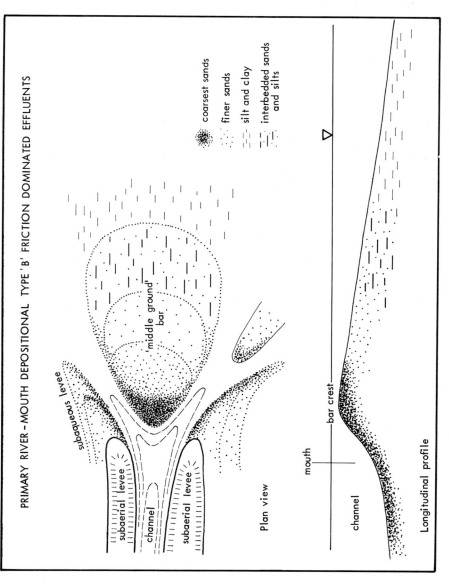

Figure 1-15. Bifurcating channel and middle-ground bar patterns associated with friction-dominated river-mouth effluents (from Wright, 1977a).

was deposited farther basinward, in the interdistributary zone between the incipient channels. A similar situation was documented by Arndorfer (1971) from a smaller crevasse.

Mikhailov (1966) found a good correlation between the morphometry of middle-ground bars and outlet geometry.

Hypopycnal Effluents and the Role of Buoyancy

At the outlets of estuaries and river mouths with low estuary numbers, E_s (Equation 1-3) ($E_s < 0.03$), stratification in and near the outlet is normally pronounced and the densimetric Froude number F_0' at the outlet can be expected to have a low value near 1.0. The buoyant forces are then pronounced relative to the turbulence-producing inertia forces. Furthermore, the condition of well-developed salt-wedge stratification in such outlets typically results in isolating the effluent from the effects of bottom friction. Buoyancy suppresses mixing, and the expansion of the effluent takes place largely in response to the buoyant forces themselves. Wright and Coleman (1971) showed, with reference to data from the highly stratified mouth of the Mississippi River, that the major features of strongly buoyant effluents could be explained to a first approximation in terms of a relatively simple theoretical model, originally proposed by Bondar (1970), which neglects turbulent diffusion. The important assumption here is that $F' = 1$ at the outlet; however, this is the case at "salt-wedge" river mouths like those of the Mississippi (e.g., Wright, 1971; Wright and Coleman, 1971), the Po (Nelson, 1970), and the Danube (Bondar, 1963, 1967, 1968).

In the (assumed) absence of turbulent diffusion, buoyant effluents (plumes) can be expected to experience lateral expansion seaward of the outlet owing to the lateral pressure gradients created by the superelevation, $\eta' = h'\gamma$ (Figure 1-16) of the lighter (freshwater) effluent relative to the ambient fluid. Lateral expansion is accompanied by vertical thinning (seaward reduction in h') in order to satisfy continuity. The corresponding expansion and thinning at x distance seaward of the outlet depend on the dynamic buoyant expansion coefficient, l_b, following:

$$\frac{b}{b_0} = (1 + l_b\, x)^{2/3} \tag{1-10}$$

and:

$$\frac{h'}{h_0'} = \frac{1}{(1 + l_b\, x)^{2/3}} \tag{1-11}$$

where h_0' is the thickness of the buoyant (upper) layer at the outlet and

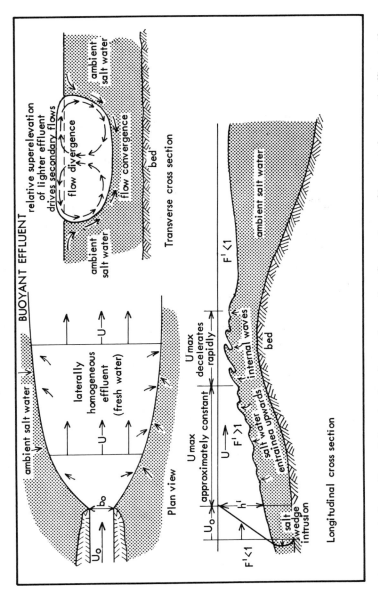

Figure 1-16. Spreading, mixing, deceleration, and secondary flow patterns of buoyant river-mouth effluents (from Wright, 1977a).

$$l_b = \frac{3}{4} \left[\frac{\frac{4}{3} (2g \; \gamma)^{1/2} \; (h_0')^{3/2} \; \left(1 - \frac{\gamma}{2}\right)}{Q_f} \right] \qquad (1\text{-}12)$$

where Q_f is the rate of freshwater discharge from the outlet (Bondar, 1970; Wright and Coleman, 1971). Figure 1-16 illustrates the resulting expansion and thinning pattern of a buoyant effluent.

The simple buoyant expansion model described above assumes that flow velocity remains constant with x owing to the suppression of turbulent mixing and momentum exchange by buoyancy. In fact, observations (Wright and Coleman, 1971, 1974) suggest that deceleration is very gradual or even that u may increase in the region between the outlet and about $x/b_0 = 4$; however, beyond this region there is a comparatively rapid decrease in effluent speed. Wright and Coleman (1971, 1974) attribute the eventual deceleration of the effluent to the upward entrainment of salt water across the density interface separating the effluent from underlying sea water. The seaward increase in vertical turbulent entertainment is at least partly attributable to the growth and eventual breaking of internal waves in response to seaward increases in F'. Seaward increases in F' to values well in excess of unity result from the seaward decrease in h' in the presence of an initially constant u (e.g., Wright and Coleman, 1971, 1974).

Visually dramatic and important features of buoyant effluents are pronounced frontal boundaries and related three-dimensional internal circulation patterns. Plume fronts are discussed by Garvine (1974, 1984), Garvine and Monk (1974), Bowman and Iverson (1978), and Wright and Coleman (1971, 1974). These fronts are exceptionally sharp as is evident from the photo in Figure 1-17. Flow (v) divergence from the centerline of the buoyant effluent near the surface converges at the frontal boundaries with inward-directed saltwater transport from outside that plunges beneath the sloping pycnocline (Figure 1-16). Flow in the lower part of the buoyant effluent is also directed inward. The net result of the combination of flow divergence near the surface and flow convergence near the pycnocline is the development of the dual helical cells illustrated qualitatively (in a somewhat idealized fashion) in Figure 1-16.

Buoyancy continues to play a role with less pronounced stratification and with increased outlet F_0' values. However, with increasing F_0' turbulent diffusion also becomes important. In intermediate cases it is common for effluent behavior to be controlled by a combination of turbulent jet diffusion, bed friction, and buoyancy (e.g., Wright and Coleman, 1974).

Figure 1-18 illustrates the typical river-mouth depositional pattern associated with buoyant effluents. This type of depositional geometry is well represented at the mouths of the Mississippi and reflects the secondary flow characteristics just described. Owing to the initially slow rate of effluent

Figure 1-17. Sharp frontal boundaries off the mouth of South Pass, Mississippi River Delta.

deceleration, the bar crest—the locus of the most rapid deposition of the coarsest material—is situated farther seaward of the outlet (4–6 outlet widths) than is the case for other effluent types. In addition, the weak flow convergence near the base of the effluent inhibits the lateral dispersion of sand and results in narrow bar deposits that ultimately prograde as the laterally restricted "bar-finger sands" described by Fisk (1961). The same secondary flow characteristics presumably also prevent the divergent flaring of the prograding subaqueous levees, thereby favoring the development of narrow, but deep, distributary channels.

The main sedimentary features of the depositional subenvironments of the Mississippi river-mouth model as described by Wright and Coleman (1974)

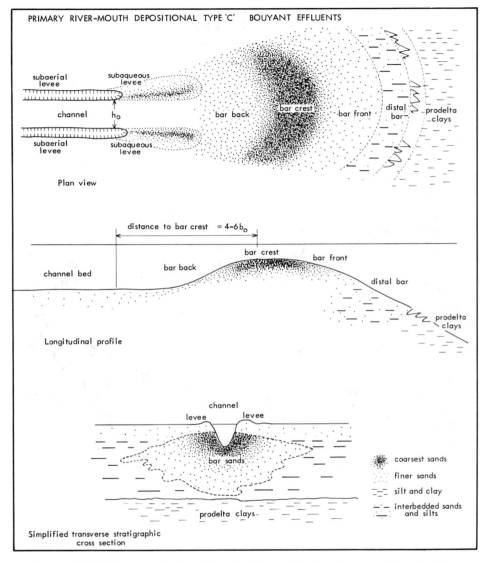

Figure 1-18. River mouth depositional pattern related to buoyant effluents (from Wright, 1977a).

are shown in Figure 1-19. The channel bed, natural levee formations, and crevasse-splay deposits are the primary features associated with the channel environment. The most significant sand accumulations in this environment are the crevasse splays. The active channels tend to scour into underlying distributary-mouth bar sands as they prograde, whereas upon abandonment

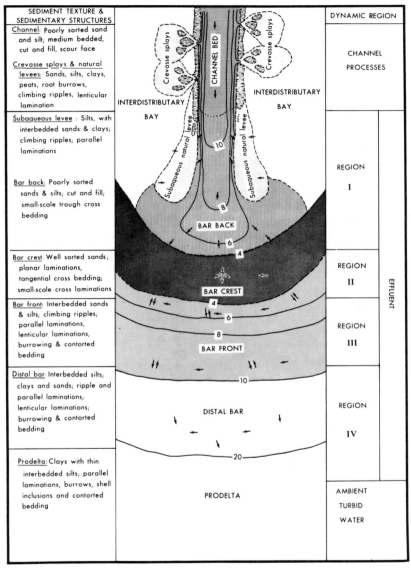

SEDIMENT TEXTURE & SEDIMENTARY STRUCTURES		DYNAMIC REGION
Channel: Poorly sorted sand and silt; medium bedded, cut and fill, scour face		CHANNEL PROCESSES
Crevasse splays & natural levees: Sands, silts, clays, peats, root burrows, climbing ripples, lenticular lamination		
Subaqueous levee : Silts, with interbedded sands & clays; climbing ripples; parallel laminations		REGION I
Bar back: Poorly sorted sands & silts; cut and fill; small-scale trough cross bedding		
Bar crest: Well sorted sands; planar laminations, tangential cross bedding; small-scale cross laminations		REGION II
Bar front: Interbedded sands & silts; climbing ripples; parallel laminations, lenticular laminations, burrowing & contorted bedding		REGION III
Distal bar: Interbedded silts; clays and sands; ripple and parallel laminations; lenticular laminations; burrowing & contorted bedding		REGION IV
Prodelta: Clays with thin interbedded silts; parallel laminations, burrows, shell inclusions and contorted bedding		AMBIENT TURBID WATER

Figure 1-19. Idealized depositional model of highly stratified "Mississippi-type" river-mouth system (from Wright and Coleman, 1974).

the channels normally fill with silts and clays; hence, accumulations of channel sands tend to be relatively unimportant. (The sparsity of sandy channel fill is primarily characteristic of highly stratified distributaries that have high depth–width ratios, and not of river mouths as a whole.) Seaward of the outlet, the distributary-mouth bar consists of the three main subcomponents shown in Figure 1-19. The salient depositional features are the bar back (the seaward ascending portion of the distributary-mouth bar) and subaqueous natural levees. The bar crest is morphologically continuous with the bar back but is situated seaward of the distal ends of the subaqueous levees. Although the bar back and bar crest receive some sediment during low stage, they are related primarily to the seaward gradients in bed-load transport that characterize flood stage. Growth of subaqueous natural levees, however, appears to continue year round and is significantly related to the near-bottom lateral convergence that takes place during low and normal river stages.

Like the bar back and bar crest, the bar front is composed primarily of bed load but is characterized by progressive seaward fining. The bar front grades seaward into the distal bar as the effects of bed-load transport become less significant and as silt becomes the predominant size fraction. The distal bar is a transitional province between the distributary-mouth bar and the prodelta. Beneath the diffuse band of turbid brackish water that normally surrounds the delta, clays are deposited from suspension to form the prodelta, the basal and most continuous of the depositional units.

Wave-Induced Effects at River Mouths

Many river mouths are influenced significantly by waves. Although the effluent mechanisms just discussed remain operative in these cases, wave processes produce important and often complex modifications to effluent behavior. With particular reference to tidal inlets, Bruun (1978) presents a relatively detailed treatment of some important wave-effluent interactions; Wright et al. (1980) describe some significant wave effects on river-mouth effluents.

Wave-effluent interactions are bidirectional. Wave modifications by seaward-flowing effluents cause increased wave shoaling and reduction of wave phase speed and length, resulting in oversteepening and often "premature" breaking of the waves. The enhanced phase speed reduction may also increase refraction coefficients around the effluent depending on underlying topography. Where wave attack is moderate to high, wave breaking in the effluent region promotes intense vertical mixing, breaking down density stratification and reducing the role of buoyancy. Wave-induced mixing and momentum exchange also increase the rate of seaward decrease in effluent velocity. Waves also cause inshore trapping and local, temporary impoundment of effluent waters and can produce abnormal setup in the vicinity of the outlet (Wright, 1977b; Wright et al., 1980). Figure 1-20 shows a cross section of flow and density characteristics as observed at a wave-dominated river mouth (Shoalhaven, Australia) during a period of high river stage and high waves and it illustrates some of the effects just described.

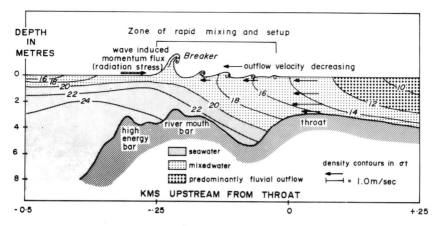

Figure 1-20. Flood-stage effluent/wave interaction and resultant density and flow structure as observed at the mouth of the Shoalhaven River (from Wright *et al.*, 1980).

The generalized depositional patterns typical of wave-dominated river mouths are illustrated in Figure 1-21 for the cases of shore-normal (Figure 1-21a) and obliquely incident (Figure 1-21b) waves. Owing to the rapid seaward diminution of effluent speeds and accompanying shoreward re-working by waves, a sandy river-mouth bar develops at a short distance seaward of the mouth. Lateral redistribution of bar sediments by wave-induced transport processes "smooths" the bar configuration into a regular arcuate shape. Rapid deposition also takes place along the flanks of the effluent producing broad, shallow subaqueous levees akin to the "ramp-margin shoals" (e.g., Oertel, 1972) of tidal inlets. Dissipative surf over these shallow platforms causes shoreward migration of sediment in the form of swash bars (Figure 1-21). These sand features ultimately weld to the shore adjacent to the mouth causing the outlet to be constricted. The same basic features are also present when waves break at an oblique angle to the shore, but they are skewed downdrift with respect to longshore currents and littoral drift.

Macrotidal River Mouths

Many of the world's deltas exhibit depositional patterns that are strongly influenced or, in many cases, dominated by extreme tides. A few notable examples include the deltas of the Ord (Australia), Shatt-al-Arab (Iraq), Amazon, Ganges–Brahmaputra (Bangladesh), Klang (Malaya), and Yangtze (China) Rivers. The process environments of tide-dominated deltas are distinguished by at least three important characteristics: (1) Mixing by tidal activity obliterates vertical density stratification so that the effects of

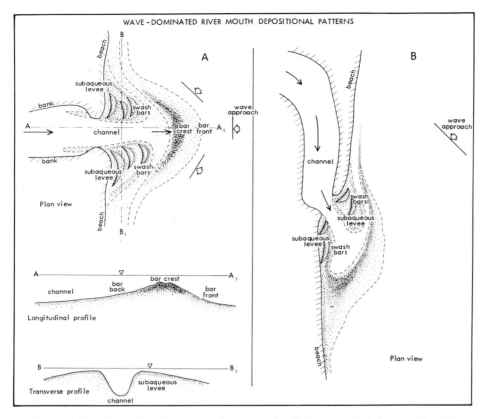

Figure 1-21. Depositional patterns characteristic of river mouths influenced by high wave energy: (a) normal wave incidence; (b) oblique wave incidence (from Wright, 1977).

buoyancy at the river mouths are negligible. (2) For at least part of the year, tides account for a greater fraction of the sediment-transporting energy than the river, and flow in and seaward of the river mouth is subject to reversals over a tidal cycle that cause bidirectional sediment transport. (3) The range of positions of the land–sea interface and of the zone of marine–riverine interactions is greatly extended both vertically and horizontally.

High tide ranges and strong tidal currents interact with the morphology of the distributary channels and delta front to drastically alter the associated depositional processes. Amplification of tides can result in swift bidirectional currents over the delta front and in river mouths. For example, observations in and near the mouth of the Ord River of Western Australia, which experiences a spring tide range of 5.9 m, revealed flood and ebb currents in excess of 3 m/s (Wright *et al.*, 1973, 1975). Where strong tidal currents of this type cause pronounced bidirectional transport of sediment, it has been found that flood- and ebb-dominated bed-load migrations tend to follow proximal but mutually evasive courses (Ludwick, 1970; Price, 1963). When this bidirectional sediment transport is approximately parallel to the direction of river outflow and sediment supply, linear subaqueous ridges commonly result. Linear tidal ridges of this type have been described by Off (1963) and Wright et al. (1975). The dynamics of tidal ridges have been studied most rigorously and intensively by Huthnance (1973, 1982). Figure 1-22 shows the form and location of tidal ridges in and seaward of a generalized macrotidal river mouth (Wright, 1977b). The crests of these ridges are often exposed at low tide, and they may range from 10 to 20 m in relief. Such ridges are well developed near the mouths of the Ord, Klang, Ganges–Brahmaputra, Shatt-al-Arab, and Purari Rivers, among numerous others in high tidal range environments. In such environments, these tidal ridges replace the normal distributary-mouth bar and distal-bar deposits as the dominant accumulation forms of the delta front. As delta progradation proceeds, the ridges, which tend to remain remarkably stable in position (Off, 1963; Wright et al., 1975), may assume subaerial expression and separate large, straight tidal channels.

In most high tidal range environments, tidal influence extends considerable distances into the lower reaches of the river channel itself. In tide-dominated river channels, the ratio of tidal amplitude to channel depth is normally high. Consequently, tides in rivers behave as finite-amplitude waves, which become increasingly deformed with distance above the mouth. The effect of this deformation is to shorten the flooding phase and extend the ebbing phase. Because of this pronounced tidal asymmetry, average flood velocities significantly exceed average ebb velocities. The result is that tide-induced (i.e., disregarding riverine contribution) bed shear stresses have a net upstream effect over the bed as a whole (Wright et al., 1973, 1975). Hence, there is appreciable upstream transport of bed load. In the lower Ord, the largest and most prominent bedforms were found to be flood oriented. The sediment budget of the Ord is balanced by ebb-dominated sediment transport in deeper channels, where ebb currents become concentrated during falling tide and, presumably, by seaward flushing of sediment during river floods (Wright et al., 1973). Upstream transport of bed load by flood-tide currents causes extensive sand accumulations within the channel (Figure 1-22) as well as bidirectional cross-bedding. If a channel becomes abandoned by the river, it will normally be filled with sand by the continued activity of tidal currents. Sand-filled channels are thus of major significance in tide-dominated deltas.

River channels subject to strong tidal influence (e.g., the Ord) are distinguished by their characteristic bell or funnel shapes, wherein channel width decreases progressively upstream. The rates of upstream convergence vary considerably among individual rivers and may be either linear or exponential. The lower Ord is an ideal example of an exponentially convergent channel in that channel width and depth decrease exponentially

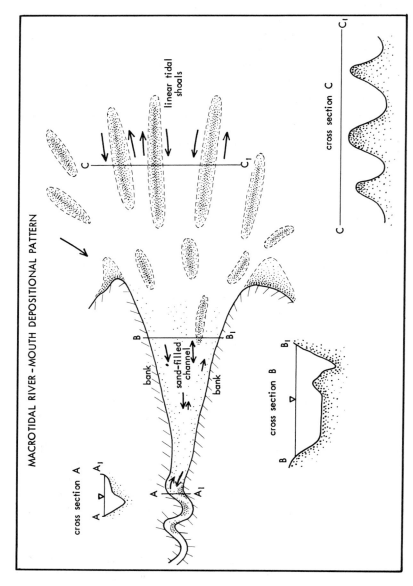

Figure 1-22. Sand-filled, "bell-shaped" channel and linear tidal ridges associated with macrotidal river mouths (from Wright, 1977).

from 9 km and 7 m, respectively, at the mouths to 90 m and 1.5 m, respectively, at 65 km upstream from the mouth. This convergence is accompanied by an upstream increase in channel sinuosity similar to that shown in Figure 1-22. Upstream convergence at exponential rates is a common attribute of many other rivers experiencing strong tidal influence (Wright *et al.*, 1973).

Post-Depositional Deformation and Mass Movement of Delta-Front Deposits

Dynamic delta-molding processes do not cease with the deposition of deltaic sediments. Subsequent to their deposition, delta-front and prodelta deposits are subject to continued deformations and flowages. These processes occur in response to high depositional rates near the river mouth, combined with differential loading by denser bar sands overlying low-density prodelta clays. Deformations are of several types, including peripheral slumping, mudflow gullying, clay diapirism, mass wasting induced by degassing of fine sediments, and shelf-edge faulting (Coleman *et al.*, 1974; Coleman and Prior, 1980). These processes have been described by Shepard (1955, 1973), Morgan (1961), Morgan *et al.* (1963, 1968), Coleman *et al.* (1974), and Coleman and Prior (1980). Although most of these studies are based on data from the Mississippi Delta, the same mechanisms are active in many other deltas.

Bathymetric maps and profiles of the delta-front deposits of the Mississippi Delta as well as the deltas of the Magdalena, Orinoco, Niger, and Amazon, among others, display a high degree of irregularity characterized by "stairstep"-like slump features as well as extensive gullies radiating from the locus of most rapid bar deposition (the latter features were described by Shepard in 1955). Recent detailed side-scan sonar mapping of delta-front and shelf-edge morphology off the mouths of the Mississippi (Prior and Coleman, 1978; Coleman and Prior, 1980) has illuminated the three-dimensional nature of the features that collectively produce the irregularities. Included are bar-front slumps, mudflow gullies, collapse depressions, faults, gas vents, and mud lobes. Some of the features are illustrated in Figure 1-23 (from Coleman and Prior, 1980). Subaqueous bar-front slumping is a particularly important process that can result in the bodily advection of large blocks of shallow water deposits (e.g., bar sands) seaward into the deeper water realm of the prodelta. The instabilities responsible for this slumping process can occur on very low gradient slopes ($\tan\beta < 0.035$; Coleman and Prior, 1980). Figure 1-24 shows the rotational character of the slumps.

The rapid localized loading of dense, incompressible distributary-mouth bar sands over the less dense plastic clays of the prodelta and marine environments can also cause rather dramatic diapiric activity. The process of

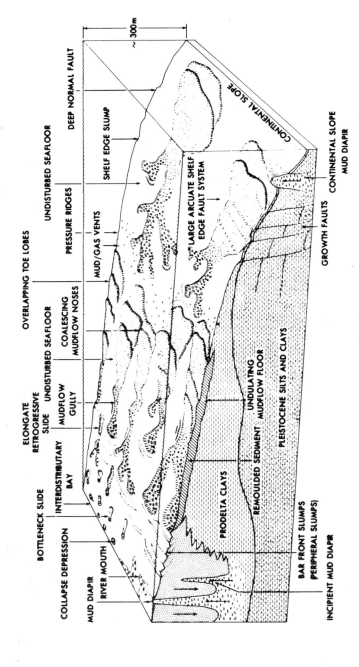

Figure 1-23. Types and distribution of subaqueous sediment failures in the Mississippi River Delta (from Coleman and Prior, 1980).

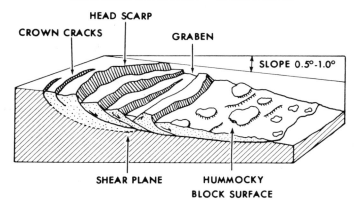

ROTATIONAL SLUMP - EARTHFLOW

HEAD SCARP

CROWN CRACKS

GRABEN

SLOPE 0.5°-1.0°

SHEAR PLANE   HUMMOCKY
                  BLOCK SURFACE

Figure 1-24. Morphology of rotational peripheral slides (from Coleman and Prior, 1980).

diapirism related to delta-front progradation is illustrated in Figure 1-25 and has been studied in detail by Morgan (1961) and Morgan *et al.* (1963, 1968). Bar accumulation creates considerable pressure on the underlying clays and causes the latter to flow toward positions of least pressure. Stress on the clays is ultimately relieved by diapiric intrusion, anticlinal folding, and faulting (Figure 1-25). Owing to these processes in the Mississippi Delta, clay diapirs or mudlumps can intrude upward through bar sands from depths as great as 200 m, and corresponding subsidence of bar deposits can lead to the accumulation of localized pods of sand as thick as 150 m. The mudlumps emerge as thin spines around the peripheries of actively accreting distributary-mouth bars. Mudlumps may appear as subaerial islands or remain wholly submerged. They are highly dynamic and can experience vertical displacements of several meters per year. Whereas peripheral slumping provides a mechanism for transporting shallow-water sediments into deep water, clay diapirism is responsible for bringing prodelta and marine sediments, along with their associated structures and faunal remains, into the shallow delta-front or interdistributary-bay environments.

Processes of Delta Plain Development

The rates and patterns of sediment transport and deposition by marine processes subsequent to the initial river mouth deposition affect the longer term evolution of delta plain plan form as well as the assemblage of depositional features that comprise the delta plain surface. Where the river is completely dominant over marine forces, the delta shape may simply express

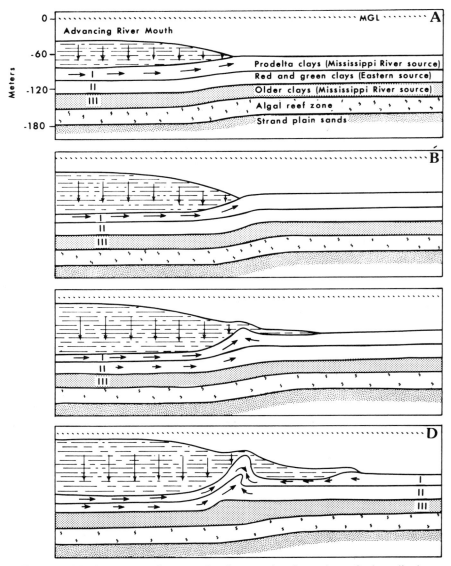

Figure 1-25. Sequence of events leading to the formation of clay diapirs or "mudlumps" in the Mississippi Delta (from Morgan *et al.*, 1968).

the pattern of distributary progradation and branching with interdistributary features consisting of open bays, marshes, or swamps (e.g., Mississippi). However, where significant marine energy in the form of waves, tides, or coastal currents is present, deltaic sediments are redistributed and remolded into features and shapes that are more nearly in equilibrium with the particular marine process. On a large scale and over the long term, gross deltaic configuration is also influenced by less dynamic constraints or controls such as receiving basin geometry, tectonic stability, and rates of subsidence because of compaction.

The Effects of Waves

In addition to the immediate roles played by waves in modifying river mouth processes as already discussed, waves attack the shores and nearshore zones of most deltas. Waves sort and redistribute the sediments debouched by rivers and remold them into wave-built shoreline features, such as beaches, barriers, and spits. As in the case of nondeltaic coasts, the extent to which waves are able to mold the sediments at the shoreline into wave-dominated configurations depends on the deep-water wave climate, as well as on the manner in which the waves are modified before reaching the shore by the processes of wave refraction, shoaling, shallow-water deformation, and dissipation by bottom friction. In deltas, however, the continued rapid introduction of sediments from a riverine source tends to negate the effects of waves to varying degrees. Consequently, the overall configuration and landform suites of delta plains reflect the relative degree to which riverine discharge of clastic sediments is able to overwhelm the sediment-reworking abilities of waves. Theoretical delta forms resulting from the interactions of riverine sediment supply and wave distribution of sediment have been treated by Komar (1973), whereas Wright and Coleman (1971, 1972, 1973) and Wright et al. (1980) have examined the relationships between actual delta morphologies and associated river discharge and wave regime.

An informative field model for illustrating the adjustment of delta configuration to wave processes is offered by the Jaba Delta on the island of Bougainville (Papua New Guinea; Wright et al., 1980). The present-day Jaba Delta is a small young delta lobe consisting almost entirely of tailings and overburden from a copper mine in the Jaba River catchment; the discharge of sediment and the development of the resulting deltaic lobe have been monitored since mining began and the delta was initiated in 1972. An average of 26×10^6 metric tons of sediment reach the coast annually, and this sediment is quickly redistributed alongshore by the 0.5–1.0 m waves to produce the regular, arcuate sediment bulge shown in Figure 1-26. The progressive evolution of the delta configuration over the period 1975–1977 is illustrated in Figure 1-27. Northward (downdrift) migration of a secondary sediment bulge and an intervening embayment is evident in the earlier stages. However, progressive equilibrium adjustment over the 2-yr period ultimately resulted in a smoother and more regularly arcuate shoreline by the end of the period.

The morphology of large-scale deltas reflects the rate at which the river introduces new sediment to the coast relative to the rate at which wave processes are able to redistribute the sediments. At a first approximation, the

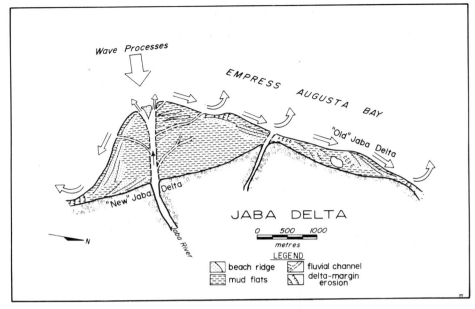

Figure 1-26. Depositional morphology and net sediment drift patterns of the Jaba Delta, August, 1977 (from Wright *et al.*, 1980).

latter rate can be assumed roughly proportional to the wave energy flux near the shore. (The nearshore energy flux or "power" is proportional to the square of the breaker height.) A wave-climate computer program that predicts the effects of wave refraction, shoaling, frictional attenuation, and associated changes in wave energy flux for multiple sets of waves was applied to several deltas by Wright and Coleman (1971, 1972, 1973). The study revealed that, in accordance with logical expectations, deltas that experience the highest nearshore wave energy flux have the straightest shorelines and the best developed interdistributary beaches and beach-ridge complexes. However, deltaic wave regimes tend to differ markedly from wave regimes affecting nondeltaic shores in that wave energy near the shore is only slightly correlative with deep-water wave climate. It is apparent that nearshore wave energy reflects deep-water wave energy only partially; it is to a greater extent a function of varying but considerable degrees of frictional attenuation over the comparatively flat offshore profiles that front the deltas. Although frictional dissipation is often of secondary importance on most nondeltaic coasts, it is the most important determinant of the wave energy flux affecting deltas owing to the flat, shallow bottom slopes that characterize delta fronts. The lowest nearshore wave energy flux values tend to occur along delta coasts that are fronted by the flattest offshore profiles and widest delta fronts.

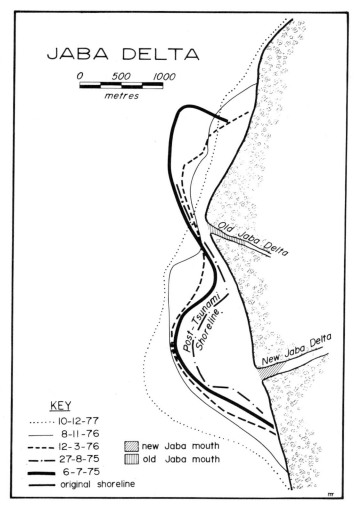

Figure 1-27. Evolution of the shoreline of the Jaba Delta over the period July, 1975–December, 1977 (from Wright *et al.*, 1980).

Among the 16 deltas compared by Wright and Coleman (1972, 1973), the Mississippi was shown to be the most "river dominated" in terms of river mouth discharge (of water and sediment) relative to nearshore wave energy flux. The Senegal Delta at the other extreme was found to be the most wave dominated. Between the two extremes, delta morphologies define a broad range of patterns. At the river-dominated end of the spectrum, deltas are highly indented and have extended distributaries; marshes, bays, or tidal flats characterize interdistributary regions (e.g., the Mississippi Delta). With increasing nearshore wave power and decreasing river discharge, delta shorelines become more regular, assuming the form of gentle, arcuate protrusions, and beach ridges become more common (e.g., the Nile Delta). The highest nearshore wave energies and lowest relative discharges are associated with deltas that exhibit wave-straightened shorelines and abundant beach ridges (e.g., the Senegal Delta). Those deltas having the highest relative discharges and associated river-dominated configurations are also the ones having the flattest delta-front slopes, whereas steep slopes front wave-dominated deltas. It is by flattening the subaqueous profile more rapidly than waves can redistribute deltaic sediments that the river is able to overcome wave effects and mold the delta to its own design.

Effects of Coastal Currents and Tides

Oceanic currents impinging on coastal waters near deltas occasionally cause sediments discharged by a river to be transported and dispersed along the coast to considerable distances from the river mouth. In a few instances, unidirectional coastal currents may be as swift as the river outflow itself. Under such circumstances, the sediment-laden river effluent may merge with the coastal current and be deflected laterally along the adjacent coastline without significant loss of sediment-transporting capacity. This process may result in arresting the seaward progradation of the delta front while causing extensive accumulation of river-derived clastics downdrift from the river outlet.

Probably the most prominent example of a delta influenced by strong coastal currents is the Amazon River mouth and delta system (Figure 1-28). Although the Amazon has filled an embayment with a delta plain encompassing nearly 500,000 km^2, the development of a seaward protrusion has been precluded by the strong Guiana Current, which flows toward the northwest over the delta front (Metcalf, 1968; Gibbs, 1970). Because of the strength of the Guiana Current and its close proximity to the coast, a large percentage of the sediment debouched by the Amazon is swept northward for hundreds of kilometers and ultimately is deposited as mudflats and low-energy beach ridges along the coasts of northern Brazil, French Guiana, and Surinam. Van Andel (1967) reports that this same current system also significantly affects the distribution of sediments over the front of the Orinoco Delta (Venezuela). Another example is the Purari Delta in the northern part of the Gulf of Papua (Papua New Guinea). Seaward of the mouths of the Purari, the mud-laden effluents merge into a band of low salinity water that remains trapped inshore by waves and winds. This band carries a heavy load of fine-grained material westward under the influence of the prevailing southeast trades. The

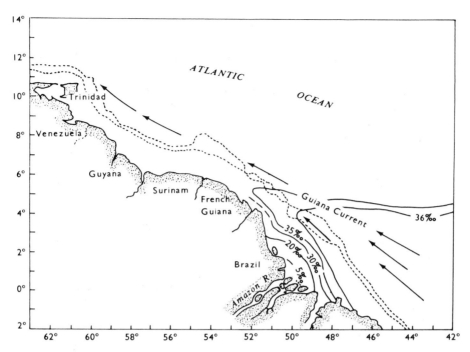

Figure 1-28. Surface currents over the Amazon Delta front (from Gibbs, 1970).

sediment is then pumped up into estuaries west of the delta by tidal currents (Thom and Wright, 1983).

Reversing tidal currents, which tend to parallel the coast in shallow water, are also effective in dispersing river-derived sediments alongshore and straightening the coast. Tidal processes are also of basic importance to depositional patterns in the interdistributary and distributary-margin regions. Intricate networks of nonriverine tidal channels and creeks, such as those that characterize the delta plain of the Ganges–Brahmaputra (Figure 1-9), are common features of many deltas in high tide range environments. Networks of this type are particularly common in humid regions, where abundant vegetation, such as mangroves, is present to stabilize channel banks. These tidal channels tend to be characterized by sand fill, bidirectional bed-load transport, and resultant bedforms and cross-bedding. In more arid macrotidal regions, interdistributary tidal channels are present but are much less abundant. Barren tidal flats composed of silts and clays deposited from suspension in overbank flows are the most abundant feature. In many cases these flats are covered or interbedded with evaporite deposits (Coleman and Wright, 1978).

Other Processes and Controls

There are numerous other passive controls as well as slower geologic processes that act as determinants of deltaic sedimentation patterns and delta plain morphology. An exhaustive review of these is not feasible; however, a few of the more salient remaining processes deserve at least brief mention. The overall large-scale configuration of a delta is affected to varying degrees by the geometry of the receiving basin in the vicinity of the delta (Coleman and Wright, 1973a, 1975). Tectonic stability of the receiving basin is a fundamental determinant of the vertical thickness and rate of seaward advance of deltaic accumulations. If subsidence rates are high, considerable thickness can result, whereas deltas prograding over flat, tectonically stable platforms tend to advance more rapidly but as thin sheets.

The patterns by which the loci of the most intense deltaic deposition change position over long periods of time are also of considerable importance. Occasionally a delta may prograde seaward without undergoing any significant lateral shift. Deltas filling the closed ends of laterally restricted structural troughs (e.g., the Shatt-al-Arab, which enters the end of the Persian Gulf) tend to follow a consistent depositional course. In many cases, however, a delta may build out in one direction for several centuries, until the channel becomes overextended and inefficient, subsequently switching to a totally new position. Over the past 4000 yr, the Mississippi Delta has changed position at least seven times. At each position the river succeeded in building a major lobe that was subsequently abandoned as the lower course of the river shifted to a new position. The chronologic sequence of delta building, as discussed by Kolb and Van Lopik (1966), is shown in Figure 1-29. The abandonment of each lobe was followed by coastal retreat under the combined influences of subsidence and marine reworking. In other situations, shifting of the location of deltaic deposition can take place as a result of changes in river course well upstream, so that the river occupies a new alluvial valley (e.g., the Hwang Ho of China). A third type of delta switching, referred to as alternate channel extension, is displayed by the Danube Delta. Two or more main courses tend to be maintained but to alternate in relative dominance. While the subdelta of one course is prograding, that of the other course may experience a period of transgression.

Biologic, chemical, and diagenetic processes, which in turn often reflect climate, also play major but highly varied roles. The luxuriant vegetation that characterizes deltas in the humid tropics is important in trapping and binding terrigenous sediment as well as in effecting vertical accumulation of peat and organic debris. For example, Coleman et al. (1970) reported that in the Klang Delta of Malaya, plant growth and organic accumulation are so rapid that peats in the interdistributary and distributary-margin regions build to elevations higher than those of adjacent natural levees.

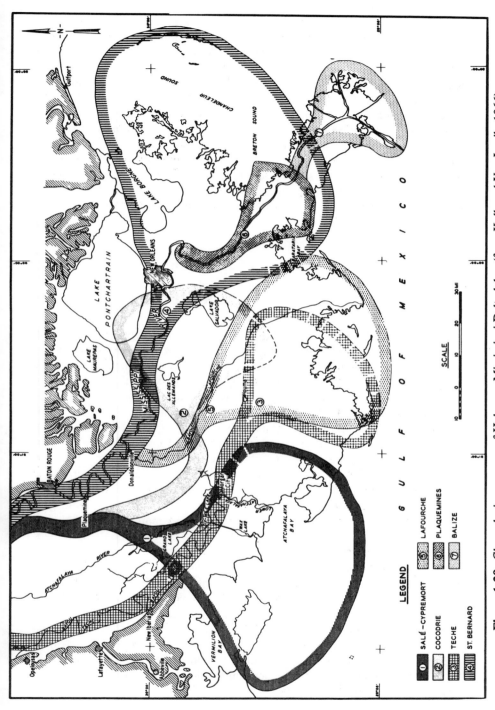

Figure 1-29. Chronologic sequence of Holocene Mississippi Delta lobes (from Kolb and Van Lopik, 1966).

The processes associated with the presence and movements of sea ice and river ice cannot be neglected, owing to the abundance of deltas along the arctic coasts of North America and the USSR. Freezing of arctic rivers causes cessation of flow for most of the year, and the result is that the total annual discharge is concentrated into a relatively short period. The resultant swift and violent spring floods transport coarse, poorly sorted material to the coast. The combination of ice rafting in channels and comparative disequilibrium between flood discharge and distributary channels results in numerous bifurcations and rejoinings that produce broad, braided, and relatively inefficient distributary networks. In many cases, river flood takes place before the breakup of sea ice near the shore. Under such circumstances, sediments may be debouched over the nearshore fast ice and ultimately deposited considerable distances offshore or alongshore from melting ice floes.

Deltaic Sediments and Sedimentary Structures

Terrigenous deltaic sediments range in size from clays to gravel, the size as well as the composition depending on the sediment yield and the nature of the drainage basin. Some rivers (such as the Burdekin of Australia) transport large quantities of sand and coarse material, whereas the sediment load of many others (e.g., the Mississippi) consists largely of fine silts and clays transported in suspension. The Mississippi debouches 4.97×10^{11} kg of sediment into the Gulf of Mexico annually (Everett, 1971), of which 45% is clay, 36% is silt, and 19% is very fine to fine sand. Fine sand is the coarsest material discharged at the river mouth. Because of the highly varied textural characteristics of the sediments made available to deltas by rivers, it is erroneous to suppose that any deltaic environment can be characterized by a particular mean or median grain size; however, relative grain sizes exhibit definite environmental relationships. The coarsest material, whether fine sand or coarse gravel, tends to accumulate nearest the river mouth or in high-energy regions, whereas the finest material is deposited farthest seaward or in the lowest energy regions. Of equal or greater importance, however, are the suites of primary sedimentary structures that characterize each environment. The sediment textural properties and structures of each deltaic environment will be described separately. Detailed descriptions of Mississippi Delta sediments and sedimentary structures can be found in Coleman and Gagliano (1964), Coleman et al. (1964), and Gould (1970), and more recently in Coleman (1976) and Coleman and Prior (1980, 1982). Some excellent color photographs of cores and sediments from deltaic environments are presented by Coleman and Prior (1982).

The Prodelta

The prodelta is the most ubiquitous of the deltaic environments and the one that varies least among different delta systems. It directly overlies the sediments of the inner shelf environment and is the basal unit of the most commonly occurring deltaic sequences. The prodelta is characteristically a blanket of clays deposited from suspension having high lateral continuity and low lithologic variation. In most instances the prodelta deposits exceed 6 m in vertical thickness; in the Mississippi Delta the prodelta ranges from 20 to 50 m thick and extends seaward to water depths of 70 m.

Within the prodelta the finest clays occur at the greatest depths, and there is often a slight increase in the percentage of silts toward the upper portions. However, clays predominate throughout. Because deposition is entirely from suspension, parallel laminations are by far the most common primary structure. Although prodelta deposits normally appear as structureless massive clays to the naked eye, the laminae are frequently evident from X-ray radiographs except where they have been eradicated by burrowing organisms. Laminae are very thin near the base of the prodelta but thicken upward. Near the uppermost parts of the prodelta, parallel and lenticular silt laminations alternate with thicker clay laminae.

The relatively slow rates of deposition that characterize the base of the prodelta permit marine fauna to flourish. Accordingly, fauna, including forams, molluscs, and echinoids, are most abundant near the seaward margins of the prodelta and decrease in abundance landward (and upward) as depositional rates increase. In the Mississippi Delta, the prodelta frequently contains pods of distributary-mouth-bar sands and associated cross-bedding, flowage structures, multiple small faults, and shallow-water fauna. These pods of sand, set in a matrix of prodelta clays, represent slump blocks transported into the prodelta environment by the slumping processes previously described.

Delta Front

Progressing upward in the deltaic sequence, variability within and between deltas increases and, in most cases, lateral continuity of depositional units decreases. Overall, the delta-front deposits are significantly coarser grained than the prodelta deposits and continue to coarsen and become better sorted upward. Such current- and wave-induced structures as various scales of cross-bedding, scour-and-fill features, planar beds, and ripple marks become more abundant. Faunal remains and burrowing are relatively rare. Depending on the process environment of any particular delta, the delta front may consist of distal bar, distributary-mouth bar, tidal ridge, or, in the case of deltas that experience high wave energy, shoreface deposits.

Distal Bars

Distal-bar deposits are normally found at the base of most delta-front sequences immediately above and transitional with the prodelta clays. They have relatively high lateral continuity and vary from 6 to 25 m in vertical thickness. The most common characteristics are alternating layers of sand, silt, and clay, with the sand and silt layers thickening upward and clay layers thinning upward. Small-scale cross-bedding is common, particularly near the top of the distal bar. Faunal content exhibits a pronounced decrease upward.

Distributary-Mouth Bars

The distal-bar deposits grade upward into distributary-mouth bar sands in the typical delta sequence. The lateral continuity of distributary-mouth bar deposits tends to be lower than that of the units below but also to vary considerably between deltas. In the case of high-bed-load rivers or rivers with friction-dominated river mouths, the frequency of bifurcations is very high and the close proximity of adjacent distributaries may result in nearly continuous sheets of coalescing distributary-mouth bars. The deltas of the Volga (USSR), Paraná (Argentina), and Colville (United States) Rivers are prominent examples. In other instances, such as off the mouth of the Amazon, the extreme size of the river mouth creates a single extensive and continuous bar. In the Mississippi Delta, however, the lateral spreading of distributionary-mouth bar sands is strongly inhibited, so that bar deposits prograde seaward in directions approximately normal to the mean coastline trend as narrow, linear sand bodies (referred to by Fisk, 1961, as bar-finger sands) having very low lateral continuity. In the Mississippi Delta these bar sands account for the major sand accumulations.

Distributary-mouth bar sands tend to be dirty (i.e., to have significant silt content) and poorly sorted near their base but to coarsen and become cleaner upward. Cross-bedding, mostly small scale, is present throughout, together with occasional cut-and-fill structures and scour features. Faunal content is minor, but transported organic debris is often present in local pockets, particularly near the top of the bar. Very thin clay beds and flaser structures are sometimes present, particularly where tidal currents are moderate.

Tidal Ridges

In macrotidal environments, linear tidal ridges often replace the distributary-mouth bar as the major sand accumulation of the delta front, although distributary-mouth bars may also be present. Unfortunately, there is very little information concerning the sedimentary properties of these features.

The tidal ridges examined by Wright *et al.* (1973) at the mouth of the Ord were composed of dirty sands and had pronounced bidirectional large- and small-scale cross-bedding. Although individual tidal ridges have low lateral continuity, their relative abundance and regular spacing cause them to be a dominant sand accumulation form. The tidal ridges may occur below, within (i.e., interfingering with), or just above distal bar deposits.

Channels and Distributary Margins

Above the delta front, sediment variability is extremely high. Every delta has channel deposits and distributary-margin deposits above certain areas of the delta front; however, lateral continuity of these deposits is very low, and their nature varies considerably among deltas. Typically, these deposits include subaqueous and subaerial levees, channel fills, and crevasse splays.

Subaqueous levees in the Mississippi Delta are characterized by dirty sands, small-scale cross-bedding in multiple directions, and occasional thin laminations of clay or organic debris (Coleman and Gagliano, 1965). The subaerial levees are vertical extensions of the subaqueous levees and result from overbank flow and the entrapment of sediment by vegetation. Intensive burrowing by vegetation and fauna and small ferric nodules produced by oxidation are present, in addition to the same structures that characterize the subaqueous levees (Coleman and Gagliano, 1965).

Channel-fill characteristics are highly dependent on dynamic environment and on the dominant river-mouth processes. Buoyancy-dominated river-mouth processes and low tidal range preclude sand fill in most of the Mississippi distributaries. Under these conditions, abandoned channels most often fill with silts, clays, and organic clays above a scour base or above beds of dirty sands. In cases where distributaries transport high percentages of bed load or where river mouths are shallow and have dominating friction effects, sandy channel fills exhibiting pronounced cross-bedding are most common. In macrotidal river channels, however, pronounced bidirectional bed shear results in extensive sand filling of channels; channel fills in these environments are among the most important sand accumulations. Large- and small-scale bidirectional cross-bedding is present throughout. In many instances the highest amplitude foreset beds dip upstream.

Extreme variability of sediments and structure over very short distances characterizes the regions immediately flanking the distributary levees. Laterally outward from and continuous with the subaqueous levees are narrow, wedge-shaped overbank deposits consisting of thin alternating stringers of sands, silts, and clays. In the Mississippi Delta, overbank sediments are capped by the more extensive crevasse-splay deposits, which generally are similar to miniature deltas. They tend to coarsen upward and toward the distributary channels.

Interdistributary Bays and Flats

Above the delta front, the most extensive depositional environments are those of the interdistributary regions. Lateral continuity is greater than that of the channel and distributary-margin deposits but less than that of the distal bar. In many cases [for example, in the Mississippi, Ebro (Spain), Danube, and Magdalena (Colombia) Deltas], the interdistributary regions are occupied by shallow bays. A combination of shallow marine depositional processes, riverine influence, and brackish-water faunal activity causes interdistributary bays to exhibit extreme lithologic and textural variability, as well as a broad range of primary structures. Some of the more common characteristics include parallel laminations of silt and clay, sandy lenticular laminations resulting from local re-sorting by waves, occasional current ripples and cut-and-fill structures, and abundant burrows and shell remains from shallow-water molluscs. Crevasse-splay deposits are frequently interbedded with the interdistributary-bay deposits.

In the interdistributary regions of deltas that experience high tide ranges, the interdistributary-bay deposits tend to be replaced by subtidal and intertidal flats. Tidal-flat deposits are usually more continuous than interdistributary-bay deposits. Typical internal features include lenses of parallel laminated clays and silts that alternate with sands displaying small-scale cross-bedding, flaser structures, thin layers of algae and evaporite deposits, and root and animal burrows. Sand-filled tidal channels characterized by bidirectional cross-bedding are often present locally.

Beach and Dune Deposits

In deltas that experience appreciable nearshore wave energy, beaches, beach-ridge plains, and eolian dune forms are normally the most abundant depositional forms adjacent to and between distributaries. These deposits are similar in most respects to nondeltaic beaches, as described in Chapter 6, except that they often are more extensive owing to rapid progradation. In contrast to other types of deltaic sediments, beach and dune sands are normally clean and well sorted. Planar beds that dip seaward at low angles are the most common primary structure of the beach, but multidirectional small-scale cross-stratification and larger scale landward-dipping cross-bedding (related to swash bar migrations) are also present. Dune deposits exhibit festoon cross-bedding, tangential cross-bedding, planar bedding, occasional distorted beds, and root burrows.

Marshes and Swamps

Extreme organic activity characterizes the marshes and swamps that occupy low, flat areas of the subaerial portions of many deltas. Except for occasional parallel laminations of silt or clay, these deposits are typically structureless.

Depending on climate and intensity of plant growth, either organic clays or peats, intercalated with fine clastics, are common. Burrowing organisms are abundant in these environments, and bioturbation is frequently extreme. Coleman (1966) found that in the upper delta plain and lower alluvial valley of the Mississippi, the freshwater swamps encompass two environments: the well-drained swamp and the poorly drained swamp. The well-drained swamp is exposed to oxidizing conditions for part of the year, and formation of iron oxide and calcium carbonate nodules results (Coleman, 1966). The poorly drained swamp is continually inundated by standing water and is characterized by reducing conditions, iron carbonate, vivianite, and pyrite nodules.

Variability of Deltaic Depositional Models

No attempt at generalizing, no matter how detailed, can adequately convey the reality of natural deltas; each has its own individual attributes that set it apart from the others. Still, comparisons of 34 major river systems by Wright *et al.* (1974) revealed that in terms of particular sets of morphologic or environmental variables, such as river-mouth morphology, delta-plain landform suites, delta morphometry, or energy regime, deltas tend to cluster together into a relatively few deltaic groups or "families" within which the individual delta members are mutually similar. By further generalization, Coleman and Wright (1973b, 1975) concluded that in terms of their gross sand body geometry, deltas can be broadly classed into at least six extreme types. Figure 1-30 illustrates the six geometric models in the form of generalized isopach maps of relative net sand thicknesses. Obviously, absolute sand thicknesses will vary appreciably, depending on absolute delta size. Each type reflects a particular combination of dominating processes and controls. Following progradation and ultimate preservation, each type should also be represented by a particular diagnostic stratigraphic sequence.

Perhaps the majority of natural deltas is intermediate between the extremes illustrated in Figure 1-30 and possess attributes of more than one type. As would be expected from the foregoing discussion of deltaic processes, deltaic depositional geometries occupy a continuum with each member expressing the relative contributions of different fluvial and marine transport processes in molding the delta. Figure 1-31 attempts to illustrate the relationships of the six extreme models and several intermediate examples to the relative dominance of fluvial, wave, and tidal processes. The relative proximity of the intermediate cases to the extreme models is also roughly indicated. It should be borne in mind that processes other than those attributable to fluvial outflow, waves, and tides are frequently important as are "passive" controls such as receiving basin geometry and shelf configuration. A few select examples of modern deltaic models from the broader

1. Low wave energy; low littoral
drift; high suspended load

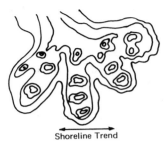

Shoreline Trend

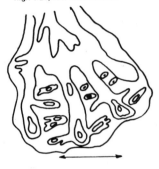

3. Intermediate wave energy;
high tide; low littoral drift

2. Low wave energy; low littoral
drift; high tide

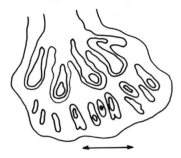

4. Intermediate wave energy;
low tide

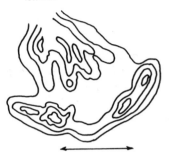

5. High wave energy; low littoral
drift; steep offshore slope

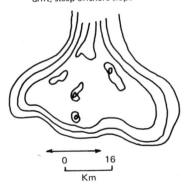

0   16

Km

6. High wave energy; high littoral
drift; steep offshore slope

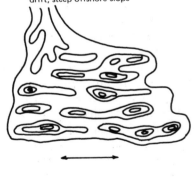

Figure 1-30. Generalized isopach maps of six extreme deltaic sand-body geometry models (from Coleman and Wright, 1973).

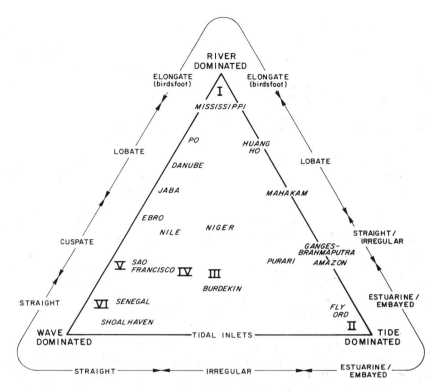

Figure 1-31. Simplified comparison of selected deltaic depositional models in terms of their relative similarity to the six extremes from Figure 1-30 (roman numerals designate the six types), relative importance of river, wave, and tide processes, and delta shape.

spectrum of models illustrated in Figure 1-31 will now be described. There are numerous other comparisons of deltaic models in the literature. The interested reader is referred to Coleman (1976), Coleman and Wright (1975), LeBlanc (1975), Galloway (1975), and Coleman and Prior (1982).

The Mississippi Model: Type I

The model of the modern Mississippi delta is the archtype of the Type I geometry (Figure 1-30). It is typical of deltaic environments having low tide range, low wave energy, flat offshore slopes, low littoral drift, and relatively high discharge of fines as suspended load. This geometric type consists primarily of distributary-mouth-bar sands that in prograded sequences compose separate elongate protrusions oriented approximately normal to the overall coastal trend. The lateral near-bottom convergence resulting from highly stratified effluent processes restricts the lateral spread of individual sand bodies. Differential compaction of prodelta clays underlying the distributary-mouth-bar sands leads to localized increases in vertical sand accumulation, creating the series of locally thick sand pods shown in Figure 1-30. In the modern Mississippi, these sand pods can attain local thicknesses as great as 120 m (Coleman and Wright, 1975).

Although the modern active delta of the Mississippi is the prime representative of the Type I model, comparisons of surficial delta morphologies suggest that other near analogies may include the Paraná, Dneiper (USSR), and Danube. The Mississippi has a mean annual discharge of 15,360 m^3/s and average maxima and minima of 57,400 m^3/s and 2,830 m^3/s, respectively. Sediment discharge averages 4.97×10^{11} kg in a year and consists largely of fine clastics. Nearshore wave power is very low. Tides are diurnal, and maximum tropic tide range is only 43 cm. The subaerial configuration and physiographic patterns of the Mississippi Delta are shown in Figure 1-32. The coastline overall is highly crenulate and has shallow bays, lakes, marshes, and swamps that occupy most of the areas between and adjacent to distributaries. The delta plain, including active and abandoned portions, occupies a total area of 28,568 km^2.

Figure 1-33 shows a generalized stratigraphic block diagram of the Mississippi (from Coleman, 1976). An important feature is the upward decrease in the lateral continuity of deposits. Whereas the relict shelf deposits and overlying basal prodelta deposits are highly continuous, the bar sands, which occur in the upper part of the sequence, are very restricted in the alongshore dimension.

The Mahakam Model: Intermediate

Allen *et al.* (1979) have recently provided an excellent and detailed description of a deltaic model intermediate between the fluvially dominated (Type I) and tide-dominated (Type II) extremes. This is the Mahakam Delta on the Eastern Coast of Kalimantan (Indonesia). Compared with the Mississippi, the Mahakam Delta is small, comprising a total area of only 5000 km^2 including subaqueous (delta front and prodelta) portions (Allen *et al.*, 1979). The present Mahakam Delta is a fan-shaped multidistributary lobe as shown in Figure 1-34; it is prograding across the continental shelf. In contrast to the Mississippi, the interdistributary regions are occupied by estuarine (tidal) channels and tidal flats giving the delta plain a much less irregular configuration.

Allen *et al.* (1979) estimate that the Mahakam discharges on the order 1000–3000 m^3/s of water and about 8×10^6 m^3/yr of sediment. Waves are usually less than 60 cm in height and, hence, probably make a small

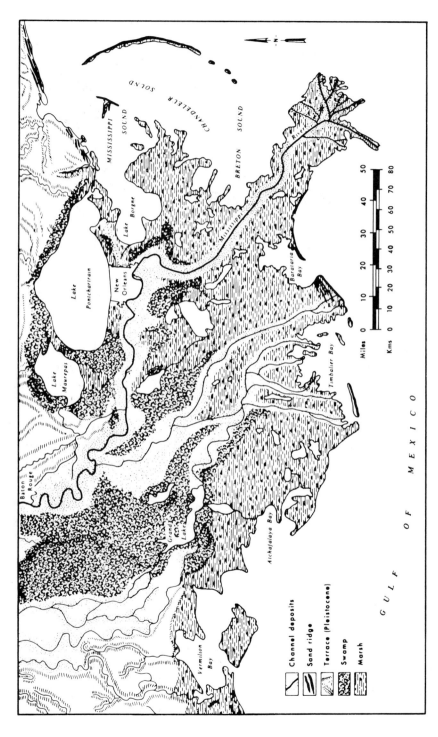

Figure 1-32. Physiography of the Mississippi Delta (from Wright and Coleman, 1973).

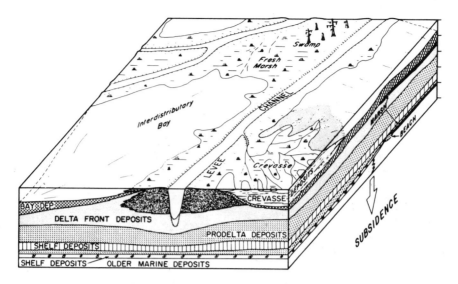

Figure 1-33. Block diagram of Mississippi Delta facies (from Coleman, 1976).

contribution to the redistribution of this sediment. Tides, however, contribute significantly although they do not totally dominate. The tides are mesotidal, semidiurnal with a mean range of 1.2 m and a spring range of 3 m. These tides are responsible for tidal currents on the order of 1 m/s at channel mouths (Allen *et al.*, 1979).

A simplified longitudinal stratigraphic cross section of the Holocene Mahakam Delta is shown in Figure 1-35 (from Allen *et al.*, 1979). As in the Mississippi example, the basal prodelta muds are widespread. Above the prodelta, the effective dispersal of sediment by the combination of multiple effluents and tidal currents has resulted in a relatively continuous delta front unit consisting of bar sands and subtidal sands and sandy muds with silts and clays predominating off interdistributary areas. The "cleanest" sand bodies are the channel sands of the deltaic plain. These sand bodies have low sinuosity, trend parallel to distributaries, and are on the order of 10 m in vertical thickness.

The Ord Model: Type II

Extreme tidal ranges and associated strong bidirectional tidal currents are responsible for the sand distribution patterns of the Type II geometry (Figure 1-30). Broad, seaward-flaring channel sands, fronted seaward by linear elongate sand bodies in the form of tidal ridges, characterize this type. The

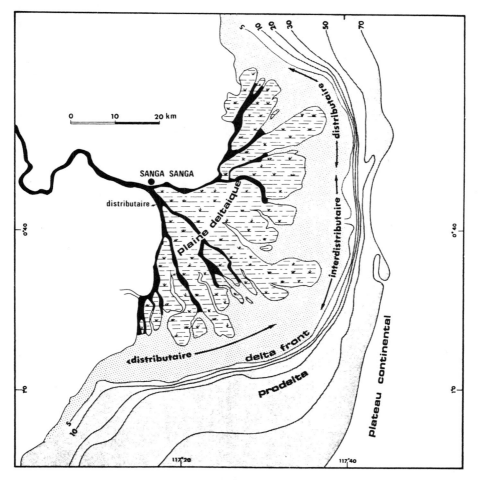

Figure 1-34. General morphology of the Mahakam Delta (from Allen *et al.*, 1979).

Ord River Delta of Western Australia (Figure 1-36) is presented here as the type example, but the deltas of the Victoria (Australia), Shatt-al-Arab, Ganges–Brahmaputra, and Fly Rivers are equally representative.

The discharge of the Ord River exhibits extreme annual variability owing to the monsoonal climate and ranges from an extreme low of only 0.07 m³/s in September to a maximum of 730.4 m³/s in January; the annual mean is 163 m³/s. In an average year the river discharges 22×10^9 kg of sediments, including appreciable quantities of medium-sand bed load as well as a high suspended load of silt and clay. These sediments have accumulated within and immediately seaward of the narrow Cambridge Gulf to form the Ord

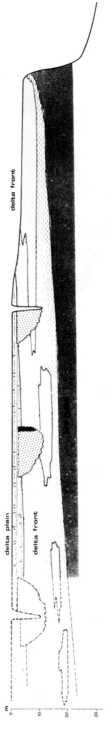

Figure 1-35. Simplified longitudinal stratigraphic cross section of the Mahakam Delta (from Allen *et al.*, 1979).

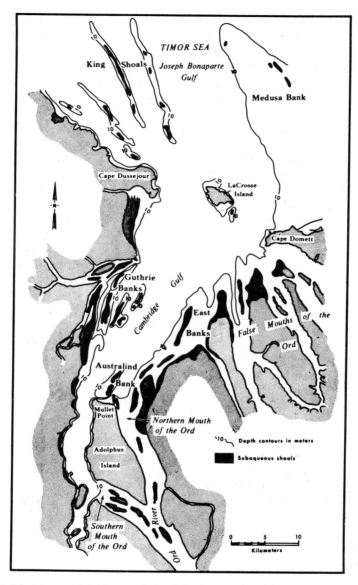

Figure 1-36. Linear tidal ridges and shoals in the mouths of the Ord River and in the entrance to the Cambridge Gulf (from Wright *et al.*, 1975).

delta complex. Semidiurnal tides at the mouth of the Ord have average spring ranges of about 6 m. Consequently, the lower course of the Ord is tide dominated; even at distances of more than 40 km above the mouth, tidal discharge rates substantially exceed the mean rate of river discharge (Wright et al., 1973). Nearshore wave power is very low because of frictional attenuation over broad intertidal shoals. Major surface morphologic features of the Ord Delta are shown in Figure 1-36. A funnel-shaped main river channel, smaller tidal creek channels, and extensive tidal-flat surfaces are the dominant features of the subaerial delta plain. Tidal ridges are the most prominent accumulation forms of the subaqueous delta within and seaward of the structural Cambridge Gulf.

Unfortunately, there is no continuous stratigraphic cross section of the Ord or of its analogs although Coleman and Wright (1975, 1978) present an idealized composite stratigraphic column extrapolated from surface sediment distributions and shallow cores. The basal unit probably consists of interbedded sands and prodelta instead of the massive clays that normally characterize prodelta deposits. The delta plain surface is a major sink for fine sediments transported in suspension and spread overbank across inter-distributary regions and behind the open coast shoreline at high tide. These fines have accumulated as a relatively thick unit of tidal flat muds capped by evaporites (Thom et al., 1975, discuss tidal flat evolution). Sandwiched between these lower and upper mud units are the major sand bodies, consisting of linear delta-front tidal ridges and sandy channel deposits.

The Burdekin Model: Type III

Intermediate wave energy, acting in combination with moderate tides, produces the third type of sand-body geometry (Figure 1-30), which is ideally represented by the Burdekin Delta of Australia (Figure 1-37; Coleman and Wright, 1975). Higher waves redistribute sands parallel to the coastline trend, remolding them into beach ridges and barriers, whereas bidirectional tidal currents produce sand-filled river channels and tidal creeks oriented approximately normal to the shoreline trend. Generally, the beach sands tend to be cleaner than the channel sands. As tide range decreases, channel sands tend to become less well developed. This type can exhibit a fairly broad range of variability, depending on relative strength of waves versus tides. This is possibly one of the most common delta types, and in addition to the Burdekin, there are numerous present-day analogies, including the deltas of the Irrawaddy (Burma), Mekong (Vietnam), and Red (Vietnam) Rivers.

The discharge of the Burdekin, like that of the Ord, is characterized by an extreme range and is highly erratic; sudden violent floods are common. Discharge rates vary from 10 m³/s to 24,286 m³/s around an annual mean of 476 m³/s. Sediment load is appreciably coarser than that of either the

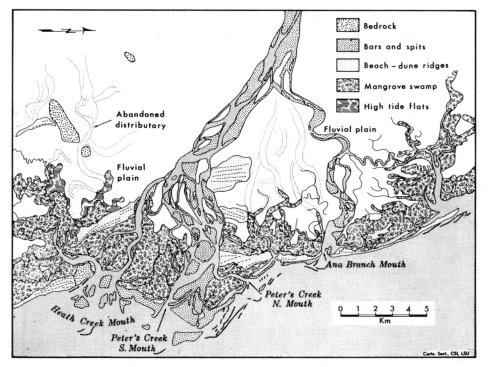

Figure 1-37. Physiography of the Burdekin Delta (from Coleman and Wright, 1975).

Mississippi or the Ord Rivers, consisting of abundant coarse sand; gravel and cobbles are present in the alluvial valley in significant quantities. Nearshore wave power is not extreme but is appreciably higher than that experienced by the Mississippi, Mahakam, or Ord. Tides are mesotidal with a mean spring range of 2.2 m. Littoral drift is weak.

Figure 1-37 shows the major morphologic features of the Burdekin Delta. In the lower delta plain the river bifurcates into numerous distributaries, which are choked with sandy shoals. Although the distributaries tend to be funnel shaped, the effect is not nearly so pronounced as in the lower Ord. Sand-filled tidal creeks are also common. At the distributary mouths, distributary-mouth-bar sands merge laterally with intertidal sand flats. Immediately above the sand flats, interdistributary areas of the lower delta plain exhibit mangrove swamps but alternate with beach ridges and dunes. Beach ridges are separated by broad swales occupied by mangroves or tidal flats. Although barren evaporite-crusted tidal flats are present locally in the lower delta plain, they are far less numerous than in the Ord. The two major sand units of the Burdekin model are: a lower unit of distributary-mouth-bar and channel sands and a higher unit of beach and dune sands. An idealized composite stratigraphic column of the Burdekin can be found in Coleman and Wright (1975) or in Coleman (1976); an intermediate example that possesses many features in common with Type III is the Niger Delta (Nigeria; Allen, 1964, 1970).

The Shoalhaven Model: Intermediate/Type IV

The fourth type of deltaic geometry (Figure 1-30) is characterized by offshore or bay-mouth barriers that shelter lagoon, bay, or estuarine environments into which a low-energy delta then progrades. Examples include the Appalachicola Delta (Florida) and the Sagavanirktok Delta of the Alaskan North Slope. The Nile Delta is probably intermediate between Types III and IV. The example selected for discussion here, the Shoalhaven Delta (Figure 1-38) of southeastern Australia is, in fact, intermediate inasmuch as it possesses features of Types IV, V, and VI. However, in terms of its stratigraphy and spatial arrangement of depositional units, the Shoalhaven is closest to Type IV. The process environment and morpho-stratigraphy of the Shoalhaven Delta have been described by Roy (1983), Roy et al. (1980), Wright et al. (1980), and Thom et al. (1981).

The Holocene deltaic plain of the Shoalhaven River is 85 km^2 in area and has infilled a shallow estuary sheltered from wave attack by a Holocene sand barrier (Figure 1-38). Although the mean discharge rate of the Shoalhaven is only 57.3 m^3/s, flows of 5000 m^3/s can accompany floods. The bed load of the river consists of fine to medium sands; suspended silts and clays are transported during floods. Tides are semidiurnal with a mean spring-tide range of 1.6 m. The open coast experiences a moderately high energy wave regime: The median wave height is 1.5 m, but storm waves frequently exceed 5 m and occasionally exceed 10 m. The persistent wave attack has precluded the development of a deltaic protrusion.

Figure 1-39 shows a stratigraphic cross section of the Shoalhaven Delta and barrier complex based on drilling by Thom et al. (1981) and Roy (1983). The most important sand unit is the shore-parallel barrier and shoreface unit. In contrast to other models, the major accumulation of "prodelta" muds occurs landward of the main sand body (the barrier) and at roughly the same elevation, within the protected estuarine environment. Although suspended fines reach the open sea, wave agitation prevents mud accumulation as a distinct unit over the inner shelf; the fines become widely dispersed alongshore and offshore. Secondary sand bodies take the form of fluvial channel sands imbedded within the deltaic muds (Figure 1-39). This type of deltaic stratigraphy is common along the embayed coast of south-eastern Australia.

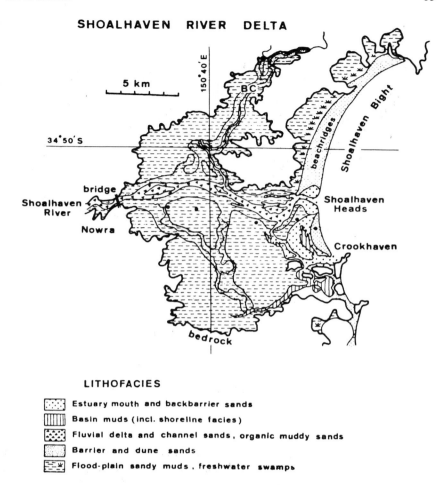

SHOALHAVEN RIVER DELTA

LITHOFACIES

Estuary mouth and backbarrier sands

Basin muds (incl. shoreline facies)

Fluvial delta and channel sands, organic muddy sands

Barrier and dune sands

Flood-plain sandy muds, freshwater swamps

Figure 1-38. Geomorphic map of the Shoalhaven Delta (from Roy, 1983).

The São Francisco Model: Type V

Persistent, moderate wave energy, acting over moderate offshore slopes, redistributes the sands debouched by the river to form the extensive sand sheets typical of Type V. The delta plain of this type consists primarily of beach ridges and dune fields, whereas shoreface sands are the prevailing deposits of the subaqueous delta. Distributary-mouth-bar deposits are restricted to the immediate vicinity of the river mouth and are quickly re-sorted and remolded by waves into the shoreface or beach. High lateral

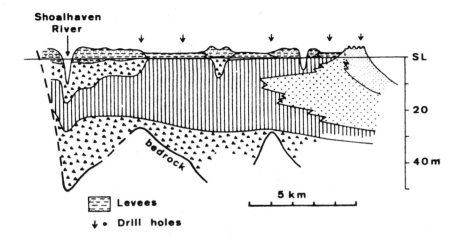

Shoalhaven River

SL

20

40 m

5 km

Levees

Drill holes

LITHOFACIES

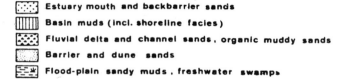

Estuary mouth and backbarrier sands

Basin muds (incl. shoreline facies)

Fluvial delta and channel sands, organic muddy sands

Barrier and dune sands

Flood-plain sandy muds, freshwater swamps

Figure 1-39. Longitudinal stratigraphic cross section of the Shoalhaven Delta (from Roy, 1983).

continuity of sands is a basic characteristic. There are numerous examples of this delta type, including the São Francisco Delta (Brazil), the Godavari (India), Tana (Kenya), and Grijalva (Mexico). The São Francisco is the ideal representative.

The Rio São Francisco del Norte is the second largest river system in Brazil (the Amazon is the largest), having a mean annual discharge rate of 3420 m³/s. The basin is within the humid tropics, and flow remains strong all year; discharge maxima and minima are 5818 m³/s and 1166 m³/s, respectively. Sediment load is large and consists of large quantities of suspended fines, as well as large amounts of sandy bed load. The river enters the South Atlantic Ocean near latitude 11°S. Semidiurnal tide range is significant, spring tides averaging 2.5 m, but tidal processes tend to be subdued relative to wave-induced processes. The delta lies within the trade winds, which blow onshore the year around. In addition, offshore slopes are moderately steep, averaging 0°15′. Consequently, the delta shoreline experiences moderate, persistent waves. However, waves attacking the delta shoreline are less powerful than those affecting the Shoalhaven Delta;

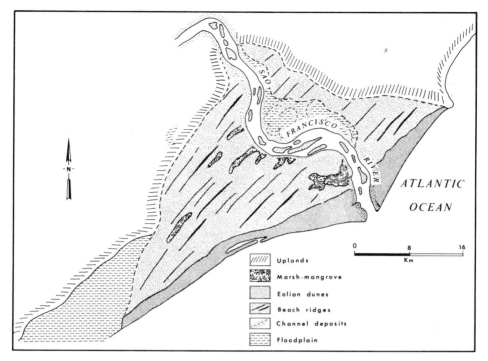

Figure 1-40. Physiography of the São Francisco Delta (from Wright and Coleman, 1973).

consequently, the São Francisco has succeeded in building a cuspate protrusion (Figure 1-40).

From Figure 1-40 it is evident that the delta shoreline is highly regular. Owing to high wave energy and strong onshore winds, broad, sandy beaches composed of clean, well-sorted sands, backed by an extensive plain of beach ridges and transgressive dunes, are the prevalent forms. Fines are removed to well offshore of the foreshore deposits. The dominant feature of the morphostratigraphy is a thick and highly continuous sheet of shoreface sands capped by eolian dune sands.

The Senegal Model: Type VI

The sixth and last type of delta geometry results from deltaic progradation in the presence of high wave energy and strong unidirectional littoral drift. The shoreline of this delta type is, like the Shoalhaven, completely straightened by waves. Long, continuous sandy barrier spits trending parallel to the coast

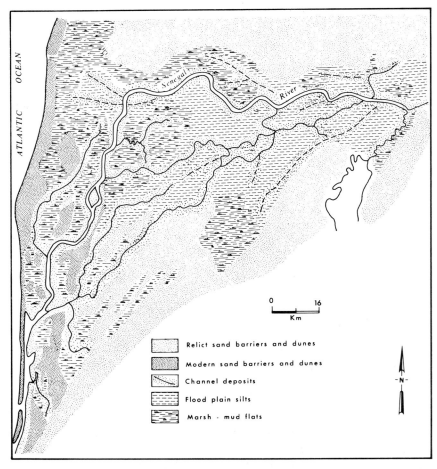

Figure 1-41. Physiography of the Senegal Delta (from Wright and Coleman, 1973).

alternate with fine-grained channel fills. The Senegal Delta of West Africa is the archetype. Another example is the Marowijne (Surinam and French Guiana).

The discharge rate of the Senegal River (Figure 1-41) averages 770 m³/s, and maxima and minima are 3460 m³/s and 20 m³/s, respectively. Spring tide range at the mouth averages 1.9 m. The high incident wave power exhibits a strong longshore component toward the south, causing pronounced littoral drift. Consequently, sands issuing from the mouth of the Senegal are immediately swept alongshore and remolded into a barrier spit. The spit continues to extend alongshore as river-mouth-bar sands merge with the distal end of the accreting spit, a process that causes extensive southerly deflection of the river. Eventually the channel becomes overextended, loses its gradient advantage, and breaches the upper end of the spit, whereupon the whole process is reinitiated. The delta thus prograds seaward as a series of sand spits and abandoned channel fills (Figure 1-41).

As in the case of the São Francisco Delta, clean quartzose sands, related to foreshore and beach-ridge deposition, are abundant and exhibit high lateral continuity. In the shore-normal dimension, the beach ridge sands alternate with organic muds that have filled the intervening abandoned channels.

Summary

Deltas result from the rapid and continued supply of sediment by rivers to coasts and upper continental shelves. They exhibit a wide range of configurations and are made up of a variety of depositional landforms. Correspondingly, the sediments and sedimentary sequences associated with deltas are highly varied. The surface forms, geometry, and sedimentary sequences of a delta depend on the discharge regime and sediment load of the associated river; on the nature of the river-mouth processes; on relative magnitudes of marine forces, particularly tides, waves, and currents; on deformational processes; and on such long-term processes as changes in river course and receiving-basin tectonics.

The depositional subenvironments of a delta include the prodelta, the delta front, the distributary channels, distributary margins, interdistributary regions, and delta shorelines. The prodelta normally consists of fine-grained sediments deposited from suspension. Distal-bar silts, distributary-mouth-bar sands, and such marine-redistributed deposits as tidal ridges and shoreface deposits may compose the delta front. Distributary-channel sediments may consist of silts and clays or of sand, depending on the dynamic environment. Natural levee and crevasse-splay sands and silts are the most common forms of channel-margin deposits. Extremely high morphologic and sedimentologic variability characterizes the interdistributary regions; depending on the particular process regime, interdistributary features may include bays, lakes, tidal flats, marshes, swamps, beach ridges, or dune fields. Delta shorelines may be highly irregular and consist of fine-grained marsh and intertidal mudflat deposits, if incident wave energy is very low, or may take the form of straight beaches, if energy is high.

The delta models described in the preceding section each reflect a particular combination of different river-mouth processes, sediment size, tidal influence, wave influence, and lateral transport by currents. Future case studies of the processes and morphologic and sedimentologic responses of other deltas will, no doubt, broaden the spectrum of delta models and will certainly lead to a more detailed understanding of the models discussed herein. In addition to studies of the global process-response variability of deltas, there is also an acute need for detailed systematic field investigations of specific mechanisms responsible for sediment transport, deposition, and redistribution in deltaic environments. Studies of this latter type are necessary to establish firmly the first principles of causality in deltaic formation and development. Some of the critical areas demanding further rigorous research include: (1) the relationships between river discharge regime and sediment transport in delta distributaries; (2) tidal transport phenomena in delta distributaries and the interactions between tidal currents and river flow; (3) effluent mechanisms and their effects on sediment dissemination at river mouths; (4) the soil mechanics of subaqueous mass movement of deltaic sediments; (5) field assessment of the relationships between riverine supply of sediment and wave redistribution of sediments; and (6) the process of temporal evolution of delta landscapes and the role of equilibrium adjustment.

References

Abramovich, G.N., 1963. *The Theory of Turbulent Jets*. M.I.T. Press, Cambridge, MA.

Albertson, M.L., Dai, Y.B., Jenson, R.A., and Rouse, H., 1950. Diffusion of submerged jets. *Trans. Amer. Soc. Civil Engr.*, **115**, 639–697.

Allen, G.P., Laurier, D., and Thouvenin, J., 1979. *Étude Sédimentologique du Delta de la Mahakam*. Notes et Mémoires No. 15, Total, Paris, Compagnie Francaise des Pétroles, 156 pp.

Allen, J.R.L., 1964. Sedimentation in the modern delta of the River Niger, West Africa. *In*: Van Straaten, L.M.J.U. (ed.), *Deltaic and Shallow Marine Sediments*. Elsevier, Amsterdam, pp. 26–34.

Allen, J.R.L., 1970. Sediments of the modern Niger delta: a summary and review. *In*: Morgan, J.P. (ed.), *Deltaic Sedimentation: Modern and Ancient*. Soc. Econ. Paleont. Mineral. Spec. Publ. 15, pp. 138–151.

Arndorfer, D., 1971. *Process and parameter interaction in Rattlesnake Crevasse, Mississippi River delta*. Unpubl. Ph.D. Dissert., Louisiana State Univ., Baton Rouge, LA.

Axelson, V., 1967. Laitaure delta. *Geografiska Annaler*, **49**, 1–127.

Bates, C.C., 1953. Rational theory of delta formation. *Amer. Assoc. Petrol. Geol. Bull.*, **37**, 2119–2161.

Bondar, C., 1963. Data concerning marine water penetration into the mouth of the Sulina Channel. *Studii de Hidraulica*, **9**(1), 293–335 (in Romanian).

Bondar, C., 1967. Contact of fluvial and sea waters at the Danube River mouths in the Black Sea. *Studii de Hidrologie*, **19**, 153–164 (in Romanian).

Bondar, C., 1968. Hydraulic and hydrological conditions of the Black Sea waters penetration into the Danube mouths. *Studii de Hidrologie*, **25**, 103–120.

Bondar, C., 1970. Considerations theoriques sur la dispersion d'un courant liquide de densite redruite et a niveau libre, dans un bassin contenant un liquide d'une plus grand densite. *Symposium on the Hydrology of Deltas*, UNESCO, **11**, 246–256.

Bonham-Carter, G.F., and Sutherland, A.J., 1968. Diffusion and settling of sediments at river mouths: a computer simulation model. *Gulf Coast Assoc. Geol. Soc. Trans.*, **17**, 326–338.

Borichansky, L.S., and Mikhailov, V.N., 1966. Interaction of river and sea water in the absence of tides. *In*: *Scientific Problems of the Humid Tropical Deltas and Their Implications*, UNESCO, pp. 175–180.

Bowman, M.J., and Iverson, R.L., 1978. Estuarine and plume fronts. *In*: Bowman and Esaias (eds.), *Oceanic Fronts in Coastal Processes*. Springer-Verlag, New York, pp. 87–104.

Bruun, P., 1978. *Stability of Tidal Inlets*. Elsevier, Amsterdam, 510 pp.

Coleman, J.M., 1966. Ecological changes in a massive fresh-water clay sequence. *Gulf Coast Assoc. Geol. Soc. Trans.*, **16**, 159–174.

Coleman, J.M., 1976. *Deltas: Processes of Deposition and Models for Exploration*. Continuing Educ. Publ. Co., Champaign, IL, 102 pp.

Coleman, J.M., and Gagliano, S.M., 1964. Cyclic sedimentation in the Mississippi River deltaic plain. *Gulf Coast Assoc. Geol. Soc. Trans.*, **14**, 67–80.

Coleman, J.M., and Gagliano, S.M., 1965. Sedimentary structures, Mississippi River deltaic plain. *In*: Middleton, G.V. (ed.), *Primary Sedimentary Structures and Their Hydrodynamic Interpretation*. Amer. Assoc. Petrol. Geol. Spec. Publ. 12, pp. 133–148.

Coleman, J.M., Gagliano, S.M., and Smith, W.G., 1970. Sedimentation in a Malaysian high tide tropical delta. *In*: Morgan, J.P. (ed.), *Deltaic Sedimentation: Modern and Ancient*. Soc. Econ. Paleont. Mineral. Spec. Publ. 15, pp. 185–197.

Coleman, J.M., Gagliano, S.M., and Webb, J.E., 1964. Minor sedimentary structures in a prograding distributary. *Mar. Geol.*, **1**, 240–258.

Coleman, J.M., and Prior, D.B., 1980. *Deltaic Sand Bodies*. Amer. Assoc. Petrol. Geol., Continuing Education Course Note Series No. 15, 171 pp.

Coleman, J.M., and Prior, D.B., 1982. Deltaic environments of deposition. *In*: Scholle, P.A. and Spearing, D. (eds.), *Sandstone Depositional Environments*. Amer. Assoc. Petrol. Geol., pp. 139–178.

Coleman, J.M., Suhayda, J.N., Whelan, T., and Wright, L.D., 1974. Mass movement of Mississippi River delta sediments. *Gulf Coast Assoc. Geol. Soc. Trans.*, **24**, 49–68.

Coleman, J.M., and Wright, L.D., 1971. *Analysis of Major River Systems and Their Deltas, Procedures and Rationale, with Two Examples*. Louisiana State Univ., Coastal Studies Inst. Tech. Rept. 95, 125 pp.

Coleman, J.M., and Wright, L.D., 1973a. Formative mechanisms in a modern depocenter. *In*: *Stratigraphy and Petroleum Potential of Northern Gulf of Mexico*, Part II. New Orleans Geol. Soc. Seminar, Jan. 22–24, 1973, pp. 90–139.

Coleman, J.M., and Wright, L.D., 1973b. Variability of modern river deltas. *Gulf Coast Assoc. Geol. Soc. Trans.*, **23**, 33–36.

Coleman, J.M., and Wright, L.D., 1975. Modern river deltas: Variability of process and sand bodies. *In*: Broussard, M.L. (ed.), *Deltas: Models for Exploration*. Houston Geol. Soc., pp. 99–149.

Coleman, J.M., and Wright, L.D., 1978. Sedimentation in an arid macro-tidal alluvial river system: Ord River, Western Australia. *J. Geol.*, **86**, 621–642.

Everett, D.K., 1971. *Hydrologic and Quality Characteristics of the Lower Mississippi River*. Louisiana Dept. Public Works, 48 pp.

Fisk, H.N., 1961. Bar finger sands of the Mississippi delta. *In*: Peterson, J.A. and Osmond, J.C. (eds.), *Geometry of Sandstone Bodies*. Amer. Assoc. Petrol. Geol., pp. 29–52.

Galloway, W.E., 1975. Process framework for describing the morphologic and stratigraphic evolution of deltaic depositional systems. *In*: Broussard, M.L. (ed.), *Deltas: Models for Exploration*. Houston Geological Soc., Houston, TX, pp. 87–98.

Garvine, R.W., 1974. Dynamics of small-scale oceanic fronts. *J. Phys. Oceanogr.*, **4**, 557–569.

Garvine, R.W., 1984. Radial spreading of buoyant, surface plumes in coastal waters. *J. Geophys. Res.* **89**, 1989–1996.

Garvine, R.W., and Monk, J.D., 1974. Frontal structure of a river plume. *J. Geophys. Res.*, **79**, 2251–2259.

Gibbs, R.J., 1970. Circulation in the Amazon River estuary and adjacent Atlantic Ocean. *J. Mar. Res.*, **28**, 113–123.

Gilbert, G.K., 1884. *The Topographical Features of Lake Shores*. U.S. Geol. Survey Annual Rept. 5, pp. 104–108.

Gould, H.R., 1970. The Mississippi delta complex. *In*: Morgan, J.P. (ed.), *Delta Sedimentation: Modern and Ancient*. Soc. Econ. Paleont. Mineral. Spec. Publ. 15, pp. 3–30.

Harleman, D.R.F., and Abraham, G., 1966. *One Dimensional Analysis of Salinity Intrusion in the Rottendam Waterway*. Delft Hydrodynamics Laboratory Publ. No. 44.

Hayashi, J., and Shuto, N., 1967. *Diffusion of warm water jets discharged horizontally at the water surface*. Int. Assoc. Hydraulic Res., 12th Congr., Proc. 4, pp. 47–59.

Hayashi, J., and Shuto, N., 1968. *Diffusion of warm cooling water discharged from a power plant*. Conf. Coastal Engr., Council Wave Res., 11th, Proc., London.

Holeman, J.N., 1968. The Sediment yield of major rivers of the world. *Water Resources Res.*, **4**, 737–747.

Huthnance, J.M., 1973. Tidal current asymmetries Over the Norfolk Sandbanks. *Est. Coastal Mar. Sci.*, **1**, 89–99.

Huthnance, J.M., 1982. On the formation of sandbanks of finite extent. *Est., Coastal, and Shelf Sci.*, **15**, 277–299.

Inman, D.L., and Nordstrom, C.E., 1971. On the tectonic and morphologic classification of coasts. *J.Geol.*, **79**, 1–21.

Jopling, A.V., 1960. *An Experimental Study on the Mechanics of Bedding*. Unpubl. Ph.D. Dissert., Harvard Univ.

Jopling, A.V., 1963. Hydraulic studies on the origins of bedding. *Sedimentology*, **2**, 115–121.

Kashiwamura, M., and Yoshida, S., 1967. Outflow pattern of fresh water issued from a river mouth. *Coastal Engr. Japan*, **10**, 109–115.

Kashiwamura, M., and Yoshida, S., 1969. *Flow Pattern of Density Current at a River Mouth*. Intl. Assoc. Hydraulic Res., 13th Congr., Proc. 3, pp. 181–190.

Kashiwamura, M., and Yoshida, S., 1971. Transient acceleration of surface flow at a river mouth. *Coastal Engr. Japan*, **14**, 135–142.

Kolb, C.R., and Van Lopik, J.R., 1966. Depositional environments of Mississippi River deltaic plain, southeastern Lousiana. *In*: Shirley, M.L. (ed.), *Deltas in Their Geological Framework*. Houston Geol. Soc., pp. 17–61.

Komar, P.D., 1973. Computer models of delta growth due to sediment input from rivers and longshore transport. *Geol. Soc. Amer. Bull.*, **84**, 2217–2226.

LeBlanc, R.J., 1975. Significant studies of modern and ancient deltaic sediments. *In*: Broussard, M.L. (ed.), *Deltas: Models for Exploration*. Houston Geol. Soc., Houston, TX, pp. 13–85.

Lisitzin, A.P., 1972. *Sedimentation in the World Ocean*. Soc. Econ. Paleont. Mineral. Spec. Publ. 17, 218 pp.

Ludwick, J.L., 1970. *Sand Waves and Tidal Channels in the Entrance to Chesapeake Bay*. Old Dominion Univ., Inst. Oceanography Tech. Rept. 1, 79 pp.

Metcalf, W.G., 1968. Shallow water currents along the northeastern coast of South America. *J. Mar. Res.*, **26**, 232–243.

Mikhailov, V.N., 1966. Hydrology and formation of river mouth bars. *In*: *Scientific Problems of the Humid Tropical Zone Deltas and Their Implications*. UNESCO, pp. 59–64.

Mikhailov, V.N., 1971. *Dynamics of the Flow and the Bed in Nontidal River Mouths*. Moscow, Div. of Hydrology, 258 pp.

Milliman, J.D., 1981. Transfer of river-borne particulate material to the oceans. *In*: Martin, J.M., Burton, J.D., and Eisma, D. (eds.), *River Inputs to Ocean Systems*. Proceedings of Scientific Committee on Oceanic Research (SCOR) Special Workshop, Rome, pp. 5–12.

Milliman, J.D., and Meade, R.H., 1983. World-wide delivery of river sediment to the oceans. *J. Geol.*, **91**, 1–21.

Morgan, J.P., 1961. Mudlumps at the mouths of the Mississippi River. *In: Genesis and Paleontology of the Mississippi Mudlumps*. Louisiana Dept. of Conserv., Geol. Bull. 35, Part I, 1–116.

Morgan, J.P., Coleman, J.M., and Gagliano, S.M., 1963. *Mudlumps at the Mouth of South Pass, Mississippi River: Sedimentology, Paleontology, Structure,*

Origin and Relation to Deltaic Processes. Louisiana State Univ., Coastal Studies Series 10, 190 pp.

Morgan, J.P., Coleman, J.M., and Gagliano, S.M., 1968. Mudlumps: Diapiric structure and Mississippi delta sediments. *In*: Braunstein, J., and O'Brien, G.D. (eds.), *Diapirism and Diapirs*. Amer. Assoc. Petrol. Geol. Memoir 8, pp. 145–161.

Nelson, B.W., 1970. Hydrography, sediment dispersal, and recent historical development of the Po River delta, Italy. *In*: Morgan, J.P. (ed.), *Deltaic Sedimentation: Modern and Ancient*. Soc. Econ. Paleont. and Mineral. Spec. Publ. 15, pp. 152–184.

Oertel, G.F., 1972. Sediment transport of estuary entrance shoals and the formation of swash platforms. *J. Sed. Petrol.*, **42**, 858–863.

Off, J., 1963. Rhythmic linear sand bodies caused by tidal currents. *Amer. Assoc. Petrol. Geol. Bull.*, **47**(2), 324–341.

Patterson, C.P., 1875. *Report of the Superintendent*. U.S. Coast Survey, 44th Congr., 1st Session, House Document 81, Sec. VIII, pp. 45–46.

Price, W.A., 1963. Patterns of flow and channeling in tidal inlets. *J. Sed. Petrol.*, **33**, 279–290.

Prior, D.B., and Coleman, J.M., 1978. Disintegrating retrogressive landslides on very-low-angle subaqueous slopes: Mississippi delta. *Mar. Geotechnology*, **3**, 37–60.

Roy, P.S., 1983. Holocene sedimentation histories of estuaries in Southeastern Australia. *In*: Hodgkin, E. (ed.), *Man's Impact on the Estuarine Environment*. Dept. of Conservation and Environment, Western Australia.

Roy, P.S., Thom, B.G., and Wright, L.D., 1980. Holocene sequences on an embayed high energy coast: an evolutionary model. *Sed. Geol.*, **26**, 1–19.

Schlichting, H., 1955. *Boundary Layer Theory*. Pergamon Press, London, 535 pp.

Scruton, P.C., 1956. Oceanography of Mississippi delta sedimentary environments. *Amer. Assoc. Petrol. Geol. Bull.*, **40**, 2864–2952.

Scruton, P.C., 1960. Delta building and the deltaic sequence. *In*: Shepard, F.P., Phleger, F.B., and Van Andel, T.H. (eds.), *Recent Sediments, Northwest Gulf of Mexico, Tulsa*. Amer. Assoc. Petrol. Geol., pp. 82–102.

Shepard, F.P., 1955. Delta front valleys bordering the Mississippi distributaries. *Geol. Soc. Amer. Bull.*, **66**(12), 1489–1498.

Shepard, F.P., 1973. Sea floor off Magdalena delta and Santa Maria area, Colombia. *Geol. Soc. Amer. Bull.*, **84**, 1955–1972.

Smart, J.S., and Moruzzi, U.L., 1972. Quantitative properties of delta channel networks. *Z. Geomorph.*, **16**(3), 268–282.

Stolzenbach, K.D., and Harleman, D.R.F., 1971. *An analytical and experimental investigation of surface discharges of heated water*. M.I.T., Ralph M. Parsons Lab., Water Resources and Hydrodynamics, Dept. of Civil Engr. Rept. 135, 212 pp.

Takano, K., 1954a. On the velocity distribution off the mouth of a river. *J. Oceanogr. Soc. Japan*, **10**, 60–64.

Takano, K., 1954b. On the salinity and velocity distributions off the mouth of a river. *J. Oceanogr. Soc. Japan*, **10**, 92–98.

Takano, K., 1955. A complementary note on the diffusion of the seaward flow off the mouth of a river. *J. Oceanogr. Soc. Japan*, **11**, 1–3.

Taylor, R.B., and Dean, R.G., 1974. *Exchange Characteristics of Tidal Inlets.* Proc. Coastal Engr. Conf., 14th, Copenhagen, pp. 2268–2289.

Thom, B.G., and Wright, L.D., 1983. The geomorphology of the Purari delta. *In*: Petr, T. (ed.), *The Purari-Tropical Environment of a High Rainfall River Basin,* Dr. Junk, Amsterdam, pp. 47–65.

Thom, B.G., Wright, L.D., and Coleman, J.M., 1975. Mangrove ecology and deltaic estuarine geomorphology: Cambridge Gulf-Ord River delta, W.A. *J. Ecology*, **63**, 203–232.

Townsend, A.A., 1976. *The Structure of Turbulent Shear Flow*, 2d ed. Cambridge Univ. Press, Cambridge, 315 pp.

Turner, J.S., 1973. *Buoyancy Effects in Fluids*. Cambridge Univ. Press, Cambridge, 367 pp.

Ünlüata, Ü.A., and Özsoy, E., 1977. *Tidal Jet Flows Near Inlets.* Proc. of Hydraulics in the Coastal Zone Conf. Amer. Soc. Civil Engrs., College Station, TX, 1977, pp. 90–98.

Van Andel, T.H., 1967. The Orinoco Delta. *J. Sed. Petrol.*, **37**(2), 297–310.

Waldrop, W.R., and Farmer, R.C., 1973. *Three-Dimensional Flow and Sediment Transport at River Mouths.* Louisiana State Univ., Coastal Studies Inst. Tech. Rept. 150, 137 pp.

Wang, F.C., 1984. The dynamics of a river-bay-delta system *J. Geophys. Res.* **89**, 8054–8060.

Welder, F.A., 1959. *Processes of Deltaic Sedimentation in the Lower Mississippi River*. Louisiana State Univ., Coastal Studies Inst. Tech. Rept. 12.

Wright, L.D., 1970. *Circulation, Effluent Diffusion, and Sediment Transport Mouth of South Pass, Mississippi River Delta.* Louisiana State Univ., Coastal Studies Inst. Tech. Rept. 84, 56 pp.

Wright, L.D., 1971. Hydrography of South Pass, Mississippi River delta. Amer. Soc. Civil Engrs., Proc., *J. Waterways, Harbors and Coastal Engr. Div.*, **97**, 491–504.

Wright, L.D., 1977a. Sediment transport and deposition at river mouths: a synthesis. *Geol. Soc. Amer. Bull.*, **88**, 857–868.

Wright, L.D., 1977b. *Morphodynamics of a Wave-Dominated River Mouth*. Proc. Coastal Eng. Conf., 15th, pp. 1732–1737.

Wright, L.D., and Coleman, J.M., 1971. Effluent expansion and interfacial mixing in the presence of a salt wedge, Mississippi River delta. *J. Geophys. Res.*, **76**(36), 8649–8661.

Wright, L.D., and Coleman, J.M., 1972. River delta morphology: wave climate and the role of the subaqueous profile. *Science*, **176**, 282–284.

Wright, L.D., and Coleman, J.M., 1973. Variations in morphology of major river deltas as functions of ocean wave and river discharge regimes. *Amer. Assoc. Petrol. Geol. Bull.*, **57**(2), 370–398.

Wright, L.D., and Coleman, J.M., 1974. Mississippi River mouth processes: Effluent dynamics and morphologic development. *J. Geol.*, **82**, 751–778.

Wright, L.D., Coleman, J.M., and Erickson, M.W. 1974. *Analysis of major river systems and their deltas: morphologic and process comparisons*. Louisiana State Univ., Coastal Studies Inst. Tech. Rept. 156, 114 p.

Wright, L.D., Coleman, J.M., and Thom, B.G., 1973. Processes of channel development in a high-tide-range environment: Cambridge Gulf-Ord River delta. *J. Geol.*, **81**, 15–41.

Wright, L.D., Coleman, J.M., and Thom, B.G., 1975. Sediment transport and deposition in a macrotidal river channel, Ord River, Western Australia. *In*: Cronin, L.E. (ed.), *Estuarine Research*, Vol. II, Academic Press, New York, pp. 309–322.

Wright, L.D., and Sonu, C.J., 1975. Processes of sediment transport and tidal delta development in a stratified tidal inlet. *In*: Cronin, L.E. (ed.), *Estuarine Research*, Vol. II, Academic Press, New York, pp. 63–76.

Wright, L.D., Thom, B.G., and Higgins, R., 1980. Sediment transport and deposition at wave-dominated river mouths: Examples from Australia and Papua New Guinea. *Est. Coastal Mar. Sci.*, **11**, 263–277.

2

Estuaries *

Maynard M. Nichols and Robert B. Biggs

Introduction

Although estuaries are one of the most well-studied entities of coastal environments, our geological understanding is still rudimentary. Why such a divergence in our level of understanding? Most inquiry has centered on human needs in an effort to resolve competing demands for use of estuaries, e.g., for shipping and waste disposal as opposed to use as a recreational outlet or food sources. Far more effort has gone into detailed engineering, fishery, and pollution studies than into generalizing and understanding estuaries as a sedimentary environment. As a result, there are few unifying models of deposition, few clues for matching modern and ancient deposits. Our geologic understanding is further tempered by Schubel and Hirschberg (1978) who note that "estuarine deposits rarely can now be delimited unequivocally from other shallow water marine deposits in the geologic record because of their limited areal extent, their emphemeral character and their lack of distinctive features." Then why bother to understand the geologic attributes and processes in estuaries?

The authors believe estuaries figure significantly in the sedimentary make-up of a coastal system, or framework of estuarine environments: lagoon-bay-inlet-tidal flat and marsh. Taken together, these environments make up 80 to 90% of U.S. Atlantic and Gulf coasts (Emery, 1967); and they are found on every continent. Estuaries also have been abundant in the recent past, and they have persisted in periods before the Quarternary when sea level was relatively constant. Estuaries figure prominently because they function either to collect or to convey material between the land and the sea,

*Contribution number 1205 from the Virginia Institute of Marine Science

an important segment of the sedimentary cycle. In vertical sequences, estuaries mark transitions between fluvial and marine environments, a key phase for tracing transgressions and regressions associated with petroleum reservoirs. Further, geologists studying ancient deposits still ask: What kind of environment was it? Under what conditions and in what manner did the sediment accumulate? These are difficult questions. To sharpen our understanding, it is necessary to examine estuaries as they now exist.

This chapter is written for those who need help to read the record of ancient coastal environments from a knowledge of modern environments. Therefore, the following sections review estuaries, their sediment characteristics, evolution, stratigraphic relationships, and processes showing how, within bounds of available data, they relate to the resultant sediments. Many questions are asked to emphasize fundamental ideas. The answers however, are often ambiguous and necessarily incomplete. The gaps in our understanding reveal where much work remains to be done; study has scarcely begun.

What is an estuary? In geomorphic terms, an estuary is "an inlet of the sea, reaching into a river valley as far as the upper limit of tidal rise" (Fairbridge, 1980). Fairbridge has proposed a physiographic classification of estuaries (Figure 2-1), modified by the regional history of sea level, morpho-tectonic factors, climatic factors, and freshwater and sediment supply.

In oceanographic terms, Pritchard (1967) has defined the estuary as "a semi-enclosed coastal body of water which has a free connection with the open sea and within which sea water is measurably diluted with fresh water derived from land drainage." The general characteristics of estuaries as classified by Pritchard are presented in Table 2-1. The distinction between the Pritchard and Fairbridge classifications is twofold. Pritchard's classification is a short-term dynamic scheme, because an estuary may change from a Class B to a Class C with a seasonal change in river discharge. Although Fairbridge recognizes short temporal effects, his classification explicitly considers long-term processes like climatic change and secular sea level rise. A second difference between the two schemes lies in the definition of the landward boundary of the estuary; Pritchard's landward boundary is a chemical one (where the chlorinity falls below 0.01 ‰ and the ratios of the major dissolved ions change radically from their ratios in sea water); Fairbridge's landward boundary is physical (the upstream limit of a measurable tide). For some estuaries, the difference is trivial, but for others the Fairbridge estuary may extend 100 km landward of the Pritchard estuary. From a geological perspective, tidally associated bidirectional sediment properties found in what might otherwise be interpreted as fluvial sediments may represent that portion of the estuary landward of Pritchard's boundary. Neither Fairbridge nor Pritchard deal with the geologically important seaward boundary in an acceptable manner. Both definitions assume that one can locate oneself inside a semienclosed basin or an inlet. This is seldom possible in geological studies with limited outcrop availability. Estuarine conditions, so far as sediment properties are concerned, can extend

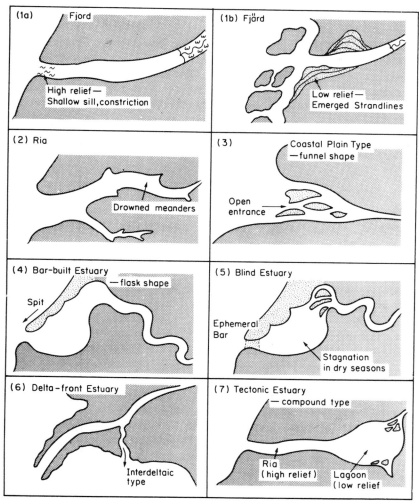

Figure 2-1. Basic estuarine physiographic types. Hydrodynamic characteristics are not considered here; discharge, tidal range, latitude (climate), and exposure all play important roles in modifying these examples, in addition to long-term secular processes such as tectonics and eustasy (schematic) (from Fairbridge, 1980).

for 100 km seaward of the geomorphic estuary mouth. Table 2-2 (Schubel and Hirschberg, 1982) illustrates how the landward estuarine boundary can fluctuate (when using Pritchard's definition) with changing freshwater discharge. By implication, the Amazon, which has zero estuarine length, discharges all of its fresh water beyond its mouth onto the continental shelf. From a sedimentological perspective, the Amazon's estuary is the continental shelf.

Table 2-1. General estuarine characteristics of Drowned River Valley estuaries (from Biggs and Cronin, 1981; reproduced by permission of Humana Press).

Estuarine type[a]	Dominant mixing force	Mixing energy	Width/depth ratio	Salinity gradient	Mixing index[b]	Turbidity	Bottom stability	Biological productivity	Example
A	River flow	Low	Low	Longitudinal vertical	≥ 1	V.high	Poor	Low	Southwest Pass Mississippi River
B	River flow, tide	Moderate	Moderate	Longitudinal vertical lateral	$<\frac{1}{10}$	Moderate	Good	V.high	Chesapeake Bay
C	Tide,wind	High	High	Longitudinal lateral	$<\frac{1}{20}$	High	Fair	High	Delaware Bay
D	Tide,wind	V.high	V.high	Longitudinal	?	High	Poor	Moderate	?

[a]Follows Pritchard's advection-diffusion classification scheme (Pritchard, 1955).

[b]Follows Schubel's definition: $\text{MI} = \dfrac{\text{vol. freshwater discharge on 1/2 tidal period}}{\text{vol. tidal prism}}$ (Schubel, 1971).

Table 2-2. Lengths of selected estuaries during periods of high river flow and low river flow (from Schubel and Hirschberg, 1982; reproduced by permission of Academic Press).[a]

Estuary	High riverflow length (km)	Low riverflow length (km)
Amazon	0	0
Chang Jiang	0	30–40
Chesapeake Bay	280	320
Congo	–	40
Huang Ho	0	0
Hudson	25	55
Long Island Sound	145	145
Mississippi	10	100
Rio de la Plata	150	300
San Francisco Bay	50	60
St. Lawrence	150	175

[a]The length of the estuary is defined as the distance from the mouth of the estuarine basin landward to the last traces of measurable sea salt in the lower layer.

Davies (1973) and Hayes (1975) classified estuaries and related coastal environments according to tidal range. Where the spring tide range is less than 2 m, the environment is termed *microtidal*; between 2 and 4 m, *mesotidal*; and more than 4 m, *macrotidal*. Microtidal ranges are typical of the U.S. Gulf coast, whereas macrotidal ranges are common on European coasts bordering the English Channel.

In this chapter, two key examples of estuaries are used, the Chesapeake Bay (United States) and the Gironde estuary (France) (Figure 2-2), which are the best studied and most significant in the world because of the information they have provided concerning sediment processes and their products.

The Chesapeake Bay is a microtidal estuary with a tidal range of 0.2–0.9 m and relatively weak currents, mostly less than 0.5 m/s. The Bay is shaped by a broad axial channel 270 km long with major branching tributary estuaries and countless small tributaries. Tides extend to the fall line, about 330 km from the sea, and salinity penetrates about 290 km. The Bay is the estuary of the Susquehanna River, which drains 72,000 km^2 and discharges an annual average of 1098 m^3/s and about 2.5 megatons of sediment. With a width of 8 to 48 km and a mean depth of 8.4 m, the Bay is essentially a shallow pan. The margins and floor are, therefore, vulnerable to significant wave stirring and erosion. Suspended sediment concentrations, however, are relatively low, 10–150 mg/liter, and the Bay is biologically rich.

The Gironde is a macrotidal estuary with a tide range exceeding 4 m and strong tidal currents reaching 3.0 m/s. The estuary is narrowly funnel-

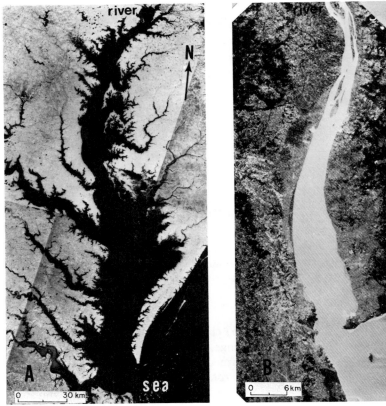

Figure 2-2. Satellite imagery of the (a) Chesapeake Bay on U.S. mid-Atlantic coast, and (b) the Gironde estuary on the southern French Atlantic coast.

shaped, very regular, and has two main channels separated by alluvial islands and shoals in landward reaches. Tides extend 165 km from the sea, and the limit of salt intrusion varies markedly, 30–70 km upstream, depending on seasonal river flow. The estuary receives a mean annual river discharge of 766 m³/s and a sediment influx of about 2.2 megatons from its 80,000 km² drainage basin. The estuary is narrowed to 2–11 km by growth of marshes and tidal flats. Channel depths range from 7 to 30 m. Suspended sediment concentrations are very high, 1–10 g/liter, and transport is vigorous.

Occurrence and Distribution of Estuaries

Estuaries exhibit diverse geomorphic, oceanographic, and sedimentologic characteristics but share several important attributes. They occur at the

ESTUARIES OF THE WORLD

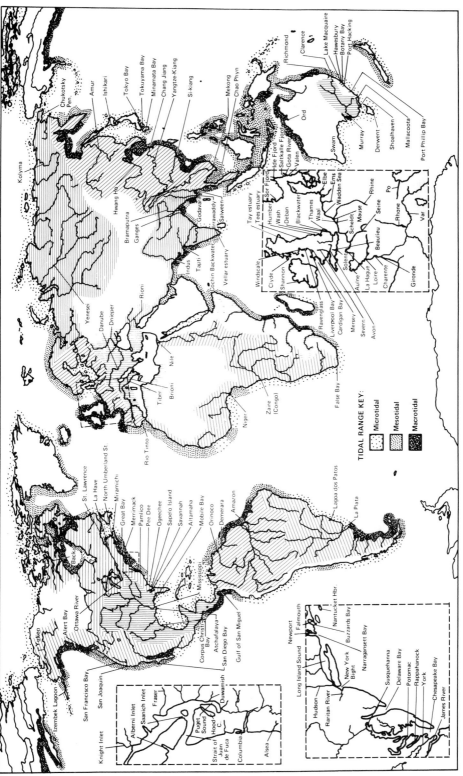

Figure 2-3. Major estuaries of the world and related deltas and bays with estuarine characteristics, (from Olausson and Cato, 1980). Hachured areas on continents represent major drainage basins. Modified and reproduced with permission of John Wiley and Sons.

mouths of rivers that have low sediment loads relative to sediment dissipative forces. Deltas, in contrast, receive high sediment loads and have had their river mouths filled in. Subsidence, at least in the short term, exceeds sediment accumulation rate and the estuary exhibits open water. Tidal currents are active and mix fresh and salt water.

Estuaries are found world-wide in all types of climate and tidal conditions (Figure 2-3). They are best developed on mid-latitude coastal plains with wide continental shelves, especially in former drowned river valleys undergoing submergence or crustal downwarping, a condition that can provide the volumetric capacity to store the influx of sediments. Other factors that promote the maintenance of estuaries include moderate to high tidal range, i.e., where fast currents can keep estuaries clear of sediments. Alternately, estuaries are well developed along coasts that have experienced recent glaciation that has overdeepened former river valleys, creating fjords. Finally, estuaries are well developed along sandy coasts that experience longshore drift that builds spits or barriers across coastal indentations or river mouths, creating estuarine lagoons. The fate of all of these estuaries is to fill with sediment. Emery and Uchupi (1972) have estimated that, if all suspended sediment discharged by rivers of the Atlantic and Gulf Coasts of the United States was deposited in their estuaries, then these basins would fill in 9500 yr, on the average, assuming constant sea level and excluding sediments from the Mississippi River. The reader is cautioned, though, that the rates of sediment supply to different estuaries can vary by many orders of magnitude. The Ganges–Brahmaputra River discharges as much suspended sediment in one day $(4.4 \times 10^6$ ton/day) as the Susquehanna River discharges to Chesapeake Bay in 1.8 years $(2.5 \times 10^6$ ton/yr).

Estuary Evolution

Estuaries form best where river valleys or coastal embayments become drowned or revived, by recent marine submergence. The initial form is a fluvial land mass, usually a river valley, though in northern areas erosion by ice sheets has modified the fluvial topography. Evidence for inundation of river valleys is shown in Figure 2-2a of Chesapeake Bay: The estuary shape resembles a river valley with branching tributaries leading into present-day streams.

Modern estuaries were formed mainly during the most recent rise of sea level, which began about 15,000 yr ago, a period of deglaciation. Sea level rose from a depth on the continental shelf of 100–130 m to its present position (Figure 2-4a). The rate of sea level rise was not constant, however, and its position on the continental shelf and coastal plain varied. At the lowest stand of sea level when the shoreline lay near the shelf edge, estuaries probably were few and confined to valleys near the shelf edge. According to

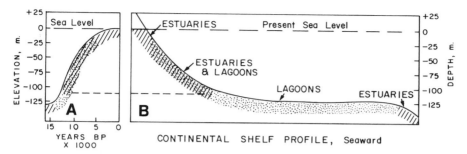

Figure 2-4. Relationship of estuary and lagoon development to (a) post-glacial sea level rise and to (b) generalized profile of continental shelf. Modified and reproduced with permission; from Emery (1967), and sea level rise curve from Milliman and Emery (1968).

Emery (1967), lagoons formed as sea level advanced slowly across the flat outer shelf. Further rise at a more rapid pace, between 12,000 to 6000 yr ago, brought the shore zone to steeper parts of the shelf profile (Figure 2-4b). In this zone estuaries developed, partly in drowned shelf channels, together with lagoons. Estuaries reached peak development in size and number as sea level rise slowed 3000 to 5000 yr ago.

Once formed, estuaries make good sediment traps. They tend to fill rapidly at first, decreasing in depth, contracting in volume and surface area until the river flows directly to the sea across a depositional plain. Progressive infilling smoothes the initial shape and closes cross-sectional areas that tend to come into equilibrium with river and tidal discharge. A regular exponential seaward increase of cross-sectional areas in the Gironde (Figure 2-5) suggests near equilibrium between geometry and flow. Enclosure by spit growth in late stages and the advance of tidal flats and marshes into the estuary tend to accelerate infilling. Why, then, are so many river mouths occupied by estuaries instead of deltas?

The formation and life span of an estuary depends first on the balance between the rise of sea level relative to the land and sediment accumulation (Figure 2-50). Infilling opposes submergence. Where the pace of sea level rise exceeds infilling, as in Chesapeake Bay and Long Island Sound, estuaries are well developed and persist. Where sea level is nearly stable, as in the Gironde, sediment infilling can catch up to or exceed sea level rise. This condition accelerates the natural reclamation of an estuary and shortens its life. One of the most sensitive zones to an imbalance of sea level rise and infilling lies at the estuary head, or river entrance, where sedimentation rates are fast and result in an estuarine delta. Late stages of infilling depend on the trapping efficiency of the circulatory regime to retain the sediment supply, a process treated in a later section of this chapter entitled *Escape or Entrapment?* Many present-day estuaries are not filled because they are geologically very young.

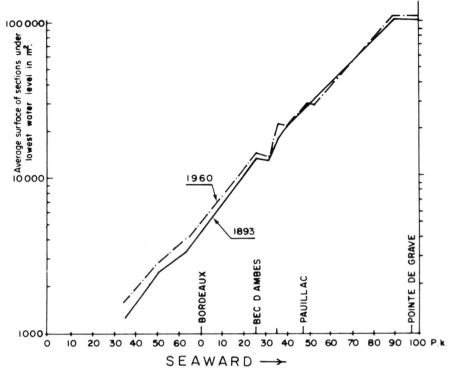

Figure 2-5. Distribution of cross-sectional areas along the Gironde estuary between 1893 and 1960 (from Migniot, 1971).

A prime example of estuary evolution during the Holocene sea level rise is provided by Allen *et al.* (1973), Feral (1970) and by Jouanneau and Latouche (1981) from many cores, dates, historic charts, and paleogeographic interpretations of the Gironde estuary, France. The detailed reconstruction shows how estuary evolution produced a sequence of lithofacies in response to eustatic sea level rise, changing climate, river flow, and sediment supply at the site of the present-day estuary. The sequence of infilling displays four phases. The stage is set by the pre-Holocene regression, a time when the Gironde River incised a deep valley into limestone beds, excavated enormous amounts of fluvial sand and gravel, and distributed it on the inner continental shelf. The terrain is marked by decalcified clays and paleosols (Figure 2-6).

Phase 1: 18,000–10,000 yr b.p.; sea level rise 120–50 m; rapid transgrassion across shelf (Figure 2-7a)

Peri-glacial braided river with relatively steep slope, high sediment loads, flooding, and accumulation of coarse sand and gravel mixed with eolian sand.

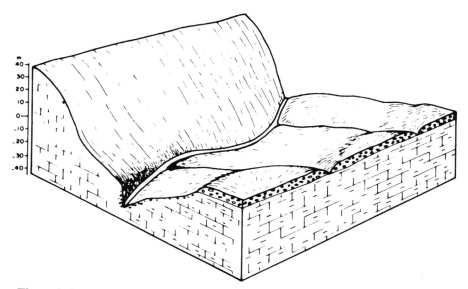

Figure 2-6. Regression phase of the Gironde estuary evolution during pre-Holocene, prior to 18,000 b.p. (from Jouanneau and Latouche, 1981).

Phase 2: 10,000–6000 yr b.p.; sea level rise rapid, 50–10 m (Figure 2-7b)

Rise of sea level raised river base-level, decreased slope, and reduced river competency. This led to a meandering-type river with accumulation of channel sands and sandy clay, alluvial levees, and swamp-marsh organic clay with plant debris. Infilling rate lags sea level rise and associated increase of volumetric capacity.

Phase 3: 6000–1000 yr b.p.; sea level rise slower; 10–0 m (Figure 2-7c)

Flooding of river valley at present-day site and formation of open estuary. Rapid accumulation and infilling accumulation of estuarine and marine channel sand bars and laminated clays rich in shell. Marine sand introduced via ocean littoral drift and landward flow in channel. (Note: U.S. mid-Atlantic estuaries are approximately at this phase today.)

Phase 4: 1000 yr b.p. to present; sea level stable at present level (Figure 2-7d)

Accretion of tidal flats and marshes and narrowing of cross section. Accumulation of alluvial mud in channel and sand in shoals and alluvial islands of upper estuary. Fluvial supply likely

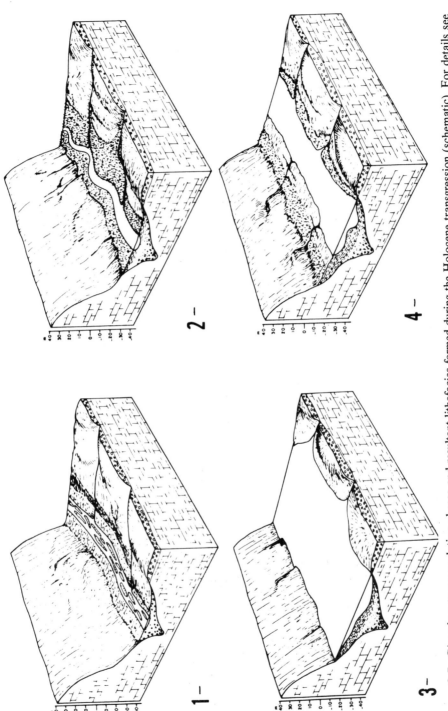

Figure 2-7. Gironde estuary evolution phases and resultant lithofacies formed during the Holocene transgression (schematic). For details see text. (a) Phase I; (b) Phase II; (c) Phase III; (d) Phase IV (from Jouanneau and Latouche, 1981).

increased by higher sediment yield. Marine sand supplied to lower estuary via ocean littoral drift and landward flow. Infilling nearly in equilibrium with sea level rise.

The last sea level rise is but one phase in a series of Pleistocene sea level fluctuations. Each rise probably reinundated the site of former valleys and former estuaries. Therefore, estuarine sediment deposits are likely to have been reworked and eroded many times.

Sources and Nature of Estuarine Sediments

Estuarine sediments are derived from a number of sources including the watershed, the continental shelf, the atmosphere, erosion of the estuarine margins and bottom, and biological activity within the system. A particular estuarine deposit usually consists of various proportions of materials from these sources. The dominance of one sediment source depends on its magnitude relative to all other sources and the dynamics of erosional, transportation, and depositional processes. Thus, lower estuary or marine diatoms may be a dominant component of the microfossils in upper estuarine sediment because the frustrules can be transported toward the estuary head by landward-flowing bottom currents. The source of zinc in one estuarine sediment may be the watershed, whereas in a nearby sediment zinc may have a strong signal from an atmospheric source. The reader should realize that a marine or fluvial source for one estuarine sedimentary component cannot be generalized to that same origin for all sedimentary components, even in the same lithofacies.

The nature and distribution of sedimentary facies in an estuary are controlled by the interaction between the kinds and quantity of available sedimentary components (such as shells, organic matter, quartz sand grains, fecal pellets, and clay minerals), hydrodynamic processes, and bottom morphology (Allen *et al.*, 1973). Examples of the relationship between morphology, facies, and the dynamic environment are presented in Figures 2-8 and 2-10 for the Gironde estuary (Allen *et al.*, 1973) and in Figures 2-9 and 2-11 for the Chesapeake Bay.

Gironde Estuary

The surface sediments consist of silt and clay with interbedded sand in the fluvial channels, sand on the channel bars and sandy muds on the intertidal flats (Figure 2-10) (Allen *et al.*, 1973). Rippled sand sheets and occasional dunes are present on the channel floor. The bars are covered by ripples and dunes with amplitudes proportional to the varying intensity of tidal currents.

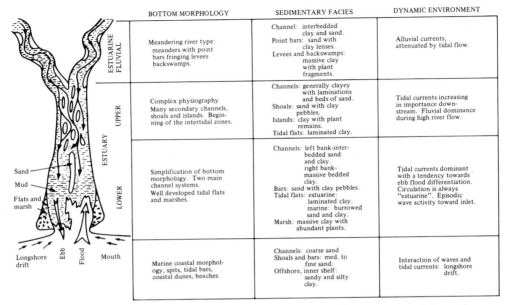

	BOTTOM MORPHOLOGY	SEDIMENTARY FACIES	DYNAMIC ENVIRONMENT
ESTUARINE FLUVIAL	Meandering river type: meanders with point bars fringing levees backswamps.	Channel: interbedded clay and sand. Point bars: sand with clay lenses. Levees and backswamps: massive clay with plant fragments.	Alluvial currents, attenuated by tidal flow.
UPPER ESTUARY	Complex physiography. Many secondary channels, shoals and islands. Beginning of the intertidal zones.	Channels: generally clayey with laminations and beds of sand. Shoals: sand with clay pebbles. Islands: clay with plant remains. Tidal flats: laminated clay.	Tidal currents increasing in importance downstream. Fluvial dominance during high river flow.
LOWER ESTUARY	Simplification of bottom morphology. Two main channel systems. Well developed tidal flats and marshes.	Channels: left bank-interbedded sand and clay. right bank-massive bedded clay. Bars: sand with clay pebbles. Tidal flats: estuarine: laminated clay. marine: burrowed sand and clay. Marsh: massive clay with abundant plants.	Tidal currents dominant with a tendency towards ebb flood differentiation. Circulation is always "estuarine". Episodic wave activity toward inlet.
	Marine coastal morphology, spits, tidal bars, coastal dunes, beaches.	Channels: coarse sand Shoals and bars: med. to fine sand. Offshore, inner shelf: sandy and silty clay.	Interaction of waves and tidal currents: longshore drift.

Figure 2-8. Relationship between bottom morphology, sediment type, and dynamic forces in the macrotidal Gironde estuary (from Allen *et al.*, 1983).

The tidal flats are composed of laminated or bedded muddy sediments, usually tan, gray, red, or brown, and no evidence of bioturbation. In the more seaward reach of the fluvial estuary, extensive broad tidal flats have developed and some have been colonized by vegetation, forming marshes with distinctive black or dark-green sediment, in which bedding structures are disturbed by roots (Allen *et al.*, 1973).

The lower Gironde exhibits a lithological contrast to the upper estuary. The entire estuary floor is covered with medium to coarse sand. Tidal flats and salt marshes are well developed and both shell and living molluscs become abundant. Few physical sedimentary structures are present because of intensive bioturbation.

At the estuary mouth, the bottom sediments vary from well-sorted fine sands to poorly sorted sands and gravels. Intertidal areas are thin, sandy beaches overlie silty bioturbated intertidal flats. On the inner shelf, just seaward of the estuary mouth, lies an extensive depositional sequence of alternately well-bedded and bioturbated silts.

Chesapeake Bay

The uppermost estuarine segment of the Chesapeake consists of a broad, shallow, subtidal delta of the Susquehanna River. Sediments consist of fine

	BOTTOM MORPHOLOGY	SEDIMENTARY FACIES	DYNAMIC ENVIRONMENT
ESTUARINE FLUVIAL / UPPER	Shallow Submarine Delta	Med. to fine moderately sorted sands interbedded with silts at downstream end.	River discharge attenuated by tide and wind waves.
	Simple with one or two main channel systems	nearshore - sands offshore - massive mud deposits with occasional sand laminations and beds	Littoral drift Tidal currents
ESTUARINE	Simple with one relict channel, occasional well developed tide marshes where pre-Holocene topography is flat, occasional offshore bars where shelf topography is gentle	nearshore - sands, multiple bars and bar fields. offshore - massive mud deposits with occasional sand laminations and beds becoming more abundant at seaward end	Littoral drift wave action Tidal currents
LOWER	Multiple channels shoals and bars	Flood tidal delta - coarse to fine sands with or without x-bedding	Waves and strong tidal currents with ebb-flood separate
MARINE ESTUARINE	Marine coastal morphology with spits, bars, dunes and well developed beaches	tidal delta grading seaward into coarse relict sands and gravels	Waves, strong tide currents and littoral drift

Figure 2-9. Relationship between bottom morphology, sediment type, and dynamic forces in the microtidal Chesapeake Bay. Interrelationships based on the data of Byrne *et al.* (1982) and Kerhin *et al.* (1983).

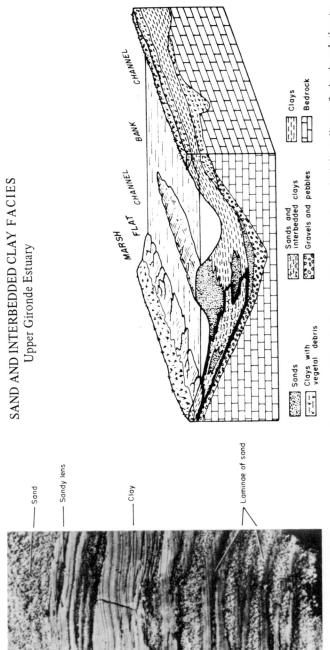

Figure 2-10. Schematic cross-section through the upper Gironde Estuary, France showing principle sediment facies in relation to channel cross-section. Core from channel sand and interbedded clay facies, left (from Allen, 1972).

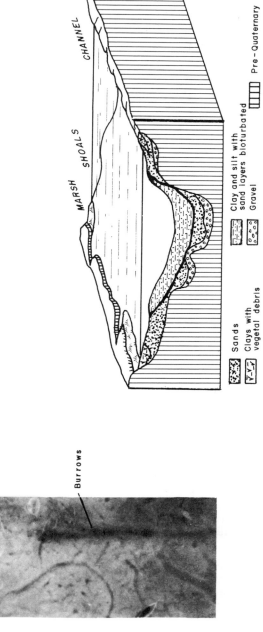

MUD FACIES, Bioturbated
Upper Chesapeake Bay

Figure 2-11. Schematic cross-section through upper Chesapeake Bay showing principle sediment facies in relation to channel cross-section. Core from channel mud facies, bioturbated, left.

to medium sands transported from the river during high discharge and winnowed by waves.

Immediately seaward of the submarine delta lies a large region characterized by gray or black muds, grading toward shallow waters into sandy muds with sands restricted to a narrow beach zone. Aside from occasional organic fragments, there are very few apparent sedimentary structures (Figure 2-11). Marshes are restricted to small protected pockets in coves and tributaries (Kerhin et al., 1983).

Along most of the length of the Chesapeake, deepwater sediments are silts and clays with rare shell or sand layers and laminations. In this region of higher wave fetch, nearshore sands extend to greater depths than farther landward. Broad shelves, cut into pre-Holocene sediments, are apparent, especially on the eastern margin. Locally, large-scale sand wave fields are present, but most of the shallow shelves are surfaced with thin, coarse, lag deposits reworked by waves. Spits and baymouth barriers are common where tributaries enter the main bay. Extensive salt marsh deposits are abundant where the transgressed surface is relatively flat and where sedimentation has, in the past, exceeded the local rate of sea level rise.

The seawardmost portion of the Chesapeake is, as is the Gironde, an estuarine flood tidal delta complex. Estuarine muds become progressively more sandy as the mouth is approached. The topography becomes complex and there is segregation of ebb and flood currents. Sands become coarser and more gravelly (Byrne et al., 1982). In contrast to the Gironde, though, the inner shelf off the Chesapeake is covered with sand that is a lag of pre-Holocene shelf materials.

In brief, the facies character and pattern in the fluvial, upper estuary, and mouth zones of the Chesapeake Bay are broadly similar to those of the Gironde estuary. In middle and lower estuarine zones, however, the facies differ in that grain size increases with depth in the Gironde but becomes finer with depth in the Chesapeake. Whereas flats of the Gironde are mainly mud, shoals of the Chesapeake are sand as a result of wave action in open water (Figures 2-10, 2-11).

Estuarine Processes

Sediments supplied to an estuary enter a remarkably complicated environment. They may undergo back-and-forth transport by ebb and flood tidal currents and recycle many times prior to deposition. Some sediments meet sharp boundaries of fresh and salty water while other sediments are eaten by animals that transform their physical and chemical form. Sediments are either "filtered" from estuarine water or flushed into the sea. Consequently, the amount and kind of sediment discharged from an estuary may differ markedly from the sediment supplied. How, then, are the sediments modified by different processes in an estuary? And how do different processes lead to retention and accumulation of some sediments while others escape into the sea? These questions are of fundamental interest because they affect the depositional history of estuaries and the character of facies that accumulate. This section examines the significant hydrodynamic, chemical, and biological factors and considers how they modify, retain, and accumulate sediment in an estuary. To understand the relationship between dynamic factors and sediment modifications, it is necessary to consider the general nature of processes and basic concepts.

Sedimentary processes in estuaries are determined by the dissipation of energy from river inflow, density gradients, the tide, waves, and meterological forces. As energy is dissipated, sediments are transported, mixed, exchanged, or accumulated and the bottom geometry is modified. These forces and processes act simultaneously or independently to supply and distribute varying kinds of amounts of sediment. They also interact with each other so their effects may be additive or they may cancel. Additionally, these processes fluctuate in intensity at time scales ranging from seconds to years or longer; some are periodic, others are random. Despite this complexity, it is possible to recognize dominant processes and show how they relate to resultant sediment characteristics—sediment structure, texture, composition, and depositional patterns and sequences.

General Concepts

The movement of estuarine sediments consists of a cycle of four processes: (1) erosion of the bed, (2) transportation, (3) deposition on the bed, (4) consolidation of deposited sediment. These processes strongly depend on the flow dynamics and on the particle properties, their size, shape, density, and composition. For cohesive sediments the bulk properties, physiochemical particle behavior, and interparticle bonding are of fundamental importance. These features make cohesive sediments respond very differently to hydrodynamic forces than do noncohesive sediments such as single sand grains. There is a substantial volume of literature on the movement of noncohesive sediments from fluvial and engineering studies, e.g., Graf (1971), Vanoni (1975), Komar (1976a, b), and Sternberg (1972); this discussion emphasizes cohesive sediments that are less well known and that dominate sediment movement in estuaries.

What is cohesion? Cohesive or so-called muddy sediments have a small size, less than about 125 μ, and a large surface area relative to mass. They are characterized by abundant clay minerals having an interlayered crystal structure and normally a net negative surface charge. Cohesion is the result of interparticle surface attractions between clay minerals. Particles cohere when repulsive surface charges are suppressed in a weak electrolyte such as sea water. Mineral cohesion is reinforced by organic cohesion, a binding of particles by mucus secretions or biogenic pelletization. Cohesion makes bed sediment more resistant to erosion than if immersed weight were the only stabilizing force.

Erosion

What conditions promote erosion? To erode sediment from a bed, fluid lift and drag forces produced by turbulence must overcome stabilizing forces of the sediment (Figure 2-12a). For noncohesive sediment, the main stabilizing force is the immersed particle weight, but for cohesive sediment with a negligible weight, the main forces are interparticle adhesion and organic binding. Normally cohesive beds are soft and layered with density and shear strength increasing downward. Sediment moves whenever the shear stress, τ, transmitted to the bed by fluid flow reaches a threshold or critical shear stress for erosion, τ_e, equal to the shear strength "holding" the sediment to the bed. The shear strength is thus equal to the critical shear stress.

In general, erodibility of sandy sediment is a direct function of grain size and shear stress on the bed (Figure 2-12b). The critical bed shear stress for erosion, τ_e, of noncohesive sediments is commonly obtained from a refinement of data obtained from Shields' (1936) diagram (Miller *et al.*, 1977), Figure 2-13. This relates dimensionless shear stress, θ_t, which employs specific weights of fluid and sediment required to erode a flat-bedded sand, to the grain Reynolds number, Re, which employs particle size and dynamic viscosity. Although these data are mainly for azoic and noncohesive freshwater sediments ($> 100 \mu$), organic detritus in silt and sand sizes generally falls within the limits of scatter about Shields' curve (Fisher *et al.*, 1979) (Figure 2-13). Fecal pellets and microbially colonized beads, however, alter the resistance to erosion (Rhoads, *et al.*, 1978). Also, the curve does not account for boundary conditions, e.g., bedforms, slope, and nonuniform grain size distributions. Despite these limitations, Shields' curve is useful for predicting the minimum competence of flow required to move a given grain size. Baker and Ritter (1975) used it for determining paleoflow. For cohesive sediment ($< 125 \mu$ grain size), there is no entirely satisfactory way at present to predict the critical shear stress or erosion rates as a function of erosion resistance for a range of flow conditions. A number of engineering laboratory experiments, however, indicate general trends.

The yield strength, τ_y, the force required to break bonds between aggregates of cohesive particles, generally correlates to the critical shear velocity for erosion according to experimental results of Migniot (1968) and Krone (1963). By using a variety of estuarine, marine, and freshwater muds and forcing suspensions through a narrow capillary tube at various driving pressures, the rate of shearing in the tube at a given pressure is proportional to the shear stress necessary to break the interparticles and aggregate bonds. Much estuarine sediment work is based on the generalized curve of Migniot

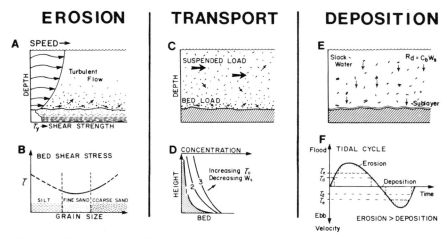

Figure 2-12. Schematic diagrams showing types of sediment movement, erosion, transport and deposition, and general relationship of sediment characteristics to hydrodynamic forces (adopted from G. P. Allen, unpublished).

(a) Velocity profile showing reduction of current speed with depth in turbulent flow field. Shear strength profile increasing downward in bed.

(b) Variation of shear stress exerted on the bed, as a function of grain size.

(c) Distribution of suspended load in relation to bed load.

(d) Vertical distribution of suspended sediment and change with increasing shear stress on the bed, and decreasing settling velocity, w_s.

(e) Deposition activity in weak flow where shear stress on the bed is less than the critical shear stress for deposition; rate of deposition, Rd, is product of concentration, C_b, and settling velocity, w_s.

(f) Tidal time-velocity curve of flood and ebb current with zones hachured (under curve) when shear stress exceeds the critical value for erosion, τ_e, and the critical value for deposition, τ_d.

(1968) that relates the drag force applied by flow U_* to yield strength, τ_y (Figure 2-14). For example, a shear velocity of 0.65 cm/s is required to erode mud having a strength, τ_y, of 0.25 dynes/cm. For flow over a flat, smooth bed, a force of that magnitude is transmitted by currents of 12 cm/s at 1 m above the bed. A shear velocity of 1.5 cm/s would correspond to about 40 cm/s, the precise value depending on roughness of the bed (McCave, 1978). The erosion rate constant for estuarine muds, i.e., that relates the rate of erosion to applied shear stress over the critical value, is in the range of 0.01–0.2 g/cm^2/s/dyne/cm^2. Migniot's relationship can provide an evaluation of the long-term balance of erosion or sedimentation along an estuary and, thus, indicate whether the bed is a source or sink (McCave, 1979).

The yield strength and, hence, erosion resistance increases with depth in the bed as overburden pressure or consolidation increase (Lambermont and Lebon, 1978). This is evident when it is recognized that aggregates

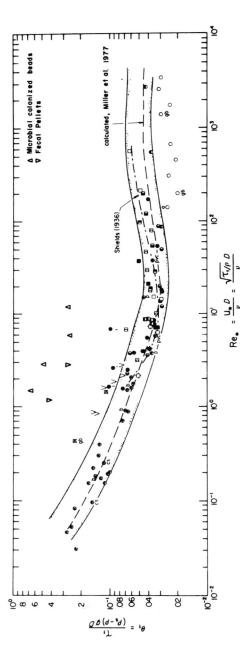

Figure 2-13. Shields diagram of dimensionless shear stress, θ_t, required to erode flat-bedded noncohesive sand with grain Reynolds number, Re_*, within the envelope dashed line represents Shields original curve and a calculated curve derived from friction velocity, U_*, versus grain diameter by Miller *et al.*, 1977. Symbols from various sources after Miller *et al.*, 1977; triangles are biogenically modified microbially colonized beads (Rhoads *et al.*, 1978) and fecal pellets (Nowell *et al.*, 1981).

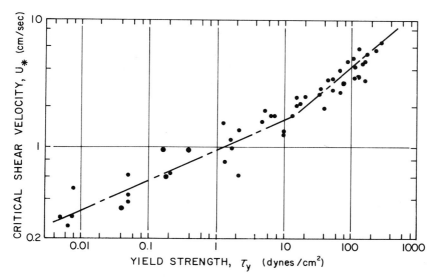

Figure 2-14. Critical shear velocity for erosion. U_* versus the yield strength of aggregated bed material, τ_y, from Migniot (1968).

(cohesively bound particles) on the surface of a fresh deposit consist of weak, "fluffy" units with a high amount of interstitial water. Increasing aggregation implies formation of aggregated units of decreasing density and shear strength. As consolidation proceeds, the aggregates are crushed by compaction, and they become comparatively stronger and more tightly packed into units (Figure 2-12a). Krone (1962) recognized five distinct layers 1–2 cm thick, each with its own structure and value of critical shear stress. Greatest downward increase in shear strength occurs in the surficial layer. An increase of erosion resistance with depth requires a decrease of erosion rate with time for a constant fluid discharge. Equations for predicting the rate of erosion are summarized by Mehta *et al.* (1982).

Most laboratory erosion studies (e.g., Partheniades, 1965; Postma, 1967) are limited to freshwater sediments and a remolded bed that has different resistance characteristics than undisturbed estuarine sediments. To avoid these difficulties, Young and Southard (1978) constructed a portable sea flume and deployed it *in situ* on a natural bed. Estimates of critical shear stress for erosion, τ_e, and threshold shear velocity, U_*, from the sea flume disclosed that values of erosional U_* were only about one-half those for corresponding sediment in a laboratory flume. This difference was ascribed to a change in sediment cohesion caused by natural bioturbation and lower bulk density of biologically aggregated sediment. But even small-scale experiments do not necessarily reproduce the large-scale phenomena of nature (Seibold and Berger, 1982).

A cohesive bed can be eroded in two distinct ways. When bed shear stress is just above a critical value, the indirect impact of fluid stresses strip off surface sediment aggregate by aggregate. At higher stress when the bulk shear strength is exceeded, direct stress causes mass erosion along a plane of failure deep in the bed and the overlying sediment is detached in lumps or instantly suspended. Despite considerable study, the details of stress and sediment response are not clear. It is difficult to predict the changing character of cohesive sediments and the complicated feedback between the sediments and water movement at the sediment–water interface.

Transport

The most fundamental case of sediment transport is the steady settling of smooth, nearly spherical particles. These particles are acted on by a downward gravity force and an opposing or upward force. For fine particles smaller than about 100 μ having a relatively small mass, the viscous resistance of the water is the main opposing force and settling velocity varies with the square of the diameter, d, following Strokes' Law (Ruby, 1933):

$$W_s = Cd^2 \qquad (2\text{-}1)$$

where W_s is the settling velocity; and C includes the various constants, particle density, fluid density, acceleration of gravity, and fluid viscosity. For very fine particles smaller than about 0.5 μ, Brownian motion in the fluid retards settling. For coarse particles larger than coarse sand (about 1000 μm), the mass is so great that viscous resistance is negligible. The particles are affected by inertial forces that generate frontal pressure and a wake. Settling velocity, therefore, varies with the square root of the diameter following the Impact Law:

$$W_s = C_2\sqrt{d} \qquad (2\text{-}2)$$

where the various constants as above are included in C_2. For particles of intermediate size, 100 to 1000 μ, there is a transition where both viscosity and mass are important. The heavy line of Figure 2-15 represents a general curve for a range of diameters. A dispersed clay size particle of 1.0 μ diameter settles very slowly, 0.0001 cm/s or 30 cm in about 3 days, while a silt particle with 20 μ diameter settles 0.04 cm/s or 30 cm in 10 minutes. Grains with high specific gravity settle faster than light ones of the same diameter such as organic-rich grains. Irregularly shaped grains (e.g., discs and flakes) settle more slowly than spheres because they offer more resistance. Shape factors and density corrections are provided by Lerman (1979) and Krumbein and Pettijohn (1938). Because measurements of natural sediment have a wide range of particle density and are affected by particle shape and by flocculation or deflocculation, size measurements are referred to as equivalent to Strokes' settling velocity based on quartz spheres of 2.65 density. For example, floccules with a true diameter of 100 μ may have an equivalent diameter of 10 μ. However, basic data for *in situ* bulk

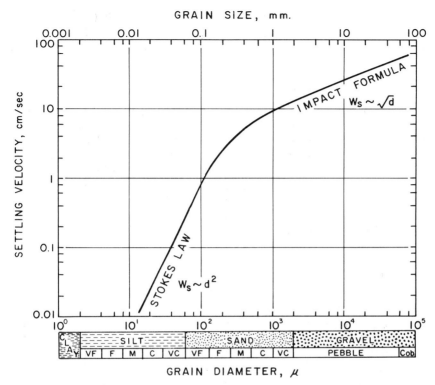

Figure 2-15. General curve of settling velocity for free-falling nearly spherical grains at 20°C in still water as a function of grain diameter (from Krumbein and Pettijohn, 1938).

density of various sized particles in estuarine suspensions are lacking (McCave, 1979).

If current speed of a sediment-transporting flow is diminished, the settling relations predict that coarse material will settle out first and fine particles will follow. By settling at different rates, sand-size material tends to separate from clay and silt-sized particles. The sediment is therefore sorted during transport. Silt and clay that settle slowly are usually distributed throughout the flow while sand is concentrated near the bed.

What are the basic transport modes? Sediment is moved in an estuary as suspended load or bed load (Figure 2-12c). The transport mode differs according to how energy is transmitted to the particles and their dependence on the bed. Bed load movement occurs intermittently in a thin layer several grain diameters thick, on or in close contact with the bed. Energy is transmitted by intergrain contact through rapid and frequent collisions manifest in rolling or discrete jumps. In suspended load, particles are borne throughout the flow field and largely independent of the bottom (Figure 2-12c). Energy is transmitted by water turbulence. Suspended load moves about as fast as the mean flow velocity, whereas bed load moves at a slower rate than that of the mean flow.

The mode of transport is affected by flow and bed characteristics. Bedforms such as ripples and sand waves modify the near-bottom flow field and feed sediment into suspension. Under turbulent flow, suspended particles are not necessarily maintained in suspension because there may be a continuous exchange between the suspension and the bed (Dyer, 1972). Details of boundary layer transport are discussed by Komar (1976a) and Yalin (1977). Although the relative importance of bed-load transport to suspended load is not well established, suspended load is usually the prominent feature in most estuaries (Officer, 1981).

The suspended load may be moved by processes of advection and diffusion. The term *advective transport* means movement of suspended material in a given water mass from one point to another without changing its spread. Usually this is a net horizontal movement. *Diffusive transport* changes the spread and transports sediment from a zone of higher concentration to one of lower concentration. Turbulence of waves and tides is mainly responsible for diffusive transport.

Can we predict the vertical distribution of suspended load? The vertical distribution of suspended sediment load within the flow depends on the relative magnitude of turbulence and on the particle settling velocity, W_s (Owen, 1977). The faster the settling velocity the more turbulence will be required to maintain sediment in transport, and hence the higher the bottom shear stress, which is a product of turbulence at the bottom (Figure 2-12d). With proportionally fast settling velocities, the sediment tends to travel near the bed, whereas with slow settling velocities the sedient is mixed throughout the depth of flow (Owen, 1977) (Figure 2-12d).

A simple one-dimensional formulation used in fluvial studies to predict the vertical variation of dilute sediment concentration, C, of a given grain size that results from a balance between the downward gravitational motion and upward buoyant effect of turbulence is:

$$W_s \, C = K_z \frac{dc}{dz} = 0 \qquad (2\text{-}3)$$

where W_s is settling velocity, K_z is the upward turbulent diffusion coefficient for sediment, and z is the height above the bed (Graf, 1971). Equation 2-3 can be solved to produce the concentration of sediment, C, of constant size and settling rate from a known reference concentration, Ca, at a height, $a = z$, just above the bed by:

$$\frac{C}{Ca} = \left(\frac{h-z}{z} \cdot \frac{a}{h-a} \right)^{w_s/ku_*} \qquad (2\text{-}4)$$

where k is the Von Karman constant (0.4 in open channel flow); u_* is the shear velocity; and h is the total flow depth. This equation, which arises from the concept of upward sediment diffusion, predicts a progressive decrease of concentration away from the bed (Kranck, 1980b). The ratio w_s/ku_* determines whether the sediment moves as bed load ($w_s/ku_* > 1$) or as suspended load ($w_s/ku_* < 1$). As the exponent w_s/ku_* decreases, either through an increase in shear velocity or a decrease in settling velocity, sediment becomes more evenly distributed with depth. Equations 2-3 and 2-4 are well founded in theory and have been verified in flumes and rivers where current is steady and unidirectional, logarithmic velocity increases toward the bed as the stress distribution is linear, and where sediment concentrations are constant. Unfortunately, these conditions are not common in estuaries, but the equations may give reasonable approximations (McCave, 1979). Measured values are preferred.

Deposition

What are the conditions for deposition? The temporary emplacement of suspended particles on the bed by deposition is largely controlled by sediment concentration, settling velocity, and fluid shear. In the simplest case, settling is still water or at slack tide, the rate of deposition, R_d, is predicted by the product of the suspended concentration, C_b, and its settling speed, W_s (Figure 2-12e). In a weak current, sediment continues to settle toward the bed but can only deposit if the shear stress exerted on the bed by the current, τ, is lower than the initial bonding strength of the particles to the bed. The shear stress below which deposition can proceed is known as the limiting shear stress for deposition, τ_d (Owen, 1977). Deposition of fine sediment occurs when the limiting or critical shear stress for deposition, τ_1, falls below a certain value. When shear stress is below τ_1, the rate of deposition is proportional to the ratio $(1 - \tau/\tau_d)$. As bottom stress increases with an increase of current velocity, the rate of deposition decreases; when τ equals τ_d deposition ceases. Neglecting factors such as waves, bed roughness, and the influence of organisms, the rate of deposition is given by:

$$R_d = C_b \, W_s \, (1 - \tau/\tau_d) \qquad (2\text{-}5)$$

Because the limiting shear stress for deposition of cohesive sediments is difficult to measure, most estuary studies use values from flume tests. These are in the range of 0.41–0.81 dynes/cm^2 (Krone, 1962; Partheniades et al. 1969; McCave, 1972), which correspond to a current speed of about 17 to 25 cm/s at 1 m above the bed assuming steady smooth boundary flow. Usually the value of τ_d is significantly less (by about 30%) than the critical shear stress for erosion, τ_e. Such a difference reflects the stronger inter-aggregate bonds in the bed surface layer than those between aggregates momentarily contacting the bed during initial stages of deposition.

According to a model of McCave and Swift (1976), deposition of suspended sediment is accomplished by particle entrapment in a viscous sublayer near the bed. Because turbulence is low or flow is laminar, even particles with slow setting speeds can settle and deposit. Sediment within the boundary layer cannot be conveyed up to higher levels by turbulent movements (Sundborg, 1967). Trapping occurs below a limiting shear stress for deposition. Presumably this is controlled by the balance between input to and ejection from the sublayer and is a function of settling velocity. The boundary layer is complicated by periodic disruptions of the sublayer whereby low velocity fluid is ejected and high velocity fluid is entrained (Gordon and Whitting, 1977). Additionally, strong interparticle bonding of high suspended concentrations can modify the near-bed flow structure by thickening the viscous sublayer causing a large shear stress reduction and a high rate of deposition (Gust, 1976).

According to a tidal model of Owen (1971a,b) as employed by Greenberg and Amos (1983), deposition of suspended sediment occurs near slack water when the shear stress falls below a certain critical value, τ_d (Figure 2-12f). By contrast, erosion occurs near the mid-stage of a tidal cycle, a time of strong current velocity, either flood or ebb, when shear stress is greater than a critical value, τ_e (Figure 2-12f). Because the value τ_d is less than that of τ_e, there is an intermediate period of the tidal cycle when neither erosion nor deposition occurs (Owen, 1971a,b). Experiments of Krone (1962) and Partheniades and Kennedy (1966), however, support simultaneous deposition and erosion. It remains to discover whether τ_d and τ_e are essentially the same or if there is a range of $\tau_d < \tau < \tau_e$. A difference between critical erosion and deposition velocities is of great importance for net transport and fate of suspended sediment in estuaries.

Consolidation

As settling and accumulation of fine sediment on the bed proceeds, buried material consolidates through "bedding down" of particles and escape of pore water. Experiments (e.g., Migniot, 1968; Owen, 1970) indicate that settling and consolidation occur in four temporal phases (Figure 2-16a): (1) 0.1–1 hour: flocculation and rapid settling of flocs forming an upper "fluff" layer of high concentrations, about 2 to 10 g/liter, and a lower "fluid mud" layer greater than 10 g per liter, with slow settling, a state called "hindered settling" because the abundance of flocs hinders the upflow of displaced water (Krone, 1962). (2) 1–10 hours: the floc structure collapses, flocs "bed down," and water escapes through interstices of deposited flocs. (3) 10–500 hours: pore water escapes slowly through drainage wells. (4) more than 800 hours: consolidation proceeds very slowly as weight of the overlying sediment forces out most remaining water. During each phase, sediment

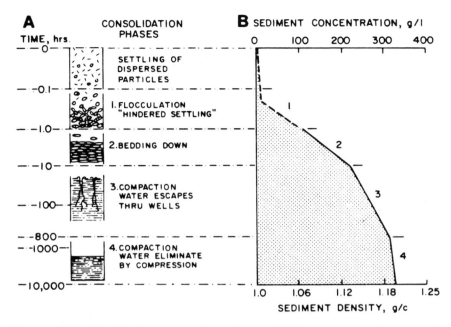

Figure 2-16. Graphical representation of consolidation stages based on experimental data (after Migniot, 1968). For details see text.

concentration and hence density and strength increase with time and burial (Figure 2-16b).

In estuaries with high suspended concentrations, the rate of deposition during decelerating tidal current ($<\tau_d$) and near slack water is too fast to permit a normally self-weight consolidated mud to form. Consequently, over many tidal cycles layers of dense suspensions (10 to 320 g per liter) called fluid mud, accumulate several meters thick. Once the mud becomes static, the suspensions slowly consolidate. Under the influence of accelerating tidal currents, upper mud layers with concentrations less than about 10 g/liter redisperse or flow slowly along the bed (Kirby and Parker, 1983).

Process Interrelationships. The response of sediments to hydrodynamic forces is illustrated in the basic "Hjulstrom Diagram" (Figure 2-17), a plot of grain size versus mean current velocity. The area above the upper curve of Figure 2-17 defines the erosion regime within which particles are set into motion. The area below the lower curve is the deposition regime in which particles in transport come to rest while the intermediate wedge-shaped area is the transportation regime.

A basic question is, What strength current is needed to move sediment? The solid curves (Figure 2-17) represent the critical velocities required for

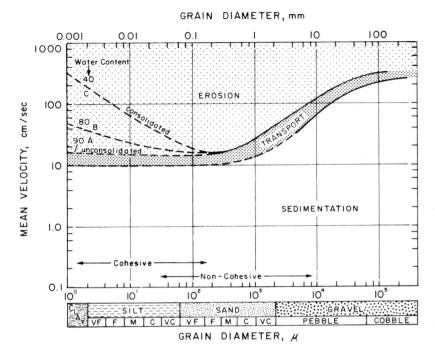

Figure 2-17. Erosion, transport, and deposition regimes for mean current velocities versus grain size. Solid lines represent critical velocities required for erosion, transport, and deposition of bed sediment while dashed lines represent extrapolated trends. Lines A-C and critical velocities for various stages of consolidation represented by decreasing water content. Adapted from Hjulstrom (1939), Postma (1967), and McDowell and O'Connor (1977).

erosion and deposition of a given grain size. The trend of the erosion curve indicates that for sediment coarser than fine sand ($>$ 125 μ), the erosion velocity increases with an increase of grain size. Coarser grains require more force to move than fine ones. However, erosion velocities are also quite high for consolidated silt and clay. Fine sand of 125–250 μ is the most erodable grain size and will be moved downstream more often than coarse grains.

Consolidated clay and silt require more force to move than fine sand. Because fine sediment is cohesive, it forms strong bonds after compaction and, thus, resists erosion. Additionally, it tends to deposit in a smooth surface that reduces turbulence and the shear velocity of moving water. Once eroded, sediment of a given grain size, e.g., clay, can be transported at velocities lower than those required for erosion. However, when the velocity is diminished to a certain level, i.e., the lowest velocity for transportation, the sediment deposits. Fine sediment of a given size that settles to the bed in areas of reduced current velocity or during slack tide will accumulate unless much higher velocities occur later. It is not surprising, therefore, that deposits accumulate slowly in macrotidal estuaries containing high suspended silt and clay loads.

Although Hjulstrom's diagram is frequently used for a broad range of particle size and for sediment movement over rippled beds, the curves are based on very limited data. For example, the data come from rivers where the velocities are not at a specific height but represent the discharge divided by channel cross-sectional area. And data for grain sizes smaller than 350 μ are either based on a questionnaire sent to engineers or extrapolated (Nowell *et al.*, 1981). Postma (1967) modified Hjulstom's diagram to include the effects of water content or degree of consolidation. As shown in Figure 2-17, the erosion velocity increases as water content of bed sediments decreases (lines A-C). Additionally, Sundborg (1967) included the effects of specific gravity, state of movement, suspended load concentration, and deposition velocity. As shown in Figure 2-18, four general areas are recognzied that represent sedimentological regimes: I. bed erosion and transport in suspension with resulting net erosion; II. transport in suspension and net deposition of suspended load; III. bed erosion and transport of bed load with resulting net erosion or net accumulation; IV. no transport with resulting deposition of

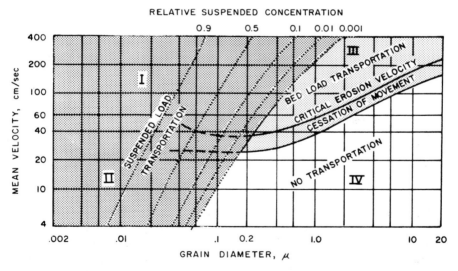

Figure 2-18. Relationship between flow velocity, grain size, and state of movement for uniform sediment of 2.65 density. The curves representing the relative concentration of suspended sediment indicate the ratio between the concentration at one-half the water depth and a reference level near the bed. Dotted lines represent diffuse boundaries. Velocity is measured at 100 cm above the bed; after Reineck and Singh (1980) and Sundborg (1967).

bedload or suspended load. Note that there is no well-defined limit of velocity sufficient to suspend sediment; instead, a continuous transition occurs from the state where no sediment of a given grain size goes into suspension and that of successive concentration increases. The diagram is most useful to show the state of movement for uniform sediment and corresponding regimes. Unsorted estuarine sediments occurring on a rough bed are much more complex. The diagram is qualitative for cohesive sediment and semiempirical for noncohesive sediment.

Hydrodynamic Factors

The characteristics of bed sediments and the accumulation of sediment in an estuary are affected by river inflow, tides, waves, wind, and meteorological forces. To understand these forces, it is necessary to examine the independent contribution of each factor and then the combined contribution of interacting factors expressed as specific mechanisms.

Effects of River Inflow

River inflow can supply a significant amount of sediment to an estuary and it can freshen estuarine water. Inflow maintains longitudinal and vertical salinity gradients that, in turn, drive a unique estuarine circulation for dispersal of sediment. Effects of inflow are most marked in near-river zones of an estuary. In times of flood, however, these effects may extend farther seaward or throughout the estuary (Gibbs, 1977).

The distance a river is capable of thrusting fresh water and sediment into an estuary is quantifiable in terms of its "flushing velocity" (Gibbs, 1977). This parameter is obtained by dividing the mean annual river discharge by the cross-sectional area at the freshwater–saltwater transition that is taken as 1 ppt salinity at the surface. The flushing velocity is also a measure of the river's ability to keep sea water from intruding landward. As shown in Figure 2-19, large rivers like the Amazon generally thrust water and sediment farther seaward to the ocean than small rivers like the Delaware, which are only capable of moving material into landward parts of their estuaries. The flushing velocity determines the position in a river-estuary system where the sedimentary environment and dispersal routes change. Just what changes take place as one passes from a fluvial to an estuarine environment?

Landward of the freshwater–saltwater transition, river flow is restricted to channels; velocities and turbulence are relatively high. Fine sediment (<16 μ) is transported seaward in suspension throughout the flow while sand (62–250 μ) is transported in near-bottom water. Coarser sediment is carried as

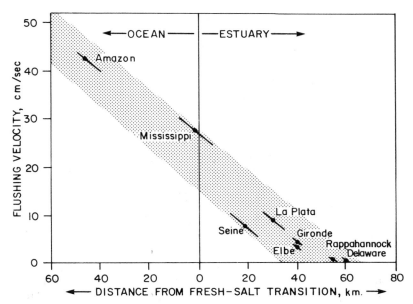

Figure 2-19. Position of the freshwater–saltwater transition, a zone of the turbidity maximum, expressed as distance from the mouth, in relation to flushing velocity for various river estuaries. Length of line represents seasonal range; the dot is average position (after Gibbs, 1977).

bed load expressed in ripples or sand waves. As the river channel widens and its cross-sectional area expands seaward, velocities decrease. Bed load and coarse sediment of the suspended load tend to drop out in freshwater zones forming elongate sand shoals and bars, some of which aggregate into point bars or ebb-flood spits. The suspended load decreases constantly during seaward transportation. It is partly diluted because the volume of water increases as the channel cross sections expands seaward. A part of the load is deposited. Because a lesser amount is spread over a larger area of the bed, accumulation rates usually decrease seaward. Sandy channel deposits are generally replaced seaward by silt and clay mixed with sand. Fractionation of the river load is not complete, however, because flow fluctuates with time. Additionally, bidirectional tidal currents, which penetrate landward from the freshwater–saltwater transition and superimpose on river flow, can mix the river-borne load.

As fresh water encounters the freshwater–saltwater transition near the bottom, river flow converges with landward-flowing estuarine water and its velocity approaches zero (Figure 2-20a). Seaward transport of bed load, therefore is arrested. The remaining suspended sediment load, by contrast, is widely dispersed either by fluctuating tidal currents or by the estuarine circulation and waves. Thus, the suspended load that is carried in near-surface water can pass farther seaward than the flow convergence as the light river flows out on top of denser salty water.

River flooding can produce exceptional hydrodynamic conditions with important sedimentological consequences. During flooding, more sediment is supplied to an estuary during a few days than during many months or years of average inflow. High-velocity currents gather up loose sediment deposited in riverine channels under previous lower discharge conditions and sweep them seaward through the upper estuary or farther seaward. Meade (1969) speculated that floods may push a salt intrusion through an estuary mouth and thus allow sediment to escape directly into the sea. Some river-borne sediment by-passes the Gironde estuary when the turbidity maximum and salt intrusion are displaced to the mouth by high river inflow (Allen, 1972). By contrast, flood-borne sediment introduced into Chesapeake Bay is retained and deposited within the Bay (Schubel, 1974). The degree to which suspended sediment is thrust through an estuary is mainly determined by the river's flushing velocity. Despite flooding, morphological changes within an estuary are relatively small except when floods reach the estuary mouth and scour out bars or enclosing spits. Most flood energy is dissipated in a large transport of suspended sediment into, within, or through an estuary (Nichols, 1977).

When the freshwater–saltwater transition is retained within an estuary, suspended sediment is dispersed by the estuarine circulation of a salt-wedge type (Figure 2-31a). The basic pattern is seaward through the upper layer, settling into the lower layer, and landward through the lower layer (Figure 2-20). As the moving mass of flood flow pushes the freshwater–saltwater transition seaward, near-bottom transport in the upper estuary changes from landward to seaward. Further, the sharp salinity gradients and high suspended concentrations favor flocculation. As the transition shifts seaward, so too does the current convergence zone, or null zone, between landward and seaward flow. Therefore, the zone for entrapment of high suspended loads and accumulation of sediment also shifts seaward in response to flooding. Even though the transition zone marks a distinct change in the sediment environment and in the sediment type, it seems unlikely that sharp boundaries develop in bed deposits because the transition zone moves up and down the estuary with fluctuations of river inflow.

Effects of Tides

In many estuaries, tides are the major energy source for mixing fresh and salt water, for resuspending sediment from the bed, and for transporting suspended sediments seaward or landward. The dynamic character of the tide is expressed as a cyclic rise and fall of the water surface accompanied by horizontal currents that vary in strength with location. Tides affect sedimentary processes through three fundamental processes: (1) tide wave

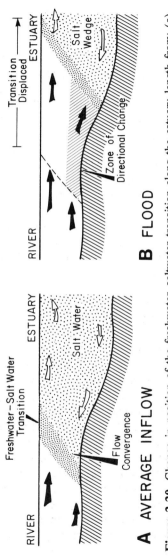

Figure 2-20. Change in position of the freshwater–saltwater transition along the estuary length from: (a) average river inflow and (b) flood. Arrows represent direction of residual flow (schematic).

deformation with distance landward; (2) tidal discharge and channel stability; and (3) cyclic tidal current fluctuations. Basically, tides are linked to astronomical forces and are shaped by basin geometry.

Tides are long waves generated by the gravitational pull of the moon and sun on mobile ocean water. These forces are coupled to the rotation of the earth and sun that varies with orbit and changes of declination, among other factors. A first approximation to tidal phenomena is given by the equilibrium theory of the tide (Russell and Macmillan, 1970; Redfield, 1980). The strongest component of force is due to the moon (M_2) and has a semidiurnal period of about 12 hours and 25 minutes. Most estuaries have two high tides and two low tides during each lunar day, a period of 24 hours and 50 minutes, which is one rotation of the earth increased by the eastward movement of the moon in orbit. The total tidal force resulting from the combined action of sun and moon is maximal once or twice a month yielding tides with a greater range (the vertical difference between high and low tide) than average, called spring tides. At alternate times in the month, the range is smaller than average, yielding neap tides. The tide range varies with estuary location depending on the global character of ocean tides and shoaling constraints on the continental shelf. For example, the mean tide range in lower Chesapeake Bay is 0.8 m, on the Georgia coast 2.5 m, and in the Gironde Estuary entrance 4.5 m. Tidal range is important because it determines the intertidal volume of water available for discharge and the magnitude of tidally induced flow.

Tidal Wave Deformation

To produce a cyclic rise and fall of the water surface, oceanic tides advance as sinusoidal progressive waves. Such waves are not subject to friction on the bottom. Tidal current velocities attain maximum values at high and low water and a value of zero at mean water level (Figure 2-21a). Duration of flood current equals ebb current, being 12 hours and 25 minutes altogether; therefore, no residual transport of water takes place.

When ocean tides are propagated into shallow estuaries, they become modified and thus produce a landward or seaward transport. Salomon and Allen (1983) recognize three distinct processes: (1) frictional damping on the bottom; (2) landward constriction or convergence in the channel; and (3) reflection on shoals or from the estuary head.

Friction with channel boundaries dissipates the energy of the tide wave and the amplitude decreases exponentially landward as a function of the relation (LeFloch, 1961):

$$A = A_0 e^{-kx} \qquad (2\text{-}6)$$

where A is the amplitude at x distance from the mouth; A_0 is the amplitude at the mouth; and k is a friction coefficient, a function of the velocity and

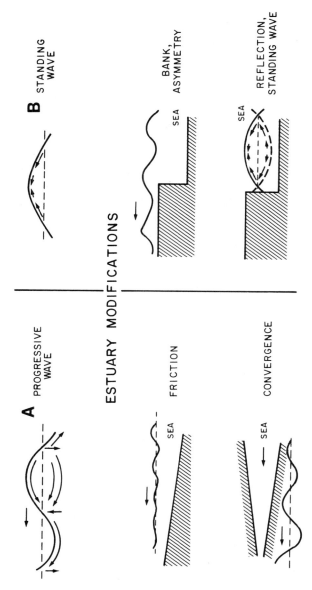

Figure 2-21. Tidal current speed and direction in relation to wave forms: (A) progressive wave, (B) standing wave, upper figures. Modifications of different tide waves in an estuary caused by friction, convergence and reflection from a bank or estuary head; schematic; after Salomon and Allen (1983). Reproduced by permission of Total-Compagnie Francaise des Petroles.

frictional resistance indexed by the Chezy coefficient (Ippen and Harleman, 1966). As amplitude decreases landward and current velocity diminishes, fine sediment is likely to be retained and deposited in landward reaches.

Frictional dissipation also affects the tide wave symmetry. In an estuary, where the water depth does not appreciably exceed the tide wave amplitude, the tide wave is propagated as a shallow water wave whereby the speed C is given by:

$$C = \sqrt{gh} + \eta \qquad (2\text{-}7)$$

where h is the water depth; g is the acceleration of gravity; and η is the local wave height of the tidal wave surface above still-water level. Because η/h increases the water depth decreases, the tide wave speed decreases. Consequently, the flood crest (high water) propagates faster than, and tends to overtake, the trough (low water). This effect deforms the tide wave whereby the duration of rise (flood) is shortened and the fall (ebb) is prolonged (Figure 2-22a). Flood velocities are increased because the front of the flood crest has a greater slope than the ebb, whereas ebb velocities are decreased. Therefore, the greater flood velocities produce greater bottom erosion and transport of sediment than the ebb, thus favoring residual movement landward (Figure 2-22b). In response to a landward increase in

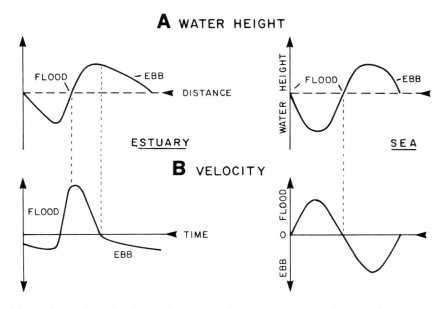

Figure 2-22. Modifications of a progressive wave: (A) tide height with distance landward from a deep water symmetrical wave (right) to shallow asymmetrical wave (left); (B) corresponding modification of current velocity from symmetrical time-velocity curve (right) to asymmetrical curve (left) (from Salomon and Allen, 1983). Reproduced by permission of Total-Compagnie Francaise des Petroles.

tide wave asymmetry in the Ord Estuary, a macrotidal estuary of Western Australia, the largest bedforms within the channel migrate landward under influence of flood currents and cause extensive sand accumulation (Wright *et al.*, 1975; Wright, Chapter 1, this volume).

Although frictional damping decreases amplitude landward, a decrease in channel cross section can cause a concentration of energy and thus an increase in amplitude. The relationship is given by Green's law, which ignores friction:

$$h = K(H^{1/4}B^{1/2}) \qquad (2\text{-}8)$$

where h is wave amplitude; H and B are the channel depth and width, respectively; and K is a constant. Therefore, if an estuary shoals or narrows landward, which is the usual case, the amplitude will increase. However, if the opposing effect of friction that reduces the amplitude is added, then there are three types of models (LeFloch, 1961; Figure 2-23):

1. Convergence exceeds frictional dissipation. If the loss of energy by friction is less than the convergence, the tidal amplitude increases landward from the mouth before decreasing toward the river, a trend characterizing a *hypersynchronous* estuary.

2. Convergence equals friction. The tidal amplitude is maintained before diminishing toward the river, a trend characterizing a *synchronous* estuary.

3. Convergence is less than friction. The tidal amplitude decreases landward from the mouth, a trend characterizing a *hyposynchronous* estuary.

Most estuaries are hypersynchronous, and tidal currents attain maximum strength in middle or landward reaches. In the Gironde estuary, intensification of tidal currents in upper reaches causes bed erosion, accompanied by bed resuspension. Consequently, a tidal turbidity maximum forms during low river inflow, a time when density circulation is reduced (Allen *et al.*, 1980).

Reflection of an advancing tide wave from an estuary shoal, bank, or estuary head can increase the amplitude as the wave changes from a progressive wave to a standing wave (Figure 2-21). Instead of moving, the wave form of a standing wave remains stationary while the water flows rapidly through it. Tidal currents are slack at high and low water, the times at which currents attain maximum velocity in a classic progressive wave (Figure 2-22). Reflection occurs when the estuary length or distance between the mouth and shoal is a multiple of one-quarter the wave length. In the case of the Ord Estuary, the entire channel length is equivalent to one-quarter the wave length, and the tide behaves as a standing wave (Wright *et al.*, 1973). High water occurs almost simultaneously throughout the estuary. In response to a landward increase of tide wave asymmetry, bedforms migrate upstream in the channel under the influence of flood currents. The reduced tidal

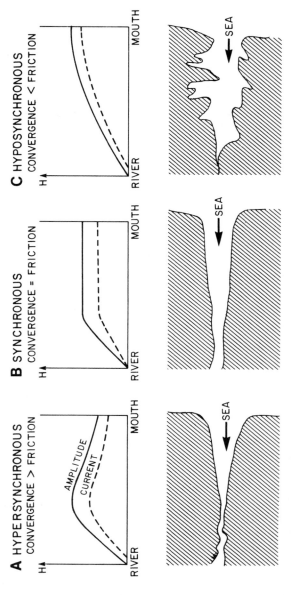

Figure 2-23. Modification of tide range in estuaries with varying ratios of convergence to friction effects (from Salomon and Allen, 1983; adapted from LeFloch, 1961).

amplitude and current velocity landward tend to retain sediment in the estuary.

In most estuaries, the tides are a combination of standing and progressive waves. In Chesapeake Bay, high water progresses landward through seaward parts while an amplified standing wave forms near the Bay head. Because of its extreme shallowness relative to its length, the Bay is one of the few estuaries that can hold a complete semidiurnal tide wave at all times.

Tidal Discharge and Channel Stability

In tide-dominated channels, the bed and banks are not static but continuously changing in response to the transporting ability of the tidal flow and the supply of sediment from external sources. Like fluvial channels, tidal channel morphology is an integral part of the hydrodynamic system, and it develops toward a state of maximum stability. To attain dynamic equilibrium, tidal discharge and bottom geometry are continuously coadjusted through erosion and deposition. The channel must be neither too deep nor too shallow for the amount of discharge and for the load of sediment it transports (Wright et al., 1973).

The tidal discharge is determined by the tidal prism, i.e., the amount of water that flows into and out of an estuary during one flood or ebb period. The discharge, Q, is a function of mean velocity, V, and cross-sectional area, A, so that at any section x:

$$Q_x = V_x A_x \qquad (2\text{-}9)$$

In a section of an estuary, the tidal prism is measured from simultaneous tide levels and corresponding surface area observations. Because estuary tidal range is controlled by the ocean tide, measurements of the prism for an entire estuary relate to flow through the entrance. Hydraulic engineers have long recognized the effects of channel stability as channel flow sections rapidly fill in following enlargement by dredging, a case demonstrated in the Delaware (U.S. Army Engineer District, 1973; Nichols, 1978) and in the Thames, (United Kingdom) (Inglis and Allen, 1957). In an investigation of sandy tidal inlets, O'Brien (1969, 1976) discovered that the entrance flow area is a unique function of the tidal prism (Figure 2-24). Despite river inflow, littoral drift, and varying sediment size and jettied entrances, the tidal prism, P, follows the linear relationship:

$$P = A_e \, 5.0 \times 10^4 \qquad (2\text{-}10)$$

where A_e is the minimum area of channel below mean sea level in square feet. Entrances following this relation are in equilibrium and stable. If sedimentation on estuary flats or marshes reduces the intertidal volume, then the entrance cross section is narrowed by deposition. Some estuaries, however, do not conform to equilibrium because cross sections are molded in "hard"

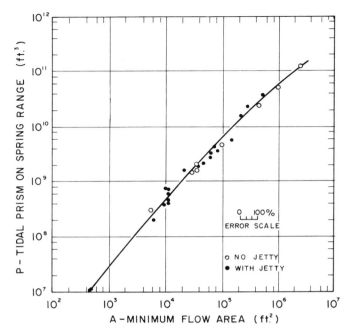

Figure 2-24. Entrance flow area versus tidal prism relationship (after O'Brien, 1969).

rock or resistant cohesive sediment, or because they have a residual sediment influx produced by unequal flood and ebb flows induced by river inflow, density gradients, or storm events.

Cyclic Tidal Current Fluctuations and Asymmetry

The simplest form of ebb and flood current is a sinusoidal and symmetrically shaped distribution of velocity versus time (Figure 2-22b). In this case, the amount of water carried past a point over the flood duration equals the amount carried over the ebb. The "ideal" symmetrical tide, however, is seldom observed (Postma, 1967). Instead, the tide wave is deformed, and ebb and flood currents are usually unequal in strength, duration, or symmetry. These contrasts produce a differential, or residual movement, either flood or ebb. Although a given parcel of water moves back and forth, it usually does not return to the same place. In narrow estuaries where currents reverse direction, the residual velocity V_r, is usually derived by averaging the individual current measurements, e.g., adding the areas under flood and ebb curves (e.g., F_1, F_2, and E_1), usually for more than eight full tidal cycles, and dividing by the total cycle duration, T; for example:

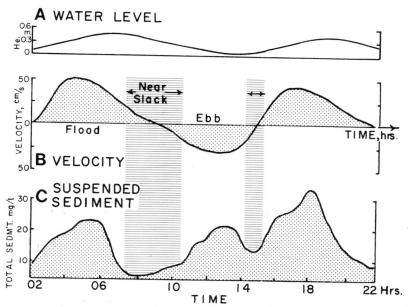

Figure 2-25. Selected time variations of water movement from an anchor station in the Rappahannock estuary channel, Virginia; (a) water level height), (b) current velocity, (c) suspended sediment concentrations. Note asymmetrical time-velocity curve with relatively long duration of low velocity near slack after flood and the corresponding low suspended sediment concentrations (at 0.5 m above the bed) (modified from Nichols and Poor, 1967).

$$V_r = \frac{(F_1 + F_2 + F \ldots - E_1 + E_2 + E \ldots)}{T} \qquad (2\text{-}11)$$

In broad estuaries, tidal currents often rotate through elliptical paths over a tidal cycle. Although such "deviations" are relatively small in one tidal cycle, over many tidal cycles they can produce a significant residual movement with important consequences for transport of suspended sediment.

Ebb-flood current asymmetry, which results from landward deformation of the tidal wave, favors landward transport when the slack after flood is longer than the ebb. For example, Figure 2-25b shows a time-velocity curve recorded at a fixed point near the bottom of the Rappahannock estuary. Asymmetry is illustrated by relatively low velocity for a longer time near the slack after flood (close to low water); i.e., the current turns slower at high water than at low water. Further, peak flood current speed exceeds peak ebb speed. Concentrations of suspended sediment with time at the same point generally rise and fall with current speed as they are resuspended from the bed (Figure 2-25c), but there is a time lag between the concentration variations and the velocity rhythm. Concentrations are lower at the slack after flood than at the slack after ebb because the settling time at flood slack is longer for a given current speed. Over many tidal cycles, such a difference favors a residual sedimentation at the end of flood and a landward shift of suspended sediment.

The lag of the time-concentration variation relative to the time-velocity curve depends on water depth, particle size, and the response of fine sediment to tidal currents. According to Postma (1980), the greatest lag effect occurs for particles with slow settling velocities, i.e., between 10 and 50 μ size. Although a lag effect has been demonstrated for fine sand (Thorn, 1975), it is most pronounced in cohesive sediment that reacts with a certain resistance to current acceleration. Because of its shear strength, higher velocities and more time is needed to erode or resuspend cohesive sediment from the bed and disperse it to a high level, than a decelerating current that requires only time to settle (Postma, 1980). Further details are given in a subsequent section of this chapter entitled, *Settling and Scour Lag Effects*.

The slack water asymmetry is enhanced by differences in flood-ebb velocities, e.g., flood velocities are higher than ebb velocities in the example, Figures 2-22b, 2-25b. The greater velocity is amplified because bed shear stress, which causes erosion, is a function of the square of the current velocity. As the tide wave becomes deformed and velocity asymmetry increases landward (Figure 2-22b), bed erosion and transport is greater during flood than ebb resulting in a residual landward movement of sediment. Flood tidal predominance of bed load is noted in the Ord estuary (Coleman and Wright, 1978) and in the Gironde where it extends landward of the salt intrusion and the turbidity maximum (Allen *et al.*, 1980).

Another effect of asymmetry is revealed by numerical modeling of tidal currents in the Gironde (Salomon and Allen, 1983). As shown in Figure 2-26, in a landward deformed tidal wave the horizontal tidal currents are accompanied by a small vertical movement that is most prominent near slack water. The vertical movements are upward at ebb slack (low water) and downward at flood slack (high water). Although the vertical motion is only a few millimeters per second, it is sufficient to deter settling at ebb slack and to enhance settling at flood slack. When this difference occurs over many tidal cycles, it favors more sedimentation at flood slack than at ebb slack. This process acts in the same direction as the normal slack asymmetry and as the flood velocity predominance. It shifts sediment landward and retains it within an estuary.

Although the type of asymmetry discussed above with a long lasting flood slack around high water is common to many estuaries, an "opposite" asymmetry whereby flood slack is short, is reported for some situations (Boon, 1975). This is the case for tidal channels cut in marshes that are

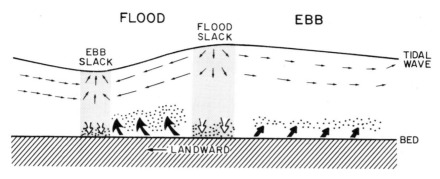

Figure 2-26. Longitudinal section of an asymmetrical tide wave illustrating a steep front that reduces flood duration but increases flood currents and resuspends much bed sediment. Further, the downward vertical motion favors more sedimentation at flood slack than at ebb slack; schematic, from Salomon and Allen (1983). Reproduced by permission of Total-Compagnie Francaise des Petroles.

scarcely flooded at high water. As a result, suspended sediment escapes seaward from the channels. Numerical modeling of channel-flow and basin storage-volume relationships by Boon and Byrne (1981) suggests that asymmetry can change from "normal" to opposite with sedimentary infilling of the estuary basin.

A change in the tide range from spring to neap can be accompanied by a marked change of tidal current intensity. Strong currents during spring tides bring more sediment into suspension than neap tides. Therefore, concentrations vary with tide range during a fortnightly cycle. They build up steeply with increasing tide range as the currents become faster and reach a peak about 4 days after maximal currents. In the Gironde, Allen *et al.* (1980) observed that during periods of decreasing tide range, when each succeeding peak tidal current velocities decrease and slack water durations increase, the ratio of sedimentation to erosion increases. This results in a residual sedimentation in the estuary during waning tidal ranges and causes pools of fluid mud to accumulate during neap tides. For 2 to 6 days during neap tide, a certain amount of sediment has time to consolidate and hence resist erosion during the following spring tides. Therefore, every 2 weeks, a thin lamination of mud accumulates permanently in the estuary. The neap–spring cycle also modulates the ratio of river flow to tidal volume, which in turn causes large variations in mixing. The Gironde, for example, a macrotidal estuary with extensive intertidal flats, changes from a well-mixed state during spring tides to a partially mixed or well-stratified state during neap tides (Allen *et al.*, 1980). Rapid mixing during spring tides combined with large-scale erosion and resuspension provides a means for seaward escape of suspended sediment from the estuary.

Effects of Waves

Although estuaries are often considered "protected environments," waves can have significant sedimentological effects, eroding shores, stripping substrates, and suspending sediment so that it can be dispersed by currents. There are essentially two genetic types of waves: (1) those generated externally in the ocean that penetrate estuary mouths, and (2) those generated internally that affect shores, marginal shoals, or the estuary floor.

Ocean Wave Impact on Entrances

Deepwater ocean waves are long low waves or swells with lengths approximately 40–60 m and periods of 5 to 8 s. As they invade shallow water where water depth becomes less than one-fourth the wave length (about 10–20 m), their circular motion of water particles becomes more elliptical and asymmetrical. When the resulting back-and-forth motions impinge on the bed and produce sufficient shear velocity, a measure of bottom shear stress (Inman, 1949), sand transport, is initiated. This transport is shoreward as the shoaling wave becomes asymmetrical and produces a shoreward advection of water. Because the bottom topography of entrances is irregular, transport of water and sediment is also affected by wave refraction. Wave crests penetrate farther landward in deeper channel water, a zone of wave divergence and low energy, than in shallow water toward margins or shoals. On shoals, waves converge and concentrate high energy. What are the resultant sediment patterns?

In the Gironde estuary, Allen (1971) found that the broad textural patterns reflect the distribution of energy that is controlled by the entrance morphology. Waves primarily dissipate in shoal areas and zones of strong refraction (Figure 2-27a), whereas tidal currents mainly flow in deeper channels below about 15 m. Wave energy is inversely proportional to water depth whereas tidal energy is directly proportional to depth (Figure 2-27b,c,d). To describe energy variations in the entrance where both waves and tides are active, Allen developed an "energy index" or ratio of wave bottom orbital velocity to the tidal bottom velocity. The mean grain size and skewness of the sands are inversely proportional, while sorting is directly proportional, to the ratio of wave to tidal activity. Tidal current zones are characterized by coarse grain sizes, relatively poor sorting, and bed-load transport. In contrast, shoal areas affected by wave energy exhibit finer grain sizes, well-sorted sands, and graded suspension transport mode. Despite the complexity of processes in estuary entrances, textural patterns, which approach dynamic equilibrium with wave and tide activity, are good indicators of energy variations.

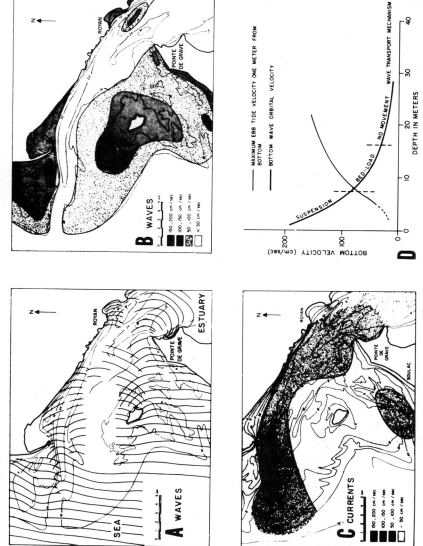

Figure 2-27. Estuary entrance dynamics by example from the Gironde estuary entrance. (a) Wave refraction pattern; (b) zones of bottom wave-orbital-velocity, cm/sec; (c) distribution of peak spring ebb-tide current velocity at one meter above the bed; (d) schematic diagram of bottom velocity as a function of depth (from Allen, 1971).

A further effect of ocean waves is to erode sand from the beach or shoreface (shoaling wave zone) and to transport it via longshore or drift currents toward the estuary entrance. Movement into the Raritan and Delaware estuaries is evidenced by migration of inlets and growth of spits (Meade, 1969). Fourier grain-shape analysis reveals a sand stream with a net influx of ocean beach sand into Chesapeake Bay (Boon *et al.*, 1983). According to Swift (1976), the intersection of estuarine tidal flows with a wave-induced littoral sand-drift system acts as a depocenter or sink for sand storage. Sand in the mouth of the Delaware estuary is blocked by a large transverse shoal that is breached by ebb and flood-dominated channels. Around tidal inlets of the Georgia coast, sand is stored in arcuate, seaward-convex shoals, which as a whole serve to focus wave energy (Swift, 1976). As upper surfaces of the shoal build upward, they become intertidal swash platforms. During storms, intersecting wave trains tend to drive sand landward forming "ramp-margin shoals" (Oertel, 1972; Swift, 1976) (Figure 2-28).

Estuary waves are shorter and less regular than ocean waves, with lengths of 15 to 25 m and periods less than 5 s. Wave size and character are determined by water depth and the direction, strength, and duration of winds blowing over the estuary. Few field observations are available; however, prediction of wave characteristics generated in shallow estuarine water is based on theoretical curves and equations of Bretschneider (1965) and the U.S. Army Corps of Engineers' Coastal Engineering Research Center (1977).

The most significant parameter expressing wave bottom effects is the maximum bottom orbital velocity. For a given wave:

$$U_{max} = \frac{H\pi}{T \sin h(2\pi h/L)} \qquad (2\text{-}12)$$

where H is the wave height, T the wave period, L the wave length, and h the water depth (Komar and Miller, 1973). Because $1/(\sin h \ 2\pi h/L)$ falls rapidly with depth, and its square falls even faster, it is evident that the maximum bed shear stress is very sensitive to water depth. Generally, strongest wave action occurs in estuaries oriented in diagonal directions along the maximum fetch. If the shores are not protected, waves approaching shore at an angle erode shores and cliffs, feed sediment to the local longshore currents, and build up spits and cusps. These can grow to such an extent that lagoonal estuaries can become segmented into a series of small estuaries linked by small channels, a process described on the Chukotsky Peninsula, USSR (Zenkovitch, 1959) and in Nantucket Harbor, Massachusetts (Rosen, 1975).

The effects of waves are both depositional and erosional. In Chesapeake Bay where fetch exceeds 30 km on more than 200 km of shoreline, the central Bay shore is marked by smooth outlines backed by steep cliffs of unconsolidated sediments and fronted by broad shoals or terraces. Shore

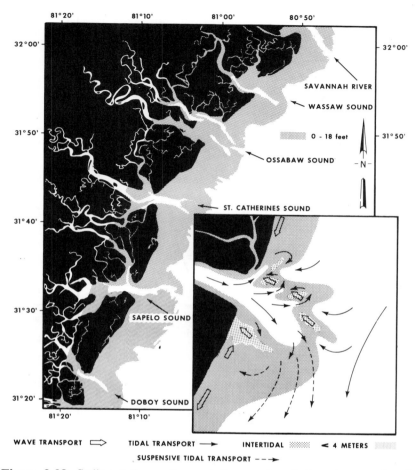

Figure 2-28. Sedimentation patterns at the mouths of Georgia estuaries (from Stanley and Swift, 1976; after Oertel, 1972).

erosion averages 0.7 m/yr and locally reaches 2.2 m/yr in Virginia (Hardaway and Anderson, 1980). Annually an estimated average of 2.5 million tons of sediment are eroded (Byrne *et al.*, 1982; Kerhin *et al.*, 1983). Where does this enormous load go? Bottom profiles and bathymetric changes during a 100-yr interval demonstrate that the eroded sand is redistributed a short distance offshore forming depositional terraces or marginal shoals at water depths of 2–5 m (Jordan, 1961; Ryan, 1953). Eroded silt and clay contribute to growth of marshes or accumulate on the channel floor (Figure 2-29). Shore erosion and local redistribution, therefore, is a self-digesting process (Rusnak, 1967) that enlarges the estuary area but decreases its total volume. Further, the erodable material can be an important term in an estuary sediment budget (Biggs, 1970; Bokuniewicz and Tanski, 1983).

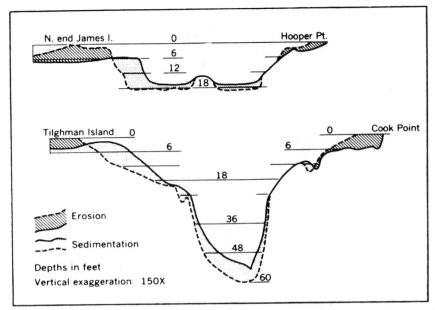

Figure 2-29. Bottom profiles from Chesapeake Bay showing changes indicative of erosion and sedimentation over a 100-year period (from Jordan, 1961).

In response to wave transport and sorting, sediment grain size of bottom sediment on shoals or tidal flats can display two opposite distributions: (1) a "normal" increase of grain size with decreasing depth and increasing distance from the tidal channel; or (2) an "opposite" trend of decreasing size with decreasing depth. The normal increase of median size occurs on sandy tidal flats between channels in the Wadden Sea (Postma, 1957), which have a convex profile form and a short distance (< 1 km) between channel and flat apex. The increase is explained by an increase in bottom stress caused by shoaling waves (Zimmerman, 1973). The opposite trend of decreasing median grain size with decreasing depth occurs on broad tidal flats along the mainland of the Wadden Sea. These have a concave profile and long distance (> 2 km) across the flat. This trend results from a decrease in wave-energy (orbital velocities) by bottom friction and hence a decrease in bottom stress (Zimmerman, 1973). Therefore, sediment distributions reflect the form of the subaqueous profile and the distance waves travel over the tidal flat.

Even small amplitude waves or "surface chop" (less than 5 cm) found in small estuaries with short fetch can have a significant effect on tidal flats. In Great Bay estuary (New Hampshire), Anderson (1972, 1983) found that such waves resuspend fine sediments, increasing the suspended sediment concentration three times more than calm conditions. In shallow embayments of San Francisco Bay, Krone (1979) reports that daily waves resuspend large amounts of fine sediment with concentrations exceeding 1.0 g/liter. Waves hold the material in suspension while slow tidal currents transport it to channels where flood tides move it landward and feed a turbidity maximum.

Meteorological Forces

Wind stress and atmospheric pressure variations create water-level changes and drift currents that affect transport of suspended sediment, particularly in shallow restricted estuaries. In Corpus Christi Bay (Texas), for example, Shideler (1984) found that contrasting onshore-offshore winds changed the direction of wind-driven currents as well as the efficiency of ocean-bay exchange and hence sediment dispersal. In the Potomac estuary, Chesapeake Bay, meteorological forces can reverse the direction and change the strength of the estuarine circulation. From long-term current measurements, Elliott (1978) discovered significant correlations of current with local winds, a downstream wind causing enhanced surface outflow and bottom inflow, whereas an upstream wind on the estuary and adjacent offshore waters caused a reversal of near-surface flow (Figure 2-30). Such fluctuations make estuarine circulation very complex. However, when meteorological forces are diminished, or their effects averaged over long periods, a basic density circulation is evident.

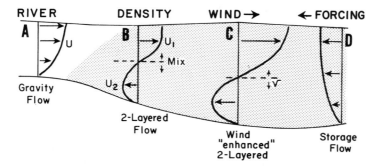

Figure 2-30. Types of flow distributions in estuaries represented by vertical current profiles: (a) river flow seaward throughout water column (left); (b) "normal" two-way estuarine circulation, seaward in near-surface, landward near-bottom, and vertical at mid-depth (middle), U_1 U_2 represent relative net longitudinal velocity, dashed line is level of no-net motion; (c) enhanced two-layered circulation driven by wind seaward; (d) off-shore meterological forcing and resulting storage flow.

Transport and Accumulation Mechanisms

The dynamic forces act on sediments either singly or in combination. This section examines the combined contribution of several forces—river flow, tides, and density circulation—to determine how they act or interact to produce distinctive sediment patterns and to retain sediment in an estuary.

Estuarine Circulation

Superimposed on the obvious back-and-forth movement of tidal currents, there is a subtle flow termed the *estuarine circulation*. What drives this circulation?

If there were no freshwater inflow into an estuary and relatively small tides, sea water would eventually fill the estuary. When fresh water is introduced from a river at the estuary head it pushes the sea water seaward down the estuary. The fresh water, which is less dense than sea water, tends to flow outward over the sea water. In contrast, the intruding sea water penetrates the estuary as a salt wedge along the bottom and underneath the seaward-flowing fresh water (Figures 2-20a, 2-31a). This creates vertical salinity stratification with a narrow zone of abrupt salinity change called a halocline between the upper and lower layer or wedge. If the seaward-flowing upper layer has a sufficiently high velocity, it can create interfacial waves on the halocline. These waves break upward, thus ejecting parcels of salt water into the seaward-flowing upper layer, a process called entrainment. In the process, no fresh water is mixed downward. Consequently, the salt content within the wedge is almost constant along its length, but the salt content of the upper layer increases as water moves seaward. Because the salt wedge loses salt water by mixing into the upper layer (while remaining in a fixed position for a given river flow), there must be a small inflow of salt water from the ocean to compensate for salt "lost" to the upper layer. Thus, the mixing of salt water from the sea with fresh water from the river provides the density distribution that drives the characteristic estuarine circulation.

In the typical coastal plain estuary, salinity increases seaward from the estuary head from nearly zero to 30 to 35 ‰. Further, the salt content varies across the estuary with lower salinity on the left side (viewed landward) and higher salinity on the right side. A major cause of this variation is the Coriolis effect (Pritchard, 1955). There is also a salinity gradient from surface to bottom that can increase 30 ‰ within a depth of 50 cm in a sharp halocline (Figure 2-31a). Consequently, salinity is a good indicator of water stability and mixing. Typically, net landward flow through the lower layer is 5–10 cm/s while the upward flow into the upper layer is about 10^{-3} cm/s. At the boundary between seaward- and landward-flowing layers, the net velocity is

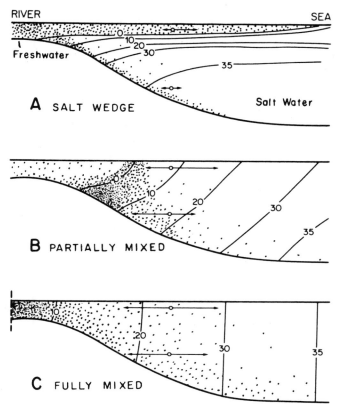

Figure 2-31. Schematic representation of types of estuarine circulation. The dots indicate relative sediment concentration and the arrows net water movements over ebb and flood. (a) Salt wedge estuary. It is assumed that most river sediment is carried to the sea with the river water in the upper layer. An example is the Mississippi; (b) Partially mixed estuary. The suspended matter is concentrated in a turbidity maximum; (c) Fully mixed estuary. The suspended matter is concentrated nearshore as in the case of a tidal mechanism without appreciable river flow (from Postma, 1980). Reproduced by permission of John Wiley and Sons.

zero, a feature called the level of no-net motion (Figure 2-30). The depth and inclination of the level and associated layer thickness changes with inflow, tides, and geometry. Just how does the circulation change as the river flow decreases, as the tide increases or as the width and depth change? Pritchard (1955) recognized that when other factors are held constant, a change in tidal current strength which dominates mixing, produces a sequence of circulation types:

1. *Type A or salt wedge*. This circulation is dominated by river flow, and tides play a negligible role. Fresh water spreads out on top of denser salt

water and thins seaward. The underlying layer is wedge shaped and thins landward (Figure 2-31a). A sharp interface, or halocline, between fresh and salty water is characterized by internal waves and upward advection of salt.

The seaward zone of the Mississippi River is a salt-wedge estuary. The wedge penetrates 150 km landward at low inflow but extends only about 1.5 km above the mouth at high inflow. Most suspended sediment in the wedge is fluviatile and disperses through the upper layer (Figure 2-31a). The bottom layer is filled with sediment that settles through the halocline; however, upstream water movement is slow. Fluvial bedload is carried to the wedge tip, where its seaward movement is arrested and accumulates in a river mouth bar.

2. *Type B or partially mixed.* In this type, the tidal currents are increased so that river flow does not dominate the circulation. Tidal mixing removes the sharp boundary between layers, and both salt water and fresh water are transferred upward as well as downward (Figure 2-31b). Landward flow is sufficiently strong to transport suspended sediment to the salt intrusion head where it accumulates in a turbidity maximum or a shoal. Sediment comes from either the river via settling from the upper layer into the lower layer, or it is derived from the sea. Chesapeake Bay is mainly a partially mixed estuary.

3. *Type C or well mixed.* In this circulation, the tide range and current velocities are large enough to break down the vertical salinity stratification completely. Therefore, the water is vertically homogeneous. Salinity varies laterally, however, being higher on the right (viewed landward) than on the left. There is little landward flow but mainly lateral movement and mixing from right to left. Suspended sediment is dispersed away from lateral or weak river sources and, also, above the bed by tidal resuspension. The Delaware and Raritan estuaries are well mixed.

If river flow and tides are constant and the estuary cross-sectional area is increased by increasing the width, e.g., by shore erosion, then the ratio of tidal volume to river flow is increased. Net velocities related to river flow alone are reduced and the circulation tends to shift from Type A toward Type C. A decrease of depth as by shoaling, other factors held equal, increases the ratio of river flow to tidal flow. Tides are more effective in vertical mixing and the circulation tends to shift from Type A toward Type C.

In summary, other factors being held equal, an estuarine circulation tends to shift from Type C through Type B to Type A by: increasing river flow, decreasing tides, decreasing width, and increasing depth. The characteristics of different estuarine mixing types are summarized in Table 2-1.

Development of the Turbidity Maximum

In the upper or middle reaches of many estuaries, particularly partially mixed estuaries, the concentrations of suspended sediment are 10 to 100 times higher than those either in the river or farther seaward in the estuary. This feature, termed the turbidity maximum, is found in many world estuaries including estuaries of different shape and size from small inlets to the world's large rivers like the Amazon, the Thames, and the Mississippi. Enormous amounts of sediment are retained in the maximum despite strong river currents that tend to flush it seaward. The maximum also persists despite dilution and rapid mixing of fresh and salty water. In a theoretical case (Figure 2-32a) fluvial suspended sediment is simply diluted by less turbid water, and it can deposit as it enters the salt intrusion head (Figure 2-32b). Instead, high suspended concentrations are found. What processes common to many estuaries retain suspended sediment and form a turbidity maximum?

A turbidity maximum can develop in different ways. Figure 2-33 illustrates three models of longitudinal suspended sediment concentrations in relation to salinity and corresponding field examples. In model A, suspended sediment reaches a peak near the landward limit of salty water. This limit corresponds to the null zone or current convergence of river flow and landward estuarine flow near the bottom (Figure 2-34). Sediment derived either from the river, the sea, or recycled by the estuarine circulation within the estuary is effectively trapped in the null zone so long as particle settling exceeds upward mixing. In higher salinity zones of the middle estuary where mixing is less vigorous, suspended sediment can settle from the upper layer into the lower layer, then join sediment entering from the sea, and travel landward to the salt intrusion head (Figure 2-34). Sediment that recycles

TURBIDITY MAXIMA MODELS

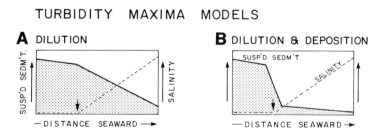

Figure 2-32. Schematic distribution of suspended sediment concentration along estuary length in relation to salinity. (A) model for simple dilution of fluvial load, (B) dilution and deposition. Arrows indicate inner limit of salty water or current null zone (modified from Buller *et al.*, 1975; Meade, 1972).

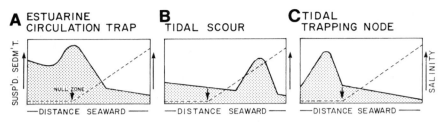

Figure 2-33. Schematic distribution of suspended concentration along estuary length showing location of the turbidity maximum in relation to salt intrusion limit, salinity. (A) model for entrapment in the estuarine circulation, an example for the Rappahannock estuary, Virginia, (Nichols, 1977); (B) model for intertidal input and tidal scour based on the Tay estuary, Scotland (Buller *et al.*, 1975); (C) model for entrapment in a tidal trapping node, an example for the Gironde estuary, France (Allen *et al.*, 1980).

many times in the estuarine circulation becomes uniform in size and mineralogy because very fine and less dense sediment is swept out to sea and relatively coarse dense sediment comes to rest. According to Postma (1967), the magnitude of a turbidity maximum depends first on the amount of suspended sediment from the river or sea and second on the strength of the estuarine circulation. Both these factors directly relate to river inflow. Evidence that a turbidity maximum depends on the estuarine circulation is provided by the fact that the maximum shifts seaward as the salt intrusion head shifts downstream during high river inflow (Nichols, 1977; Dobereiner and McManus, 1983). Additionally, the salt intrusion head is often the site of rapid shoaling. This site migrates landward when channel deepening

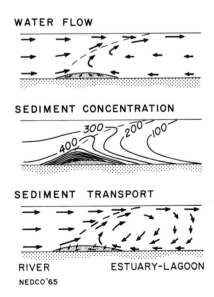

Figure 2-34. Schematic representation of the relations between water (density) circulation (top), suspended sediment concentration with turbidity maximum and suspended sediment transport (bottom) showing convergence and accumulation at the inner limit of salty water. (modified from NEDECO, 1965; Meade, 1972).

induces upstream penetration of the salt intrusion (Simmons, 1972). Numerical simulations by Festa and Hansen (1978) confirm the importance of both river and marine influx as well as the strength of the estuarine circulation.

The effect of flocculation or agglomeration on formation or maintenance of the maximum is controversial. Flocculation increases particle settling and promotes deposition at the salt intrusion limit; but the maximum contains relatively high concentrations, not less. Krone (1972) shows that flocculation enhances formation of the maximum. When flocs form, they sink into flow of the lower layer that readily recylces them back upstream to the salt limit where they can disintegrate under rigorous mixing conditions (see section on *Particle Dynamics*). Kranck (1981) indicates that flocculation increases the magnitude of the maximum by increasing the settling, as well as resuspension rates, and increasing the size range of the trapped particles. Most suspended particles in the Miramich estuary (Canada) are so small that without flocculation they probably would be carried to sea (Kranck, 1981).

Turbidity maxima can form independently of the salt intrusion limit when tidal scour resuspends fine sediment from the bed. Therefore, Model B (Figure 2-33) shows a turbidity maximum seaward of the inner salt limit. In the Tay estuary (Scotland), suspended sediment loads are derived from intertidal input during ebbing tides (Buller *et al.*, 1975). In the Gulf of San Miguel (Panama), a turbidity maximum is generated by tidal currents that develop a turbulent boundary layer over mud shoals (Swift and Pirie, 1970). Turbidity maxima can also form in a tidal trapping node, Model C (Figure 2-33), whereby the maximum lies landward of the salt intrusion at the head of tides.

Tidal Turbidity Maximum

Recent studies in the Gironde and Aulne estuaries (Allen *et al.*, 1980) reveal that tidal transport of suspended sediment is as important as the density circulation in trapping fine sediment. Concentrations in the maximum reach 1–10 g/liter, and its position may center either landward or seaward of the salt intrusion limit. Because tidal currents are strong, more than 3 m/s on the surface, the turbidity maximum grows and decays as concentrations vary during a semidiurnal tidal cycle (Figure 2-35). Also, the core of the maximum oscillates 10–20 km to and fro with ebb and flood of the tide. Such changes are linked to tidal scour, resuspension, and deposition. At high slack tide, the core of the maximum is small; concentrations are less than 1 g/liter (Figure 2-35). When ebb currents accelerate, bed mud erodes and causes the maximum to grow. But near low slack tide, part of the mud settles and the maximum decays. Similar trends follow on the flood tide (Figure 2-35). Because the tidal time-velocity curve is asymmetrical, the effect over many tidal cycles is to erode sediment preferentially during flood tide and bring about an upstream movement of sediment.

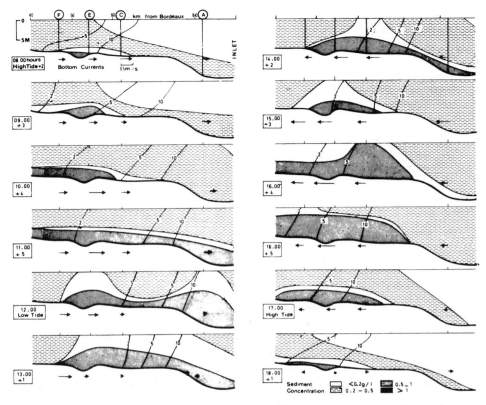

Figure 2-35. Longitudinal distribution of suspended sediment concentrations, g/l, in relation to isohalines (salinity), ppt, Gironde estuary during a semi-diurnal tidal cycle; spring tide and high river flow. Note high resuspended concentrations during peak currents (+ 4 and + 5 hour) and sedimentation near slack water, particularly at high slack tide (+ 1 and + 2 hour), (from Allen *et al.*, 1977). Reproduced by permission of Academic Press.

When current strength changes fortnightly between neap and spring tide, the concentration and position of the maximum also change. From studies in the Gironde, Allen *et al.* (1980) disclosed that during decreasing tidal amplitude, i.e., from spring to neap, and hence decreasing peak current velocities and longer slack durations, net sedimentation exceeds erosion (Figure 2-36a). Pools of fluid mud accumulate during neap tide (Figure 2-36b). The reverse process occurs when tide range increases; erosion exceeds sedimentation (Figure 2-36c). After a complete neap–spring cycle, however, a certain amount of sediment consolidates during neap conditions and, hence, resists erosion on subsequent spring tides. Therefore, every 2 weeks a thin lamination of mud accumulates permanently in the estuary. Additionally, Allen *et al.* (1980) found that when river floods push the

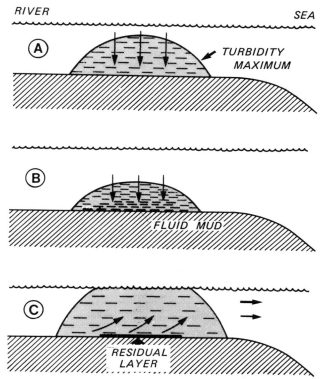

Figure 2-36. Neap-spring cycle of suspended sediment deposition and erosion in the Gironde estuary. (a) Waning tide range, beginning of deposition and reduction of maximum; (b) neap tide, rapid deposition and depletion of maximum with accumulation of fluid mud; (c) spring tide, erosion, resuspension and seaward escape; net sediment accumulation deposited in laminae, (from Salomon and Allen, 1983). Reproduced by permission of Total-compagnie Francaise des Petroles.

turbidity maximum near the estuary mouth (Figure 2-37) during a time of spring tide, large-scale resuspension and mixing promote escape of suspended sediment from the estuary into the sea. In response to similar fortnightly and seasonal fluctuations of river inflow, the turbidity maximum in the Columbia River, a mesotidal estuary, shifts landward and seaward with an excursion of about 20 km (Gelfenbaum, 1983).

In some estuaries, accumulation of marine mud is observed in rivers upstream of the innermost limit of salt intrusion. For example, marine diatom remains are recorded in freshwater deposits (Brockmann, 1929). Postma (1967) notes that the turbidity maximum in the Demerara River estuary (British Guiana) extends upstream against the residual river flow, a process that occurs by diffusion of suspended material induced by the tide. Allen *et al.* (1980) found similar trends in macrotidal hypersynchronous estuaries,

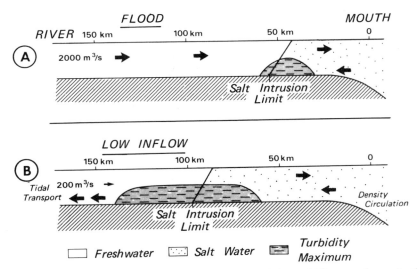

Figure 2-37. Location and longitudinal extension of the turbidity maximum in the Gironde estuary during (a) river flood, and (b) low inflow. Note landward extension of maximum landward of salt intrusion limit during low inflow, (from Salomon and Allen, 1983). Reproduced by permission of Total-Compagnie Francaise des Petroles.

the Aulne and Gironde. During low river flow, a well-developed maximum is maintained and extends about 40 km upstream of the salt intrusion reaching the limit of tides. From numerical simulations, Salomon and Allen (1983) showed that tidal amplitude and power dissipation can attain a peak in the estuary and produce high resuspensions. The tide wave is deformed and becomes more asymmetrical as it propagates upstream (see prior section, *Tidal Wave Deformation*) (Figure 2-22). Tidal processes can extend or form a turbidity maximum landward of the salt intrusion, both by creating high tidal velocities and thus greater bed resuspension and by producing flood velocity predominance near the bottom. As shown in Figure 2-38, a tidal turbidity maximum can form in the trapping node of current convergence between landward tidal flow and seaward river flow. In this zone, the tide wave and currents damp out. Therefore, during the low river inflow, tidal processes control trapping and formation of the maximum at the head of tides, whereas during high river inflow, density circulation dominates at the head of the salt intrusion.

Settling and Scour Lag Effects

In some estuarine environments like tidal flats and lagoons of the Dutch Wadden Sea, high concentrations of suspended material are observed in

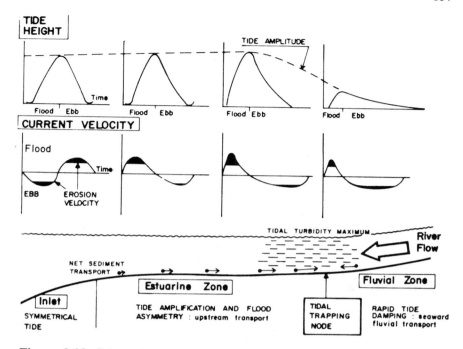

Figure 2-38. Schematic diagram illustrating the effect of tide asymmetry in macrotidal estuaries resulting in landward transport of suspended sediment and a tidal turbidity maximum toward the landward limit of tidal currents, (from Allen *et al.*, 1980). Reproduced by permission of Elsevier Scientific Publishing Co.

landward reaches despite the lack of significant freshwater inflow. The increase takes place also despite rapid water renewal with the sea, which favors transport to the sea. Instead, measurements show a residual landward transport into the estuary and an excess deposition over erosion with predominately tidal flat mud extending downward more than 12 m. Mineralogical and petrographical analyses clearly show that the silt and clay, as well as sand, in the Wadden Sea are derived from the North Sea (Van Straaten and Kuenen, 1957). How are fine sediments transported landward and retained?

An explanation of landward transport and accumulation on tidal flats is provided by Van Straaten and Kuenen (1957, 1958) and Postma (1961, 1967) in terms of a settling lag and scour lag effect. The settling lag is the time taken for a particle to reach the bed after the tidal velocity has diminished to the point at which the particle can no longer be held in suspension. That is, during decreasing current velocity, some time is needed for suspended sediment to settle (Postma, 1954). Hence, there is a time lag between slack water when current velocity is zero and the time of minimal

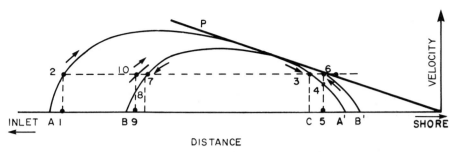

Figure 2-39. *Settling lag effect.* The diagram shows the velocities with which different water masses move with the tides at each point along a section through a tidal inlet (left) to the shore (right). Although the tide at fixed points is assumed to be symmetrical, the distance-velocity curves are asymmetrical. A water mass moves in and out along one such curve. The tangent (P) represents the maximum current velocity in each point and meets each curve at a point attained by the water mass at half tide. The curves apply only to idealized average conditions, and scour lag is neglected.

A particle at point 1 is taken into suspension by a flood current (water mass at A) of increasing velocity and starts to settle toward the bottom at point C, when the current still has a velocity equal to 2. While settling, the particle is carried farther inland by the still flooding currents and reaches the bottom at point 5 while the water has a velocity at point 4.

Scour lag effect. After the turn of the tide, the particle cannot be eroded by the same water mass (AA′) because this water parcel attains the required velocity later at a point beyond the particle toward the inlet. The sediment particle is therefore eroded by a more landward water mass (B′) and is transported toward the inlet to point B. At 7, it starts to drop out of suspension and reaches the bottom at point 9. During one tidal cycle, the particle has therefore been transported landward from point 1 to 9. After a number of these landward transport cycles, the particle may reach a point where it cannot be entrained by subsequent ebb flow currents because of the landward decrease of the average velocity of the tidal current (after Postma, 1967, and Van Straaten and Kuenen, 1957).

suspended sediment concentration. Because of this lag and the inert reaction of sediment to current changes, sediment is carried some distance beyond the point where it starts to sink. Thus, for flood current near high water, a certain amount of sediment settles toward landward areas. Because the average and maximum tidal currents also decrease landward, it follows that at the place where the sediment finally settles to the bed, the successive ebb currents available for erosion are weaker at that place than the original flood currents. After the turn of the tide, it takes time for ebb current to reach a critical erosion velocity required to resuspend the sediment. Therefore, the returning water mass contains less sediment than the incoming water mass. A certain amount is left behind on the bed. The basic mechanism is shown diagrammatically in Figure 2-39.

The scour lag is the additional velocity required to move or suspend a particle in excess of that at which it settled to the bed. The difference in critical erosion velocity and lower depositional velocity is shown in Figure 2-39. Because of the scour lag, the deposited sediment will not be resuspended after the turn of the tide; i.e., not until the currents have attained a higher velocity than when it deposited. The time elapsing between the turn of the tide and the moment of resuspension is longer than in the preceding period between the moment of deposition and turn of the tide. The constant shifting of sediment from one (flood) water mass to the other (ebb) with settling and scour lag, and the landward diminishing velocity trend, results in a distance velocity asymmetry whereby the sediment has a net landward transport with each tidal cycle (Figure 2-39). Because the process repeats itself every tide, landward zones develop high concentrations of suspended sediment and accumulate fine sediment less than 64 or 125 μ in size. This contrasts with the usual case where fine sediment increases away from shore, away from its source, or in deep water below the level of wave action. action.

The settling scour lag effect is independent of river inflow and is most clearly observed in lagoonal tide flat systems dissected by deep tidal inlets that expand landward into broad flats or shoals. The landward decrease of average water depth enhances the effect because settling distance for a given grain size is less landward than seaward. The effect is simply based on vertical turbulent exchange and settling velocity; it does not require a certain cohesion of sediments (Postma, 1980).

Estuary–Shelf Exchange

In the zone where estuaries meet the sea, transport of suspended sediment is modulated by river flow, tidal currents, and coastal drift. Greatest seaward escape of sediment in the Gironde estuary takes place place when river flow pushes the turbidity maximum near the mouth and the resuspension of bed sediments is great during spring tide (Figures 2-36, 2-37) (Castaing and Allen, 1981). During these periods, the increased seaward surface density flow supplies large quantities of suspended sediment to the shelf, a net seaward flux more than 100 times greater than during conditions of low river flow. One of the sinks for dispersed suspended sediment is a large mud field on the shelf floor adjacent to the Gironde estuary. Landsat and thermal imagery analyzed by Castaing and Allen (1981) reveal that part of the dispersed load moves northward in the coastal drift, mixes with water of adjacent inlets, and recycles into coastal lagoons. Kelley (1983) observed a similar dispersal from mineralogic evidence and landsat imagery of Delaware

Bay. Suspended sediment resuspended from the Bay floor escapes through the mouth, disperses northward around Cape May, mixes with shelf suspended sediment, and recycles into lagoons on the southern New Jersey coast. However, the bulk of the fluvial load supplied to northeast U.S. estuaries is retained within these systems (Biggs and Howell, 1984) because of their relatively high volumetric capacity, low sediment influx, and active circulatory entrapment. In contrast, the Gironde is a relatively "mature" estuary that has largely filled its drowned Holocene valley (Castaing and Allen, 1981). Sedimentation rates are faster than the present rate of transgression. A gradual reduction in flow cross section has shifted the salt intrusion head as well as the turbidity maximum, seaward particularly during high river flow. As a result of this "aging," tidal processes probably become relatively more important than density processes and a greater proportion of fine sediment escapes seaward (Castaing and Allen, 1981).

Particle Dynamics and Behavior

Fine particles in estuaries are either single particles dispersed in suspension, which are primary particles, or they may coalesce and form composite particles, referred to variously as flocs, aggregates, or agglomerates. How are the composite particles distinguished? According to Schubel (1982) and Syvitski and Murray (1981), the nature or strength of the forces that bind the component particles together are distinctive features. *Floccules* are inorganic particles, i.e., mineral or nonliving biogenic material, bound by electro-chemical forces. *Agglomerates* consist of organic or inorganic matter bound by relatively weak forces like surface tension or sticky organic matter. *Aggregates* are inorganic particles strongly bound by intermolecular, intramolecular, or cohesive atomic forces. Generally, the total *number* of suspended particles in estuaries is accounted for by individual particles while the bulk of the total *volume* or *mass* of particles consists of composite particles, either organic or inorganic material (Schubel, 1971, 1982).

Composite particles form by different processes: (1) physiochemical flocculation, (2) biological processing and binding, or (3) other processes including action of bubbles and enmeshment in precipitates (Table 2-3). Although these processes can act in conjunction, they are treated separately for discussion.

Fine-suspended particles in estuaries present some fundamentally different dynamic problems from those of sand grains. Clay-size particles have a high ratio of surface area to mass and therefore surface forces acting on the particles control its dynamic behavior in suspension. Whereas a sand grain has virtually constant properties, those of a floc change as a function of the local chemical and turbulent forces acting on the particles. As flocs become progressively larger, they also become less stable. Probably no stage is

Table 2-3. Factors affecting flocculation and aggregation of fine particles.

1. *Physicochemical processes*
   - Cohesion (cause)
     Van der Waals force
     Electrochemical force, salinity > 0.5 ‰
     Clay mineral composition
   - Collision (processes)
     Brownian motion
     Fluid shear in a velocity gradient
     Differential settling and scavenging
2. *Biological processing*
   - Biodeposition
   - Organic binding
3. *Other processes*
   - Bubble action
   - Enmeshment in precipitates

reached in which all particles have the same dynamic transport velocities (Kranck, 1975). Settling rates are difficult to measure and to predict because floc size is sensitive to local as well as to prevailing chemical and turbulent forces. Settling can vary with clay mineral type and suspended sediment concentration. The usual laboratory methods of particle size analysis employing peptizing agents yield little information about the particle behavior in the natural state, i.e., the aggregate size distribution associated with transport processes. Further, laboratory experiments fail to reproduce the natural turbulence and shear stresses important to floc formation. Also, microscopic examination is hampered by the fragile nature of flocs and the need to remove them from the natural hydrodynamic environment for analyses. Recently, some of these difficulties have been overcome by examining natural flocs and agglomerates *in situ* through laser holographic techniques (Carder, 1979; Carder and Meyers, 1979; Gibbs and Heltzel, 1982).

Because floccules settle many times faster than their dispersed component particles (Figure 2-40), sediment is transported at quite different rates and along different routes than dispersed sediment. Whereas flocculent sediment tends to sink and accumulate in an estuary, dispersed particles tend to remain in suspension and be swept out to the sea.

A complicated situation arises when fine-suspended particles discharged from a river first encounter the salty water of an estuary. Transportation and deposition are affected not only by a change in particle properties and by flocculation and aggregation processes, but by the mixing of fresh and salty water and associated estuarine circulation. Whereas flocs tend to settle by gravity, upward mixing because of salt exchange tends to resuspend them. The situation is further complicated because physicochemical flocculation is

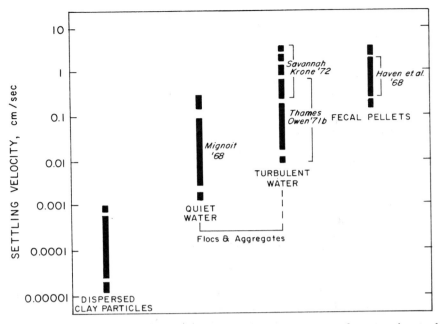

Figure 2-40. Range of settling velocities for different states of aggregation and dynamic conditions; adapted from G.P. Allen (unpublished) with data from various sources as noted.

a reversible process. Deflocculation can occur when flocs are transported from saline water into fresh water (Postma, 1967; Duinker, 1980).

The importance of "salt flocculation" has been questioned by Meade (1972, p. 97) who pointed out that while laboratory experiments demonstrate flocculation, "there is little direct evidence ... that salt flocculation has much influence on how and where fine-suspended sediments are deposited in estuaries." Meade concluded that the estuarine circulation is the most important process affecting transport and deposition of fine sediments while discounting flocculation as having been exaggerated. Although the question of whether or not flocculation plays an important role in estuarine sedimentation has not been answered, the question serves as a focus for discussion and study along different lines of research inquiry.

Flocculation and Aggregation

One line of research whereby flocculation and aggregation processes can be detected employs quantitative models of aggregation kinetics to predict the aggregation of particles when particle collisions are induced by fluid motion (Smoluchowski, 1917), or when particles are destabilized by surface charge differences and ionic content, a phenomena termed the VODL theory, (Verwey-Overbeek and Derjaquin-Landau, 1948). The relation between electrostatic particles and particle fabric is demonstrated by conceptual models (Lambe, 1953; Van Olphen, 1977). Another approach is to simulate the effects of varying sediment concentration, particle size, mineralogy, and salinity on the rate of flocculation in laboratory flumes or still-water experiments. Flocculation is observed by a change in concentration of the suspension or in the number of particles or aggregates in various size classes. In some studies flocculation is recognized by a change in suspended sediment settling rate relative to the initial concentrations or by changes in accumulation rate on the bed (Migniot, 1968, 1977; Krone, 1962; Edzwald and O'Melia, 1975; Kranck, 1980a). Electron and optical microscopy display the floccular structure and state of aggregation (grain to grain attachment) together with the particle composition (e.g., inorganic or organic) and the abundance of particles in different size classes (Harris *et al.*, 1972; Schubel, 1969; Kranck, 1973; Zabawa, 1978a, 1978b; Gibbs *et al.*, 1983). Reportedly (Schubel, 1982), direct field observations of flocculation are inconclusive, in part because *in situ* measurements are difficult. In the Savannah estuary however, Krone (1972) recognized flocculation effects by a rapid decrease of suspended sediment concentrations with decreasing tidal current speed. Although the precise nature of the process is seldom delineated, others observe flocculation effects by increasing particle size and aggregation state through the freshwater–saltwater transition (Kranck, 1979; Gibbs and Heltzel, 1982; Zabawa, 1978a). The processes and the particle behavior are best studied by combined laboratory experiments, microscopy, and field observations.

Estuarine sediments exhibit a full range of particle types from single clay minerals to groups of flocs. How, then, are flocs recognized? As shown in Figure 2-41, plate-like clay minerals in suspension can coalesce into chain-like or slightly rounded flocs with a cardhouse structure (Figure 2-41b). Individual flocs also unite into porous floc groups that may be subrounded or chain-like (Figures 2-41c, 2-42a). Such a composite particle of about 62 μ diameter and 25% mineral content, commonly will incorporate nearly 1000 component particles. Larger and more complex arrangements consist of honeycomb structures with a veil of bacteria that form a "fluff" on the surface of bed deposits (Figures 2-41d, 2-42b). Because of their loose porous structure, such structures tend to collapse with burial and compaction. As pore water is lost, bulk density increases. Although most flocs lose their identity, some flocs retain their original floccule fabric as shown by Krone (1962), by clay fabric analyses (Bennett and Bryant, 1977), and by examination of deposits (Pryor and VanWie, 1971; Pryor, 1975). When floccule fabric is not retained, the effects of flocculation can be recognized in

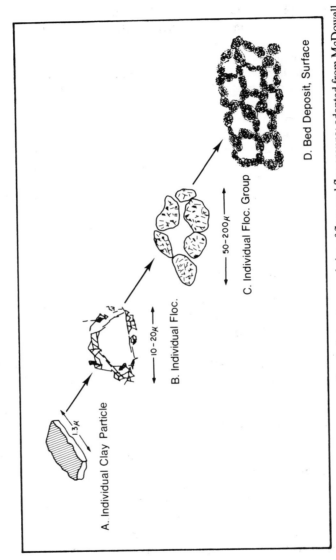

Figure 2-41. Schematic sequence of typical structures and size of flocs and floc groups; adapted from McDowell and O'Connor (1977). Reproduced by permission of John Wiley and Sons.

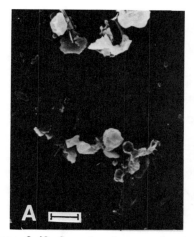

Figure 2-42. Scanning electron micrographs of (a) natural suspended agglomerate in an angular net-work or chain-like group; and (b) large floc in form of a "raft" prepared from dispersed suspension. Bar is 5 μm long. From Chesapeake Bay; Zabawa, 1978a.

ancient deposits by uniform grain size composition and constant ratios between size subfractions less than 16 μ as demonstrated in the Dutch Holocene (Favejee, 1960; Van Stratten, 1963).

The process of flocculation, whereby a suspension of dispersed grains of varying sizes is transformed into a stable flocculated distribution, is described by Kranck (1975, p. 116).

In salinities above about 3‰ fine particles in a suspension are unstable and flocculate readily on contact. Abundant particle collision will occur in the initial unflocculated suspension as gravity and turbulent forces cause a random variation in transport speed and direction of particles of different sizes. Smaller particles have a large relative surface area and will flocculate most readily (Van Olphen, 1966) whereas the larger grains are not sufficiently surface-active to flocculate with other single grains but only adhere to flocs composed of many smaller grains. As progressively larger flocs are formed the transport velocities become more uniform and particle collisions less frequent. Eventually, a state is reached in which the settling velocities of the largest flocs equal the velocities of the largest grains and further flocculation ceases.

When flocs grow to a large size, they are subject to more shear stress and become unstable. According to Kranck (1975), a greater increase in primary grain mode is required to produce a given increase in floc mode. Therefore, the ratio of floc size mode to primary grain size mode decreases with an increase in particle size (Figure 2-43). This is because the large particles have a smaller total area for the same mass and hence they are less susceptible to flocculation.

When flocs reach the bottom, small shear forces may break them apart, and the single particles or light material is resuspended or diffused upward

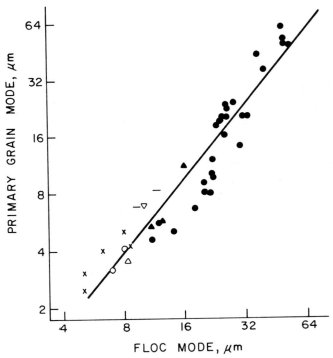

Figure 2-43. Relation between primary grain size mode and floc size mode. Dots and triangles from Petpeswick Inlet and Northumberland Strait, Nova Scotia; others from various estuaries and Scotia Shelf, circles; adapted from Kranck (1975). Reproduced by permission of Blackwell Scientific Publications.

into the water column (Kranck, 1979; Postma, 1980). This process repeats itself and forms a dense turbid bottom layer, termed a "lutocline" (Kirby and Parker, 1983). The layer acts as a reservoir from which mobile fine sediment is resuspended, mixed upward during accelerating tidal current, and to which it returns during decelerating phases. Within the layer, final settling takes place when reflocculated material becomes strong enough to withstand boundary shear forces (Kranck, 1980b). The sediment may also be permanently removed from suspension when currents disperse it to less energetic areas. The end product of accumulation is called "fluid mud," which is a highly concentrated mud suspension containing 10 to 480 g/liter.

Particle flocculation and aggregation require two essential conditions: (1) there must be numerous collisions of suspended particles or aggregates, and (2) the colliding particles must stick to each other (Krone, 1978). The following section, therefore, considers the *cause* for particles to coalesce and the processes that facilitate collisions.

Cohesion is largely determined by the balance of attractive Van der Waals forces and the repulsive electrochemical forces between particles. Clay minerals have platelike crystals with a layered lattice structure; they exhibit ion-exchange properties and normally carry a high negative surface change. In water, the total negative charge is balanced by a double layer of positive ions, and particles of like charge repel each other and thus remain dispersed. However, when the repulsive forces are destablisized by a high concentration of cations, such as Ca^{++}, Mg^{++}, and Na^+, which are contained in an electrolyte like sea water, the clay particles approach each other under the influence of ever-present Van der Waals force and flocculate. The flocs are held together by electrochemical bonding forces between the various atoms in each crystal (Overbeek, 1952; Strumm and Morgan, 1970). Additionally, they can be draped by veils of bacterial mucus. Organic particles that are electrically charged can also act as nuclei and can attract clay minerals. This simplified cohesion process is complicated by the quantity and type of cations present in solution, by pH (Packham, 1962), by the type of clay mineral, and by the particle size. These factors are complex where river-borne suspended sediment contacts salty estuarine water, but some of their effects are easily demonstrated by laboratory experiments.

The experiments of Whitehouse et al. (1960) are of special interest because they demonstrate the effects of chlorinity (salinity) and of differential flocculation of clay minerals. The results show that pure clay suspensions of illite and kaolinite begin to flocculate quickly with an increase in chlorinity between 1 and 2‰ (Figure 2-44a), as shown by an increase of settling velocity of larger particles. Thus, a small amount of electrolyte promotes a large increase in settling rate. With an increase of only 2‰ chlorinity, illite flocculates abruptly whereas montmorillonite flocculates over a much wider range.

The experimental results of Whitehouse et al. (1960) suggest that differential flocculation would take place where river water first encounters sea water, especially at high suspended concentrations (Figure 2-44b). Illite and kaolinite flocculate first and deposit followed by montmorillonite farther seaward. This is partly confirmed by longitudinal clay distributions in Ishikari Bay, Japan (Shizowa, 1970), and in the Pamlico River estuary, North Carolina (Edzwald and O'Melia, 1975). In Pamlico estuary bed sediments, however, illite dominates seaward while kaolinite dominates landward. Whitehouse's experiments apply to settling velocities of pure clay in quiescent sea water, which essentially eliminate effects of turbulence and the time required for particles to aggregate. Differential flocculation seems to occur in some estuaries, but similar effects can be attained by mixing two or more different clay populations, river-borne and marine, as in the James estuary, Virginia (Feuillet and Fleischer, 1980).

The collisions necessary to bring particles close enough for cohesive forces to act are facilitated by three processes: (1) Brownian motion, (2) fluid shearing produced by velocity gradients in the fluid, and (3) differential

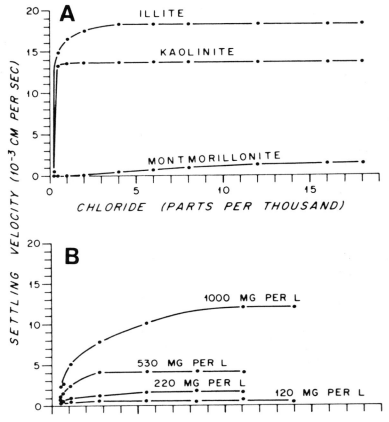

Figure 2-44. (A) Settling rates of clay floccules with changing chlorinity. Data based on Whitehouse *et al.* (1960); figure from Meade (1972). (B) Variation in clay particle settling rates with chlorinity and sediment concentration. Data based on Krone (1962); figure from Meade (1972).

settling and scavenging of different size particles. All of these processes depend on particle concentration.

Random Brownian motion results from thermal motions of water molecules. The rate of successful collisions is directly related to the second power of the particle concentration. Flocs formed by Brownian motion have a lace-like structure and are weak; therefore, they are easily dispersed by shearing or crushed in a deposit (Ariathurai *et al.*, 1977).

In fluid shearing or turbulence, particles moving at different speeds overtake one another, collide, and form flocs. The frequency of collision produced by shearing of the flow not only depends on the particle volume but on the velocity gradient as well as the particle concentration (by number). The frequency of collision is directly proportional to the number concentration of suspended particles (Krone, 1978). As larger flocs occur, the frequency of collision increases rapidly, but flocs cannot grow in size indefinitely because they are limited by the maximum fluid shear stress they can withstand. At high shear rates, interparticle bonds are broken and the flocs are torn apart. Thus, fluid shearing and turbulence can either promote or hinder flocculation depending on the relative strength of flocs and the shear forces. According to Krone (1962), flocs and aggregates produced by fluid shear tend to be spherical, relatively dense, and strong because only particles with strong interparticle cohesive bonds survive the local fluid stresses.

Collisions between particles also can occur when particles settle at unequal speeds. In still water, collisions between grains occur when faster settling grains overtake and run into slower settling grains that are of the same density. There is little stress on particles colliding by differential settling, and the resulting flocs are usually of low density, ragged, and relatively weak.

The relative efficiency of all three processes—Brownian motion, fluid shear, and differential settling—is measured by a collision frequency function (Lerman, 1979, p. 295). Brownian motion is important for submicron particles (Krone, 1978), while differential settling is probably most important when aggregation is far advanced and when currents are weak near slack water (Krone, 1978; Hawley, 1982). Most of the time, however, aggregation by fluid shear is by far the dominant collision process (Ariathurai et al., 1977).

In summary, flocculation requires that particles be brought together (collide) and stay together (cohere). The frequency of particle collisions is directly proportional to the suspended concentration (by number) (Krone, 1978); it depends very strongly on the heterogeneity of particle size and density as well as on the local fluid shear rate. Mild turbulence promotes flocculation by increasing particle collision, but severe turbulence inhibits flocculation by disrupting large flocs. Cohesion depends on the composition of the particles and on salinity of the suspensate. Increasing salinity reduces particle repulsion and thereby enhances cohesion.

Biological Processing

Because estuaries are generally more productive than other coastal environments, they contain large numbers of planktonic and benthic suspension-feeding animals that can have a strong influence on sedimentary processes. Suspension-feeding animals like oysters, barnacles, tunicates, and copepods obtain food by pumping water through gills or other particle-retention organs. Large quantities of fine-suspended sediment as small as 1 μ, inorganic and organic, are removed in the process (Galtsoff, 1928). Some material is rejected as loosely bound clumps called pseudofeces. The rest is ingested, compacted in the gut, and egested as fecal pellets or strings. The sediments, therefore, are no longer discrete grains but components of larger pellets. The process of filtration, compaction in the gut, and subsequent deposition is called *biodeposition* (Haven and Morales-Alamo, 1966, 1972).

Properties of the pellets are entirely different from those of their component particles. They exhibit a wide range of composition, size, and shape (Kraeuter and Haven, 1970) consisting of rods, pellets, ribbons, and discs. The particles are bound by mucus and the pellets are often very resistant (Humby and Dunn, 1973). Depending on size, shape, and density, fecal pellets settle from 0.4 to 2 cm/s, rates common to coarse silt and fine sand particles (Haven and Morales-Alamo, 1968) (Figure 2-40). Once deposited, pellets can be either resuspended or eaten by deposit-feeders and reworked into the bottom deposits (Haven and Morales-Alamo, 1966). Mixing of fine-grained pelletized and coarse material results in a bioturbated and poorly sorted deposit (Carriker, 1967).

Biological processing is important as a sedimentary process. Experiments reveal that the oyster, *Crassostrea virginica*, filters, eats, and voids enormous quantities of fine suspended particles, mainly in the 1–3 μm size range, and then excretes them as agglomerated fecal pellets, 50–3000 μm long. Lund (1957) found that oysters increased the volume of material settled from suspension by eightfold. Haven and Morales-Alamo (1972) showed that oysters on a typical commercial oyster bed in the Chesapeake region can agglomerate and deposit about 1–2 metric tons (dry weight) of sediment per hectare per week, an amount equal to about 6 mm of sedimentation annually. Further, Verwey (1952) calculated that mussels in the Wadden Sea deposit between 25,000 and 175,000 tons of fecal detritus each year. Additionally, Haven and Morales-Alamo (1972) found that biodeposition rates of the tunicate (*Molgula manhattensis*) were three times faster than the oyster and even greater than soft clams, ribbed mussels, and barnacles.

Filter-feeding zooplankton like copepods that are widely dispersed in estuarine waters can process and agglomerate more fine sediment than benthic invertebrates (Schubel, 1982). From experiments on Chesapeake Bay water, Schubel and Kana (1972) estimated that copepods and other zooplankton filter a volume of water equivalent to that of the entire Bay in a few days. Some copepods ingest clay particles, e.g., *Tigriopus californicus*, and perform chemical and mineralogical transformation (Syvitski and Lewis, 1980). A mineralogic change by digestive processes may account for some anomalous geochemical or clay mineral distributions such as reported in the Pamlico River estuary (Edzwald and O'Melia, 1975). Because of their increased bulk density, egested pellets settle several orders of magnitude faster than individual component mineral particles. Additionally, many filter-feeding organisms have sophisticated mechanisms for selecting and sorting fine-suspended material. Generally, these mechanisms are active in estuarine zones of low suspended concentrations, less than 5 mg/liter, where zooplankton can be abundant.

Growth of bacteria and forms of algae and fungi can bind microbes to suspended particles or attach particles to each other. By draping veils of mucal slime over agglomerated sediments (Figure 2-45), bacteria not only hold particles together but strengthen particle bonding so that the aggregate structure is strong enough to withstand considerable shear forces (Kranck, 1980a). Growth of microbes can increase the rate and size of sediment agglomerates larger than that due solely to physiochemical flocculation (Paerl, 1973, 1975). Diatoms also aid agglomeration of particles through a mucilage coat or siliceous web to which particles adhere (Ernissee and Abbott, 1975).

Other Particle Processes

Most sedimentological interest has centered on physicochemical flocculation while other aggregation processes have received less attention. In addition to biological processing, aggregates of fine-suspended material can form in a variety of other ways through:

1. The action of bubbles in water that adsorb dissolved or colloidal organic carbon. Aggregation is probably aided by bacterial colonization (Riley, 1963; Burton, 1976; Meade, 1972).

2. Adsorption of dissolved molecules onto immersed mineral surfaces (Bader *et al.*, 1960; Chang and Anderson, 1968).

20 Mμ

Figure 2-45. Scanning electron micrograph of suspended sediment from Pee Dee River estuary, S.C. showing microbial structures and slime webbing on and between agglomerates; from Zabawa (1980).

3. Enmeshment in metal hydroxides precipitated or flocculated from dissolved or colloidal organic material when river-borne iron or humic acids mix with salty water in an estuary (Coonley *et al.*, 1971; Aston and Chester, 1973; Boyle *et al.*, 1977; Sholkovitz, 1976, 1978; Mayer, 1982).

Importance of Flocculation and Biological Processing

The importance of flocculation and biological processing relative to circulatory factors in estuarine transport and deposition is not clear. It is evident, however, that flocs and bioagglomerates are found in a large number of world estuaries. The processes proceed in quite different ways and at different rates, and their effects vary widely with time and space.

Physiochemical flocculation is likely important where concentrations of suspended sediment are high, 200 to more than 5000 mg/liter, such as zones of turbidity maxima, near-bottom lutocline layers, and during river floods, a time when haline stratification is also intense. Without a large fluvial supply or a mechanism like the estuarine circulation and bed resuspension to elevate suspended concentrations, flocculation will be relatively less important (Kranck, 1981). Where concentrations are low, less than about 60 mg/liter, such as middle and seaward estuarine zones, biological processing may account for most composite particles.

The proportions of bioagglomerates and flocs in suspension, either inorganic or organic, can change seaward through the freshwater–estuarine water transition. These trends reflect the properties of flocs and their interaction with hydrodynamic conditions, as well as the biological production of agglomerates in the estuarine zone. In the turbidity maximum of the St. Lawrence River estuary, the proportions of inorganic-organic flocs change as a result of differential settling of large-size organic aggregates leaving single grains and very small flocs in suspension (Kranck, 1979). Large low-density flocs can be culled out of a heterogeneous population in favor of smaller, more compact particles according to their resistance to shear forces (Krone, 1962). Whereas inorganic flocs "grow" and disintegrate in a few hours, production of bioagglomerates like fecal pellets usually takes longer: hours to a day. However, pellets are generally more stable than flocs and settle an order of magnitude faster. Flocs may "regenerate" in suspension, or in fluid mud, many times prior to final deposition. The ultimate effect of pellet production, however, is to increase stability and cohesion of estuarine deposits (Postma, 1980).

What are the significant consequences? Flocculation and biological processing enhance settling velocity, which ranges up to several orders of magnitude greater than primary particles (Figure 2-40). Consequently, these processes remove suspended material from the water and decrease turbidity. They tend to increase the net sedimentation rate and thus retain sediment in an estuary. When settling is accelerated, the average residence time of particles at any one level is reduced (Kranck, 1980b), an important consideration for contaminated sediment. When suspended sediment is removed from the water, it is "uncoupled" from its water mass or prevailing current and accumulates with coarser sediment (e.g., coarse silt or very fine sand), which settles as single grains. The resulting grain-size distribution is poorly sorted. It is a function of aggregation history and local energy level rather than a function of dispersal or transport mode either of traction or suspension.

If flocculation leads to differential settling according to variations in the inorganic/organic composition, then estuarine sediments are potentially fractionated. This is the case documented by Kranck (1979) in the St. Lawrence River estuary turbidity maximum where sedimentation removes proportionally more flocculated organic matter than very small flocs and single mineral grains that are selectively retained in suspension. Without flocculation and bioagglomeration, fine-suspended sediment probably would be carried out of an estuary or simply settle in zones where flow is diminished.

From an evaluation of filter feeding organisms in Delaware, Biggs and Howell (1984) concluded that biological mediated sedimentation exceeds other sedimentation processes in accumulation of fine sediment. For example, the combined filter feeders, zooplankton, oysters, and mussels fix an estimated 4×10^{11} kg/yr, more than 200 times the annual input of fluvial fine sediment to the estuary.

Effects of Biological Processes on Bed Sediments

Estuarine bed sediments are continually processed and recycled by enormous numbers of benthic organisms. How, then, do organisms modify the sediments and affect transport and retention of sediments? Consider two opposing biological processes: stabilization and destabilization, which describe respectively either a net increase or decrease in the critical erosion velocity (Rhoads et al., 1978). Stabilization processes bind particles to the bed through secretion of mucus or organic films over the sediment surface. The tracks and feeding trails left by gastropods or bivalves are often floored with mucal binding and are, thus, more resistant to erosion than to surrounding sediments (Nowell et al., 1981). Diatoms stabilize mud by migrating up and down through the sediment leaving a trail of mucus (Hopkins, 1967). In The Wash (United Kingdom), Coles (1979) maintains that benthic microalgae play a key role in promoting deposition by both trapping and binding sediment. If the algae were killed by pollution, tidal flats could be expected to erode rapidly and thereby induce sedimentation in the channels (Frostick and McCave, 1979). Despite important effects, biological processes are not included in existing suspended or bedload models (Young, 1982).

When growths of algae, worm tubes, mussels, and vascular plants cover a large area, they usually stabilize the bed by cementing mud or entrapping sediment between anchor structures providing a tight fabric (Rhoads, 1974). Such features also alter the turbulent boundary layer structure and elevate the viscous sublayer thus favoring sedimentation. However, when high density cover is broken by grazers, predators, or mechanical action of storms, low density cover and individual mounds, beds, or banks can accelerate flow and cause scour (Young, 1982).

Destabilization processes reduce bed cohesiveness and increase suspended sediment concentrations above the bed. Mainly, they result from feeding activities of suspension and deposit feeders. Pelletization of fine sediment by mobile deposit feeders like worms and bivalves may have either positive or negative effects. Rhoads (1974) reports that the pelletal surfaces and sediments have an open fabric with a high near-surface porosity and low compaction and cohesion. Such a surface mud is easily eroded and resuspended. Because settling velocities are high, however, pellets tend to move as bed load (Nowell et al., 1981). Additionally, any sort of biogenic surface roughening like projecting tubes and fecal mounds enhances resuspension. In a study of bed roughening, Boyer (1980) found that meiofauna lower critical erosion velocities by as much as 70% relative to an unprocessed bed.

Bioturbation, the reworking of sediments by burrowing or deposit-feeders mainly in the upper 10–15 cm of the substrate (Figure 2-46), increases sediment pore space and, thus, water content (Rhoads, 1974; Rhoads and Young, 1970). In turn, bioturbation deters compaction and supports formation of fluid mud in zones of fast sedimentation. If organic binding is negligible, bioturbation tends to lower erosion resistance of the surface and, thus, to destablize the sediment. It tends to produce preferential resuspension of surface sediment (Rhoads and Young, 1970). Vertical sorting also takes place when deposit feeders reject particles they are unable to ingest and they tend fall to the bottom of their burrows (Rhoads and Stanley, 1965).

The effects of bioturbation go far beyond modification of sediment characteristics. By continually reworking and recycling sediment and fecal material many times, the upper 10–15 cm of sediment is exposed to overlying water and keeps exchanging matter through resuspension or chemical reactions (Figure 2-47). Therefore, sediment constituents and organic material remain available for some time before permanent burial. As bed sediment is reworked and resuspended, pore water and nutrients are released back into overlying water. This may explain why estuaries are so chemically reactive and biologically productive.

The effects of bioturbation are recorded in the sedimentary structures (Figure 2-46). Bioturbation mixes sediment and prevents preservation of layers. It tends to smooth the estuarine record and average conditions at the time of accumulation. In general, the rate of bioturbation is lower in landward, less saline estuary zones and where the sedimentation rate is rela-

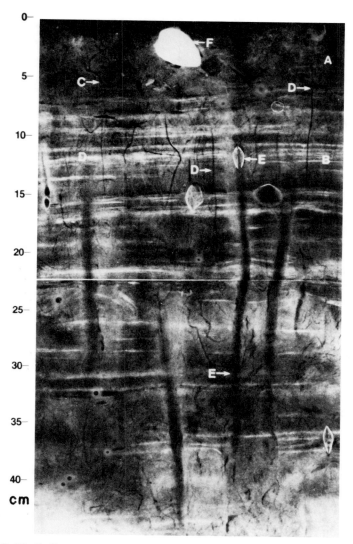

Figure 2-46. Sedimentary structures in X-ray radiograph negative, of a box core from middle Chesapeake Bay. Depth scale on left in cm. Laminations of fine sand (light) interbedded with clay and silt (dark) and cut by burrows and tubes representing bioturbation activity along length of core. Special features are: (A) Homogenous mud bed; (B) Interlaminated mud and fine sand bed; (C) Juvenile *Nereis succinea* burrow (active); (D) *Scolecolepides viridus* tube traces (some are active); (E) *Macoma balthica* shell and burrow with methane bubble concentration (inactive); (F) *Rangia cuneata* shell (disarticulated). Plate 3 from Reinharz *et al.* 1982).

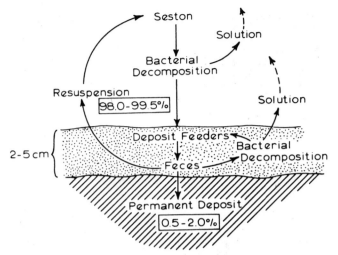

Figure 2-47. Schematic diagram showing the cycling of organic material at the benthic boundary layer. In Buzzards Bay, Mass. approximately 98–99.5 percent of the annual sediment influx to the bay floor is resuspended and cycled within the upper 2–3 cm, a bioturbated near-surface layer (from Young, 1971).

tively fast, e.g., in a turbidity maximum. Howard (1975) explores the degree of bioturbation as a tool in recognizing estuarine environments.

Geochemical Processes

Estuaries are like great chemical reaction vessels in which solutions with different chemistries are mixed in the presence of reactive particles (Goldberg, 1978). The fate of conservative chemicals supplied in solution from rivers or the ocean is usually predictable from a linear relation between the chemical concentration changes and salinity (Biggs and Cronin, 1981). Most chemicals, however, are highly reactive in the unstable estuarine environment. They may be assimilated by organisms or partition in particulate form through processes like ion exchange, precipitation, complexation with organic substances, coprecipitation with iron and manganese, and incorporation into mineral lattices and fecal material. They are strongly adsorbed on fine particles. Consequently, their movement and behavior is largely governed by sedimentary processes. By altering the flux of river-borne material to the oceans, estuary sediments can play an important geochemical role and affect global chemical budgets. The dynamics and details are beyond the scope of this chapter but treated in Olausson and Cato (1980), Burton and Liss (1976), and Martin *et al.* (1979).

Although geochemical characteristics of estuarine sediments are sensitive to environmental conditions, they are also susceptible to diagenetic changes. The classification of Berner (1981) based on authigenic mineralogy, i.e., oxic–anoxic, sulfidic–nonsulfidic, methanic, is applicable to estuaries. However, indicators of the original chemical environment also have been used to reconstruct environmental conditions. Of note for paleosalinity are evaporite beds precipitated from estuaries in an arid climate with an anti-estuarine circulation (Morris and Dickey, 1957). Additionally, the boron content of clay minerals and the calcium-to-iron ratio in phosphorite (Muller, 1969; Nelson, 1967) provide an indication of paleosalinity within broad limits.

Because mixing is an essential feature of estuaries, chemical characteristics can provide useful tracers of different sediment sources as fluvial and marine. They are useful provided the tracers have distinctive source end-members and behave conservatively; i.e., the amount of tracer per unit weight of sediment in a specific source is not subject to temporal variations during transport and deposition (Salomons and Eysink, 1981). Whereas radioactive or activable tracers and contaminants can indicate short-term movement, natural tracers are useful for long-term transport. For natural tracers of fine sediment in estuaries along the North Sea, Rhine-Meuse, Ems, Elbe, and Scheldt, Salomons and Eysink (1981) utilized differences in mineralogy, chemical composition, and stable isotopic composition of carbonates. The mixing ratios revealed that physical mixing of fluvial and marine sediments is a dominant process affecting trace metal distributions. Secondary deviations from conservative mixing are caused by addition or removal processes depending on chemical reactions and environmental conditions.

Estuarine Dispersal and Depositional System

It has been shown that different energy agents move sediment acting either independently or in combination, and the various sediment sources and types of sediment found on an estuary floor have been reviewed. It is now necessary to consider how sediments travel from their dominant source to their sink and hence give rise to characteristic facies distributions.

The complex dispersal system of an estuary is best approached by distinguishing areas dominated by fluvial, estuarine, and marine activity (Figures 2-48a, 48b). Additionally, a zone of wind wave action is recognized in shallow areas. The boundaries between areas are transition zones that shift according to river discharge, tides, and meteorological forces.

Dispersal in the fluvial area is controlled by river flow, the amount of sediment transported, and the dispersal pattern, depending on whether or not the

DISPERSAL ZONES & ROUTES

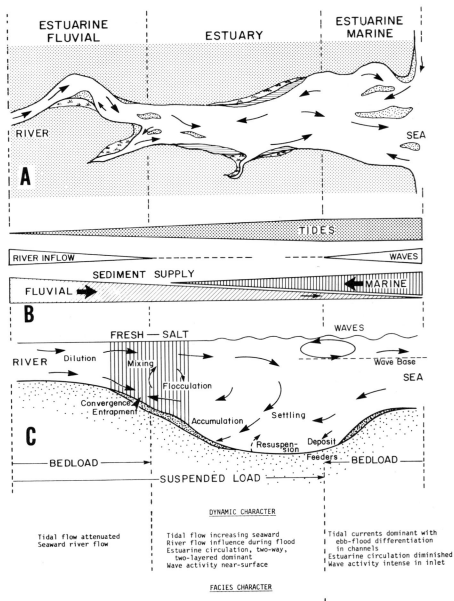

ESTUARINE FLUVIAL | **ESTUARY** | **ESTUARINE MARINE**

RIVER

SEA

A

TIDES

RIVER INFLOW | WAVES

SEDIMENT SUPPLY

FLUVIAL | MARINE

B

FRESH — SALT | WAVES

RIVER — Dilution | Mixing | Wave Base

SEA

Flocculation

Convergence
Entrapment | Accumulation | Settling

C | Resuspension | Deposit
Feeders

BEDLOAD | BEDLOAD

SUSPENDED LOAD

DYNAMIC CHARACTER

Tidal flow attenuated	Tidal flow increasing seaward	Tidal currents dominant with
Seaward river flow	River flow influence during flood	ebb-flood differentiation
	Estuarine circulation, two-way,	in channels
	two-layered dominant	Estuarine circulation diminished
	Wave activity near-surface	Wave activity intense in inlet

FACIES CHARACTER

Accumulation in meander	Shoals: rippled sand with	Shoals: ripple cross-bedded
bars dominate; cross-	clay pebbles and erosional	medium to fine sand
bedded sand and clay	contacts	Channel: ripple cross-bedded
lenses	Flats: laminated clay or sand	coarse sand
Channel: interbedded sand	and clay, bioturbated	
and clay; gravelly	Marsh: massive clay with	
Marsh and swamps: organic	plant fragments	
rich clay and silt	Channel: clay and silt laminated	
	and interbedded with sand	
	lenses and shells, massive	
	silt and clay, bioturbated, seaward	

river floods. Tidal flow is attenuated, especially during high river discharge, and net nontidal flow is seaward throughout the water column. Suspended sediment concentrations are diluted with distance downstream and partly sedimented in adjacent marshes, resulting in massive clay and silt that is rich in organic matter. Another part of the load is deposited in less energetic parts of the channel and on bars or is swept seaward into the main estuary. Bed load is directed seaward into the upper estuary. Some sand accumulates in meander bars but can be transposed from one bar to another during river floods as exemplified in the Gironde (Allen, 1972). Sand can also accumulate on the channel floor following floods, whereas mud accumulates during low river discharge. The resulting deposit consists of interbedded sand and clay. Longitudinal bars are fed by sand bed load and consist of cross-bedded sand with clay lenses or pebbles overlain by megaripples. Where fluvial supply is relatively high or tides are weak, the river builds a deltaic sediment mass in the estuary head or estuarine fluvial zone, e.g., Mobile Bay (Ryan and Goodell, 1972). Bed load of sand or flocculated fine sediment tends to accumulate where the river debouches into the open estuary forming a delta front shoal. Dispersed fine sediment is spread unequally into adjacent swamps and marshes of the estuarine fluvial zone or swept through the main channel and spread seaward over a broad area of the main estuary.

Dispersal of fine sediment in the main estuary follows the estuarine circulation. Where the ratio of river to tidal discharge is intermediate, 0.05– 1.0, the river-borne suspended load that remains in suspension for a long time is carried: (1) seaward through the upper layer, (2) downward by gravitational settling enhanced by agglomeration, (3) into less energetic zones possibly after repeated cycles of settling and tidal resuspension, or (4) back upstream in the lower layer to the convergence of landward and seaward flow where it accumulates or recycles upward and back downstream in the upper layer (Figure 2-48c). As the whole disperal process repeats itself, very fine-grained or flocculant sediment accumulates in the convergence zone and can form masses of fluid mud. The channel facies in the upper estuary is commonly clayey, laminated, or wavy-bedded with occasional lenses or beds of sand. The lower estuary displays similar deposits except that they are often bioturbated and hence are massive silt or clay and shelly. Marginal shallows that are attacked by wind waves are zones of bed erosion and of transport alongshore or into deeper water. Resultant

---

Figure 2-48. Conceptual model of dispersal zones and routes in a hypothetical estuary; (a) plan view, submerged sand shoals and spits dotted, intertidal flats hachured, and marsh. Arrows show direction of water and sediment movement. (b) Relative intensity of dominant dispersal agents, tides, river inflow and waves along estuary length in relation to dominant sediment supply, either fluvial or marine. (c) Vertical-longitudinal section of main dispersal routes. Lower table summarizes dynamic factors and associated facies character.

shoals are rippled sand with occasional clay pebbles and erosional contacts, while in protected areas tidal flats accumulate laminated clay or sand and clay burrowed by organisms. Bordering marshes hold massive clay rich in plant fragments.

Dispersal in the estuarine marine zone at the mouth is driven by the estuarine circulation and superimposed tides, waves, and longshore drift. Interactions of the latter are provided in Chapter 7 of this volume and in Figure 2-27. Generally, tides and waves are strongest over the shallow entrance bars, whereas estuarine flow brings in suspended load through the deeper channels. Terwindt (1963) describes directional properties of megaripples and current ripples in the lower Rhine estuary showing steep faces upstream in response to net landward sand transport. In the convergence between fluvial and estuary zones, ripples face both landward and seaward. Similar directional current structures are reported in the Elbe estuary (Reineck, 1963) and the Ord estuary (Wright et al., 1975).

In summary, dispersal is dominated by the estuarine circulation and superimposed bidirectional tidal transport. Dispersal directions are bimodal (Klein, 1967) and follow the channel morphology. Rates of dispersal vary with intensity of estuarine mixing and follow an energy format of reduced current competence away from zones of strong tidal currents in the river and estuary mouth, toward less energetic areas in deep water or protected shallows.

Dispersal not only mixes sediment from multiple sources but "filters" and fractionates it. Some sediment may pass through an estuary to the ocean while the rest is trapped within it. What conditions determine whether sediment is trapped or by-passed and how much?

Escape or Entrapment?

The trapping efficiency of an estuary, i.e., the fraction of the total mass of sediment input that is retained in an estuary, depends in the first place on its effective volumetric capacity (manifest in water depth) in relation to the rate of sedimentation and the energy available to transport the sediment supplied. If supply and energy are not in balance, then transport processes act to establish equilibrium by either trapping or by-passing the sediment supply. When entrapment results in accumulation, volumetric capacity and, hence, water depth below an equilibrium depth is reduced and the sediment surface becomes more dynamic. Retained sediment is subject to greater resuspension by waves, currents, and more rapid cycling and exchange with the ocean. As demonstrated by hydraulic and numerical models (Nichols, 1972; Simmons, 1965; Festa and Hansen, 1976), water depth changes estuarine mixing. As an estuary shoals and its volumetric capacity decreases, near-bottom flow from the ocean is reduced, vertical velocity increases, and the two-layered estuarine circulation weakens, i.e., the circulation type shifts from a Pritchard type B (partially mixed) toward type

C (well mixed). As a result retention of suspended sediment in the turbidity maximum is diminished and the probability of deposition is reduced. As the head of an estuary fills in with sediment, the estuary length decreases, particularly the length of the salt intrusion from the mouth to the landward limit of salty water. High river flow can shift the circulation from a Pitchard type B (partially mixed) toward type A (salt wedge) or thrust the entire salt wedge from its basin. As a result, trapping efficiency is greatly reduced and much suspended sediment escapes to the ocean. In general, filtering efficiency is low in short or shallow estuaries with either well mixed or highly stratified regimes or with high flushing velocity.

Three combinations of cycling are possible (Burns, 1963) (Figure 2-49): (1) The sediment moves directly through the estuary and escapes, either by the force of river floods or by intense wave and circulatory mixing (Figure 2-49a); alternately the sediment temporarily deposits and moves through in progressive steps, a step with each flood or storm. (2) The sediment is partially entrapped in an essentially closed circulation system, recycled many times, or resuspended from the bed, prior to accumulation (Figure 2-49b). (3) The sediment settles, deposits, and accumulates in low-energy sites below an equilibrium depth (Figure 2-49c).

The status of escape or entrapment is indicated by a sediment budget or mass balance of supply and loss. Assuming steady state and no net additions or losses, then the input flux, Mi, plus the sediment produced in the system, P, must equal the output flux, Me, plus the amount consumed in the estuary, C, and the flux to the bed, Ms.

Thus:

$$Mi + P = Me + C + Ms \qquad (2\text{-}13)$$
$$\underbrace{}_{\text{sources}} \quad \underbrace{}_{\text{losses}}$$

Then, the trapping efficiency, i.e., fraction retained, can be expressed as an index:

$$T_i = \frac{Ms}{Mi + P - C} \qquad (2\text{-}14)$$

For example, in northern Chesapeake Bay, of the total river input, an estimated 1.3 M tons/yr, 0.15 M tons/yr is exported, while 0.9 tons/yr is lost to the bed, an amount representing 71% entrapped ($T_1 = 0.11$) (Biggs, 1970). Consider how the escape–entrapment status changes with estuary evolution.

On a geologic time scale, submergence by rising sea level increases estuary water depth, length and, hence, volumetric capacity; consequently, entrapment is favored (Figure 2-49c). As a greater proportion of the influx mass, Mi, is trapped than exported, Me, then accumulation, Ms, increases and the ratio of rise, R, to depth, H below an equilibrium depth, increases (Figure 2-50). The cycling regime then favors less entrapment and more escape (Figure 2-49a). As the volumetric capacity of an estuary decreases, its trapping efficiency

CYCLING MODES

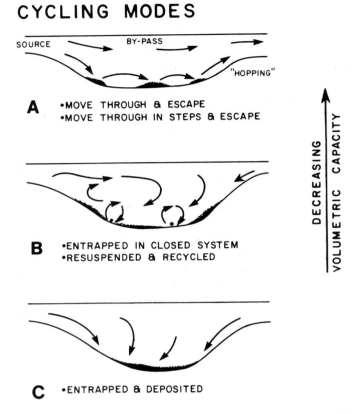

Figure 2-49. Cycling modes affecting escape and entrapment as a function of volumetric capacity; schematic. For explanation see text.

also decreases, and faster than its capacity decreases. During progress of submergence and sediment infilling, the locus of accumulation probably shifts from less energetic basins landward to the current convergence zone (landward salt limit). In turn, this zone moves landward with rising sea level and up the river valley with rising river base level and transgression. In late stages, when estuary channels are shoaled and convey much fluvial sediment directly to sea, accumulation shifts laterally into remaining backswamps, flats, or marshes. With lateral infilling the intertidal volume decreases, and hence the relative importance of river discharge compared to tidal flow, increases. Thus, with progressive infilling, the geologic function of an estuary changes from a sink for fluvial and marine sediment to a source of fluvial sediment for the ocean. Peterson *et al.* (1984) in a detailed study of the depositional

INFILLING EFFECT

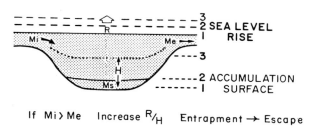

If Mi > Me   Increase R/H Entrapment → Escape

Figure 2-50. Schematic diagram showing effect of sediment infilling on the status of entrapment and escape of an estuary in longitudinal section. As the sediment influx, Mi, exceeds the export, Me, the accumulation rate Ms, may increase at a faster rate than the sea level rise, R. This trend is reflected by diminished water depth, H, below equilibrium depth, in relation to sea level rise, R, or increased R/H ratio.

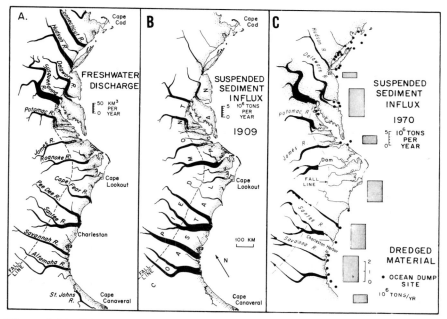

Figure 2-51. Water and sediment discharge by major rivers of the U.S. east coast between Cape Cod and Cape Canaveral. (a) Fresh water discharge, based mainly on 1931-60 U.S.G.S. stream records; (b) Suspended sediment discharge based on Dole and Stabler, 1909; (c) Suspended sediment influx affected by dams about 1970 and average annual quantities of dredged material taken from major estuaries. Figure and data modified from Meade (1969), Meade and Trimble (1974) and Nichols (1978); dredge data from U.S. Army Corps of Engineers annual reports 1970–74; and ocean disposal data from Gross (1975) and U.S.C.E.Q. (1970).

evolution of Alsea Bay, Oregon, document a change in sediment accumulation rate as a function of depth, volume and trapping efficiency.

Although most modern estuaries formed at about the same time, the stage of infilling and the escape–entrapment status, vary widely in different estuaries. Estuaries with a large fluvial supply like Atchafalaya Bay and the Yangtze are mostly filled and now discharge their sediment load on to the adjacent continental shelf. In contrast, estuaries like Chesapeake Bay and Long Island Sound with a small sediment yield and fluvial influx compared to volumetric capacity, trap nearly all the fluvial input (Biggs and Howell, 1984). Volumetric capacity in these estuaries is large because of glacial overdeepening (Gordon, 1980). Sediment influx is low despite relatively high river water discharge, as a result of the erosion-resistant glaciated terrain (Figure 2-51a,b) (Meade, 1969). On the other hand, rivers of low discharge on the U.S. southeast coast, which drain weathered Piedmont terrain, bring much more sediment into their estuaries. Consequently, the estuaries were filled nearly to capacity as fast as the advancing sea level rise drowned their valleys. Most fluvial supply accumulates in extensive marshes (Meade, 1982) or in dredged channels. However, large amounts of dredged material now by-pass the estuaries to the shelf or are recycled through dredging and disposal (Figure 2-51c).

Depositional Sequences in Estuaries

Estuarine deposits are spatially confined to topographic lows along coasts. These depressions may be river or glacial valleys or direct tectonic features. Regardless of their origin, the depressions can receive sediments from fluvial or marine sources. Thus, Rusnak (1967) has defined:

1. Positive filled basins, in which sediments accumulate principally by the longitudinal distribution of suspended river sediment and biological material.
2. Inverse filled basins, in which sediments accumulate principally from the longitudinal distribution of sediments entering from the sea.
3. Neutral filled basins, in which there may be redistribution of basin sediment from shore erosion and in which there is no change in basin volume.

Sedimentary Lithofacies

The largest estuaries are seldom more than 25–50 km wide and 300 km long, and within these spatial confines, all of the sedimentary lithofacies, characteristic of estuaries, must be deposited. The estuarine sedimentary lithofacies are usually associated with submergence and transgression, and it is useful to apply Walther's Law to the sedimentary sequence. Walther's Law, restated by Middleton (1973, p. 979) asserts:

> The various deposits of the same facies area, and, similarly, the sum of the rocks of different facies areas, are formed beside each other in space, though, in cross section, we see them lying on top of each other. As with biotopes, it is a basic statement of far reaching significance that only those facies and facies areas can be superimposed primarily, which can be observed beside each other at the present time.

Regardless of the source of sediment to an estuary, the axial sequence of lithofacies is: (1) estuarine fluvial, (2) estuarine, and (3) estuarine marine. The vertical sequence in a transgressive environment is, from the bottom up: (1) estuarine fluvial, (2) estuarine, and (3) estuarine marine. The complete axial sequence is shown in Table 2-4a, while the lateral sequence is illustrated in Table 2-4b. The complexity of the pattern of adjacent sedimentary types, as illustrated by Folk textural classes for the Gironde estuary, Chesapeake Bay and Delaware Bay, can be used as a predictor of potential vertical sequences and their variability (Figures 2-10, 2-11, 2-52).

The sedimentary lithofacies that are present beneath a particular estuary are controlled by dynamic processes (short-term like waves and tides as well as long-term processes like the rate of sea level rise) as well as the nature and quantity of available sediment. If the local relative sea level change is exceeded by the net rate of sedimentation, an estuary will fill and the estuarine environment might prograde over adjacent marine environments. Clearly, though, the vertical thickness of the entire estuarine sediment sequence cannot exceed the local relative sea level rise during the latest transgression. Roy *et al.* (1980) have proposed models for the evolution of estuarine embayments in which sedimentation rate exceeds sea level rise (Figure 2-53). These models all predict eventual filling of the estuary and direct injection of fluvial materials into the marine environment.

As observed by Clifton (1982, 1983) and Demarest *et al.* (1981), estuarine deposits rarely consist of a single fill complex formed at only one high sea stand. At least during the Pleistocene and Holocene, it seems that the estuarine environment reoccupies the same places during various transgressions. Because of the similarity of the deposited lithofacies, the time-stratigraphy of any one outcrop or drill hole is extremely complex. Any lithologic change may represent a wide range of nondepositional time, from minutes to tens of thousands of years. It is rarely clear whether an erosional surface lies within a depositional unit or between superjacent units. Clifton (1982) has developed a schematic diagram (Figure 2-54) to illustrate the superposition of estuarine fill deposits under multiple fluctuating sea level conditions. During the Pleistocene, the time between transgressions is typically 60,000–200,000 years.

Table 2-4. Generalized Sequence of Sedimentary Lithofacies in a Transgressive Estuarine Environment. A. Axial and Vertical Trend. B. Lateral and Vertical Trend

A. Axial and Vertical Sequence of Sedimentary Lithofacies in the Estuarine Environment

	River ESTUARINE FLUVIAL	Seaward ⟶ ESTUARINE	Sea ESTUARINE MARINE
V E R T			Coarse marine sands, massive or with abundant X-bedding, tidal current ridges with low angle X-bedding in fine sands with silt laminae
I C		Silt and clay with sandy lenses and laminae, massive silt and clay deposits	
A L	Massive silt and clay with abundant plant fragments and roots, sandy lenses, and laminations, grading downward into sand gravel and cobble		

B. Lateral and Vertical Sequence of Sedimentary Lithofacies in the Lower Estuary

	Shore SHORELINE DEPOSITS	⟶ SUB-TIDAL FLATS	Mid-Channel ESTUARINE MARINE
V E R T			Coarse marine sands, massive or with abundant X-bedding (as above)
I C		Laminated and massive muddy sands and sandy muds	
A L	Sand, gravel, and shell with or without washover complex and muds with plant fragments and basal peat		

Sedimentary Structures

Sedimentary structures in the beach complex, tidal flats, marshes, and deltas are thoroughly discussed in succeeding chapters of this volume and discussion will be restricted to the open water estuarine environment.

Modern sediments in estuaries are usually black, brown, green, or gray. Color banding and mottling occur frequently. Sometimes the color banding is controlled by the environment of deposition. This is especially true of black estuarine sediments in which the color is caused by metastable iron sulfides ($FeS \cdot nH_2O$), produced under reducing conditions. These sulfides convert to pyrite with time, and the sediment color changes from black to gray or green. Under oxidizing depositional conditions, lighter colors of brown or gray are encountered frequently. Oxidizing conditions prevail in the uppermost fine-grained sediments near the sediment–water interface, whereas reducing conditions are usually achieved at some small distance below the interface. Sediment color reflects this chemical variation and may change from light gray or brown under oxidizing conditions to black under reducing conditions. Color banding, although present in modern estuarine sediments, may not persist in the geologic record or may develop diagenetically.

Parallel laminations consisting of alternating of coarse and fine material are apparent in many low-energy estuarine deposits, particularly in deep oxygen depleted zones where benthic fauna and bioturbation are absent. These may be caused by seasonal fluctuations in suspended sediment load or character, or by variations of the amounts and kinds of organic matter supplied by the estuary. Sometimes these laminations are not apparent to the eye but are evident on radiographs of cores. Laminations can also occur in high energy estuarine deposits as a product of normal tidal periodic processes or aperiodic extreme events.

Sand deposits 1–10 cm thick occur sporadically through otherwise monotonous sequences of silts and clays. These usually are present as lenticular layers, a product of concentration and reworking of the sands by currents. Ripple cross-stratification is rarely preserved. Occasionally scour-and-fill structures are present.

Shell and plant fragments are common, both scattered throughout estuarine sequences and concentrated in discrete beds. Extreme events, such as hurricanes, floods, or ice conditions, may concentrate plant fragments in the deeper, lower energy portions of estuaries. Molluscs may colonize bottom areas, grow to adults, or be wiped out while still juveniles because of changing environmental conditions, leaving behind layers of shells in life positions with evidence of extensive burrowing or reworking in the under-lying sediment.

Moore and Scruton (1957) have provided an excellent conceptual summary of the kinds of sedimentary structure that might be present in estuarine and coastal systems. Primary structures are formed at the time of initial deposition from fluctuations in the competence of the transporting

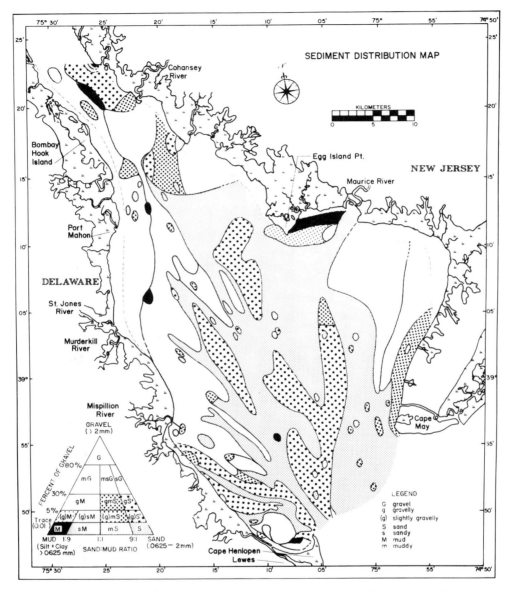

Figure 2-52. Distribution of Folk texture classes in Delaware Bay bottom sediments (from Weil, C.B., Sediments, structural framework, and evolution of Delaware Bay, Sea Grant Report DEL-SG-4-77, Weil, 1977, 199 p.)

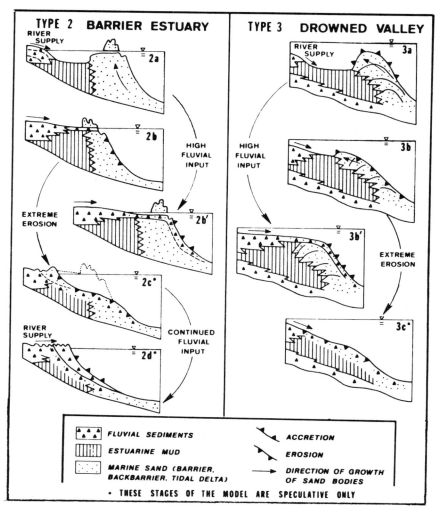

Figure 2-53. Schematic sections showing progressive stages in evolution of two estuarine types (modified from Roy *et al.*, 1980).

medium, the type of material available for deposition, or both. Secondary structures are produced after initial deposition by partial or complete destruction of primary structures and re-sorting or unmixing of homogenous sediments. Secondary structures are caused by the activity of organisms as well as waves, currents, and chemical processes. Moore and Scruton (1957) identified three factors that relate sedimentary structure to environment of formation: (1) sediment source, (2) primary sedimentation and secondary alteration intensity, and (3) sedimentation rate. The interrelation between these factors is illustrated in Figure 2-55.

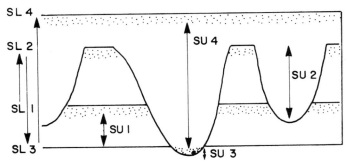

Figure 2-54. Schematic diagram illustrating a hypothetical estuarine-fill complex deposited under fluctuating sea level (SL1, SL2, SL3, SL4 indicate different levels). Under conditions of SL1, stillstand units SU1 and SU2 may be highly dissected as depicted here, or depending on position within the estuary, they may be completely eroded. Stillstand unit 3 may only be preserved as localized channel fill deposits (from Clifton, 1982).

Summary

Estuaries form in drowned river mouths and are best developed where the rate of coastal subsidence exceeds sediment accumulation. As they evolve, configurations change from deep branching, little modified by infilling, to shallow funnel-shaped approaching an equilibrium configuration between tidal discharge and sediment accumulation, and finally, reshaping by accretion of tidal flats, marshes, alluvial islands, and spits. Estuaries mainly function as sediment sinks collecting sediments from multiple sources, i.e., rivers, shores, ocean, as well as from biological production within the system.

Fine sediments are retained in estuaries by different mechanisms. Fluvial sediment is entrapped by the estuarine circulation or tidal current nodes. Accumulation is enhanced by reduced current competence, or physico-chemical flocculation and biological aggregation. Marine sediment is brought in and retained by tidal asymmetry, flood current predominance, and settling and scour lag effects. Sediments can escape estuaries during extreme river flooding or when strong tidal currents promote estuary-shelf exchange. Retained sediments are extensively reworked, mixed, and modified in particle size and composition by filter and deposit feeding organisms.

Although sediment supply, morphology, and dynamic processes fluctuate widely, the resultant deposits form recognizable and interrelated facies: (1) estuarine fluvial, (2) estuarine, and (3) estuarine marine. These constitute transgressive sequences interfingering between fluvial and marine deposits. Estuary deposits have limited extent, are shaped to the original river valley

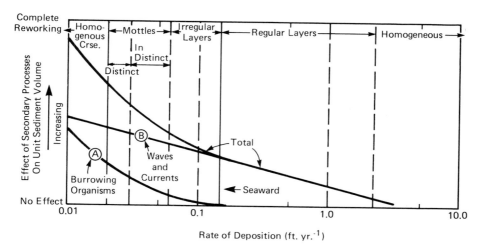

Figure 2-55. Relations of minor internal structures to rate of deposition and processes off the Mississippi Delta. Decreasing rate of deposition causes effects of secondary processes to be greatest offshore in deeper water, although intensity of physical processes such as waves is greatest in shallow water nearshore (from Moore and Scruton, 1957).

configuration, and lie transverse to the coastal depositional trend. They exhibit marked horizontal and vertical lithologic variations, being most pronounced at seaward and landward margins, which reflect variable tidal and fluvial processes. Physical and lithologic characteristics may resemble other coastal deposits. However, estuary deposits hold a mixed assemblage of fresh-brackish and marine fossil fauna with some brackish forms persisting through the section

Unsolved Questions

The review of estuarine processes and products in this chapter reveals that our geologic understanding of estuaries is incomplete. The processes *per se* are usually better known than the processes and responses together. Even less is known about the process–response together with the resultant "permanent" end product incorporated into the sedimentary record. Given an estuarinelike deposit, it is still difficult to hindcast with certainty under what conditions and in what manner the sediment accumulated. This gap arises because estuarine processes and their interaction with sediments are so complex and because they are not confined to river estuary environments.

Further, the "transient" characteristics of modern sediments are difficult to distinguish from the "lasting" characteristics. And there is little indication of how representative the present conditions are of the long-term sedimentary behavior.

Estuarine processes are complex because they are highly variable with time and place. The hydrodynamic regime, turbidity, and salinity gradients continually change as inflow, wind, and tide change, daily, fortnightly, and seasonally. Some variations are periodic, others nonperiodic or episodic. The bed geometry and composition may be ephemeral as the sediments undergo alternate erosion and deposition. At one time, the bed may be a source of sediment while at another time it is a sink. Superimposed on these variations is the multiplicity of sediment inputs; i.e., external source materials mixed with internal-produced materials like shells and detritus. Whereas fluvial and beach environments usually have one or a few distinctive source end-members, estuaries have multiple end-members. And these are constantly shifting with time. Faced with the complexities of processes and sources, an open question remains: How can we detect the depositional conditions from the lasting characteristics of a sedimentary deposit?

The recognition of estuarine deposits is bounded by the preservation potential of modern features. Modern sediments probably undergo many episodes of deposition and removal from the bed before they are "permanently" incorporated into the stratigraphic record; i.e., buried below a position of physical and biological reworking. Even thousands or millions of years after burial, estuarine deposits are susceptible to erosion, reworking, and deposition as regressive and transgressive seas invade the same river valley. What, then, are the chances of preservation as a function of the intensity and frequency of energy dissipation, reworking, and subsidence? It is of special interest to compare the end-products of *normal* estuarine processes with those of *episodic* processes like storms and floods to determine what process leaves the greatest impact on the sediment record.

Other open questions requiring priority research study are: (1) the basic nature of cohesive sediment erosion and deposition as a function of turbulence, shear stress, combined wave and current action, and biogenic activity; (2) the behavior and movement of fluid mud and its role in resuspension of bed sediment; (3) the effects of meteorologically forced flows on escape or entrapment of sediment in estuaries; (4) the effects of biological processing on particle transport and bed stabilization and its importance relative to other mechanisms as physicochemical flocculation in retaining fine sediment in estuaries; (5) the trapping efficiency of estuaries as a function of subsidence, infilling, hydrodynamic regime, and particle aggregation processes.

Estuaries offer unique opportunities for sedimentological study because of their great diversity and extreme variability. Most estuaries are readily accessible and require much fewer resources than the open ocean. A great deal of research study remains to be done.

References

Allen, G. P., 1971. Relationship between grain size parameter distribution and current patterns in the Gironde Estuary (France). *J. Sed. Petrol.*, **41**, 74–88.

Allen, G. P., 1972. *Étude des processes sedimentaires dans l'estuaire de la Gironde.* D.S. These, L'Univ. de Bordeaux, 314 pp.

Allen, G. P., 1973. Suspended sediment transport and desposition in the Gironde estuary and adjacent shelf. *In*: Proc. Int. Sympos. on Interrelationships of Estuarine and Continental Shelf sedimentation. *Mem. Inst. Geol. Bassin d'Aquitane*, Bordeaux, 7, 27–36.

Allen, G. P., Bouchet, J. M., Cabronnel, P., Castaing, P., Gayet, J., Gonthier, E., Jouanneau, J. M., Klingebiel, A., Latouche, C., Legigan, P., Ogeron, C., Pujos, M., Tesson, M., and Vernette, G., 1973. *Environments and Sedimentary Processes of the North Aquitaine Coast.* Guidebook, Inst. Geol. Bassin d'Aquitaine. 183 pp.

Allen, G. P., Salmon, J. C., Bassoulet, P., Du Penhoat, Y., and De Grandpre, C., 1980. Effects of tides on mixing and suspended sediment transport in macrotidal estuaries. *Sed. Geol.*, **26**, 69–90.

Allen, G. P., Sauzay, G., Castaing, P., and Jouanneau, J. M., 1977. Transport and deposition of suspended sediment in the Gironde Estuary, France. *In*: Wiley, M. (ed.), *Estuarine Processes*, Vol. II. Academic Press, New York, pp. 63–81.

Anderson, F. E., 1972. Resuspension of estuarine sediments by small amplitude waves. *J. Sed. Petrol.*, **42**, 602–607.

Anderson, F. E., 1983. The northern muddy intertidal: a seasonally changing sources of suspended sediments to estuarine waters—a review. *Can. J. Fish. Aquatic Sci.*, **40**, Supplement I, 143–159.

Ariathurai, R., MacArthur, R. C., and Krone, R. B., 1977. *Mathematical Model of Estuarial Sediment Transport.* U.S. Army Engineers Waterways-Experiment Station, Dredged Material Research Program, Tech. Rept. D-77-12, 79 pp.

Aston, S. R. and Chester, R., 1973. The influence of suspended particles on the precipitation of iron in natural waters. *Est. Coastal Mar. Sci.*, 1, 225–231.

Bader, R. G., Hood, D. W., and Smith, J. B., 1960. Recovery of dissolved organic matter in seawater and organic sorption by particulate material. *Geochim. Cosmochim. Acta*, **19**, 236–243.

Baker, V. R., and Ritter, D. F., 1975. Competence of rivers to transport coarse bedload material. *Geol. Soc. Amer. Bull.*, **86**, 975–978.

Bennett, R. H., and Bryant, W. R., 1977. *Clay fabric and geotechnical properties of selected submarine sediment cores from the Mississippi Delta.* Natl. Oceanic Atmos. Admin. Prof. Paper 9, Miami, FL. 86 pp.

Berner, R. A., 1981. A new geochemical classification of sedimentary environments. *J. Sed. Petrol.*, **51**, 359–365.

Biggs, R. B., 1970. Sources of distribution of suspended sediment in northern Chesapeake Bay. *Mar. Geol.*, **9**, 187–201.

Biggs, R., and Cronin, E., 1981. Special characteristics of estuaries. *In*: Neilson, B., and Cronin, L. E. (eds.), *Estuaries and Nutrients*. Humana Press, Clifton, NJ, pp. 3–24.

Biggs, R. B., and Howell, B. A., 1984. The estuary as a sediment trap; alternate approaches to estimating its filtering efficiency. *In*: Kennedy, V.S., *The Estuary as a Filter*. Academic Press, New York.

Bokuniewicz, H., and Tanski, J. J. 1983. Sediment partitioning at an eroding coastal bluff. *Northeastern Geol.*, **5**, 73–81.

Boon, J. D., 1975. Tidal discharge asymmetry in a salt marsh drainage system. *Limnol. Oceanogr.*, **20**, 71–80.

Boon, J. D., and Byrne, R. J., 1981. On basin hypsometry and the morphodynamic response of coastal inlet systems. *Mar. Geol.*, **40**, 27–48.

Boon, J. D., Frisch, A. A., and Hennigar, H., 1983. Sand transport pathways in the entrance to Chesapeake Bay: Evidence from Fourier grain-shape analysis. *Estuaries*, **6**, 317.

Boyer, L. F., 1980. *Production and Preservation of Surface Traces in the Intertidal Zone*. Ph.D. Dissert. Dept. of Geophysical Sciences, Univ. of Chicago, Chicago, IL.

Boyle, E. A., Edmond, J. M., and Sholkovitz, E. R., 1977. The mechanism of iron removal in estuaries. *Geochim. Cosmochim. Acta*, **41**, 1313–1324.

Bretschneider, C. L., 1965. *The Generation of Waves by Wind: State of the Art*. Natl. Engr. Sci. Co., Office of Naval Res., SN 134-6, 96 pp.

Brockmann, C., 1929. Das Brackwasser der Flussmundungen als Heimat und Vernichter des Lebens. Senckenberg am Meer, 29. *Natur und Museum*, **59**, 401–414.

Buller, A. T., Green, C. P., and McManus, J. 1975. Dynamics and sedimentation: the Tay in comparison with other estuaries. *In*: Hails, J., and Carr, A. (eds.), *Nearshore Sediment Dynamics and Sedimentation*. John Wiley and Sons, London, pp. 201–49.

Burns, R. E., 1963. *Importance of Marine Influences in Estuarine Sedimentation*. Proc. Federal Inter-Agency Sed. Conf., V.S.A.D. Misc. Pub. 970, pp. 593–598.

Burton, J. D., 1976. Basic properties and processes in estuarine chemistry. *In*: Burton, J. D., and Liss, P. S. (eds.), *Estuarine Chemistry*. Academic Press, New York, pp. 1–36.

Burton, J. D., and Liss, P. S., 1976. *Estuarine Chemistry*. Academic Press, New York, 183 pp.

Byrne, R. J., Hobbs, C. H., and Carron, M. J., 1982. *Baseline sediment studies to determine distribution, physical properties, sedimentation budgets and rates in the Virginia portion of the Chesapeake Bay*. Final grant report to U.S.E.P.A., 155 pp.

Carder, K. L., 1979. Holographic microvelocimeter for use in studying ocean particle dynamics. *Optical Engineering*, **18**, 524–525.

Carder, K. L., and Meyers, D. J., 1979. *New optical techniques for particle studies in the bottom boundary layer*. Proc. Soc. Photo. Opt. Instrum. Engr., **208**, 151–158.

Carriker, M. R., 1967. Ecology of estuarine benthic invertebrates: a perspective. *In*: Lauff, G. H. (ed.), *Estuaries*. Amer. Assoc. Adv. Sci. Publ. 83, Washington, DC, pp. 442–487.

Castaing, P., and Allen, G. P., 1981. Mechanisms controlling seaward escape of suspended sediment from the Gironde: a macrotidal estuary in France. *Mar. Geol.*, **40**, 101–118.

Chang, C. W., and Anderson, J. U., 1968. Flocculation of clays and soils by organic compounds. *Soil Sci. Soc. Amer. Proc.*, **32**, 23–27.

Clifton, H. E., 1982. Estuarine deposits. *In*: Scholle and Speauing (eds.), *Sandstone Depositional Environments*. Amer. Assoc. Petrol. Geol. Memoir 31, pp. 179–188.

Clifton, H. E., 1983. Discrimination between subtidal and intertidal facies in Pleistocene deposits, Willapa Bay, Washington. *J. Sed. Petrol.*, 53, 353–369.

Coleman, J. M., and Wright, L. C., 1978. Sedimentation in an arid macrotidal alluvial river system: Ord river, Western Australia. *J. Geol.*, **86**, 621–642.

Coles, S. M., 1979. Benthic microalgal populations on inter-tidal sediments and their role as precursors to salt marsh development. *In*: Jefferies, R. L., and Davy, A. J. (eds.), *Ecological Processes in Coastal Environments*. Blackwell, Oxford, U.K., pp. 25–42.

Coonley, L. S., Baker, E. B., and Holland, H. D., 1971. Iron in the Mullica River and in Great Bay, New Jersey. *Chem. Geol.*, **7**, 51–63.

Davies, J. L., 1973. *Geographical Variation in Coastal Development*. Hafner, New York, 204 pp.

Demarest, J. M., Biggs, R. B., and Kraft, J. C. 1981. Time stratigraphic aspects of a formation: interpretation of surficial Pleistocene deposits, southeastern Delaware. *Geology*, **9**, 360–365.

Dobereiner, C., and McManus, J., 1983. Turbidity maximum migration and harbor siltation in the Tay Estuary. *Amer. J. Fish. Aquatic Sci. Supplement*, 40, 117–129.

Dole, R. B., and Stabler, H., 1909. Denudation. *U.S. Geol. Survey, Water Supply Paper*, **234**, 78–93.

Duinker, J. C., 1980. Suspended matter in estuaries: adsorption and desorption processes. *In*: Olausson, E., and Cato, I. (eds.), *Chemistry and Biochemistry of Estuaries*. Wiley, Chichester, pp. 121–151.

Dyer, K. R., 1972. Sedimentation in estuaries. *In*: Barnes, R. S. R., and Green, J. (eds.), *The Estuarine Environment*. Applied Science Publ., London, pp. 10–32.

Edzwald, J. K., and O'Melia, C. R., 1975. Clay distributions in recent estuarine sediments. *Clays and Clay Minerals*, **23**, 39–44.

Elliott, A. J., 1978. Observations of the meteorologically induced circulation in the Potomac estuary. *Est. Coastal Mar. Sci.*, **6**, 285–299.

Emery, K. O., 1967. Estuaries and lagoons in relation to continental shelves. *In*: Lauff, G. H. (ed.), *Estuaries*. Amer. Assoc. Adv. Sci. Publ. 83, Washington, DC, pp. 9–11.

Emery, K. O., and Uchupi, E., 1972. *Western North Atlantic Ocean: Topography, Rocks, Structure, Water, Life and Sediments*. Amer. Assoc. Petrol. Geol. Memoir. 17, 532 pp.

Ernissee, J. J., and Abbott, W. H., 1975. Binding of mineral grains by a species of Thalassiosira. *In*: Simonsen, R. (ed.), *Third Symposium of Recent and Fossil Marine Diatoms*. Occ. Publ. of Nova Hedwigia, von Cramer, Bremerhaven, pp. 241–248.

Fairbridge, R. W., 1980. The estuary: its definition and geodynamic cycle. *In*: Olausson, E., and Cato, I. (eds.), *Chemistry and Biogeochemistry of Estuaries*. pp. 1–36.

Favejee, J. C. H. L., 1960. On the origin of the mud deposits in the Ems-Estuary. *In*: Voorthuysen, J. H., and Kuenen, P. H. (eds.), *Das Ems-Estuarium (Nordsee)*. Verhandl. Ron. Ned. Geol. Mignb. Gen. Geol. Ser., XIX, 300 pp.

Feral, A., 1970. *Interprétation sédimentologique et paléogéographique des formations alluviales de l'estuaire de la Gironde et de ses dépendances marines. Thèse 3e cycle*, Fac. Sc., Bordeaux, n°806, 158 pp.

Festa, J. F., and Hansen, D. V. 1976. A two dimensional numerical model of estuarine circulation: the effects of altering depth and river discharge. *Est. Coastal Mar. Sci.*, **4**, 309–323.

Festa, J. F., and Hansen, D. V., 1978. Turbidity maxima in partially mixed estuaries: a two-dimensional numerical model. *Est. Coastal Mar. Sci.*, **7**, 347–359.

Feuillet, J. P., and Fleischer, P. 1980. Estuarine circulation: controlling factor of clay mineralogy distribution in James River Estuary, Virginia. *J. Sed. Petrol.*, **50**, 267–279.

Fisher, J. S., Pickral, J., and Odum, W. E., 1979. Organic detritus particles, initiation of motion criteria. *Limnol. Oceanogr.*, **24**, 529–532.

Frostick, L. E., and McCave, I. N., 1979. Seasonal shifts of sediment within an estuary mediated by algal growth. *Est. Coastal Mar. Sci.*, **9**, 569–576.

Galtsoff, P. S., 1928. Experimental study of the function of the oyster gills and its bearing on the problems of oyster culture and sanitary control of the oyster industry. *Bull. U.S. Bur. Fish.*, **44**, 1–39.

Gelfenbaum, G., 1983. Suspended-sediment response to semidiurnal and fortnightly tidal variations in a mesotidal estuary: Columbia River, U.S.A. *Mar. Geol.*, **52**, 39–57.

Gibbs, R., 1977. Suspended sediment transport and the turbidity maximum. *In*: Officer, C. (ed.), *Estuaries, Geophysics, and the Environment*. National Academy of Sciences, Washington, DC, pp. 104–109.

Gibbs, R., and Heltzel, S., 1982. Coagulation and the deposition of mud. *Abs. Internal. Assoc. Sed.*, Hamilton, Canada, p. 4.

Gibbs, R., Konwar, L., and Terchurian, A., 1983. Size of flocs suspended in Delaware Bay. *Can. J. Fish. Aquatic Sci.* **40** (Suppl. No. 1), 102–104.

Goldberg, E. D., 1978. Concentration variations of dissolved elements in estuaries. *In*: Goldberg, E. D. (ed.), *Biogeochemistry of Estuarine Sediments*. UNESCO, Paris, pp. 19–28.

Gordon, C. M., and Witting, J., 1977. Turbulent structure in a benthic boundary layer. *In*: Nihoul, J. C. J. (ed.), *Bottom Turbulence*. Proceedings, Eighth Intl. Liege Colloquim on Ocean Hydrodynamics, Elsevier Oceanography Series, 19, Elsevier, New York, 306 pp.

Gordon, R. B., 1980. The sedimentary system of Long Island Sound. *Advances in Geophysics*, 22, 1–39.

Graf, W. H., 1971. *Hydraulics of Sediment Transport*. McGraw-Hill, New York, 509 pp.

Greenberg, D. A., and Amos, C. L., 1983. Suspended sediment transport and deposition modeling in the Bay of Fundy, Nova Scotia—a region of potential tidal power development. *Can. J. Fish. Aquatic Sci. Supplement*, **40**, 20–34.

Gross, M. G., 1975. Trends in waste solid disposal in U.S. coastal waters, 1968–1974. *In*: Church, T. M. (ed.), *Marine Chemistry in the Coastal Environment*. Am. Chem. Soc. Sympos. Series, 18, pp. 394–405.

Gust, G., 1976. Observations of turbulent-drag reduction in a dilute suspension of clay in sea water. *J. Fluid. Mech.*, 75, 29–47.

Hardaway, S., and Anderson, G., 1980. *Shoreline Erosion in Virginia*. VIMS Sea Grant Educational Series No. 31, 25 pp.

Harris, J. E., McKee, T. R., Wilson, Jr., R. C., and Whitehouse, U. G., 1972. Preparation of membrane filter samples for direct examination with an electron microscope. *Limnol. Oceanogr.*, **17**, 784–787.

Haven, D. S., and Morales-Alamo, R., 1966. Aspects of biodeposition by oysters and other invertebrate filter feeders. *Limnol. Oceanogr.* **11**, 487–498.

Haven, D. S., and Morales-Alamo, R., 1968. Occurrence and transport of fecal pellets in suspension in tidal estuary. *Sed. Geol.*, **2**, 141–151.

Haven, D. S., and Morales-Alamo, R., 1972. Biodeposition as a factor in sedimentation of fine suspended solids in estuaries. *In*: Nelson, B. W. (ed.), *Environmental Framework of Coastal Plain Estuaries*. Geol. Soc. Amer. Memoir. 133, Boulder, CO, pp. 121–130.

Hawley, N., 1982. Settling velocity distribution of natural aggregates. *J. Geophys. Res.*, **87**, 9489–9498.

Hayes, M. O., 1975. Morphology of sand accumulations in estuaries. *In*: Cronin, L. E., (ed.), *Estuarine Research*, Vol. 2, *Geology and Engineering*. Academic Press, New York, pp. 3–22.

Hjulstrom, F., 1939. Transportation of detritus by moving water. *In*: Trask, D. D. (ed.), *Recent Marine Sediments, A Symposium*. Soc. Econ. Paleont. Mineral. Spec. Publ. 4, pp. 5–31.

Hopkins, J. T., 1967. The diatom trail. *J. Quekett Microsc. Club*, 30, 209–217.

Howard, J. D., 1975. Estuaries of the Georgia coast, USA: Sedimentology and biology. IX Conclusions. *Senckenberg. Marit.* **7**, 297–305.

Humby, E. J., and Dunn, J. N., 1973. Sedimentary processes within estuaries and tidal inlets. *In*: Helliwell, P. R., and Bossanyi, J. (eds.), *Pollution Criteria for Estuaries*. Pentech Press, London, pp. 6.1–6.22.

Inglis, C. C., and Allen, M. A., 1957. The regimen of the Thames Estuary as affected by currents, salinities, and river flow. *Proc. Inst. Civil Engr.*, **7**, 827–878.

Inman, D. L., 1949. Sorting of sediments in the light of fluid mechanics. *J. Sed. Petrol.*, **19**, 51–70.

Ippen, A. T., and Harleman, D. R. F., 1966. Tidal dynamics in estuaries. *In*: Ippen, A. T. (ed.), *Estuary and Coastline Hydrodynamics*. McGraw-Hill, New York, pp. 493–545.

Jordan, G. F., 1961. *Erosion and sedimentation Eastern Chesapeake Bay at the Choptank River*. U.S. Coast and Geodetic Survey Tech. Bull. 16, 8 pp.

Jouanneau, J. M., and Latouche, C., 1981. *The Gironde Estuary*. Contributions to Sedimentology 10, E. Schweizerbart'sche Verlagsbuchhandlung (Nagele u. Obermiller), Stuttgart, 115 pp.

Kelley, T. T., 1983. Composition and origin of the inorganic fraction of southern New Jersey coastal mud deposits. *Geol. Soc. Amer. Bull.*, **94**, 689–699.

Kerhin, R. T., Halka, J. P., Hennessee, E. L., Blakeslee, P. J., Wells, D. V., Zoltan, N., and Cuthbertson, R. H., 1983. *Physical characteristics and sediment budget for bottom sediments in the Maryland portion of Chesapeake Bay*. Final Report to U.S. Environmental Protection Agency, Grant #R805965, Maryland Geol. Survey, Johns Hopkins Univ., Baltimore, MD, 190 pp.

Kirby, R., and Parker, W. E., 1983. Distribution and behavior of fine sediment in the Severn Estuary and Inner Bristol Channel, U.K. *Can. J. Fish. Aquatic Sci.*, **40**, 83–95.

Klein, G. D. V., 1967. Paleocurrent analysis in relation to modern marine sediment dispersal patterns. *Amer. Assoc. Petrol. Geol.*, **51**, 366–382.

Komar, P. D., 1976a. Boundary layer flow under steady unidirectional currents. *In*: Stanley, D. J., and Swift, D. J. (eds.), *Marine Sediment Transport and Environmental Management*. John Wiley, New York, pp. 91–106.

Komar, P. D., 1976b. The transport of cohesionless sediments on continental shelves. *In*: Stanley, D. J., and Swift, D. J. (eds.), *Marine Sediment Transport and Environmental Management*. John Wiley, New York, pp. 107–126.

Komar, P. D., and Miller, M. C. 1973. The threshold of sediment movement under oscillatory water waves. *J. Sed. Petrol.*, **43**, 1101–1110.

Kraeuter, J., and Haven, D. S., 1970. Fecal pellets of common invertebrates of lower York River and Lower Chesapeake Bay, Virginia. *Chesapeake Sci.*, **11**, 159–173.

Kranck, K., 1973. Flocculation of suspended sediments in the sea. *Nature*, **246**, 348–350.

Kranck, K., 1975. Sediment deposition from flocculated suspensions. *Sedimentology*, **22**, 111–123.

Kranck, K., 1979. Dynamics and distribution of suspended particulate matter in the St. Lawrence Estuary. *Nat. Can.*, **106**, 163–179.

Kranck, K., 1980a. Experiments on the significance of flocculation in the setting of fine-grained sediment in still water. *Can. J. Earth Sci.*, **17**, 1517–1526.

Kranck, K., 1980b. Sedimentation processes in the sea. *In*: Hutzinger, O. (ed.), *The Handbook of Environmental Chemistry*. Springer-Verlag, New York, 2:A, pp. 61–75.

Kranck, K., 1981. Particulate matter grain-size characteristics and flocculation in a partially mixed estuary. *Sedimentology*, **28**, 107–114.

Krone, R. B., 1962. *Flume Studies of the Transport of Sediment in Estuarial Shoaling Processes, Final Report*. Hydraulic Engr. Lab. and Sanitary Engr. Res. Lab., Univ. of California, Berkeley, 110 pp.

Krone, R. B., 1963. *A study of rheologic properties of estuarial sediments*. Tech. Bull. 7, Committee on Tidal Hydraulics, U.S. Army Corps of Engr. WES, Vicksburg, MI.

Krone, R. B., 1972. *A Field Study of Flocculation as a factor in estuarial shoaling processes*. Tech. Bull. 19, Committee on Tidal Hydraulics, U.S. Army Corps of Engineers, 62 pp.

Krone, R. B., 1978. Aggregation of suspended particles in estuaries. *In*: Kjerfve, B. (ed.), *Estuarine Transport Processes*. Univ. of South Carolina Press, Columbia, SC, pp. 171–190.

Krone, R. B., 1979. Sedimentation in the San Francisco Bay system. *In*: Conomos, T. J. (ed.), *San Francisco Bay: The Urbanized Estuary*. Pacific Division, Amer. Assoc. Adv. Science, San Francisco, CA pp. 85–96.

Krumbein, W. C., and Pettijohn, F. J., 1938. *Manual of Sedimentary Petrography*. Appleton-Century-Crofts, New York, 549 pp.

Lambe, T. W., 1953. The structure of inorganic soil. *Amer. Soc. Civil Engr. Separate 315*, **79**, 1–49.

Lambermont, J., and Lebon, G., 1978. Erosion of cohesive soils. *J. Hydr. Res.*, **16**, 27–44.

LeFloch, 1961. *Propagation de la maree dans l'estuarie de la seine et en seine— maritime*. These D. S., Univ. Paris, 507 pp.

Lerman, A., 1979. *Geocheical Processes, Water and Sediment Environments*. John Wiley and Sons, New York, 481 pp.

Lund, E. J., 1957. A quantitative study of clearance of a turbid median and feeding of the oyster. *Publ. Inst. Mar. Sci. Univ. Texas*, **4**, 296–312.

Martin, J. M., Burton, J. D., and Eisma, D., 1979. *River Inputs to Ocean Systems*. UNEP, IOC, SCOR, 369 pp.

Mayer, L., 1982. Retention of riverine iron in estuaries. *Geochim. Cosmochim. Acta*, **46**, 1003–1009.

McCave, I. N., 1972. Transport and escape of fine-grained sediment from shelf areas. *In*: Swift, D. J. P., Duane, D. B., and Pilkey, O. H. (eds.), *Shelf Sediment Transport: Process and Pattern*. Dowden, Hutchinson and Ross, Stroudsburg, PA, pp. 225–248.

McCave, I. N., 1978. Sediments in the abyssal boundary layer. *Oceanus*, **21**, 27–33.

McCave, I. N., 1979. Suspended sediment. *In*: Dyer, K. R. (ed.), *Estuarine Hydrography and Sedimentation, a Handbook*. Cambridge Univ. Press, Cambridge Univ. Press, Cambridge, pp. 131–185.

McCave, I. N., and Swift, S. A., 1976. A physical model for the rate of deposition of fine-grained sediments in the deep sea. *Geol. Soc. Amer. Bull.*, **87**, 541–546.

McDowell, D. M., and O'Connor, B. A., 1977. *Hydraulic Behavior of Estuaries*. John Wiley and Sons, New York. 292 pp.

Meade, R. H., 1969. Landward transport of bottom sediments in estuaries of the Atlantic Coastal Plain. *J. Sed. Petrol.*, **39**, 222–234.

Meade, R. H., 1972. Transport and deposition of sediments in estuaries. *In*: Nelson, B. W. (ed.), *Environmental Framework of Coastal Plain Estuaries*. Geol. Soc. Amer. Memoir 133, Boulder, CO. pp. 91–120.

Meade, R. H., 1982. Sources, sinks and storage of river sediments in the Atlantic drainage of the United States. *J. Geol.*, **90**, 235–252.

Meade, R. H., and Trimble, S. W., 1974. Changes in sediment loads in rivers of the Atlantic drainage of the United States since 1900. *Proc. Paris Symposium IAHS-AISH Publ.*, 113, 99–104.

Mehta, A.J., Parchure, T. M., Dixit, J. G. and Ariathurai, R., 1982. Resuspension potential of deposited cohesive sediment beds. *In*: Kennedy, V.S. (ed.), *Estuarine Comparisons*. Academic Press, New York, pp. 591–609.

Middleton, G. W., 1973. Johannes Walther's law of correlation of facies. *Geol. Soc. Amer. Bull.*, **84**, 979–988.

Migniot, C., 1968. A study of the physical properties of various forms of very fine sediments and their behaviour under hydrodynamic action. *La Houille Blanche*, **7**, 591–620.

Migniot, C., 1971. L'evolution de la Gironde au cours des temps. *Bull. Inst. Geol. Bassin d'Aquitaine, Bordeaux*, **11**, 221–281.

Migniot, C., 1977. effect of currents, waves and wind on sediment. *La Houille Blanche*, **32**, 9–47.

Miller, M. C., McCave, I. N., and Komar, P. D., 1977. Threshold of sediment motion under unidirectional currents. *Sedimentology*, **24**, 507–527.

Milliman, J. D., and Emery, K. O. 1968. Sea levels during the past 35,000 years. *Science*, **162**, 1121–1123.

Moore, D. G., and P. C. Scruton, 1957. Minor internal structures of some recent unconsolidated sediments. *Bull. Amer. Assoc. Petrol. Geol.*, **41**(12), 2723–2751.

Morris, R. C., and Dickey, P. A., 1957. Modern evaporite deposition in Peru. *Amer. Assoc. Petrol. Geol. Bull.*, **41**, 2467–2474.

Muller, G., 1969. Sedimentary phosphate method for estimating paleosalinity: limited applicability. *Science*, **163**, 812–813.

NEDECO, 1965., *A study of the siltation of the Bangkok Port Channel. Vol. 2, The field investigation*. Netherlands Engineering Consultants, The Hague, 474 pp.

Nelson, B. W., 1967. Sedimentary phosphate method for estimating paleosalinity. *Science*, **148**, 917–920.

Nichols, M., 1972. Effect of increasing depth on salinity in the James River Estuary. *In*: Nelson, B. W. (ed.), *Environmental Framework of Coastal Plain Estuaries*. Geol. Soc. Amer. Memoir 133, pp. 571–589.

Nichols, M., 1977. Response and recovery of an estuary following a river flood. *J. Sed. Petrol.*, **47**, 1171–1186.

Nichols, M., 1978. The problem of misplaced sediment. *In*: Palmer and Gross (eds.), *Ocean Dumping and Marine Pollution*. Dowden, Hutchinson, and Ross, Stroudsburg, PA, pp. 147–161.

Nichols, M., and Poor, G., 1967. Sediment transport in a coastal plain estuary; *J. Waterways and Harbors*, Proc. Amer. Soc. Civil Engr., No. WW4, Paper 5571, **93**, 83–95.

Nowell, A. R. M., Jumars, P. A., and Eckman, J. E., 1981. Effects of biological activity on the entrainment of marine sediments. *Mar. Geol.*, **42**, 133–153.

O'Brien, M. P., 1969. Dynamics of tidal inlets. *In*: Castañares, A., and Phleger, F. (eds.), *Coastal Lagoons, A Symposium*. UNAM-UNESCO, Memoir Symposium Intl. Langunas Costeras, pp. 397–406.

O'Brien, M. P., 1976. *Notes on Tidal Inlets on Sandy Shores*. U.S. Army Coastal Engr. Res. Center GITI Rept. 5, 20 pp.

Oertel, G. F., 1972. Sediment transport of estuary entranced shoals and formation of swash platforms. *J. Sed. Petrol.*, **42**, 858–863.

Officer, C. B., 1981. Physical dynamics of estuarine suspended sediments. *Mar. Geol.*, **40**, 1–14.

Olausson, E., and Cato, I., 1980. *Chemistry and Biogeochemistry of Estuaries*. John Wiley and Sons, New York, 452 pp.

Overbeek, J. T. G., 1952. Electrokinetic phenomena. *In*: Krurgt, H. R. (ed.), *Colloid Chemistry*, Vol. 1, Chapter 7. Elsevier, Amsterdam.

Owen, M. W., 1970. *Properties of a consolidating mud*. Rept. INT 83, Hydraulics Research Station, Wallingford, U.K.

Owen, M. W., 1971a. *The Effect of Turbulence on Floc Settling Velocities*. Paper D4, Proc. 14th Congr. Intl. Assoc. Hydraulic Res., Paris, D4-1-D4-6.

Owen, M. W., 1971b. *The Effect of Turbulence on the Settling Velocities of Silt Flocs*. Paper D4, Proc. 14th Congress, Intl. Assoc. Hydraulic Res. Paris, pp. 27–32.

Owen, M. W., 1977. Problems in the modeling of transport erosion, and deposition of cohesive sediments. *In*: Goldberg, E. D., McCave, I. N., O'Brien, J. J., and Steel, J. J. (eds.), *The Sea*, Vol. 6. John Wiley and Sons, New York, pp. 515–537.

Packham, R. F. 1962. The coagulation process. 1. The effect of pH and the nature of the turbidity. *J. Applied Chem.*, **12**, 556–564.

Paerl, H. W., 1973. Detritus in Lake Tahoe: structural modification by attached microflora. *Science*, **180**, 496–498.

Paerl, H. W., 1975. Microbial attachment to particles in marine and fresh water ecosystems. *Microbial Ecology*, 2, 73–83.

Partheniades, E., 1965. Erosion and deposition of cohesive soils, *J. Hydr. Div.*, Proc. Amer. Soc. Civil Engr., **91** (HY1), 105–138.

Partheniades, E., Cross, R. H., and Avora, A., 1969. Further results on the deposition of cohesive sediments. *Proc. 11th Conf. Coastal Engr.*, Amer. Soc. Civil Engr., London, **1**, 723–142.

Partheniades, E., and Kennedy, J. F., 1966. *Depositional behavior of fine sediment in a turbulent fluid motion.* Proc. 10th Intl. Conf. on Costal Engr., Tokyo, pp. 707–729.

Peterson, K., Scheidegger, K., and Komar, P. 1982. Sand dispersal patterns in an active margin estuary of the northwestern United States, as indicated by sand composition, texture, and bedforms. *Mar. Geol.* **50**, 77–96.

Peterson, C. D., Scheidegger, K. F., and Schrader, H. J., 1984. Holocene depositional evolution of a small active-margin estuary of the northwestern United States. *Mar. Geol.*, **59**, 51–83.

Postma, H., 1954. Hydrography of the Dutch Wadden Sea. *Arch. Neerl. Zool.*, **10**, 406–511.

Postma, H., 1957. Size frequency distribution of sands in the Dutch Wadden Sea. *Arch. Neerl. Zool.*, **13**, 319–349.

Postma, H., 1961. Transport and accumulation of suspended matter in the Dutch Wadden Sea. *Neth. J. Sea Res.*, **1**, 148–190.

Postma, H., 1967. Sediment transport and sedimentation in the marine environment. *In*: Lauff, G. H., (ed.), *Estuaries*. Amer. Assoc. Adv. Sci. Publ. 83, Washington, DC, pp. 158–179.

Postma, H., 1980. Sediment transport and sedimentation. *In*: Olausson, E., and Cato, I. (ed.), *Chemistry and Biogeochemistry of Estuaries*. Wiley, Chichester, pp. 153–186.

Pritchard, P. W., 1955. Estuarine circulation patterns. *Proc. Amer. Soc. Civil Engr.*, **81**, 717–1–717–11.

Pritchard, D. W., 1967. Observations of circulation in coastal plain estuaries. *In*: Lauff, G. H. (ed.), *Estuaries*, Amer. Assoc. Adv. Sci., Publ. 83, Washington, DC, pp. 3–5.

Pryor, W. A., 1975. Biogenic sedimentation and alteration of argillaceous sediments in shallow marine environments. *Geol. Soc. Amer. Bull.*, **86**, 1244–1254.

Pryor, W. A., and VanWie, W. A., 1971. The "Sawdust Sand"—an Eocene sediment of floccule origin. *J. Sed. Petrol.*, **41**, 763–769.

Redfield, A. C., 1980. *Introduction to Tides*. Marine Science International, Woods Hole, MA, 108 pp.

Reineck, H. E., 1963. Sedimentgefuge im Bereich der sudlichen Nordsee: *Abhandl. Sencken. Nat. Gesell.*, **505**, 1–138.

Reineck, H. E., and Singh, I. B., 1980. *Depositional Sedimentary Environments*. Springer-Verlag, New York-Heidelberg-Berlin, 549 pp.

Rhoads, D. C., 1974. Organism-sediment relations on the muddy sea floor. *Oceanogr. Mar. Biol. Ann. Rev.*, **12**, 263–300.

Rhoads, D. C., and Stanley, D. J., 1965. Biogenic graded bedding. *J. Sed. Petrol.*, **35**, 956–963.

Rhoads, D. C., Yingst, J. Y., and Ullman, W. J., 1978. Seafloor stability in central Long Island Sound, Part I: Temporal changes in erodibility of fine-grain sediment. *In*: Wiley, M. L. (ed.), *Estuarine Interactions*. Academic Press, New York, pp. 221–244.

Rhoads, D. C., and Young, D. K., 1970. The influence of deposit-feeding organisms on sediment stability and community trophic structure. *J. Mar. Res.*, **28**, 150–178.

Riley, G. A., 1963. Organic aggregates in seawater and the dynamics of their formation and utilization. *Limnol. Oceanogr.*, **8**, 372–381.

Rosen, P. S., 1975. Origin and processes of cuspate spit shorelines. *In*: Cronin, L. E. (ed.), *Estuarine Research*, Vol. 2. Academic Press, New York, pp. 77–92.

Roy, P. S., Thom, B. G., and Wright, L. D., 1980. Holocene sequences on an embayed high-energy coast: an evolutionary model. *Sed. Geol.* **26**, 1–19.

Ruby, W. W., 1933. Settling velocities of gravel sand and silt particles. *Amer. J. Sci. 5th Series*, **25**, 325–338.

Rusnak, G. A., 1967. Rates of sediment accumulation in modern estuaries. *In*: Lauff, G. H. (ed.), *Estuaries*. Amer. Assoc. Adv. Sci. Publ. Washington, DC 83, pp. 180–184.

Russell, R. C. H., and Macmillan, D. H., 1970. *Waves and Tides*. Greenwood Press, Westport, CT, 348 pp.

Ryan, J. D., 1953. *The Sediments of Chesapeake Bay*. Maryland Dept. of Geology, Mines and Water Resources Bull. 12, 120 pp.

Ryan, J. J., and Goodell, G., 1972. Marine geology and estuarine history of Mobile Bay, Alabama. *In*: Nelson, B. W. (ed.), *Environmental Framework of Coastal Plain Estuaries*. Geol. Soc. Amer. Memoir 133, pp. 517–554.

Salomon, J. C., and Allen, G. P., 1983. Role sedimentologique de la maree dans les estuaires a fort marnage. Compagnie Francais des Petroles. *Notes and Memoires*, **18**, 35–44.

Salomons, W., and Eysink, W. D., 1981. Pathways of mud and particulate trace metals from rivers to the southern North Sea. *Intl. Assoc. Sed. Spec. Publ.*, 5, pp. 429–450.

Schubel, J. R., 1969. Size distributions of the suspended particles of the Chesapeake Bay turbidity maximum. *Neth. J. Sea Res.*, 4, 283–309.

Schubel, J. R., 1971. Estuarine circulation and sedimentation. *In*: Schubel, J. R. (ed.), *The Estuarine Environment: Estuaries and Estuarine Sedimentation*. Amer. Geol. Inst. Short Course Lecture Notes, Amer. Geol. Inst., Washington, DC.

Schubel, J. R., 1974. Effects of Tropical Storm Agnes on the suspended solids of the Northern Chesapeake Bay. *In*: Gibbs, R. (ed.), *Suspended Solids in Water*. Plenum Press, New York, pp. 113–132.

Schubel, J., 1982. An eclectic look at fine particles in the coastal ocean. *In*: Kimrey, L., and Burns, R., eds., *Proc. of a Pollutant Transfer by Particulates Workshop*. Seattle, WA, pp. 53–142.

Schubel, J. R., and Hirschberg, D. J., 1978. Estuarine graveyard and climatic change. *In*: Wiley, M., ed., *Estuarine Processes*, Vol. 1. p. 285–303.

Schubel, J. R., and Hirschberg, D. J., 1982. The Chang Jiang (Yangtze) Estuary: Establishing its place in the community of estuaries. *In*: Kennedy, V. S. (ed.), *Estuarine Comparisons*. Academic Press, New York, pp. 649–666.

Schubel, J. R., and Kana, T. W., 1972. Agglomeration of fine-grained suspended sediment in northern Chesapeake Bay. *Power Tech.*, **6**, 9–16.

Seibold, E., and Berger, W. H., 1982. *The Sea Floor, an Introduction to Marine Geology*. Springer-Verlag, New York, 288 pp.

Shideler, G. L., 1984. Suspended sediment responses in a wind-dominated estuary. *J. Sed. Petrol.*, **54**, 731–745.

Shields, A., 1936. *Anwendung der Aehnlickkeits-Mechanik und der Turbulenzforschung auf die Geshiebebewegung*. Mitlerlungen der Preussische Versuchsanstalt fur Wasserbau, Erd, Schiffbau, Berlin, No. 26.

Shizowa, T., 1970. The experimental study and differential flocculation of clay minerals—one application of its results to recent sediments in Ishikari Bay. *Japanese Assoc. Mineral. Petrol. Econ. Geol. J.*, **63**, 75–84.

Sholkovitz, E. R., 1976. Flocculation of dissolved organic and inorganic matter during the mixing of river water and seawater. *Geochim. Cosmochim. Acta*, **40**, 831–845.

Sholkovitz, E. R., 1978. The flocculation of dissolved Fe, Mn, Al, Cu, Ni, Co and Cd during estuarine mixing. *Earth Planet. Sci. Letters*, **41**, 77–86.

Simmons, H. B., 1965. *Channel Depth as a Factor in Estuarine Sedimentation*. U.S. Army Committee on Tidal Hydraulics, Tech.-Bull. 8, 15 p.

Simmons, H. B., 1972. Effects of man-made works on the hydraulic, salinity, and shoaling regimens of estuaries. *In*: Nelson, B. W. (ed.), *Environmental Framework of Coastal Plain Estuaries*. Geol. Soc. Amer. Memoir 133, pp. 555–570.

Smoluchowski, M., 1917. Versuch einer mathematischen theorie der koagulationskinetik kolloider losungen: *Z. Phys. Chem.*, **92**, 129–168.

Stanley, D. J., and Swift, D. J. P., 1976. *Marine Sediment Transport and Environmental Management*. John Wiley and Sons, New York, 602 pp.

Sternberg, R. W., 1972. Predicting initial motion and bed load transport of sediment particles in the shallow marine environment. *In*: Swift, D. J., Duane, D. B., and Pilkey, O. H. (eds.), *Shelf Sediment Transport*. Dowden, Hutchinson, and Ross, Stroudsburg, PA pp. 61–82.

Strumm, W., and Morgan, J., 1970. *Aquatic Chemistry*. Wiley Interscience, New York, 583 pp.

Sundborg, A., 1967. Some aspects on fluvial sediments and fluvial morphology. I. General views and graphic methods. *Geogr. Ann*, **49**, 333–343.

Swift, D. J. D., 1976. Coastal sedimentation. *In*: Stanley, D., and Swift, D. (eds.), *Marine Sediment Transport and Environmental Management*. John Wiley, New York, pp. 255–310.

Swift, D. J. P., and Pirie, R. G., 1970. Fine-sediment dispersal in the Gulf of San Miguel, western Gulf of Panama: a reconnaissance. *J. Mar. Res.*, **28**, 69–95.

Syvitski, J. P., and Lewis, A. G., 1980. Interaction of zooplankton and suspended sediments. *J. Sed. Petrol.*, **50**, 0869–0880.

Syvitski, J. P., and Murray, J. W., 1981. Particle interaction in Fjord suspended sediment. *Mar. Geol.*, **39**, 215–242.

Terwindt, J. H. J., de Jong, D., and Van der Wilk, E., 1963. *Sediment Movement and Sediment Properties in the Tidal Area of the Lower Rhine (Rotterdam Waterway).* Trans. Jubilee Convention, Pt. 2, Kon. Ned. Geol. Mijnbowk, Genoots. Verhand. Geol. Ser., Vol. 21, Part 2, pp. 243–258.

Thorn, M. F. C., 1975. *Deep Tidal Flow Over a Fine Sand Bed.* Proc. 16th Congress Intl. Assoc. Hydraulic Res., São Paulo.

U.S. Army Corps of Engineers, 1977. *Shore Protection Manual.* Coastal Engineering Research Center, No. 008–022–00113–1, 1262 pp.

U.S. Army Engineer District, Philadelphia, Corps of Engineers, 1973. *Long Range Spoil Disposal Study.* Part III, 140 pp.

U.S. Council on Environmental Quality, 1970. Ocean dumping—A national policy A Report to the President. Council on Environmental Quality, Washington, D.C., 45 pp.

Van Olphen, H., 1977. *An Introduction to Clay Colloidal Chemistry*, 2d ed. John Wiley and Sons, New York, 318 pp.

Vanoni, V. A., 1975. *Sedimentation Engineering.* Amer. Soc. Civil Engr., New York, 745 pp.

Van Straaten, L. M. J. U., 1963. Aspects of Holocene sedimentation in the Netherlands. *Verhandel. Koninkl. Ned. Geol. Mignbouwk. Genool. Geol. Ser.*, **21–1**, 149–172.

Van Straaten, L. M. J. U., and Kuenen, P. H., 1957. Accumulation of fine grained sediments in the Dutch Wadden Sea. *Geol. en Mijnbouw (NW. Ser.)*, **19**, 329–354.

Van Straaten, L. M. J. U., and Kuenen, P. H., 1958. Tidal action as a cause of clay accumulation. *J. Sed. Petrol.*, **28**, 406–413.

Verwey, E. J. W., 1952. On the ecology of distribution of cockle and mussel in the Dutch Waddensea, their role in sedimentation and the source of their food supply. *Arch. Neerl. Zool.*, **10**, 172–239.

Verwey, E. J. W., and Overbeek, J. T. G., 1948. *Theory of the Stability of Hydrophobic Colloids.* Elsevier, New York.

Weil, C. B., 1977. *Sediment, structural framework, and evolution of Delaware Bay; a transgressive estuarine delta.* Univ. of Delaware Sea Grant Tech. Rept. No. DEL-SG-4-77, 199 pp.

Whitehouse, U. G., Jeffrey, L. M., and Debbrecht, J. D., 1960. Differential settling tendencies of clay minerals in saline waters. *In*: Swineford, A. (ed.), *Clays and Clay Minerals.* Proc. National Conf. on Clays and Clay Minerals, 7th, Washington, DC, 1958, Pergamon Press, New York, pp. 1–179.

Wright, L. D., Coleman, J. M., and Thom, B. G., 1973. Processes of channel development in a high-tide-range environment. Cambridge Gulf-Ord River Delta. *J. Geol.*, **81**, 15–41.

Wright, L. D., Coleman, J. M., and Thom, B. G., 1975. Sediment transport and deposition in a macrotidal river channel, Ord River, Western Australia. *In*: Cronin, L. E. (ed.), *Estuarine Research*, Vol. 2. Academic Press, New York, pp. 309–322.

Yalin, M. S., 1977. *Mechanics of Sediment Transport*, 2d ed. Pergamon Press, New York, 298 pp.

Young, D. K., 1971. Effects of infauna on the sediment and seston of a subtidal environment. *Vie et Milieu. Supp.*, **22**, 557–571.

Young, R., 1982. Mechanisms of erosion, deposition, and transport of cohesive sediments in the boundary layer. *In*: Kim, L., and Burns, R. (eds.), *Proc. of a Pollutant Transfer by Particulates Workshop*, Seattle, WA, pp. 193–223.

Young, R. A., and Southard, J. B., 1978. Erosion of fine-grained marine sediments: Seafloor and laboratory experiments. *Geol. Soc. Amer. Bull.*, **89**, 663–672.

Zabawa, C. F., 1978a. Flocculation in the turbidity maximum of northern Chesapeake Bay: University of South Carolina, Columbia, S.C., Ph.D. dissertation, 123 pp.

Zabawa, C. F., 1978b. Microstructure of agglomerated suspended sediments in Northern Chesapeake Bay Estuary. *Science*, **202**, 49–51.

Zabawa, C. F., 1980. *Estuarine Sediments and Sedimentary Processes in Winyah Bay, SC*. Geologic Notes, Dept. of Geology, Univ. South Carolina, Columbia, SC, 40 pp.

Zenkovitch, V. P., 1959. On the genesis of cuspate spits along lagoon shores. *J. Geol.*, **67**, 267–277.

Zimmerman, J. T. F., 1973. The influence of the subaqueous profile on wave induced bottom stress. *Neth. J. Sea Res.*, **6**, 542–549.

3

Intertidal Flats and Intertidal Sand Bodies

George deVries Klein

Introduction

The intertidal zone represents a diverse depositional surface demarcated by the elevation of both high and low tide stage. Intertidal flats are low-sloping features within this interval that are exposed at low-tide stage. Intertidal sand bodies represent linear shoals or bars deposited by tidal currents that occur in the lower portions of intertidal flats and the nonvegetated intertidal zone. They are exposed also at low tide, in contrast to tidal current sand ridges, which are always subtidal.

Intertidal environments have fascinated sedimentologists for some time. Within them, a variety of sedimentary features is known, many of which are common to ancient sedimentary rocks. Because of geologists' urge to understand the origin, depositional processes, environment, and predictive trends of sedimentary rocks, they have examined modern systems of sediment transport and deposition to interpret the ancient sedimentary rock record. A uniformitarian approach has underlain many of the interpretations common to practioners of sedimentology, and the nonvegetated intertidal zone is no exception to such analysis, largely because of its accessibility.

Earliest studied tidal flats occur on the coasts of northwestern Europe along the North Sea. Much of the earlier and descriptive work done there was summarized in a literature inaccessible to most of the English-speaking segment of the geological community. The earlier papers were small-scale studies focusing on specific features. Hantzschel (1939) provided the first English-language synthesis on German work.

Intertidal flat sedimentation studies exploded immediately following World War II. The first major work was by Van Straaten in The Netherlands who presented a series of papers focusing first on specific sediment features and later expanding to include comparative studies of other areas and a synthesis of depositional processes (Van Straaten, 1952, 1953, 1954, 1959, 1961; Van Straaten and Kuenen, 1957). These studies were followed by an extensive on-going program of research along the northwest coast of Germany by Reineck (1963, 1967, 1972) and Reineck and Wunderlich (1968a,b); these are summarized in a chapter in Reineck and Singh (1973, 1980). Evans' (1965, 1975) work along the coast of eastern England in The Wash and work by Bajard (1966) in France comprise other earlier European work. In North America, intertidal flat sedimentary structures were first described from the Bay of Fundy by Kindle (1917); these were followed by a regional summary from there by Klein (1963), a comparative study of Bay of Fundy and western European intertidal flats by Klein and Sanders (1964), and by detailed work on intertidal sand bodies by Klein (1970) and Knight (1980), including their bedform migration (Klein and Whaley, 1972; Dalrymple et al., 1978). Thompson (1968) provided a most detailed summary of intertidal flats in an arid setting along the Gulf of California, and Pestrong (1972) completed a distributional sedimentological study of tidal flats along San Francisco Bay. In Australia, Gellatly (1970) described sedimentary structures from intertidal sand bodies in King Sound; the regional intertidal flats there were described more recently by Semeniuk (1981).

These studies provided a baseline for comparison and development of facies models. Facies models on intertidal flats were provided by Klein (1971), whereas facies models for intertidal sand bodies were proposed first by Knight and Dalrymple (1975).

As a result of these studies, many geologists have assumed that tidal sedimentation research focuses only on intertidal flat environments (see discussion by Klein, 1976; 1977a). This assumption (derived from accessibility) has overlooked the significance of tidal sedimentation processes in the subtidal zone and deepwater marine settings. Thus, in subtidal settings, particularly where an increase in shelf width tends to enhance tidal current intensity (Cram, 1979; Klein, 1977a,b; Klein and Ryer, 1978), large tidal current sand ridges are common and they show evidence of tide-dominated deposition (Off, 1963; Houbolt, 1968; Boggs, 1974; Stride, 1963, 1982; Belderson, 1964; Klein et al., 1982; Belderson et al., 1972; McCave and Langhorne, 1982; Kenyon et al., 1981). Tidal processes are common also to many other coastal environments including deltas (Coleman, 1980), barrier islands, and tidal marshes other than those associated with intertidal flats and intertidal sand bodies (see Klein, 1977a, Table 1). The intertidal flats and intertidal sand bodies that are the topic of this chapter occur most prominently along tide-dominated coastlines, especially macrotidal and mesotidal coasts (Davies, 1964), whereas tidal action in other coastal settings tends to be subordinate to other processes. Tidal sedimentation processes are active also in deepwater settings in water depths ranging from 2000 to 2500 m (Keller et al., 1973; Shepard et al., 1969; Shepard and

Marshall, 1973; Lonsdale *et al.*, 1972; Lonsdale and Malfait, 1974), and the sedimentary features so produced show some element of commonality in type to features produced in coastal and relatively shallow subtidal conditions.

Global Distribution

It has long been known that tidal ranges and tidal current velocities along coastlines tend to vary with shelf width and the associated relative intensity of wave action (Davies, 1964; Hayes, 1975, 1979; Cram, 1979; Klein, 1977a,b; Klein and Ryer, 1978). Tidal ranges have been used to classify coastlines according to the following scheme of Davies (1964): microtidal (< 2 m), mesotidal (2–4 m), and macrotidal (> 4 m). Hayes (1979) has discussed the variation in coastline morphology and sedimentary systems in each of these settings. It is clear from his summaries (Hayes, 1975, 1979) that extensive intertidal flats occur in macrotidal areas, whereas narrower but significant intertidal flats occur also in mesotidal areas. Figure 3-1 shows the global distribution of both mesotidal and macrotidal coast where such intertidal flats occur.

In actual fact, very few of these places have been studied in detail. Perhaps the most spectacular intertidal flats are those from the Yellow Sea of Korea, a macrotidal coast. These have been studied in reconnaissance by Chung and Park (1977) and Wells and Huh (1979), who demonstrated a textural

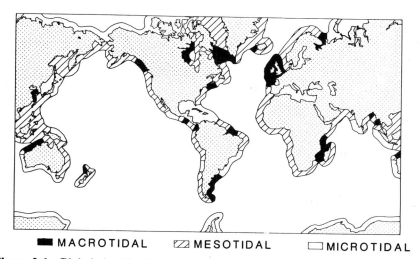

■ MACROTIDAL   ▨ MESOTIDAL   ▢ MICROTIDAL

Figure 3-1. Global classification of coastlines by tidal range (redrawn from Davies, 1964).

sediment distribution there similar to intertidal flats from the North Sea (see below). Intertidal sand bodies have been observed from the Yellow Sea also, but these have not been described at all. The Bay of Fundy intertidal flats of the Minas Basin are the only intertidal flats and intertidal sand bodies from a macrotidal coast that have been examined in any degree of detail (Klein, 1963, 1970; Knight, 1980; Knight and Dalrymple, 1975; Dalrymple *et al.*, 1978; Lambiase, 1980). The Baie du Mt. Saint Michel of France is an area of well-described intertidal flats occurring also in a macrotidal domain (Larsonnieur, 1975; Bajard, 1966). The Wash of eastern England (Evans, 1965; McCave and Geiser, 1979) is also well described and it too occurs in a macrotidal setting (Davies, 1964).

Intertidal flats have been described from many mesotidal coasts. These include the North Sea coastline of The Netherlands (Van Straaten, 1952, 1959, 1961), and of Germany (Reineck, 1963, 1967, 1972). Mesotidal-range intertidal flats along the Gulf of California were described by Thompson (1968) and those of San Francisco Bay by Pestrong (1972). Similarly, intertidal sand bodies (Gellatly, 1970) and intertidal flats (Semeniuk, 1981) from the mesotidal domain of King Sound, Australia, are noteworthy because of their combined semitropic and local arid setting. Muddy intertidal flats have been reported and analyzed in detail from mesotidal coasts in New Hampshire (Anderson *et al.*, 1981) and the north coast of Surinam (Wells and Coleman, 1981a,b).

Sedimentation Processes

Sediment transport processes in the intertidal zone are distributed in a contour-parallel fashion from high-tide level to low-tide level (Figure 3-2). These transport process zones include a bed-load tidal current sediment transport in combination with late-stage, sheet-like runoff prior to exposure, a second zone where bed-load tidal current processes alternate with suspension deposition (transitional zone), and a third zone dominated by suspension sedimentation. These zones are superimposed by processes and features caused by exposure to air, tidal scour, and bioturbation. These sediment transport zones also control the textural distribution of sediment.

Suspension Processes

Suspension processes of sediment transport occur over intertidal flats during periods of submergence. Deposition of sediment from suspension occurs only during periods of negligible velocity associated with slack water periods around the time of peak submergence or high tide. Such sediment, however,

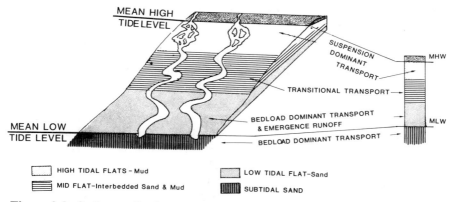

Figure 3-2. Sediment distribution, and sediment transport and depositional zones across an intertidal flat based on data from The Netherlands, West German, and eastern English coastline of North Sea based on work by Van Straaten (1959, 1961), Reineck (1963), and Evans (1975). Columnar section on right shows vertical sequence produced by progradational intertidal flat coastline (redrawn from Klein, 1972).

is resuspended as tidal current velocities increase, by periodic wave action or when storms occur.

The source of suspended sediment appears to be twofold: offshore continental shelf zones and resuspended material from the intertidal flats themselves (Postma, 1954, 1961; Van Straaten and Kuenen, 1957; Groen, 1967; Anderson *et al.*, 1981; Wells and Coleman, 1981a,b). Suspended sediment from offshore areas is brought landward by tidal currents containing relatively large concentrations of sediment. During peak submergence of the intertidal flat, the slack water stage is characterized by a negligible bottom current velocity and minimal turbulence so that suspended material settles to the intertidal zones' seabed. This slack water period may last as long as 2 hours (Postma, 1961), permitting sufficient time for suspended material to settle to the bottom because the settling velocity of the particulate matter exceeds bottom current velocities capable of maintaining material in suspension. In the landward zone of intertidal flats (known as high-tidal flat; Figure 3-2), water levels begin to fall in the early stages of the ebb tidal cycle and as it does so, the suspended sediment that settled to the intertidal flat surface becomes exposed and remains there. Some of this fine-grained sediment is resuspended by ebb tidal currents of slightly increased velocity. Because this velocity is relatively small, the seaward distance of sediment transport off the intertidal zone or downslope on the intertidal flat is less than the combination of resuspension, additional yield of suspended sediment, and increased relative distance of landward transport associated with the succeeding phase of flood-dominated deposition. This process

favors sediment accumulation on the high intertidal flats and is termed
settling lag (Postma, 1954, 1961).

A second factor that aids in concentrating relatively larger volumes of
suspended sediment is the time-velocity asymmetry of tidal currents
(Postma, 1961; Groen, 1967). Along the North Sea Coast of The Netherlands, tidal currents on some tidal flats and offshore zones are characterized
by a flood-dominated time-velocity asymmetry which means that current
velocities are greater during the flood stage of a tidal cycle than the ebb stage
of a tidal cycle (Figure 3-3). These greater tidal current velocities erode fine-
grained sediment and also are the cause of a larger competence of tidal
currents resulting in transport of relatively large concentrations of sediment.
Because flow directions are landward, such sediment is dispersed landward
also and accumulates during the high-water, slack-water stage of the tidal
cycle.

Van Straaten and Kuenen (1957) pointed out also that a scour lag exists
under these suspension-dominated processes. This scour lag involves the
resuspension of fine-grained sediment from the intertidal flat surface. Net
accumulation of fine-grained sediment from suspension on high-tidal flats
comes about also because with exposure and subsequent desiccation,
material cannot be eroded by tidal currents. However, in the submerged part
of intertidal flats, mud can be and is resuspended and thus net mud
accumulation diminishes in the middle portion of intertidal flats. Mud is
seldom preserved on lower intertidal flats because of this process.

Most of the resuspension of mud from tidal flats occurs during periods of
storm activity along a coast or during relatively moderate to large wave
energy expended along coasts (Anderson *et al.*, 1981; Wells and Coleman,

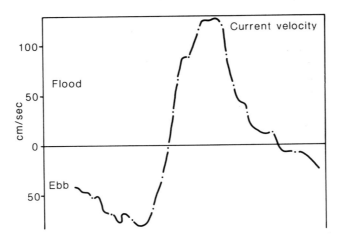

Figure 3-3. Time-velocity curve showing a typical asymmetric velocity pattern
common to intertidal and subtidal areas (redrawn from Postma, 1961).

1981b). Nevertheless, on many intertidal flats, it is not uncommon to observe water-saturated muds showing a gel-like character (Wells and Coleman, 1981b). This material tends to move downslope as a slurry or a slide with associated small-scale slump scars, but it acts to baffle wave action along an intertidal flat (Wells et al., 1980). This baffling effect damps out wave action along intertidal flats unless wave energy increases. When wave energy increases, resuspension of fine-grained sediment occurs. A seasonal periodicity to resuspension during winter months coupled with net sediment accumulation during fair-weather spring and summer months has been documented along the intertidal flats of New Hampshire and elsewhere along the northeastern coast of North America (Anderson et al., 1981) and southwestern Korea (Wells et al., in press).

Bedload Processes

Transport and deposition of sediment by bedload processes occurs by means of tidal current systems. Tidal current systems operate, obviously, in the intertidal zone only during periods of submergence. When such tidal currents flow across the intertidal zone, sand-sized sediment is moved by bed shear when bottom current velocities exceed at least 10 cm/s (Reineck and Wunderlich, 1968a). With increased bottom tidal current velocity, sandy zones are deformed into large and small bedforms, with current ripples moving with an average threshold velocity of 50 cm/s, dunes with an average threshold velocity of 40 cm/s, and sand waves with average threshold velocities of 55 to 60 cm/s (Dalrymple et al., 1978). Minimal threshold bottom-current velocities for migration of dunes and sand waves of 45–47 cm/s, respectively, were observed by Klein (1970) and Klein and Whaley (1972).

The internal nature and orientation of cross-stratification, cross-strata set boundaries, and other discontinuities are controlled by the time-velocity asymmetry of tidal currents, and the longer term alternation of neap and spring tides (Reineck, 1963; Klein, 1970; Visser, 1980; Boersma and Terwindt, 1981). When maximum tidal current velocities tend to be of nearly equal intensity, a vertical stacking of units of herringbone cross-stratification can be developed by accumulation of sediment coupled with opposite migration of dunes and sand waves with each reversing tidal phase (Reineck, 1963). Moreover, the depth of scour and reworking of sediment will change from the neap to the spring stage; and as Klein (1970) demonstrated, during the change from the spring to the neap stage of a lunar tidal cycle, the depth of scour decreases, the velocity spectrum changes, and vertically stacked herringbone cross-stratification is preserved. When time-velocity asymmetry is a dominant characteristic of tidal currents, truncation surfaces, termed *reactivation surfaces* (Figure 3-4) (Collinson, 1969; Klein, 1970), are formed during the subordinate-velocity phase. As a consequence of this

Figure 3-4. Excavation in dune in ebb-dominant zone in tidal current time-velocity asymmetry showing avalanche cross-stratification and reactivation surfaces (arrows), Economy Point, Minas Basin, Bay of Fundy, Nova Scotia, Canada (from Klein, 1970; republished with permission of the Society of Economic Paleontologists and Mineralogists).

velocity asymmetry, a unimodal orientation of cross-strata is observed, with bundles of such cross-strata bounded by reactivation surfaces dipping in the same direction, but truncating avalanche cross-stratification at a lower angle. Figure 3-5 shows the mechanism for formation of such reactivation surfaces.

Superimposed on these reversals of flow directions of tidal currents during a tidal cycle are longer term cycles of changes in magnitude of bottom-current velocities during the change from neap to spring tide and back to neap tide. DeRaaf and Boersma (1971), Boothroyd and Hubbard (1975), Allen and Friend (1976), Visser (1980), and Boersma and Terwindt (1981) all observed that the distance of migration of bedforms is greater during spring tides and minimal, approaching zero, during neap tides. More recently, Boersma and Terwindt (1981) demonstrated that on one intertidal sand body examined in detail, tidal current velocities, characterized by time-velocity asymmetry, show larger mean and maximum bottom-current velocities and sediment transport rates during the spring tide phase of the lunar tidal cycle than occurs during the neap phase (Figure 3-6). Within dunes and sand wave complexes in a flood-dominated area, where bottom-current velocity measurements were obtained, accretionary bundles of cross-strata with distinct bounding surfaces were observed. These could be correlated to both

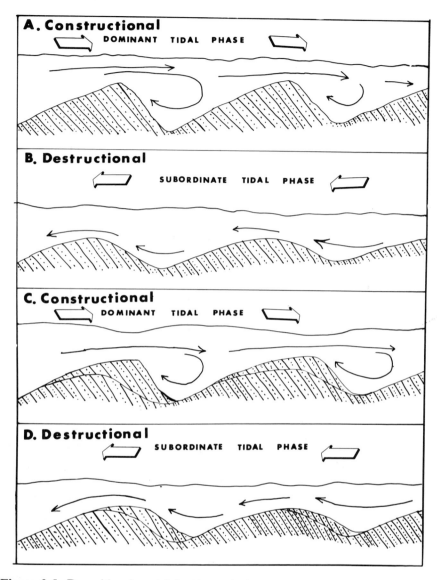

Figure 3-5. Depositional model for time-velocity asymmetry control for development of reactivation surfaces. During dominant constructional phase, dunes developed on seabed (A). When tide reverses into subordinate velocity phase, dune is scoured in part, leaving a reactivation surface (B) dipping at lower angle but with same orientation as dune slip face truncating avalanche cross-strata. When tide turns into next dominant phase, second dune forms and buries original reactivation surface (C). The next turn of tide into another subordinate velocity phase (D) develops a second reactivation surface (reprinted by permission of International Human Resources Development Corporation from *Clastic Tidal Facies* by George deVries Klein).

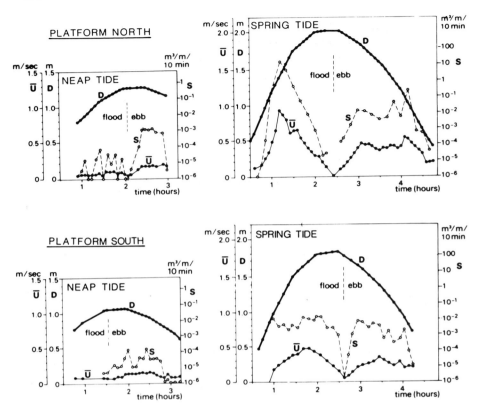

Figure 3-6. Variation during a single tidal cycle in water depth (D) in meters, mean current velocity (Ū) in m/s, and sand transport rate (S) in m³/m/10 minutes from two observation stations on intertidal sand body, Westerschelde Estuary, The Netherlands, during both spring and neap tide. Tidal current velocity curves show both time-velocity asymmetry and noticeable change in velocity magnitude between spring and neap tide (redrawn from Boersma and Terwindt, 1981).

neap and spring tide sediment transport. Reactivation surfaces (which they termed *pause planes* because of lack of migration of bedforms) and associated structures graded laterally into a bundle of thick avalanche cross-strata organized into distinct laminae. (These cross-strata are termed *vortex structures* by Boersma and Terwindt.) Laterally, these structures grade into a terminal interval of cross-strata that contain less well-sorted sand with the angle of repose decreasing downcurrent. These are termed *slackening structures* (Figure 3-7). The entire sequence is overlain by an "ebb cap" of sediments showing ebb-oriented cross-strata. This lateral change in structures records deposition during a single flood-dominated tidal phase with the reactivation structure representing nondeposition during a subordinate phase

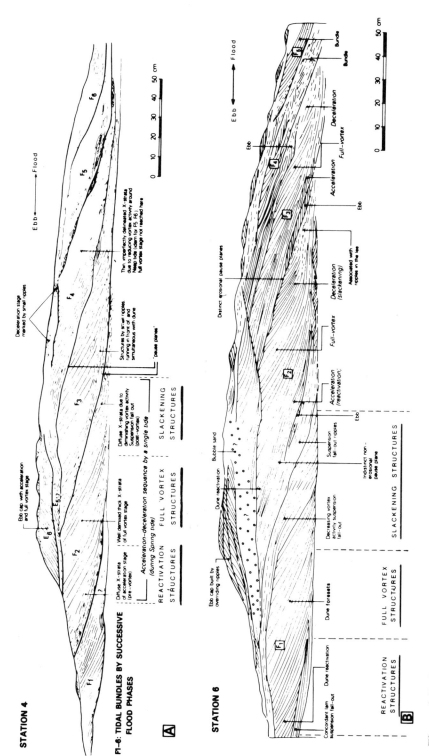

STATION 4

F1–6: TIDAL BUNDLES BY SUCCESSIVE FLOOD PHASES

Ebb ⟷ Flood

50 cm

0 10 20 30 40 50 cm

Deceleration stage marked by small ripples

Then, imperfectly delineated x-strata due to reducing vortex activity around Suspension tail-out (post-vortex). Neap tide idem for F5, F6. Full vortex stage not reached here

Structures by small ripples running in front of and simultaneous with dune

'pause planes'

Diffuse X-strata of acceleration stage (pre-vortex)

Ebb cap with acceleration and full vortex stage

Diffuse X-strata due to diminishing vortex activity Suspension tail-out (post-vortex)

Well demised thick X-strata of full vortex stage

Acceleration–deceleration sequence by a single tide

(during Spring tide)

REACTIVATION STRUCTURES | FULL VORTEX STRUCTURES | SLACKENING STRUCTURES

A

STATION 6

Ebb ⟷ Flood

0 10 20 30 40 50 cm

Distinct erosional pause planes

Ebb

Ebb

Bundle

Bundle

Deceleration

Acceleration | Full-vortex

Full-vortex

Deceleration (slackening)

Associated with ripples in the lee

Ebb

Acceleration (reactivation)

Full-vortex

Suspension tail-out ripples

Indistinct non-erosional pause plane

Bubble sand

Decreasing vortex activity suspension tail-out

Dune reactivation

Ebb cap built by overriding ripples

Dune foresets

Concordant (sm) suspension tail-out

Dune reactivation

REACTIVATION STRUCTURES | FULL VORTEX STRUCTURES | SLACKENING STRUCTURES

B

Figure 3-7. Cross-sections through dunes from intertidal sand body, Westerschelde Estuary, The Netherlands, showing series of tidal bundles containing reactivation surfaces, full vortex and slackening cross-stratification, and other structures in response to changes from spring tide to neap tide as explained in text. Diagram based on lacquer peels (redrawn from Boersma and Terwindt, 1981).

and the avalanche cross-strata representing the active phase of dune migration during the dominant flood phase. The slackening phase (Figure 3-7) represents the diminishing of flow velocities toward the end of a tidal cycle. The internal organization of cross-strata and reactivation surfaces differ between the neap and spring phase because the neap phase shows thinner bundles and thinner cross-strata sets, reflecting smaller bottom current velocities, whereas the spring tidal phase shows thicker sets of cross-strata and longer bundles, reflecting greater velocities and greater sand transport rates (Figure 3-8). Thus the lateral dimensions and sediment volume permit recognition of spring and neap tidal phases in present-day intertidal and sand bodies. This interpretation can be applied also to ancient counterparts where the neap pause planes may be represented instead by mud flaser beds (Allen, 1982).

Late-Stage Emergence Runoff

As the tidal elevation falls during the ebb-phase of a tidal cycle, the intertidal zone becomes progressively exposed. This changing water depth is coupled with changes in flow directions and changing bottom-current velocities. These three combined processes define late-stage emergence runoff.

On both the lower reaches of intertidal flats and intertidal sand bodies, the direction of flow of tidal currents will change as both the crests of dunes and sand waves, and as the crest of intertidal sand bodies, become emergent. Flow directions change from shore parallel to downslope; some of this flow becomes confined to the troughs of large-scale bedforms and moves in a manner similar to open-channel flow. Some of this downslope flow may parallel the main direction of ebb current flow, or flow at right angles or opposite to such flow, depending on the direction of slope locally and regionally depending on the depositional surface. Because of the continuing reduction in water depths, the height and size of bedforms that are migrating will decrease. Some of the larger bedforms stop migrating completely. Because of changing directions in flow, it is normal that smaller scale bedforms will become superimposed on larger scale bedforms and the orientation of the superimposed bedforms will be oblique, at right angles or opposite to the larger bedforms developed during the main ebb or flood current flow. These superimposed smaller scale features involve several sizes of diminishing bedforms. Thus, current ripples may be superimposed on dunes or sand waves (Figure 3-9), or small current ripples may be superimposed on larger current ripples (Figure 3-10). As the flow velocity is maintained while water depth decreases, double-crested current ripples (Figure 3-11) will develop as the bed shear is maintained, although depth of migration is reduced on the crest but continues in a slightly deeper trough (McMullen, 1964; Klein, 1970). Locally, scour pits may develop, and these too would show such superimposed features. Reduction of water level during

STATION 7

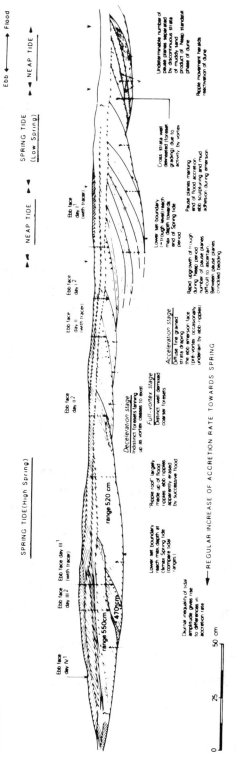

Figure 3-8. Drawing from lacquer peel obtained from intertidal sand body, Westerschelde Estuary, The Netherlands, showing record of difference in volume and types of sedimentary structures generated over a lunar month showing changes from spring tide to neap tide sediment styles (redrawn from Boersma and Terwindt, 1981).

Figure 3-9. Ebb-dominated dune with superimposed current ripples on both slip-face and crest, intertidal sand body, Pinnacle Flats, Minas Basin, Bay of Fundy, Nova Scotia, Canada. Scale in centimeters and decimeters (reprinted by permission of International Human Resources Development Corporation from *Clastic Tidal Facies* by George deVries Klein).

Figure 3-10. Superimposed smaller current ripples on larger current ripples, Girdwood Bar, Turnagain Arm, Alaska. Scale in centimeters (reprinted by permission of International Human Resources Development Corporation from *Clastic Tidal Facies* by George deVries Klein).

Figure 3-11. Double-crested current ripples, Big Bar, Minas Basin, Bay of Fundy, Nova Scotia, Canada (reprinted by permission of International Human Resources Development Corporation from *Clastic Tidal Facies* by George deVries Klein).

this late-stage emergence runoff may also produce horizontal steplike hachure marks. If wind-driven small waves move over the water surface, small currents with scouring capability may be generated. These currents partially destroy current ripple geometries and leave behind plane beds that truncate these ripples. These features have been termed *washout structures* by Van Straaten (1954, 1959) and are shown in Figure 3-12.

Local air entrapment is also common during both emergence runoff (Emery, 1945; Stewart, 1956) and later submergence. This process gives rise to internal air holes (Figure 3-13). During submergence, this air is observed bubbling off into the water. The air holes are similar to birds-eye structures in carbonates (Shinn, 1968).

Alternation of Bedload and Suspension Deposition

Within the middle portion of tidal flats, the dominant process of sedimentation is the alternation of both bedload and suspension transport and deposition. As a consequence, sedimentary features there show a preserved record of such processes. Suspension deposition occurs under the conditions discussed above involving both submergence and negligible velocity of transport by tidal currents during high-water slack tide. Bedload transport

Figure 3-12. Washout structure truncating current ripples, intertidal flats in Swansea Bay, Wales, United Kingdom. Scale is 30 cm long (reprinted by permission of International Human Resources Development Corporation from *Clastic Tidal Facies* by George deVries Klein).

Figure 3-13. Air holes in intertidal sand body, Big Bar, Minas Basin, Bay of Fundy, Nova Scotia, Canada (reprinted by permission of International Human Resources Development Corporation from *Clastic Tidal Facies* by George deVries Klein).

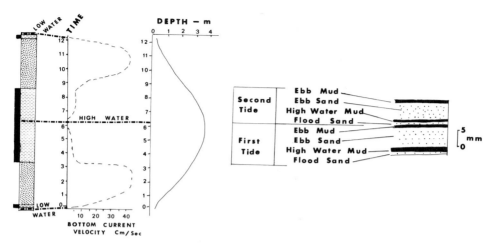

Figure 3-14. Development of tidal bedding according to Reineck and Wunderlich (1968b). Marker beds emplaced at high and low tide bracketed bed-load deposition of sand (stippled) and suspension deposition of mud (dashed; also black bar). Velocity changes during tidal cycle controlled type of deposition. During two tidal cycles, four couples of sand and mud comprising a tidal bed are deposited (on right) (reprinted by permission of International Human Resources Development Corporation from *Clastic Tidal Facies* by George deVries Klein).

and deposition typically begins under bottom-current velocities in excess of 10 cm/s (Figure 3-14; Reineck and Wunderlich, 1968a,b).

Because the volume of available sand is relatively small, the type of sedimentary features that develop in response to this alternation of bedload and suspension deposition includes small-scale dunes and current ripples with their internal cross-stratification and thin lamina of clay organized as flaser bedding (Figure 3-15). A large variety of lenticular, wavy, and flaser bedding (Figure 3-15) is characteristic of this depositional process (Reineck and Wunderlich, 1968a,b). The differentiation of these types of structures is controlled, however, by the relative volume of sand and mud, by the relative duration of both the bedload and suspension mode of deposition, and by current velocities. If sand exceeds mud, ripples form with isolated clay drapes in ripple troughs and crests (Figure 3-16). If mud and sand content is nearly equal, the volume of flaser bedding increases (Figure 3-16). When the relative volume of mud increases significantly with respect to sand, current ripples become isolated and are preserved as lenticular beds, some with internal flaser bedding (Figure 3-16). Wavy bedding occurs if the clay layers are draped continuously over both symmetrical and asymmetrical ripples. It should be observed, however, that in subtidal areas, similar flasers form by storms (McCave, 1970).

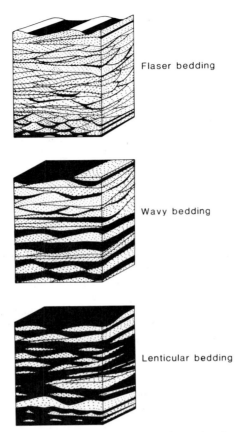

Figure 3-15. Three block diagrams showing three-dimensional organization of (from top down) flaser bedding, wavy bedding, and lenticular bedding (redrawn after Reineck and Wunderlich, 1968a).

This process of alternation of bedload and suspension sedimentation also gives rise to thin, parallel-layered beds of alternating and interbedded sands and muds, termed *tidal bedding* (Figure 3-17) by Reineck and Wunderlich (1968b). They monitored the duration, the volume, and the type of sedimentation under this regime more precisely (Figure 3-14) and showed that deposition of alternating suspension muds and parallel-layered bedload sands was controlled by critical bottom current velocities. Sand accumulated whenever velocities exceeded 10 cm/s (Figure 3-14), whereas mud accumulated when velocities were less than 10 cm/s. Using a series of color markers (Figure 3-14), suspension processes were observed to coincide with slack-water high tide and negligible-velocity low tide stage (Reineck and Wunderlich, 1968b).

Figure 3-16. Simple and bifurcated flaser bedding (lower and upper third), lenticular bedding, and wavy bedding (middle third and below knife scale), Schelde Estuary, Haringvliet, The Netherlands (reprinted by permission of International Human Resources Development Corporation from *Clastic Tidal Facies* by George deVries Klein).

Other Processes

Exposure

The twice-daily rise and fall of tide level in the nonvegetated intertidal zone causes the intertidal seabed to be exposed for moderate to long periods; up to as much as 10 hours. The effect of this exposure to the atmosphere is to induce heating by the sun's radiation, water evaporation, water runoff (mentioned already), and water escape. The most common sedimentary features that develop in response to exposure are mudcracks and runzel marks. Mudcracks are common and have been described extensively.

Figure 3-17. Tidal bedding, intertidal flats, northwest Germany. Scale in centimeters (photo by Friederich Wunderlich; reprinted by permission of International Human Resources Development Corporation from *Clastic Tidal Facies* by George deVries Klein).

Runzel marks (Reineck, 1963; Hantzschel and Reineck, 1968) have been described from intertidal flats, and their origin, in terms of exposure, involves at least two processes. First, foam is often observed at the waterline, and, as the tide goes out, it becomes stranded. Cohesive action by the foam on sediments causes a suction effect. When the foam is blown by offshore winds, it removes some surface sediment with it, leaving a wrinkled surface; these wrinkled surfaces comprise the runzel marks.

A second mode of origin was observed (see Klein, 1977a) at low tide on a barrier island beach off South Pass, Louisiana, on the Mississippi Delta. There, raindrop marks were observed on the beach surface the day after a rainstorm had occurred. The beach was reworked by swash and backswash action and it disfigured and destroyed most of the raindrop marks. The disfigured markings appeared identical to the runzel marks Reineck (1963) and Hantzschel and Reineck (1968) have described.

Reworking of mudcracked surfaces by tidal currents during submergence causes the mudcracked polygons to be eroded from the seabed; these fragments are redeposited as a mud-chip breccia or so-called dessication breccia.

Tidal Scour

Scouring action and erosion by tidal currents is common in the intertidal zone, particularly in tidal channels where erosion of the thalweg produces

Figure 3-18. Shell lag concentrate at base of tidal channel, Wadden Sea, north of Groningen, The Netherlands (reprinted by permission of International Human Resources Development Corporation from *Clastic Tidal Facies* by George deVries Klein.

slump blocks of channel wall sediments in the channel floor, and related lateral sedimentation associated with tidal channel meandering produces shell beds of disarticulated molluscs on the channel floor as a lag concentrate (Van Straaten, 1952, 1954, 1959, 1961; Klein, 1963; Reineck, 1963). These shell lags are oriented typically convex-side up (Figure 3-18) and are buried by migration of point bar sediments. In addition, advection of storm-generated currents and tidal currents produces local erosional remnants on tidal flats that stand in relief with respect to surrounding tidal flats. These have been termed *ilots* by Macar and Ek (1965). On the intertidal zone seabed, local turbulent scour also may produce flute marks and current crescents on sandy bottoms (Klein, 1970, 1977a).

Soft-Sediment Deformation

A variety of compactional and soft-sediment deformation features is common on intertidal flats. Much of the sediment on intertidal flats is highly water-saturated and subject to excess pore-pressure. Failure of these materials is not uncommon and sediment moves as a slurry, debris flow and slump, leaving small-scale slump scars and slump folds in their wake (Wells *et al.*, 1980). Because of the variable water content and density contrasts of

Figure 3-19. Pseudonodules in tidal flats channel wall, The Wash, near Boston, England (reprinted by permission of International Human Resources Development Corporation from *Clastic Tidal Facies* by George deVries Klein).

interbedded sediment lithologies, pseudonodules (ball-and-pillow structure) have been observed on intertidal flats (Figure 3-19). In the case illustrated in Figure 3-19, a sudden rapid depositional event produced locally was responsible for the differential compaction and soft-sediment deformation features (Graham Evans, personal communication, November, 1969).

Intertidal Flats

Intertidal flats have been examined by sedimentologists since the early 1930s (see *Introduction*). The earliest studies focused on the intertidal flats of the North Sea and because of integrative and detailed studies by Van Straaten (1959, 1961), Reineck (1963), and Evans (1965), the intertidal flats of The Netherlands (Wadden Sea), northwest Germany (Wadden Sea), and eastern England (The Wash) became type areas from which comparisons were made.

In the intertidal zone of these three areas in the North Sea, sediment processes of transport and deposition are zoned in a contour-parallel fashion (Figure 3-2). Above the mean high water line is a supratidal zone that is dominated by submergence only during periods of elevated spring tide and storms. There, tidal marshes with *Spartina* grass are commonplace (see

Chapter 4). They develop typically on a substrate of mud; however, these marsh taxa show little preference for sediment texture. With additional accumulation of sediment, these tidal marshes encroach seaward as the position of high tide also progrades seaward.

This motif of supratidal sedimentation occurs in other areas. *Spartina* marshes have been reported from the intertidal zone of the Bay of Fundy (Klein, 1963, 1970), San Francisco Bay (Pestrong, 1972), and Boundary Bay (British Columbia) (Kellerhals and Murray, 1969) and the coast of Massachusetts (Hayes, 1969). These all occur in a temperate–humid region. As climates become more tropical, this zone tends to give way to mangrove swamps such as those reported from the tidal flats associated with the tide-dominated Klang-Langat Delta of Malaysia (Coleman *et al.*, 1970) and from the Niger Delta (Oomkens, 1974). Along the northwest coast of Australia, Semeniuk (1981) reported the presence of a mangrove supratidal zone grading laterally into a salt-pan evaporitic zone. Evaporite salt flats have also been described from arid supratidal zones adjoining intertidal flats around the Gulf of California (Thompson, 1968).

The sediment distribution of the main intertidal flat between mean high and mean low water coarsens in a seaward direction. The intertidal flat is zoned into three sections consisting of a high-tidal flat, a mid flat, and a low-tidal flat (Figure 3-2). These zones are distinguished on the basis of sediment distribution and dominant process of sedimentation. The high-tidal flats are areas dominated by deposition of fine-grained silts and clays. There, submergence lasts for less than one-third of a tidal cycle associated with the high-water level stage when velocities are negligible. Mud deposition is favored by suspension processes (as reviewed earlier). Sedimentary structures preserved here include mudcracks, silty current ripples, and bioturbation features. Bioturbation tends to be fairly strong in this setting (Figure 3-20) because the fine-grained sediment appears to serve as a suitable substrate for local infauna, and because of elevation and long exposure, many of the organisms burrow fairly deeply (Rhoads, 1967).

High intertidal flats at most locations contain mud and are, in fact, similar to the general case reviewed above from the North Sea. Thus the high-tidal flats of the Baie du Mt. Saint Michel (Bajard, 1966; Larsonnieur, 1975), San Francisco Bay (Pestrong, 1972), the Bay of Fundy (Klein, 1963, 1970), northwest Australia (Semeniuk, 1981), and the Gulf of California (Thompson, 1968) all fit into this mode of deposition.

The mid-flat environment occurs in the central portions of intertidal flats, is inundated for approximately 50% of a tidal cycle, and, therefore, experiences a near equal period of time of both suspension and bed-load sedimentation. This nearly equal alternation of bed-load and suspension deposition generates a mixed lithology (DeRaaf and Boersma, 1971) of interbedded sand and mud. Bedforms that develop include lower regime plane beds and current and interference ripples. Symmetrical ripples also occur, all within sand. Exposure features also are common. Internally, the

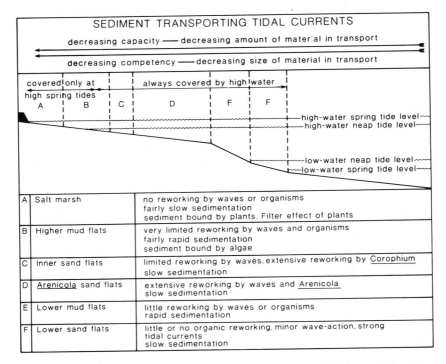

Figure 3-20. Surface sediment facies distribution of intertidal flats of The Wash, England (redrawn from Evans, 1965).

alternating lithologies are organized into wavy bedding depending on the ratio of sand to mud. Thus if the sand-mud ratio is large, flaser bedding is favored, whereas lenticular bedding is favored if the sand-mud ratio is less than unity. Tidal bedding and wavy bedding are favored where the sand-mud ratio approaches unity. These general observations pertain in particular to the North Sea, but they have also been reported from Northwestern Australia (Semeniuk, 1981), the Gulf of California (Thompson, 1968), and Boundary Bay (Kellerhals and Murray, 1969).

The low-tidal flat zone consists almost completely of sand of various size-ranges that is fashioned into bedforms of variable size, including ripples, dunes, and sand waves. Current ripples are superimposed often on the surfaces of larger bedforms and are oriented parallel, oblique, or opposite to the larger bedform. Internal cross-stratification, reactivation surfaces, micro-cross-laminae, and parallel laminae are the most common sedimentary structures observed in this setting. Bioturbation features are rare, primarily because of the high degree of instability of the sand substrate. The dominant mode of sediment transport in the low-tidal flat setting is bed-load transport by tidal currents that are characterized commonly by time-velocity asym-

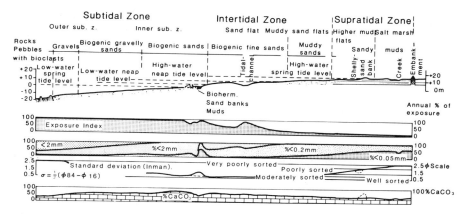

Figure 3-21. Zonation of intertidal flats in the Baie du M. St. Michel showing lateral variation in texture, sorting, carbonate content and exposure (redrawn from Larsonnieur, 1975).

metry. This transport takes place during the period of submergence of almost two-thirds of a tidal cycle or more. Emergence runoff processes also are characteristic of this setting.

The seaward-coarsening textural distribution and associated bedforms are diagnostic of the intertidal flat environment (Figures 3-20, 3-21). It owes its origin to the combination of differential time of inundation and submergence of intertidal flats during a tidal cycle, associated changes in bottom-current velocities of tidal currents during a tidal cycle (being therefore concentrated over the low-tidal flats), and the dominance of suspension processes near the time of high tide and slack water, which favors preservation of mud in high-tidal flats.

The idealized presentation (Figure 3-2) of a seaward-coarsening textural distribution on intertidal flats (Figures 3-20, 3-21) is not always identical to the North Sea (Van Straaten, 1961; Reineck, 1963; Evans, 1965), the Bay of Fundy (Klein, 1963, 1970), or the northwest coast of Australia (Semeniuk, 1981). Instead, muddy intertidal flats from high to low water are common. Wells and Coleman (1981a,b) and Wells *et al.* (1980) have demonstrated that the intertidal zone off the coast of Surinam is mud-dominated, with almost no sand occurring at all. High sediment yields there favor intertidal flat development, but the sediment yield comes mostly from the Amazon River and the continental shelf off Surinam, both of which provide sediment consisting of silt and clay. The intertidal flats of New Hampshire also comprise a mud-dominated system with almost no sand present (Anderson *et al.*, 1981). Much of the sand there appears to have been confined to estuaries or transported offshore. Broad sandy intertidal flats are also known, and the writer has observed such a broad intertidal sand flat in Swansea Bay (Wales), just east of Swansea (see Figure 3-12). In

Turnagain Arm (Alaska). Ovenshine *et al.* (1975, 1976) reported muddy intertidal flats separated from large intertidal sand bodies by a channel, although some seaward coarsening of sediment sizes in the intertidal zones there was observed.

Development of Intertidal Flat Sequences

Progradation of intertidal flats in a seaward direction depends largely on a moderate to high sediment yield. Many large tidal flats are known where they adjoin point sources (most riverine) or line sources (continental shelves) where a large volume of sediment is available. Thus, along the coast of western Korea, extensive intertidal flats occur with widths ranging from 5 to 25 km (Wells and Huh, 1979) because of sediment yield from the Hwang Ho and Yangtze Rivers of China and smaller rivers in Korea (Chung and Park, 1977; Wells *et al.*, in press). The intertidal flats of northwest Europe appear to owe their sediment availability to the high yields of the Rhine River and associated coastal current transport systems.

Rapid rates of intertidal flat progradation are documented directly from only two locations. Along the coast of northern France, LeFournier and Friedman (1974) determined an intertidal flat progradation rate of 1 km per century, based on survey records of several centuries. Along the coast of Turnagain Arm, coastal subsidence of 1.5–2.0 m occurred following the 1964 Alaska earthqake. Since the earthquake, an additional 2.0-m thick sequence of intertidal flat sediments has prograded over a depositional zone that is 1.8 km wide during the succeeding decade (Table 3-1). The high sediment yield comes from the melting of the Portage Glacier.

Along the coast of The Netherlands, the Holocene stratigraphy is well known (DeJong, 1965; Hageman, 1972), and within it, four intertidal flat successions are preserved, each separated by a ravinement surface associated with a transgression. The horizontal distance and the time interval of successive progradational events are also known and indicate a progradational rate that averages 4.9 m/yr. In comparison to northwestern France and Alaska, these rates fall midway between low (1 m/yr) for France, to extreme (12 m/yr) for Alaska (Table 3-1).

The consequences of progradation of tidal flats in the North Sea of western Europe is not only to shift shorelines in a seaward direction and thus displace laterally the components of intertidal flats, but also to generate a progradational vertical sequence of all components of the intertidal zone distributed vertically in a consistent and predictable way (Klein, 1971, 1972). When progradation occurs, each of the intertidal flat subenvironments oversteps the adjoining seaward-most subenvironment. Thus, high-tidal flat muds are observed to prograde over mid-flat interbedded mixed lithologies of sand and mud, which in turn prograde laterally over low-tidal flat sands. The supratidal zone will, in turn, prograde over the high-tidal flat. Continued

Table 1. Holocene rates of intertidal flat progradation

Location	Rate	Reference
The Netherlands	4.9 m/yr	Hageman, 1972
Northwest France	1.0 m/yr	DeJong, 1965 LeFournier and Friedman, 1974
Turnagain Arm, Alaska	12.0 m/yr	Ovenshine et al., 1975, 1976

progradation will generate a vertical sequence that fines upward (Figure 3-2). From the base upward, this vertical sequence (Klein, 1971, 1972) consists of lower intertidal flat sands, overlain by mid-flat interbedded sands and muds, and high-tidal flat muds. Such sequences may be overlain by supratidal marshes, mangrove roots, or evaporites (cf. Semeniuk, 1981). The thickness of these sequences (see Figure 3-2) coincides with Holocene tidal range (Figure 3-2; Klein, 1971, 1972) and may be used to approximate paleotidal range (see Klein, 1977a). Such a sequence fits the North Sea very closely. The nature of sequences in muddy intertidal flats, sand intertidal flats, or silty intertidal flats is less well known. In Turnagain Arm, Ovenshine et al. (1975, 1976) demonstrated that although the lithologies differ, a fining-upward grain-size trend was observed, fitting the generation zonation of depositional processes.

Intertidal Sand Bodies

The intertidal sand body system tends to occur along mesotidal and macrotidal coasts and also within some estuarine complexes (see Chapter 7). Some of these intertidal sand bodies are isolated from associated intertidal flats such as in parts of the Bay of Fundy (Klein, 1970; Knight, 1980) and in Turnagain Arm (Ovenshine et al., 1975, 1976), whereas in other locations they are welded onto intertidal flats and comprise part of the low-tidal flat zone. Examples of such cases include part of the Bay of Fundy (Klein, 1970; Knight and Dalrymple, 1975), the coast of western Korea (Chung and Park, 1977; Wells and Huh, 1979), and several locations in the North Sea.

The intertidal sand body systems have accumulated in areas of relatively strong bottom tidal current velocities, subjecting the seabed to a large rate of sand transport by bedload processes of deposition. Many of these current systems show evidence of the property of time-velocity asymmetry (Figures 3-3, 3-4) during which bottom current velocities are characterized by larger velocity magnitudes during a dominant phase of the tidal cycle and lower magnitudes during the subordinate phase of the tidal cycle (Klein,

1970; Boersma and Terwindt, 1981; Knight, 1980; Dalrymple *et al.*, 1978). Bedform migration and orientation of bedforms, topography of sand bodies, distribution of sediment facies, and dispersal patterns of sand are all controlled largely by time-velocity asymmetry on these sand bodies (Klein, 1970; Boersma and Terwindt, 1981; Balazs and Klein, 1972). The distance of migration, sand transport rates, and volume of preserved sediment also tend to be favored during times of spring tides in comparison to times of neap tides (Visser, 1980; Boersma and Terwindt, 1981), as discussed earlier (Figures 3-7, 3-8).

Intertidal sand bodies are linear in plan and asymmetrical in cross-section. Their orientation is parallel to the main flow of tidal currents. Sand on these sedimentary bodies ranges in size from very fine to very coarse, depending on availability and sorting. Bedforms developed on the sand bodies include current ripples, dunes, and sand waves (Figure 3-22). Migration of sand waves and dunes occurs only during times of dominant tidal flow (Klein and Whaley, 1972; Dalrymple *et al.*, 1978), and much of it is accomplished

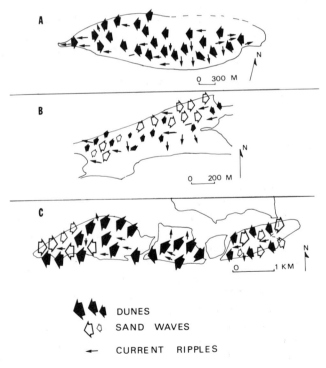

Figure 3-22. Orientation of dunes, sand waves, and current ripples on intertidal sand bars, Minas Basin, Bay of Fundy, Nova Scotia, Canada. (A) Big Bar, (B) Pinnacle Flats, (C) Economy Point (from Klein, 1970; reprinted with permission of the Society of Economic Paleontologists and Mineralogists).

during very short periods of time, perhaps as little as an hour. The orientation of internal cross-stratification and of dunes and sand waves is in agreement with sand body alignment and dominant flow direction of tidal currents (Figure 3-23). Thus, unimodal orientations of cross-strata are in agreement with dominant tidal current flow and sand body alignment. Such cross-stratification is truncated by reactivation surfaces (see earlier discussion on bedload processes). Bundles of unimodally cross-stratified zones are grouped laterally into thicker units and thinner units, representing spring and neap tide sedimentation, respectively (Boersma and Terwindt, 1981; see also Figures 3-7, 3-8).

The distribution of different zones of time-velocity asymmetry over intertidal sand bodies appears to be dependent in part on sand body

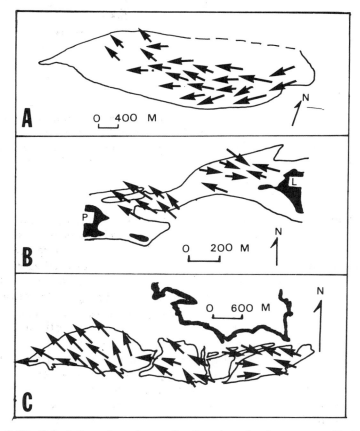

Figure 3-23. Orientation of maximum dip direction of cross-strata at Big Bar (A), Pinnacle Flats (B), and Economy Point (C) intertidal sand bodies, Minas Basin, Bay of Fundy, Canada (from Klein, 1970; republished with permission of the Society of Economic Paleontologists and Mineralogists).

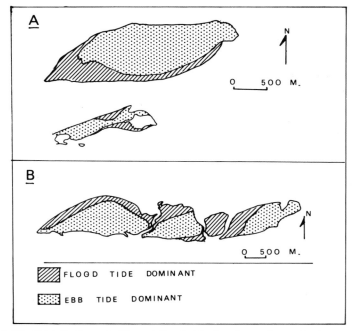

Figure 3-24. Distribution of ebb and flood tide dominant time-velocity asymmetry zones on three intertidal sand bodies, Minas Basin, Bay of Fundy, Nova Scotia, Canada. (A) Big Bar and Pinnacle flats, (B) Economy Point (from Klein, 1970; republished with permission of the Society of Economic Paleontologists and Mineralogists).

topography. As shown in Figure 3-24, from the Minas Basin, Bay of Fundy, relatively narrow and steep zones of intertidal sand bodies are dominated by tidal currents characterized by a flood-tide, time-velocity asymmetry, whereas gentler sloping surfaces, showing a wider area of exposure, are shaped and reworked by ebb-dominated tidal currents. Consequently, sediment is dispersed around the sand body through alternating flood-dominated and ebb-dominated time-velocity asymmetry zones (Klein, 1970, Figure 3-25). This elliptical dispersal pattern produces well-rounded sands on the sand bodies, as demonstrated by Balazs and Klein (1972).

Most of the Holocene intertidal sand bodies that have been described appear to be in areas of strong reworking (Klein, 1970) rather than in areas of active progradation. Some such sand bodies welded to the intertidal zone do comprise the lower parts (low-tidal flat) of fining-upward intertidal flat sequences (Klein, 1971; see also Figure 3-2). However, the nature of such sequences in macrotidal settings remains unknown simply because no areas of macrotidal settings that also are characterized by a progradational history have been documented. This situation exists also for intertidal sand bodies that are free-standing from intertidal flats.

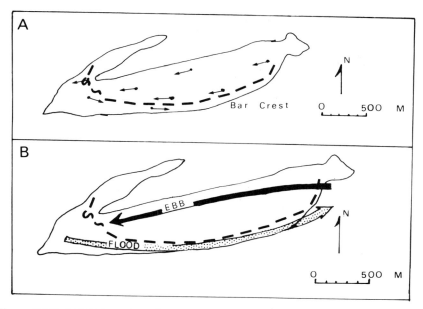

Figure 3-25. (A) Orientation of mean and maximum direction of sand grain dispersal (based on tracers), and (B) interpreted mode of grain dispersal, Big Bar, Minas Basin, Bay of Fundy, Canada (from Klein, 1970; republished with permission of the Society of Economic Paleontologists and Mineralogists).

Nevertheless, one hypothetical vertical sequence has been suggested assuming coastline progradation. It is based on Bay of Fundy observations by Knight and Dalrymple (1975) and is shown in Figure 3-26. In this hypothetical sequence, there is a vertical succession overlying subtidal sand, gravel, or bedrock overlain by an intertidal sand body. It in turn is overlain by a braided bar tidal channel deposit, which grades upward into intertidal mudflats, capped by supratidal salt marshes. Overall, this sequence contains a thicker section of sand than the fining-upward intertidal flat sequence shown in Figure 3-2 and includes braid-bar deposits. This intertidal sand body–macrotidal coastline sequence should be readily distinguishable from the normal intertidal flat sequence shown in Figure 3-2.

Summary

The nonvegetated intertidal zone can be subdivided into two sedimentary domains: (1) intertidal flats, and (2) intertidal sand bodies. Sediment deposition and transport across the intertidal flat setting is zoned shore-parallel with suspension-dominant deposition confined to high intertidal flat

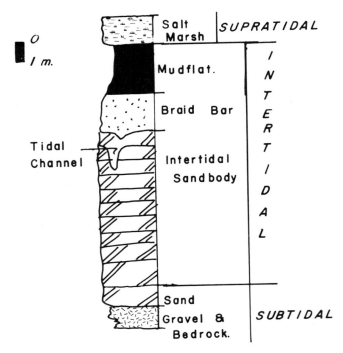

Figure 3-26. Hypothetical vertical sequence, macrotidal coastline, and intertidal sand body complex based on sediment distribution in Minas Basin, Bay of Fundy (redrawn after Knight and Dalrymple, 1975; reprinted by permission of International Human Resources Development Corporation from *Clastic Tidal Facies* by George deVries Klein).

zones, nearly equal periods of time, and intensity of alternating bedload and suspension deposition confined to the mid-tidal flat zone. Bedload–dominant transport and deposition is confined to the low-tidal flat zone. Coupling of these distinct modes of sediment transport and deposition with duration of periods of submergence favors deposition of muds in the high-tidal flats (shortest submergence associated with negligible velocities at high tide), interbedded mixed lithologies of mud and sand in the mid-flat zone, and sand in the low-tidal flat zone (longest period of submergence favoring bedload deposition). In areas where intertidal flats prograde, rates of coastline progradation have been documented ranging from 1 to 12 m/yr. Such progradation generates a fining-upward sequence that, from the base upward, develops low-tidal flat sands, overlain by mid-flat mixed lithologies of sand and mud, and capped by high-tidal flat muds. The thickness of the sequence coincides with mean tidal range.

The supratidal zone of deposition consists dominantly of marshes in temperate–humid areas, mangroves in semitropical to tropical areas, and salt flats in more arid domains. Exclusively muddy intertidal flats are known, and these appear to be dominated not only by mud suspension because of a high muddy sediment yield, but also by downslope movement by debris flow and slurry of water-saturated muds.

The intertidal sand body system is dominated by bedload deposition by tidal currents. The sand transport rate, time-velocity asymmetry, and bottom velocity variation between neap and spring tide controls the rate of bedform migration and depositional rate and volume of sedimentation. Thus, during spring tides on intertidal sand bodies, greater distances of migration are characteristic of bedforms, and these are enclosed by reactivation surfaces. Such bundles are, volumetrically, larger than similar bundles of sediments accumulating during neap tide phases of the lunar tidal cycle. These sand bodies are linear in plan also, and the topographic shape and associated time-velocity asymmetry distribution controls the orientation of bedforms, sediment facies, and grain dispersal on the sand bodies. Although no areas of coastal progradation that incorporate such sand bodies are known, a hypothetical progradational fining-upward sequence has been proposed consisting of a thick intertidal sand body at the base, overlain by braided stream channel deposits, and capped by a high-tidal flat, thin mud.

References

Allen, J. R. L., 1982. Mud drapes in sand-wave deposits: a physical model with application to the Folkestone Beds (Early Cretaceous, Southeast England). *Roy. Soc. London, Phil. Trans., Ser. A*, **306**, 291–345.

Allen, J. R. L., and Friend, P. F., 1976. Changes in intertidal dunes during two spring-neap cycles, Lifeboat Station Bank, Wells-next-the-Sea, Norfolk (England). *Sedimentology*, **23**, 329–347.

Anderson, F. E., Black, L., Watling, L. E., Mook, W., and Mayer, L. M., 1981. A temporal and spatial study of mudflat erosion and deposition. *J. Sed. Petrol.*, **51**, 729–736.

Bajard, J., 1966. Figures et structures sedimentaires dans la zone intertidale de la partie orientale de la Baie du Mont-Saint-Michel. *Rev. de Geog. Physique et de Geol. Dynamique*, **9**, 39–111.

Balazs, R. J., and Klein, G. deV., 1972. Roundness-mineralogical relations of some intertidal sands. *J. Sed. Petrol.*, **42**, 425–433.

Belderson, R. H., 1964. Holocene sedimentation in the western half of the Irish Sea. *Mar. Geol.*, 2, 147–163.

Belderson, R. H., Kenyon, N. H., Stride, A. H., and Stubbs, A. R., 1972. *Sonographs of the Sea Floor*. Elsevier, Amsterdam, 185 pp.

Boersma, J. R., and Terwindt, J. H. J., 1981. Neap-spring tide sequences of intertidal shoal deposits in a mesotidal estuary. *Sedimentology*, **28**, 151–170.

Boggs, Jr., S. 1974. Sand-wave fields in Taiwan Strait. *Geology*, **2**, 251–253.

Boothroyd, J. C., and Hubbard, D. K., 1975. Genesis of bedforms in mesotidal estuarines. *In*: Cronin, L. E. (ed.), *Estuarine Research*, **2**, 217–234.

Chung, G. S., and Park, Y. A., 1977. Sedimentological properties of the Recent intertidal flat environment, southern Nam Yang Bay, west coast of Korea. *J. Oceanogr. Soc. Korea*, **13**, 9–18.

Coleman, J. M., 1980. *Deltas: Processes of Deposition and Models for Exploration*, 2d ed. Burgess Pub. Co., Minneapolis, 124 pp.

Coleman, J. M., Gagliano, S. M., and Smith, W. G., 1970. Sedimentation in a Malaysian high tide tropical delta. *In*: Morgan, J. P., and Shaver, R. H. (eds.), *Deltaic Sedimentation*. Soc. Econ. Paleont. Mineral. Spec. Publ. 15, pp. 185–197.

Collinson, J. D., 1969. Bedforms of the Tana River, Norway. *Geog. Annaler*, **52**, 31–56.

Cram, J., 1979. The influence of continental shelf width on tidal range: paleoceanographic implications. *J. Geol.*, **87**, 441–447.

Dalrymple, R. W., Knight, R. J., and Lambiase, J. J., 1978. Bedforms and their hydraulic stability relationships in a tidal environment, Bay of Fundy, Canada. *Nature*, **275**, 100–104.

Davies, J. L., 1964. A morphogenic approach to world shorelines. *Z. Geomorph.* **8**, 127*–142*.

DeJong, J. D., 1965. Quaternary sedimentation in the Netherlands. *In*: Wright, H. E., and Frey, D. G. (eds.), *International Studies on the Quaternary*. Geol. Soc. Amer. Spec. Paper 84, pp. 95–124.

DeRaaf, J. F. M., and Boersma, J. R., 1971. Tidal deposits and their sedimentary structures. *Geol. en Mijnbouw*, **50**, 479–504.

Emery, K. O., 1945. Entrapment of air in beach sand. *J. Sed. Petrol.*, **15**, 39–49.

Evans, G., 1965. Intertidal flat sediments and their environments of deposition in The Wash. *Geol. Soc. London Quart. J.*, **121**, 209–241.

Evans, G., 1975. Intertidal flat deposits of The Wash, western margin of the North Sea. *In*: Ginsburg, R. N. (ed.), *Tidal Deposits*. Springer-Verlag, New York, pp. 13–20.

Gellatly, D. C., 1970. Cross-bedded tidal megaripples from King Sound. *Sed. Geol.*, **4**, 185–192.

Groen, P., 1967. On the residual transport of suspended matter by an alternating tidal current. *Neth. J. Sea Res.*, **3**, 564–574.

Hageman, B. P., 1972. Sedimentation in the lowest part of river systems in relation to the post-glacial sea level rise in the Netherlands. *XXIV Int. Geol. Cong.*, **12**, 37–47.

Hantzschel, W., 1939. Tidal flat deposits (Wattenschlick). *In*: Trask, P. D. (ed.), *Recent Marine Sediments*. Soc. Econ. Paleont. Mineral., Tulsa, OK, 195–206.

Hantzschel, W., and Reineck, H. E., 1968. Faziesuntersuchungen im Hettangium von Helstedt (Niedersachsen). *Geol. Staatsinst. Hamburg Mitt.*, **37**, 5–39.

Hayes, M. O. (ed.), 1969. *Coastal environments of northeastern Massachusetts and New Hampshire. Eastern Section Guidebook*. Soc. Econ. Paleont. and Mineral., 462 pp.

Hayes, M. O., 1975. Morphology of sand accumulation in estuaries. *In*: Cronin, L. E. (ed.), *Estuarine Research*, **2**, 3–22.

Hayes, M. O., 1979. Barrier island morphology as a function of tidal and wave regime. *In*: Leatherman, S. P. (ed.), *Barrier Islands*. Academic Press, New York, pp. 1–27.

Houbolt, J. J. C., 1968. Recent sediments in the southern bight of the North Sea. *Geol. en Mijnbouw*, **47**, 245–273.

Keller, G. H., Lambert, D., Rowe, G., and Staresinic, N., 1973. Bottom currents in the Hudson Canyon. *Science*, **180**, 181–183.

Kellerhals, P., and Murray, J. W., 1969. Tidal flats at Boundary Bay, Fraser River Delta, British Columbia. *Canadian Petrol. Geol. Bull.*, **17**, 67–91.

Kenyon, N. H., Belderson, R. H., Stride, A. H., and Johnson, M. A., 1981. Offshore tidal sand banks as indicators of net sand transport and as potential deposits. *In*: Nio, S. D., Schuttenheim, R. T. E., and van Weering, T. C. E. (eds.), 1981. *Holocene Marine Sedimentation in the North Sea Basin*. Intl. Assoc. Sed. Spec. Pub. 5, pp. 257–268.

Kindle, E. M., 1917. Recent and fossil ripple marks. *Geol. Survey Canada Mus. Bull.*, **25**, 56 pp.

Klein, G. deV., 1963. Bay of Fundy intertidal zone sediments. *J. Sed. Petrol.*, **33**, 844–854.

Klein, G. deV., 1970. Depositional and dispersal dynamics of intertidal sand bars. *J. Sed. Petrol.*, **40**, 1095–1127.

Klein, G. deV., 1971. A sedimentary model for determining paleotidal range. *Geol. Soc. Amer. Bull.*, **82**, 2585–2592.

Klein, G. deV., 1972. Determination of paleotidal range in clastic sedimentary rocks: *XXIV Int. Geol. Cong.*, **6**, 397–405.

Klein, G. DeV. (ed.), 1976. *Holocene Tidal Sedimentation*. Dowden, Hutchinson and Ross, Inc., Stroudsburg, PA, 423 pp.

Klein, G. deV., 1977a. *Clastic Tidal Facies*. Continuing Educ. Publ. Co., Champaign, IL, 149 pp.

Klein, G. deV., 1977b. Tidal circulation model for deposition of clastic sediments in epeiric and mioclinal shelf seas. *Sed. Geol.*, **19**, 1–12.

Klein, G. deV., Park, Y. A., Chang, J. H., and Kim, C. S., 1982. Sedimentology of a sub-tidal, tide-dominated sand body in the Yellow Sea, southwest Korea. *Mar. Geol.*, **50**, 221–240.

Klein, G. deV., and Ryer, T. A., 1978. Tidal circulation patterns in Precambrain, Paleozoic and Cretaceous epeiric and mioclinal shelf seas. *Geol. Soc. Amer. Bull.*, **89**, 1050–1058.

Klein, G. deV., and Sanders, J. E., 1964. Comparison of sediments in tidal flats in the Bay of Fundy and the Dutch Wadden Zee. *J. Sed. Petrol.*, **34**, 18–24.

Klein, G. deV., and Whaley, M. L., 1972. Hydraulic parameters controlling bedform migration on an intertidal sand body. *Geol. Soc. Amer. Bull.*, **83**, 3465–3470.

Knight, R. J., 1980. Linear sand bar development and tidal current flow in Cobequid Bay, Bay of Fundy, Nova Scotia. *In*: McCann, S. B. (ed.), *The Coastline of Canada*. Geol. Survey of Canada Paper 80-10: 123–152.

Knight, R. J., and Dalrymple, R. W., 1975. Intertidal sediments from the south shore of Cobequid Bay, Bay of Fundy, Nova Scotia, Canada: *In*: Ginsburg, R. N. (ed.), *Tidal Deposits*. Springer-Verlag, New York, pp. 47–56.

Lambiase, J. J., 1980. Sediment dynamics in the macrotidal Avon River estuary, Bay of Fundy, Nova Scotia. *Can. J. Earth Sci.*, **17**, 1628–1641.

Larsonnieur, C., 1975. Tidal deposits, Mont-Saint-Michel Bay, France. *In*: Ginsburg, R. N. (ed.), *Tidal Deposits*. Springer-Verlag, New York, pp. 21–30.

LeFournier, J., and Friedman, G. M., 1974. Rate of lateral migration of adjoining sea-margin sedimentary environments shown by historical records, Authie Bay, France. *Geology*, **2**, 497–498.

Lonsdale, P., and Malfait, B., 1974. Abyssal dunes of foraminiferal sand on the Carnegie Ridge. *Geol. Soc. Amer. Bull.*, 85, 1697–1712.

Lonsdale, P., Normark, W. R., and Newman, W. A., 1972. Sedimentation and erosion on Horizon Guyot. *Geol. Soc. Amer. Bull.*, **83**, 289–316.

Macar, P., and Ek, C., 1965. Un curieux phenomene d'erosion fammeinnienne: Le "Pains de gres" de Chambralles (Ardenne, Belge). *Sedimentology*, **4**, 53–64.

McCave, I. N., 1970. Deposition of fine-grained suspended sediment from tidal currents. *J. Geophys. Res.*, **75**, 4151–4159.

McCave, I. N., and Geiser, A. C., 1979. Megaripples, ridges and runnels on intertidal flats of The Wash, England. *Sedimentology*, **26**, 353–369.

McCave, I. N., and Langhorne, D. N., 1982. Sand waves and sediment transport around the end of a tidal sand bank. *Sedimentology*, **29**, 95–110.

McMullen, R. M., 1964. *Modern sedimentation in the Mawddach Estuary, Barmouth, North Wales*. Unpubl. Ph.D. Dissert., Univ. of Reading (UK), 399 pp.

Off, T., 1963. Rhythmic linear sand bodies caused by tidal currents. *Amer. Assoc. Petrol. Geol. Bull.*, **47**, 324–341.

Oomkens, E., 1974. Lithofacies relations in the Late Quaternary Niger Delta Complex. *Sedimentology*, **21**, 195–221.

Ovenshine, A. T., Bartsch-Winkler, S. R., O'Brien, N. R., and Lawson, D. E., 1975. Sediments of the high tidal range environment of Upper Turnagain Arm, Alaska. *In: Recent and Ancient Sedimentary Environments in Alaska*. Alaska Geol. Soc., pp. 1–40.

Ovenshine, A. T., Lawson, D. E., and Bartsch-Winkler, S. R., 1976. The Placer River Silt—an intertidal deposit caused by the 1964 Alaska Earthquake. *U.S. Geol. Survey J. Res.*, **4**, 151–162.

Pestrong, R., 1972. Tidal flat sedimentation at Cooley Landing, southwest San Francisco Bay. *Sed. Geol.*, **8**, 251–288.

Postma, H., 1954. *Hydrography of the Dutch Wadden Sea*. Unpubl. Ph.D. Dissert., Univ. of Groningen. 106 pp.

Postma, H., 1961. Transport and accumulation of suspended matter in the Dutch Wadden Sea. *Neth. J. Sea Res.*, **1**, 148–190.

Reineck, H. E., 1963. Sedimentgefuge im Bereich der Sudliche Nordsee. *Abhandl. Sencken. Nat. Gesell.*, **505**, 1–138.

Reineck, H. E., 1967. Layered sediments of tidal flats, beaches and shelf bottoms of the North Sea. *In*: Lauff, G. H. (ed.), *Estuaries*. Amer. Assoc. Adv. Sci. Spec. Publ. 83, pp. 191–206.

Reineck, H. E., 1972. Tidal flats. *In*: Rigby, J. K., and Hamblin, W. K., (eds.), *Recognition of Ancient Sedimentary Environments*. Soc. Econ. Paleont. Mineral. Spec. Publ. 16, pp. 146–159.

Reineck, H. E. and Singh, I. B., 1973. *Depositional Sedimentary Environments*. Springer-Verlag, New York, 439 pp.

Reineck, H. E., and Singh, I. B., 1980. *Depositional Sedimentary Environments*, 2nd ed. Springer-Verlag, New York, 549 pp.

Reineck, H. E., and Wunderlich, F., 1968a. Classification and origin of flaser and lenticular bedding. *Sedimentology*, **11**, 99–104.

Reineck, H. E., and Wunderlich, F., 1968b. Zeitmessungen und Bezeitenschichten. *Natur und Museum*, **97**, 193–197.

Rhoads, D. C., 1967. Biogenic reworking of intertidal and subtidal sediments in Barnstable Harbor and Buzzards Bay, Massachusetts. *J. Geol.*, **75**, 461–476.

Semeniuk, V., 1981. Sedimentology and the stratigraphic sequence of a tropical tidal flat, northwestern Australia. *Sed. Geol.*, **29**, 195–221.

Shepard, F. P., Dill, R. F., and Von Rad, U., 1969. Physiography and sedimentary processes of LaJolla submarine fan and fan valley, California. *Amer. Assoc. Petrol. Geol. Bull.*, **53**, 420.

Shepard, F. P., and Marshall, N. F., 1973. Currents along floors of submarine canyons. *Amer. Assoc. Petrol. Geol. Bull.*, **57**, 244–264.

Shinn, E. A., 1968. Practical significance of birds-eye structures in carbonate rocks. *J. Sed. Petrol.*, **38**, 215–223.

Stewart, Jr., H. B., 1956. Contorted sediments in modern coastal lagoon explained by laboratory experiments. *Amer. Assoc. Petrol. Geol. Bull.*, **40**, 153–161.

Stride, A. H., 1963. Current-swept sea floors near the southern half of Great Britain. *Geol. Soc. London Quart. J.*, **119**, 175–197.

Stride, A. H., 1982. *Offshore Tidal Sands*. Chapman and Hall, London, 222 pp.

Thompson, R. W., 1968. *Tidal flat sedimentation on the Colorado River Delta, northwest Gulf of California*. Geol. Soc. Amer. Memoir 107, 133 pp.

Van Straaten, L. M. J. U., 1952. Biogene textures and the formation of shell beds in the Dutch Wadden Sea. *Kininkl. Nederlandse Akad. Wetensch. Proc., Ser. B.*, **55**, 500–516.

Van Straaten, L. M. J. U., 1953. Megaripples in the Dutch Wadden Sea and in the Basin of Archachon (France). *Geol. en Mijnbouw*, **15**, 1–11.

Van Straaten, L. M. J. U., 1954. Sedimentology of recent tidal flat deposits and the Psammites du Condroz. *Geol. en Mijnbouw*, **16**, 25–47.

Van Straaten, L. M. J. U., 1959. Minor structures of some Recent littoral and neritic sediments. *Geol. en Mijnbouw*, **21**, 197–216.

Van Straaten, L. M. J. U., 1961. Sedimentation in tidal flat areas. *J. Alberta Soc. Petrol. Geol.*, **9**, 203–226.

Van Straaten, L. M. J. U., and Kuenen, P.D., 1957. Accumulation of fine-grained sediments in the Dutch Wadden Sea. *Geol. en Mijnbouw*, **19**, 329–354.

Visser, M. J., 1980. Neap-spring cycles reflected in Holocene subtidal large-scale bedform deposits: a preliminary note. *Geology*, **8**, 543–546.

Wells, J. T., and Coleman, J. M., 1981a. Periodic mudflat progradation, northeastern coast of South America: a hypothesis. *J. Sed. Petrol.*, **51**, 1069–1075.

Wells, J. T., and Coleman, J. M., 1981b. Physical processes and fine-grained sediment dynamics, coast of Surinam, South America: *J. Sed. Petrol.*, **51**, 1053–1068.

Wells, J. T., and Huh, O. L., 1979. Tidal flat muds in the Republic of Korea: Chinhae to Inchon. *Office of Nav. Res. Sci. Bull.*, **4**, 21–30.

Wells, J. T., Huh, O. L., and Park, Y. A., in press. *Dispersal of Silts and Clays by Winter Monsoon Surges in the Southeastern Yellow Sea*. Proc. U.S.-PRC Symposium on Shelf Sedimentation Dynamics.

Wells, J. T., Prior, D. B., and Coleman, J. M., 1980. Flowslides in muds on extremely low angle tidal flats, northeastern South America. *Geology*, **8**, 272–275.

4

Coastal Salt Marshes*

Robert W. Frey and Paul B. Basan

Introduction

Salt marshes, which represent the final stage in the leveling of marine delta plains or the filling of depressions, embayments, and other irregularities along coasts, are to some extent a measure of coastal stability or equilibrium. The overall sedimentary sequence is therefore a potential record of coastal history; it may reveal complete successions from original estuary, delta, lagoon, or bay floors to the highest intertidal flat, including lateral variations in contemporaneous facies or subfacies. Associated mineral suites are equally important indicators of both sources and possible recycling of coastal sediments. As habitable dwelling space for numerous organisms, some of which are uniquely adapted to stressful conditions, salt marsh substrates record many details of significance in paleoecology, ichnology, and environmental reconstruction.

The recognition of ancient marsh deposits is important for the sake of documenting not only a major depositional environment but also that of key phases in marine transgressions or regressions. Ancient coastal systems, in addition, contain facies that provide excellent reservoirs for hydrocarbons. In that context, the recognition of marsh facies may help indicate both the kinds of reservoirs to be expected and the spatial distribution of adjacent depositional environments. The marshes associated with a particular coastal system, whether barrier islands, saline to brackish estuaries or deltas, embayments, or "zero-energy" coasts, can be of fundamental importance in the search for new oil and gas reserves and probably for cerain coals as well.

*Contribution Number 499, University of Georgia Marine Institute, Sapelo Island, Georgia.

Environmental geology and biology are further concerns. In a world where human population expansion, residential and commercial development, and industrial pollution threaten our coastal systems, the highly sensitive marsh zone frequently is endangered. Many marshes occur in positions that help buffer the force of hurricanes before they strike the populous coastal mainland, for example. What happens when marshes are destroyed by dredging or by artificial infill, or when pollution defoliates the marshes and leaves them barren?*

Marshes provide nurseries for important fishery species, such as shrimp and fish, and are a major link in the food web of nearly all adjacent coastal animals. What happens to these organisms when the marshes are destroyed, and to the persons employed in coastal fisheries? Similarly, marshes and their associated estuarine or lagoonal substrates ordinarily are replete with incompletely oxidized organic matter. What happens when mining or dredging operations introduce this reduced matter into normally oxygenated waters?

The search for an understanding of marshes and marsh processes thus is not merely an academic exercise in science; instead, such studies have important, far-reaching implications for the quality of our own lives. This chapter, therefore, is meant not only to introduce students to an integral part of the interesting, valuable, and at times fragile coastal system, but also to encourage and stimulate research in an area where most of the work remains unfinished. Indeed, some of it has hardly begun.

The State of the Art

An understanding of complex depositional systems, such as those along coasts, depends upon detailed information from all environments that make up the systems. Salt marshes are one of these parts. Yet few marshes, anywhere, have been studied sufficiently from a geologic point of view. In fact, although interest currently is accelerating, salt marshes remain the least studied of all intertidal environments.

With the possible exception of coral reefs and similar ecosystems, salt marshes exhibit the most striking interplay between chemical, hydrological, geological, and biological processes found in any marine environment. In-depth studies of salt marshes thus require at least some consideration for all these facets.

Extremely few marshes have been studied chemically or hydrographically, and no general summaries are available. Biologically, the best syntheses are

---

*Marshes in San Francisco Bay already have been reduced by 60%, and those in San Diego Bay by 85% (Macdonald, 1977b); numerous other examples could be listed (Cooper, 1974).

Figure 4-1. Salt marsh along Blackbeard Creek, Sapelo Island, Georgia (see Figure 4-3). Person stands on levee marsh; another levee marsh occurs at upper right, along course of minor tidal creek. Between person and forest is an expanse of meadow marsh. Both tall and short grasses are ecophenotypes of *Spartina alterniflora*.

those by Reimold and Queen (1974), Chapman (1974, 1976, 1977), and Pomeroy and Wiegert (1981). Geologically, an approach such as that employed by Coleman and Wright (1975) and Coleman (1976) in the study of deltas still is needed, to discern characteristic elements of marshes that (1) occur in different climates, (2) are subjected to different tidal ranges and energy regimes, and (3) are developed on different terranes or have different sediment sources. Marsh elements common to these various areas, in turn, could be translated into detailed spatial and temporal models for this part of the coastal system. At present, however, any summary discussion of marshes must be viewed more as an introduction to the widespread literature than as a well of knowledge on the essence of salt marshes.

Our experience with marshes is limited mostly to the Georgia coast, although we have observed others in Louisiana, Florida, the Bahamas, the Carolinas, Massachusetts, the Bay of Fundy, and California. Consequently, in this chapter Georgia marshes (Figure 4-1) are extensively used as examples. Similarly, literature review was confined mainly to the marshes of North America; those of other continents are comparable in some respects but quite different in others (Chapman, 1977).

Finally, although mangrove stands are lumped together with salt marshes by some workers, the two are fundamentally different in many respects. Mangroves therefore are excluded from this discussion.

Definition and Occurrences of Coastal Salt Marshes

Salt marshes occur along the coasts of most continents and typically occupy zones that border either strictly freshwater environments (Kolb and van Lopik, 1966) or marine to brackish bays and estuaries (Phleger, 1969a); where low-energy conditions persist, as along the "zero-energy" coast of Florida (Tanner, 1960), marshes may front the open sea. All such marshes are defined as well-vegetated saline intertidal flats (Cooper, 1974; Basan and Frey, 1977). Implicit in the definition is colonization by and ecologic successions among various halophytes.* Specifically excluded are unvegetated intertidal areas, such as ordinary mud and sand flats, as well as subtidal vegetated substrates, such as *Zostera* and *Thalassia* beds.

Equally important, salt marshes thus defined consist only of a veneer of sediments deposited intertidally; underlying low-intertidal to subtidal sediments (e.g., mud flat, lagoon, or estuary) are excluded, although much of the contemporary literature fails to acknowledge this distinction. Of course, subsidence or sea level rise may yield a final marsh sequence much thicker than the original veneer (Bloom, 1964; Redfield, 1972, Figure 7). Nevertheless, the sequence represents sedimentation high in the intertidal zone.

Although the general characteristics of salt marshes are somewhat similar on a world-wide basis, several individual marsh types may be distinguished. Differences between them are related chiefly to one or more of the following variables: (1) character and diversity of the indigenous flora; (2) effects of climatic, hydrographic, and edaphic factors upon this flora; (3) availability, composition, mode of deposition, and compaction of sediments, both organic and inorganic; (4) organism-substrate interrelationships, including burrowing animals and the prowess of plants in affecting marsh growth; (5) topography and areal extent of the depositional surface; (6) range of tides; (7) wave and current energy; and (8) tectonic and eustatic stability of the coastal area.

Other than arctic marshes, which have been studied very little, three major salt marsh groups commonly are recognized in North America (Figure 4-2; Chapman, 1974; Cooper, 1974): (1) Bay of Fundy and New England marshes, (2) Atlantic and Gulf coastal plain marshes, and (3) Pacific marshes. The geologic setting for east and west coast marshes is strikingly different. By recent estimates (Inman and Nordstrom, 1971; Gross, 1972), 80 to 90% of the Atlantic–Gulf coast is comprised of lagoons, estuaries, and deltas suitable for marsh development, whereas only 10 to 20% of the Pacific coast is suitable.

---

*The presence of intertidal halophytes—salt-tolerant plants (Reimold and Queen, 1974)—is a necessary part of the definition. As pointed out long ago by Johnson (1925), the tidal influence commonly extends much farther up estuaries than does the salinity influence; hence, some freshwater marshes experience daily changes in water levels because of tidal ebb and flood but remain beyond reach of the salt wedge itself.

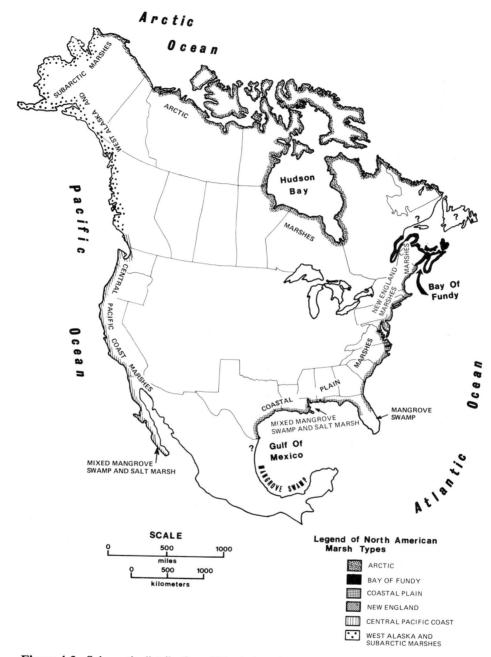

Figure 4-2. Schematic distribution of North American coastal salt marshes (adapted from Hoese, 1967; Chapman, 1974; Macdonald, 1977a).

The Pacific coast is active tectonically, and mountain building and coastal uplift have severely limited the amount of space available for marshes (Macdonald, 1977a,b). Pacific areas best suited for marsh growth, at least physiographically, include the coastal plain of the Alaskan arctic slope and that of Baja California below about 29°N latitude. Elsewhere on this coast, marshes typically are restricted to relatively narrow fringes around the few small protected bays and lagoons scattered widely along reentrants of the rocky shore, or in estuaries; north of Puget Sound the coast is even more irregular because of intricate networks of fjords and coastal archipelagos. Sedimentologically, these marshes are not unlike some in the Atlantic–Gulf coastal plain, although floristically, they are distinct from other North American marsh types.

Unlike individual Pacific marshes, extremely few of which attain sizes of more than a few thousand acres, the Atlantic–Gulf coastal plain marshes are essentially continuous over many broad regions of the coastline, are abundant locally from Texas to New York (Figure 4-2), and originally covered 5 million acres (Hoese, 1967). On the Atlantic coast, the largest marsh areas occur between Cedar Island (South Carolina) and Cumberland Island (Georgia) (Hayden and Dolan, 1979). Georgia alone has some 358,200 acres (Linton, 1969; Reimold, 1977), although this area is dwarfed by the saline wetlands of Louisiana (Day et al., 1973). Except for major estuaries, scattered sedimentary islands, and such features as Cape Cod, the New England—Canadian rocky coasts are more like those of the Pacific coast.

These marsh areas (Figure 4-2) presently are defined mainly on the basis of characteristic plant associations (Chapman, 1974; Cooper, 1974). Further differentiation of North American marshes may prove desirable in the future, particularly if geologic criteria are integrated with the floristic ones. West coast marshes are by no means homogeneous from arctic Alaska to arid Mexico (Macdonald, 1977a), and marshes along the St. Lawrence estuary (Dionne, 1972) and the Gaspé peninsula differ from ones in the Bay of Fundy; these, in turn, differ from the peat marshes of New England.

As a final example, the comparatively firm marsh substrates of the Atlantic–Gulf coastal plain differ from the "flotants" of the Mississippi delta. Such marshes consist of a mere mat of vegetation floating upon an organic ooze, grading to clay with depth (Mackin, 1962; Kolb and van Lopik, 1966). Some of these marshes literally flounder and drown in the ooze.

Environmental Conditions and Marsh Zonation

General Characteristics

Regional Settings

Regional temperature data indicate that marsh sites along the Pacific coast generally are warmer and more equable than sites at similar latitudes on the

Atlantic coast (Macdonald and Barbour, 1974, Figure 3). Although the Pacific coast extends from arctic tundra to the subtropical fringe of the Sonoran Desert, the maritime influence is more stable than the terrestrial one; the Aleutians, southern Alaska, and British Columbia are swept by the relatively warm North Pacific current, whereas the southern coastline is bathed by the relatively cold California current.

Comparable data on mean annual precipitation (Macdonald and Barbour, 1974, Figure 2) reveal a general, progressive, latitudinal increase in precipitation along the Atlantic seaboard, from a low of about 25 cm on Baffin Island to a high of 150 cm at Miami, Florida. The Alaskan arctic slope is drier than Baffin Island, yet precipitation increases spectacularly along the Gulf of Alaska (as much as 330 cm); annual precipitation remains high, although variable, from there to north-central California, then falls dramatically from that point southward, reaching a low of about 5 cm in Baja California (see also Macdonald, 1977a, Figure 8.2).

Marsh Conditions

Salt marshes are a tide-stressed environment where numerous variables exert strong selective control upon the kinds and abundance of organisms present. The diversity of marine species is low relative to that of adjacent aquatic environments because of these stressful conditions. However, some organisms may be renewed or compensated by tidal flux (Odum, 1974), and certain populations are dense.

Ecologic conditions in salt marshes grade from nearly marine to nearly terrestrial. Hence, a gradual or abrupt change in these conditions causes a corresponding change, or zonation, of organisms and their habitats. These gradients in ecologic conditions affect both plants (Reimold and Queen, 1974; Chapman, 1977) and animals (Bradshaw, 1968; Phleger, 1970; Day et al., 1973; Kraeuter and Wolf, 1974; Shanholtzer, 1974; Daiber, 1977; Macdonald, 1977a).

Obvious environmental factors that help govern the density and distribution of organisms within salt marshes include energy flow and nutrient budgets; inter- and intraspecific relationships among organisms and their potential for dispersal; area and duration of subaerial exposure caused by tidal fluctuations and topography; wave or current energy; physiochemical conditions, such as Eh, pH, salinity, and temperature; and sedimentological factors, such as sediment composition and grain size, and substrate coherence or consistency. Tide-dependent factors, such as tidal range and length of time the substrate is under water, seem to be most critical. Other important parameters, such as current velocity, salinity, and nutrient levels, also are related directly to the tidal regime and water quality.

Furthermore, in consequence of plant successions, sediment accumulation, and marsh expansion, some parts of marshes are older and better developed than other parts. Such marshes ordinarily may be divided into two

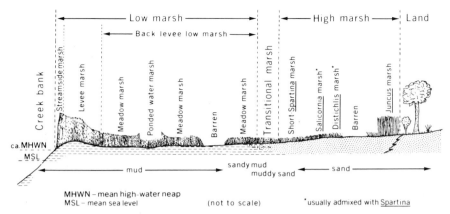

Figure 4-3. Diagrammatic classification of Georgia salt marshes. *Spartina alterniflora* is characteristic of low and transitional marshes (adapted from Edwards and Frey, 1977).

fundamental zones (Figure 4-3; Ranwell, 1972): low marsh and high marsh. Low marshes are younger, lower topographically, and more subject to adjacent estuarine or marine conditions. High marshes are older, occupy a higher topographic position, and are more influenced by terrestrial conditions. Low marshes, on the average, are subjected to more than 360 tidal submersions per annum; high marshes undergo substantially fewer submersions and may remain continuously exposed to air for periods exceeding 10 days, between tidal inundations.

Nevertheless, boundaries for these zones, or their relationship to a given tidal datum (Figure 4-4), may differ from one coast to another (Reimold, 1977, Table 7.3). The low marsh corresponds to the upper-middle intertidal zone, beginning between half-tide level and mean high-water neap in Massachusetts (Redfield, 1972) and about mean high-water neap on the Pacific coast (Phleger, 1969a; Macdonald, 1977a,b). Similarly, high marshes begin at about mean high water in Massachusetts and about mean higher high water on the Pacific coast. High marshes therefore occupy the uppermost part of the intertidal zone, terminating landward at, or slightly above, mean high-water spring. The relationships of various plant zones to these tidal datum levels were reviewed by Fornes and Reimold (1973).*

Differences in marsh boundaries with respect to a given tidal datum seem to be related primarily to two factors: tidal regularity and substrate composition. Atlantic tides are regular and nearly equal in semidiurnal range (Gross, 1972); hence, many Atlantic marshes have boundaries similar to

---

*The legal and commercial implications of these zones, and the problems associated with them, are formidable (Fornes and Reimold, 1973).

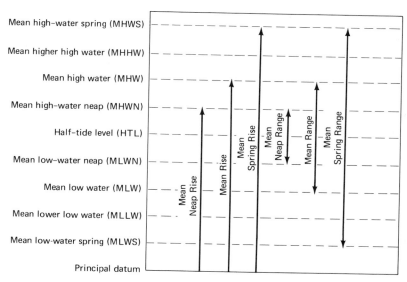

Figure 4-4. Tidal datum levels useful as reference planes in coastal hydrography. Mean high-water spring: average level of high waters occurring during spring tides. Mean higher high water: average level of higher high waters of each day. Mean high water: average level of all high waters. Mean high-water neap: average level of high waters occurring during neap tides. Half-tide level: level midway between mean high and mean low water. Mean low-water neap: average level of low waters occurring during neap tides. Mean low water: average level of all low waters. Mean lower low water: average level of lower low waters of each day. Mean low-water spring: average level of low waters occurring during spring tides. "Lower low" and "higher high" refer to coasts characterized by semidiurnal tides of unequal amplitude (adapted from Fornes and Reimold, 1973).

those in Massachusetts. Pacific tides are markedly unequal in semidiurnal range, apparently accounting for most of the difference between Pacific and Atlantic marsh zones. In contrast, Gulf Coast tides are irregular and of small amplitude; thus, the distinction between high and low marshes is less significant. Here, the general marsh surface approximates mean high water (Kolb and van Lopik, 1966).

In addition to the tidal regime, which probably exerts the main control upon marsh zones, the composition or consistency of the substrate is important on a local or regional basis. Generally, low-marsh plants tend to favor substrates containing abundant interstitial water, which is more prevalent in mud than in sand. High-marsh plants tend to favor drier substrates, such as permeable sands. In extremely muddy marshes, such as those in Georgia, the low marsh therefore may extend somewhat higher into the intertidal zone than it otherwise would, i.e., at a level approaching mean higher high water. The "flotants" of Mississippi delta marshes, which may subside because of the nearly liquid subsurface, are an even more striking example of the influence of substrate upon marsh development.

On most coasts, unvegetated tidal flats or tidal stream banks occupy the zone between mean low-water spring and the low-marsh fringe (Guilcher, 1963). Locally, this lower intertidal zone may be colonized by typically subtidal halophytes; but these areas are not considered as marshes *sensu stricto*. At the opposite extreme, the influence of marshes may extend to the lowermost supratidal zone (Figure 4-3). Here the boundaries may be extremely gradational or may be delimited by a "debris line" of tide-rafted plant litter and/or by a conspicuous change in slope. The transition from low to high marsh in some Pacific areas, especially Oregon, may be marked by a low, steep terrace (Macdonald, 1977a).

New England marshes are covered predominantly by high-marsh mead-ow*; low-marsh plants occur mostly along the borders of tidal creeks (Miller and Egler, 1950). From Chesapeake Bay southward, east coast marshes contain larger percentages of low marsh (Cooper, 1974). On the average, about 60% of the marsh area in Georgia is streamside–levee marsh and low-marsh meadow. In addition to the typical low and high areas, some southern California marshes exhibit a conspicuous middle-marsh zone (Warme, 1971; Macdonald, 1977b).

Low marshes and high marshes may be subdivided further (e.g., Figure 4-3). The smaller zones within low marshes are defined primarily by physiographic or geomorphic criteria, whereas those in high marshes are defined mainly by floristic criteria. Low-marsh zones may repeat in a rather complex mosaic (Figure 4-5) created by the repetition of individual sets of physical or ecological conditions. Tidal creeks, in general, are reference boundaries. Marshes that are intricately dissected by tidal creeks repeat marsh divisions behind each creek bank (Figures 4-1, 4-3, and 4-5). Because high marshes have few tidal creeks to replicate given sets of ecological or physical conditions, the zonation is rather straightforward. These zones ideally develop in a linear fashion, parallel to one another.

Marsh Maturation

Hypothetically, end-member marshes in an evolutionary sequence should be those in which the entire area is covered either by high marsh or low marsh. Coastal regimes seldom, if ever, are stable for periods sufficient to achieve such purity, however, and marshes normally consist of some mixture of low- and high-marsh areas. Many variables influence the distribution of these two zones, but the primary factor is the amount of area subject to regular tidal inundation. In turn, this area is a product of topography, tidal range, and

---

*Not to be confused with the low-marsh meadow of southeastern marshes (Figure 4-3).

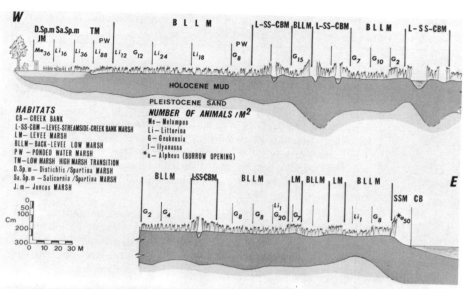

Figure 4-5. Actual sampling transect across part of marsh, showing repetition of habitats because of intervening tidal drainages (see Figures 4-3 and 4-6). Distribution of plants and selected animals also illustrated (from Basan and Frey, 1977).

length or duration of the sedimentation process, as well as the stability of the coastal area (from the standpoint of both crustal and eustatic stability and subsidence caused by compaction, soft-sediment deformation, etc.).

A related and important aspect of this "time stability" concept is: How much sediment is supplied, and how fast are the coastal bays, lagoons, estuaries, and interdistributary bays filled with sediment to form high intertidal flats (Rusnak, 1967)? Salt marshes normally develop contemporaneously or penecontemporaneously with the infilling process. Although many studies are based on gross coastal stratigraphy (Fischer, 1961; Kraft, 1971), we are not aware of any that truly document the finite differences in stratigraphy between the primary infill (bay, lagoon, estuary) and the ultimate marsh. Furthermore, marshes themselves represent a sedimentary succession, albeit subtle, from the lowest, initially colonized substrate, through various intermediate stages, to the highest, quasiterrestrial substrate (Basan and Frey, 1977). Studies of both aspects of salt marsh sequences are necessary to understand marsh processes.

The succession from low-marsh to high-marsh substrates has been observed in 500- to 1000-yr-old marsh muds cropping out on erosional beaches in South Carolina and Georgia (Frey and Basan, 1981). The respective marsh subenvironments (Figure 4-3) are analogous to a "facies tract" (or Walther's Law): the present lateral succession should also represent

Table 4-1. Idealized stages in marsh maturation, barrier islands of Georgia[a] (compare with Johnson, 1925; Chapman, 1974).

Stage	Characteristics
Youth	A. Substantially more than 50% of area consists of low marsh. In early youth, total area may consist of low marsh, vegetated exclusively by *Spartina alterniflora*, initiated either as small marsh islands along higher parts of tidal flats or as narrow fringing marshes around sound or estuary margins. Zonation patterns absent to very simple in early youth, giving way to a complex of repetitive low-marsh zones governed by the density and distribution of tidal drainages (e.g., Figure 4-5). High-marsh vegetation restricted mainly to terrestrial fringes.
	B. Drainage systems well developed. Pronounced meandering and crevasse splaying, with possible headward erosion of individual tributaries, during early youth. Intensity of these processes declines progressively during middle and late youth and the positions of drainage channels become correspondingly more stable.
	C. Relatively rapid sedimentation, especially during early youth. Marsh substrates actively accrete, both vertically and laterally, until lateral growth is inhibited by margins of sounds or estuaries (e.g., Figure 4-6). During middle and late youth, accretion is mostly vertical, and the rate of deposition decelerates as the marsh increasingly approaches an equilibrium among topography, tidal hydraulics, and sediment supply.
	D. Stratigraphic record consists predominantly of low-marsh environments, as modified by channel migrations. In many places the vertical sequence consists only of thin veneers of Holocene sediment spread over shallow basements of Pleistocene sand (e.g., Figure 4-5), rather than the fill of open lagoons or estuaries; these sand platforms are remnants of old barrier islands.
Maturity	A. Low- and high-marsh areas approximately equal in size (e.g., Figure 4-3). Low marsh and lower edge of high marsh vegetated by ecophenotypes of *Spartina alterniflora*. In early maturity the remaining high-marsh zones may consist of mixtures of *S. alterniflora, Salicornia* spp., or *Distichlis spicata*, followed by isolated stands of *Juncus roemerianus*, as is true in late youth, whereas in later maturity these plants typically occur either in mosaic clumps or in narrow, concentric zones.
	B. Good drainage system remains, especially in low-marsh areas; but many tidal creeks are partially or totally infilled in the high marsh. In late youth and early maturity much lateral erosion and rotational slumping occur along creek banks; yet erosion in one place tends to be compensated by deposition in another, so that little net difference in channels results.

Table 4-1. *(Continued.)*

Stage	Characteristics

C. Relatively slow deposition. Tidal sedimentation restricted mainly to low-marsh areas; extremely slow rates of deposition in the high marsh, except where barrier washovers occur or torrential rains sweep sands off adjacent Pleistocene or Holocene barrier island remnants. Where supplies of washover or terrestrial sediments are not readily available and tidal hydraulics prevent significant deposition of clays and silts, the transition from mature to old-age marshes may be exceedingly slow.

D. Stratigraphic records are variable. Those in low-marsh areas are similar to ones from youthful marshes, whereas ones in high-marsh areas may depict the succession from lowest to highest marsh, including numerous channel migrations and fills.

Old age   A. Substantially more than 50% of area consists of high marsh. In late old age, virtually all of the area may consist of high marsh, including the encroachment of quasiterrestrial or terrestrial vegetation upon the marsh surface. In early and middle old age the zonation consists of concentric bands of short *Spartina alterniflora*, *Salicornia* spp., *Distichlis spicata*, and *Juncus roemerianus*, followed by or admixed with such plants as *Sporobolus*, *Borrichia*, and *Batis*. Taller forms of *S. alterniflora* are restricted to the few drainage channels remaining.

B. Drainage mostly by surface runoff; most tidal channels are filled, and the marsh substrate is more or less uniform in elevation. Aeolian processes are correspondingly more important in the distribution and reworking of sediments.

C. Extremely slow rates of deposition. Tidal processes are largely ineffectual, and the transition to a terrestrial environment depends upon the availability of terrestrial sediments and mechanisms for their dispersal.

D. Stratigraphic records should show complete sequences, from underlying subtidal or low intertidal sediments, through the earliest marsh stages, to the oldest marsh or quasiterrestrial environments. Old marshes probably are not obtained without the deposition of washover or terrestrial sands; hence, the final record would be a coarsening-upward sequence.

[a]This model, which requires considerable testing, applies only to the more seaward marshes in Georgia, such as those associated with dissected barrier islands. More landward marshes, particularly those adjoining freshwater drainages, differ not only with respect to discrete modes of growth but also by plant zonations. In brackish marshes, for example, *Spartina alterniflora* may be largely or totally replaced by *S. cynosuroides*, and *Juncus roemerianus* may occupy most of the marsh area. (See also Howard and Frey, 1975a, p. 9; Dörjes and Howard, 1975, Figure 9).

the eventual vertical or laterovertical succession (Redfield, 1972, Figure 8).

This process, referred to as *marsh maturation*, is reflected by the stage of plant and animal succession (Gould and Morgan, 1962, Figure 3; Clarke and Hannon, 1969, Figure 1; Chapman, 1974; Macdonald, 1977b) and by the size, position, and differentiation of the marsh within the coastal complex. In geologic terms, marsh maturation is a by-product of simple sediment progradation whereby the high marsh has displaced the low marsh by both vertical and lateral succession. In spite of a variety of conditions that might induce submaturity in a marsh, a simple scale of relative maturation can be established (Table 4-1). A youthful marsh is one where low-marsh environments constitute most of the total area. In a mature marsh, the areas of high and low marsh are approximately equal. In an old marsh, the high-marsh environment comprises most of the total area. The emphasis here clearly is upon physiographic "stage" and not finite "age" of the marsh in question. Numerous Pacific coast marshes are no more than 100 to 200 years old (Macdonald and Barbour, 1974; Macdonald, 1977a), whereas thick peat accumulations at places in Massachusetts are approximately 4000 years old (Redfield, 1972); some may be more than 5500 years old (Kaye and Barghoorn, 1964).

In general, high rates of tidal deposition and consequent marsh expansion tend to favor low-marsh development whereas conditions approaching an equilibrium tend to favor high-marsh development, especially if growth of adjacent tidal flats is minimal (Macdonald, 1969, 1977b). However, rapid deposition of "colluvium" derived from adjacent terrestrial sources tends to favor high-marsh development, even in places where low marshes otherwise would predominate. Furthermore, a submature to supermature marsh (rather than an old marsh) may be the stable phase in areas where (1) the rate of deposition of terrigenous detritus is low; (2) tidal range and tidal energy are pronounced; (3) substrate accretion is affected mostly by tidal sedimentation (including physical and biogenic sediment trapping); and (4) the area available for marsh growth is restricted by the confines of bay or estuary margins (Figure 4-6).

Even within a youthful, mature, or old marsh, relative differences in maturity are apparent. The marshes shown in Figure 4-6 are entirely youthful, for example, yet the marsh in the center of the area exhibits a considerably more youthful drainage pattern than do the peripheral marshes (cf. Ragotzkie, 1959, Figure 7, Table 1).

Characteristic Plants

Vegetation is such a conspicuous and fundamental part of salt marshes, as much so as the sedimentological, mineralogical, and geochemical aspects, that geologists cannot hope to comprehend marshes and marsh processes

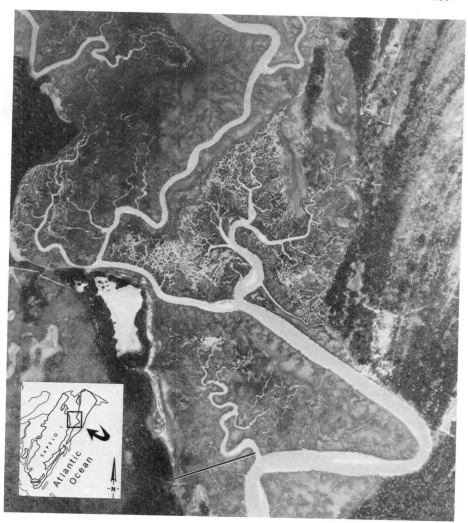

Figure 4-6. Tidal drainage system, Blackbeard Marsh, Georgia. The marsh complex, although extremely youthful in places, has filled in all intertidal areas between Sapelo Island (left; a Pleistocene barrier island remnant) and Blackbeard Island (right; a Holocene barrier remnant characterized by profusions of old beach ridges). Eroded barrier remnants are major sources of sand in both channel point bars and the marshes; all large bars are ebb oriented. Line indicates position of transect in Figure 4-5 (photo courtesy U.S. Navy).

Figure 4-7. Dense stand of cordgrass *Spartina alterniflora* along tidal creek bank, Sapelo Island. Adjoining small tidal bar colonized by polychaete *Diopatra cuprea*; its dwelling tubes commonly are aligned in rows perpendicular to bidirectional tidal flow, and they aid in sediment trapping and substrate stabilization.

without some basic understanding of the various floras. Moreover, the occurrence and distribution of vegetation in marshes provides part of the framework for characterizing and classifying subenvironments or habitats, both sedimentologically and biologically (Basan, 1979). A comprehensive list of Atlantic and Gulf marsh plants was compiled by Duncan (1974) and of Pacific marsh plants by Macdonald and Barbour (1974) and Macdonald (1977a).

As mentioned previously, the salt marsh vegetation of eastern Canada is generally similar to that of New England (Chapman, 1974, 1977). Along both the Atlantic and Gulf coasts of the United States, low marshes are characterized by different ecophenotypes (growth forms) of the smooth cordgrass,* *Spartina alterniflora* (Miller and Egler, 1950; Kurz and Wagner, 1957; Gould and Morgan, 1962; Adams, 1963; Kerwin and Pedigo, 1971; Redfield, 1972; Nixon and Oviatt, 1973; Gabriel and de la Cruz, 1974; Kirby and Gosselink, 1976; Reimold, 1977; Cleary *et al.*, 1979). Growth forms of this species are tallest and most luxuriant along the edge of tidal creeks (Figure 4-7; Valiela and Teal, 1974); the stems become progressively shorter toward the high marsh (Figures 4-1, 4-3, and 4-5). The main difference between New England and southern low marshes is the smaller areal extent of the former. Along brackish tidal streams in southeastern marshes, however, *S. alterniflora* is mostly replaced by the big cordgrass *S. cynosuroides*. Productivity of these plants in Georgia is indicated in Table 4-2 (cf. Turner, 1976).

*Common names of marsh halophytes used here are those recommended by Duncan (1974).

Table 4-2. Annual net[a] productivity of cordgrasses in Georgia salt marshes (see Figure 4-3) (adapted from Odum, 1973).

Plant species/environment	Dry weight	
	g/m^2	tons/acre
Spartina alterniflora:		
Streamside–levee marshes	4000	17.8
Meadow marshes	2300	10.2
Short-*Spartina* high marshes	750	3.3
Spartina cynosuroides:		
Brackish streamside–levee marshes	2000	8.9

[a]Organic matter available for consumption or export after plant metabolic requirements have been satisfied.

High-marsh floras of the Atlantic coast exhibit more diversity than do low-marsh floras; the former are characterized by a variety of grasses, succulents, rushes, and other plants adapted to more terrestrial-like habitats (Duncan, 1974). In Cape Cod marshes, exemplified by well-known ones at Barnstable Harbor (Redfield, 1972), the main high marsh is dominated by salt meadow cordgrass (*Spartina patens*) and spike grass (*Distichlis spicata*), inter-mingled with sea-lavender (*Limonium carolinianum*), seaside plantago (*Plantago juncoides*), aster (*Aster subulatus*), seaside goldenrod (*Solidago sempervirens*), salt-bush (*Atriplex patula*), sea-blite (*Suaeda maritima*), and glasswort (*Salicornia* sp.). At higher levels, especially along marsh margins, black-grass (*Juncus gerardi*) is common. Sampshire (*Salicornia europaea*) and glasswort (*S. bigelovii*) surround bare pannes. Where ground water emerges along marsh fringes, the regular halophytes are replaced by reed (*Phragmites communis*), saltmarsh bulrush (*Scirpus robustus*), and cat-tail (*Typha* sp.). Information on other U.S. north Atlantic marsh plants may be found in references cited by Miller and Egler (1950), Redfield (1972), and Nixon and Oviatt (1973).

High marshes along the U.S. south Altantic coast are dominated, in addition to dimunitive *Spartina alterniflora*, by glasswort (*Salicornia bigelovii*), sampshires (*S. europaea, S. virginica*), spike grass (*Distichlis spicata*), and marsh rush (*Juncus roemerianus*). In mature marshes the plants are distributed in zones that tend to repeat, in a landward direction, one behind the other (Figure 4-3), whereas in submature marshes they may be distributed in irregular mixtures or in mosaic patterns. Locally, other halophytes may be common on the terrestrial fringe; in Georgia, these especially include saltwort (*Batis maritima*), sea ox-eye (*Borrichia frutescens*), and Virginia dropseed (*Sporobolus virginicus*). Salt meadow cordgrass (*Spartina patens*) is present but is nowhere as abundant here as in marshes farther north. Other plants are more typical of the middle and upper reaches of riverine estuaries (Gallagher and Reimold, 1973; Dörjes and Howard, 1975). Information on high-marsh vegetation elsewhere along the U.S. south Atlantic and Gulf coasts may be found in references cited previously.

In south Florida and along parts of the Gulf coast, typical marsh halophytes become admixed with or replaced by mangroves (see Figure 4-10b; Rützler, 1969; Day et al., 1973; Cooper, 1974). Where abundant mangroves and salt marsh plants occur together, the mangroves typically lie seaward of the marsh plants because the former can root in deeper water. However, Egler (1952) suggested that in some places marsh halophytes may be pioneer species in the succession to mangrove stands.

Among ancient marshes, some of the above relationships may be reconstructed by means of peat petography (Cohen and Spackman, 1972; Allen and Cohen, 1977) or by pollen analyses (Sears, 1963; Meyerson, 1972). These subjects are beyond the scope of this chapter, however.

Marsh floras typically are more diverse on the Pacific coast than on the Atlantic coast, and the resulting zonations or successions are correspondingly more complex (Macdonald, 1977a, Figures 8.3–8.7). About 40 species are considered to be characteristic of marshes north of the Canadian border, whereas some 54 species are considered characteristic of marshes south of the border (Macdonald and Barbour, 1974, Tables 3–4). Of these halophytes, most fall into one of three fidelity patterns: (1) species occuring with great frequency at almost every locality examined within their known latitudinal range; (2) species occurring with high frequency but which may be conspicuously absent at a few specific localities; and (3) species presently known from relatively few but widely scattered sites, throughout their recorded ranges.

Latitudinal gradients in Pacific marsh plant distributions (phytogeographic provinces) are not pronounced, yet five climatic groups have been recognized (Macdonald, 1977a): (1) arctic, including the north and west coasts of Alaska; (2) subarctic, between Anchorage, Alaska, and the Queen Charlotte islands; (3) temperate, from British Columbia to central California; (4) dry Mediterranean, in southern California; and (5) arid, extending throughout Baja California, Mexico. Associated climatic factors were summarized by Macdonald (1977a, Figure 8.2), who suggested that productivity of Pacific marshes is generally comparable to that of Atlantic and Gulf marshes (Turner, 1976).

Interestingly, of the characteristic Pacific marsh plants listed by Macdonald and Barbour (1974) and Macdonald (1977a), five also are characteristic U.S. south Atlantic marsh plants: Batis maritima, Distichlis spicata, Salicornia bigelovii, S. europaea, and S. virginica. Their ecologic roles are not necessarily the same on the two coasts, nor for that matter at different sites along the Pacific coast. Distichlis spicata and S. europaea may be the pioneer species on certain northern California sand spits, for example, and farther south *D. spicata* may be restricted to dune vegetation. (*Distichlis spicata* also is common locally on Georgia coastal dunes.) The dominant Atlantic species, *Spartina alterniflora*, artificially introduced into Willapa Bay, Washington, has now successfully colonized sand flats up to 80 m beyond the fringes of the native marsh.

Characteristic Sediments

In a sense, salt marsh deposits may be termed either *soil* or *sediment* because of their many ambivalent characteristics (Edelman and van Staveren, 1958; Cotnoir, 1974). Marsh substrates commonly exhibit inherent pedologic structures (in addition to sedimentary structures) and identifiable soil types and horizons. The term *soil* also has some standing in engineering geology (Kolb and van Lopik, 1966). Nevertheless, *soil* also implies a direct genetic relationship with some underlying bedrock, regolith, or residuum from which it has been derived. *Sediment*, whether allochthonous or autochthonous, carries no connotation of an underlying parent material and therefore seems to be the preferable term.

Bouma (1963), Macdonald (1977b), and others have suggested that salt marsh substrates contain finer, more poorly sorted sediments than any other intertidal environment. Such generalities, however, tend to disregard the subfacies developed within salt marshes as well as the relationship between marsh substrates and those in closely adjacent habitats. The texture of many low-marsh sediments is virtually indistinguishable from that of adjoining sand or mud flats, for example, and high marshes may grade directly into terrestrial soil or regolith (Figure 4-3).

Nevertheless, the finite details of marsh sediments—their precise mineralogic or biogenic composition, biogeochemical attributes, original and intermediate sources (recycling), and specific modes of transport and deposition—remain virtually unknown in the great majority of coastal areas. Mechanisms of accumulation of sediments on marsh surfaces, equally problematical, are discussed subsequently.

Atlantic Marsh Sediments

Brief descriptions of marsh sediments in eastern Canada may be found in publications by Dionne (1972), Amos (1980), and Yeo and Risk (1981), and references cited by them. Perhaps most striking among maritime deposits are the red sediments of the Bay of Fundy, derived from adjacent Triassic red beds; subsurface marsh sediments are brown or gray, however, and consist of silt and clay overlain by a thin veneer of peaty detritus.

Along the Atlantic coast and shelf of the United States, two distinct clay-mineral facies have been suggested (Hathaway, 1972). The southern clay-mineral facies, which extends from Florida to Chesapeake Bay, is composed primarily of kaolinite and montmorillonite (or a complex of mixed-layer clays), whereas the northern clay-mineral facies, north of the Bay, is composed primarily of illite, chlorite, and traces of feldspar and hornblende. Although considerable overlap exists between components of the two facies, this scheme provides a convenient frame of reference for comparing northeastern and southeastern marshes in general.

New England Marsh Sediments

Over most of their course, rivers draining New England terranes (Figure 4-2) are incised into more resistant bedrock (mainly Paleozoic, Lower Mesozoic) than are southern rivers (mainly Upper Mesozoic, Cenozoic). Furthermore, the thick, well-developed saprolites of southern terranes provide abundant clay for redistribution along the coast. During times of heavy rainfall, the streams are literally red with this clay.* Most silts and clays originally present in the glacial deposits of New England already have been winnowed away, however, either to be trapped in numerous lakes and swamps inherited from the glacial topography or swept out to sea. Less inorganic material, therefore, is available for deposition in northern coastal environments (Meade, 1969). Instead, thick beds of peat have accumulated under the present marsh surface (Redfield and Rubin, 1962; Redfield, 1965, 1972).

Peat forms from the degradation of roots, stems, or leaves of marsh plants, especially the grass *Spartina* (Kerwin and Pedigo, 1971; Nixon and Oviatt, 1973). Thick ice cover shears off stands of *Spartina* during winter and mechanically degrades the grass into smaller pieces during spring thaw. In addition, freezing of the marsh surface retards biological decomposition and physical erosion of the substrate, which also enhances peat accumulation. Because peat is the most compressible of all natural sediments and may even compress under its own weight (autocompaction; see Bloom, 1964; Kaye and Barghoorn, 1964), thick accumulations may build up over a relatively short period of time. Even this peat contains considerable inorganic sediment, however (Redfield, 1972, Figure 6). Some of the major properties of Massachusetts peat are indicated in Table 4-3.

In addition to peat and fine-grained inorganic sediment, most of which has received meager study, ice-rafted pebbles and cobbles occur locally. In places, large glacial erratics may be seen. Sand is abundant in many of the tidal creeks within marshes and in some high marshes. Moreover, marshes associated with reworked glacial deposits may contain abundant and diverse

---

*Judging by the reports of early colonists, however, Georgia streams carried relatively small amounts of clay prior to human exploitation of the land.

Table 4-3. Characteristics of salt marsh peat, Barnstable Harbor, Massachusetts (adapted from Redfield, 1965).

Depth (m)	Composition (weight%)			Specific gravity (g/ml)
	Water	Ash	Organic matter	
0.03–0.06	74.0	20.2	5.8	1.11
0.94–1.03	84.3	11.5	4.2	1.15
1.68–1.76	89.0	5.1	5.9	1.19
2.40–2.50	70.4	24.4	5.2	1.13
3.14–3.22	90.4	2.3	7.3	1.05
3.87–3.96	70.0	25.1	4.9	1.16
4.78–5.00	67.0	29.6	3.4	1.23
Mean:	77.9	16.9	5.2	1.15

inorganic sediments, such as clay, silt, sand, gravel, cobbles, and boulders (Redfield, 1972).

Atlantic Coastal Plain Marsh Sediments

Marsh sediments between New England and Chesapeake Bay were described by Fischer (1961), Kraft (1971), Belknap and Kraft (1977), and workers cited by them. Below the bay, sediments are increasingly more southern in aspect.

Southern coastal plain marshes are in many respects the antithesis of New England marshes. Peat is not a significant constituent of sediments (although peats do occur in Louisiana, South Carolina, and Florida; Kolb and van Lopik, 1966; Cohen and Spackman, 1972; Staub and Cohen, 1979). Inorganic constituents are correspondingly more abundant. Here, stems of *Spartina alterniflora* remain standing throughout the relatively mild winters. Some grass is reworked by tides and may form small local lenses of peaty material (Edwards and Frey, 1977); but thick, continuous peat beds do not accumulate because of tidal flushing (Ragotzkie and Bryson, 1955; Kirby and Gosselink, 1976), rapid degradation of plant material by intense biologic activity (May, 1974), and extremely slow rates of coastal warping or submergence (Winker and Howard, 1977; DePratter and Howard, 1977). Thus, in spite of the luxuriance of Georgia marsh plants (Table 4-2), approximately 90% of the aerial (above-ground) primary production is dispersed as detritus. By one estimate, only 1 g of dry matter per square centimeter every 10 yr accumulates in marshes around Sapelo Island (Smalley, 1959).

These marsh substrates, like numerous others (Kurz and Wagner, 1957), contain large percentages of total carbon (see Table 4-4; Teal and Kanwisher, 1961). Where intact roots are removed prior to analysis of sediments, however, the percentage is considerably lower (Edwards and Frey, 1977, Figure 5). These observations suggest that most of the carbon in older marsh substrates is derived from *in situ* root matter, whether or not the material has been degraded beyond recognition.* Other major sources of organic carbon include benthic microbes, phytoplankton, animal remains, fecal and pseudofecal pellets, and "organoclays" (Pomeroy, 1959; Odum and de la Cruz, 1967; Kraeuter, 1976; Hanson and Wiebe, 1977; Pinet and Frey, 1977).

In a representative southern Coastal Plain marsh near Sapelo Island (Figure 4-1), the substrate is composed of approximately constant proportions of silt- and clay-size sediment (Figure 4-8). The maximum proportion of silt observed is 60% and that of clay is 55% (Edwards and Frey, 1977). The greatest variations occur along bare creek banks and in ponded water marshes, each of which is influenced by tidal drainages unlike those in the remainder of the marsh. Sand content of low marshes is uniformly rather low, but clean to slightly muddy sands predominate in high marshes. Surface and subsurface samples are generally very similar.

The coarser sediment in this marsh is composed essentially of quartz, potassium feldspar, and plagioclase. Muscovite, hornblende, epidote, monazite, magnetite, ilmenite, and garnet are minor constituents. In closely adjacent marshes, kaolinite is the preponderant clay mineral (35 to 40%), followed by mixed layer clays (20%), whereas smectite, vermiculite, and illite make up smaller (10 to 15%) but nearly equal parts of the clay fraction (Edwards and Frey, 1977, Table 1).

Gulf Coast Marsh Sediments

Although the Gulf coast contains important marsh areas (Figure 4-2), few have been studied sufficiently geologically. In general, these marshes seem to be much like those along the southeastern Atlantic coast (Kurz and Wagner, 1957; Bernard et al., 1962; Eleuterius, 1972). Marshes of the Mississippi delta complex are a major exception, however; here, in addition to purely freshwater marshes (Coleman, 1966), three major marsh types are recognized (Kolb and van Lopik, 1966):

1. Floating marsh or "flotant": a mat of roots or other plant parts admixed with finely divided "mucky" materials 10 to 35 cm thick, underlain by 1 to 5 m of organic ooze that grades with depth to clay. Consolidation of the

---

Spartina decomposes much less rapidly than do several other marine grasses (Wood et al., 1969; Gosselink and Kirby, 1974).

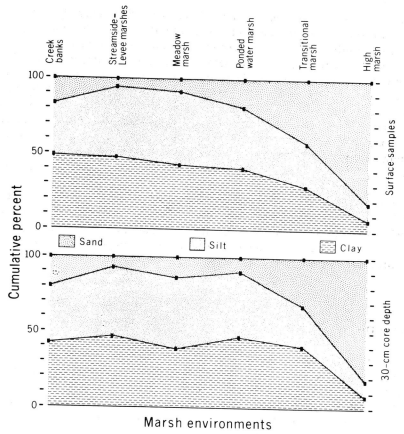

Figure 4-8. Proportions of sand, silt, and clay in salt marsh sediments, Sapelo Island, Georgia (see Figure 4-3) (adapted from Edwards and Frey, 1977).

ooze results in a black "organic clay" or peaty layer in which the organics generally exceed 50% of the deposit.

2. Brackish water–freshwater marsh: vegetative mat and muck 10 to 20 cm thick, underlain by 0.3 to 3.1 m of coarse- to medium-textured fibrous peat, underlain, in turn, by fairly firm, blue-gray clay or silty clay containing lenses of dark clays rich in organic matter. Only 10 to 20% of these deposits are thought to consist of inorganic sediments.

3. Saline–brackish water marsh: a mat of roots, stems, and leaves 5 to 20 cm thick, underlain by rather firm blue-gray clay containing few plant parts. Tiny organic particles are disseminated throughout. Clays are increasingly firmer and freer of organics with depth. The silt to fine sand content may range as high as 30%; clays rich in organics average 50% of the deposit, and peat typically ranges between 15 and 30%.

In these marshes, the distribution of live vegetation generally correlates with substrate type and therefore is a useful guide sedimentologically. Because the substrates are so soft (have such low shear strength), the marsh surfaces subside readily; in a real sense, these marsh sediments accumulate downward rather than upward. The peats are a mixture of allochthonous and autochthonous components; terrigenous clastics are delivered both by river flood and storms and are redistributed by the tides. Additional general information, including carbon budgets and the occurrence of natural hydrocarbons, may be gleaned from other works on Louisiana marshes (Penfound and Hathaway, 1938; Gould and Morgan, 1962; Mackin and Hopkins, 1962; Coleman and Gagliano, 1965; Day et al., 1973; Broussard, 1975; Coleman, 1976).

Unlike other Gulf coast marshes, many of the bays have been studied in considerable detail. At least indirectly, this information is of interest to marsh research. Montmorillonite is the most abundant clay mineral carried by rivers that enter the Gulf of Mexico, for example; illite is found in subordinate amounts, and kaolinite is a rather rare constituent (van Andel, 1960).

California Marsh Sediments

Thin peats or peatlike deposits have been observed in Alaska and Oregon marshes, but thick accumulations of peat are rare to absent along most parts of the Pacific coast (Macdonald, 1977a). Indeed, although plant litter is abundant in many southern California high marshes, relatively little of it becomes incorporated into the substrate (Warme, 1971); even the carbon content of these sediments is low.

Among the better known west coast marshes are those in southern San Francisco Bay, Mugu Lagoon, and Newport Bay. Sediments in the Newport marshes were described by Stevenson and Emery (1958); those of the other two areas are summarized below. Additional information on Pacific coast marsh sediments may be found among references cited by Macdonald (1977a,b).

Marshes near San Francisco are composed primarily of clay (45 to 65%) and silt (~ 40%) (Pestrong, 1972a,b; see also Rumboltz et al., 1976; Knebel et al., 1977). Within the clay fraction, montmorillonite dominates (50 to 60%); mica and chlorite constitute smaller amounts (25 to 35%). The silt fraction is composed mainly of quartz and plagioclase feldspars; other constituents include potassium feldspars, pyritized diatoms, iron sulfides, glaucophene, jadite, sphene, and garnet. Only 5 to 10% of the sediment is composed of sand-size particles.

Most of the coarser material in the bay–tidal flat–marsh system is found in channels (Pestrong, 1972a,b). This material nevertheless comprises only 35% of the channel sediment. A typical marsh channel contains sand, shell fragments, and organic debris in the coarse fraction. Channels at the bayward edge of the marsh contain 20% sand and coarse shell debris, 15% silt, and 65% clay, whereas at the landward edge, channel deposits typically contain no sand or shell debris, 15% silt-size particles (mostly in the form of clay "floccules"), and 85% clay.

Marsh sediments here exhibit a spatial sorting as well. The mean grain diameter is approximately 9 ϕ at the seaward edge of the tidal flat, grading to 10.5 ϕ at the landward edge of the marsh (Pestrong, 1972a, 1972b, Table 1). In general, the organic carbon content of *Spartina* marshes is about 14% and that of *Salicornia* marshes is about 18%.

Mineralogic investigations of marshes in southern California mainly stress the gross percentages of sand, silt, and clay (Stevenson and Emery, 1958). Several investigators, however, have documented the size grading of sediments from lower to higher parts of the marsh. Similarly, marshes that border Mugu Lagoon, near Ventura, California, exhibit three distinct sedimentologic zones (Warme, 1971, Figure 3). The lower marsh substrate, which is established approximately 1 m above mean sea level, contains clean sand, mud, and mixtures of both. The sand content ranges between zero and 92% and averages about 50%. Quartz is the dominant mineral; other particles in the coarse fraction are shell fragments, valves of ostracods and foraminifers, and muddy and limonitic aggregates. The carbon content of sediments ranges from 0.4 to 3.0%, which is appreciably lower than that of southern San Francisco Bay marshes (Pestrong, 1972b, Table 1).

The middle marsh, which occurs approximately 1.5 m above mean sea level (Warme, 1971), contains considerably more mud; sand here, as in the upper marsh, is present only in the lee of beach ridges, etc. The coarse fraction otherwise is composed largely of plant debris. Tests of foraminifers and gastropods also are common. The organic carbon content is as high as 5.8%. The upper marsh—between approximately 1.75 m and 2.0 m above mean sea level—however, is composed almost entirely of silt ($\sim$ 82%) and clay (18%). The coarse fraction here mostly contains plant debris, some gastropod shells, and muddy aggregates of sand and gravel size.

Salt pans, which also occur in some New England marshes (Chapman, 1974) but are absent in Coastal Plain marshes (Basan and Frey, 1977; Edwards and Frey, 1977), are present in the highest parts of the Mugu Lagoon marsh (higher than 2.0 m above mean sea level; Warme, 1971). Pans contain 46 to 82% silt; the remainder of the sediment is predominantly clay-size particles. The carbon content is less than 1%. The coarse fraction contains some sand-size mica, gastropod shells, and authigenic gypsum.

Miscellaneous Biogenic Sediments

Perhaps the most obvious biogenic component of marsh sediments is peat, which is one of the precursors of coal. The distinction between so-called rooted (autochthonous) and transported (allochthonous) coals is especially important in interpreting ancient coastal swamps and marshes (Swain, 1971; Cohen and Spackman, 1972; Staub and Cohen, 1979). In general, rooted coals are associated with low-energy environments, such as bays, lagoons, estuaries, and the adjacent marshes, whereas transported coals are associated with fluviomarine or fluviodeltaic systems. Fecal and pseudofecal pellets are less conspicuous but, in numerous marshes, are more important than peat. Characteristics of these pellets are discussed subsequently (see also Kraeuter and Haven, 1970).

Animal skeletons, especially carbonate tests and shells, are another potentially important constituent of marsh sediments. Inorganic precipitation of calcite is not important along most North American coasts or in coastal marshes, although aragonite occurs with other evaporites in Ojo de Liebre lagoon, Baja California (Phleger, 1969b; cf. Horodyski et al., 1977). Local concentrations of biogenic carbonate may be found on marsh flats, however (Macdonald, 1969; Warme, 1971; Basan and Frey, 1977; Edwards and Frey, 1977). Contributions of preservable shell debris to the sediment of most marshes nevertheless are small, because animals having durable tests (e.g., molluscs) are relatively rare. Moreover, acid conditions within marsh substrates tend to cause dissolution of much of the shell material, as do algal and fungal microborers (Stevenson and Emery, 1958; Wiedemann, 1972a,b). In some cases the organic matrix of shells decomposes, leaving a paste or slurry of disaggregated fine crystals. Similarly, the exoskeletons of arthropods, whether totally proteinaceous or impregnated with calcium carbonate, tend to decompose rapidly where microbes are abundant. Cores of marsh sediments in Georgia typically yield only scattered fragmentary carapaces or claws of crabs.

The diversity of shelled animals along most marsh tidal creeks is low, although certain populations, such as oysters or mussels, may be dense (Pestrong, 1972a, Figure 12; 1972b, Figure 6). In Georgia, oysters form only small isolated clumps on the marsh flat (Edwards and Frey, 1977, Plate 9, Figure 42), whereas they build large, discontinuous beds along the margin of large creeks (Figure 4-9a); in small tributaries the oysters may floor the channel (Figure 4-9b). The shape and distribution of various colonies may thus provide clues to the interpretation of ancient analogs. Subfossil oyster beds commonly are buried and exhumed along creeks as a result of channel migration (Frey et al., 1975). Many colonies are reworked, however, and may form local, discontinuous shell pavements (lag deposits) on creek banks or channel bottoms (Wiedemann, 1972a). Some of these shells may go through several cycles of exhumation and redeposition as marsh creeks migrate.

Shell beds such a those along Louisiana coastlines also may underlie marsh deposits (Figure 4-10a). These shell beds may be laterally continuous for 20 mi and are up to 60 ft thick (J.M. Coleman, 1977, personal communication). Although the main shell beds occur in an open bay

Figure 4-9. Beds of the oyster *Crassostrea virginica* along banks of Blackbeard Creek, Georgia (see Figure 4-6). (a) Beds elongated parallel to channel trend, in this case more massive on the seaward end (left); midtide. (b) Beds flooring the mouth of a small tributary, causing oceanward (leftward) shift of channel and silting-in and elevation of the old floor to a "hanging valley"; at confluence, oyster bed also shifts left; ebb tide. (Hanging valley confluences lacking oysters also occur in Georgia marshes.)

(Atchafalaya Bay), the sedimentary facies are not unlike those found in some marshes. As a result, bay-fill sequences in the fossil record may appear similar to tidal stream–salt marsh sequences. Bay fills, indeed, may grade vertically into true marsh sequences. Thus, the lateral continuity of bay oyster beds versus the lateral discontinuity of marsh oyster beds may help provide a means by which the two environments can be distinguished in the geologic past.

Deposits of transported shells may cap marshes in a transgressive sequence (Kolb and van Lopik, 1966); typical marsh sediments are overlain by 0.3 to 2 m of shells, shell fragments, silt, and sand. Shells of bay oysters and associated molluscs are reworked and accumulate as a transgressive sheet (J. M. Coleman, 1976, personal communication). Sable Island, east of the Mississippi delta, is an example of such a deposit (Figure 4-10b).

Some marsh sediments in southern California are composed of a combination of skeletal and terrigenous material (Warme, 1971). Other

(a)

(b)

Figure 4-10. Relations of selected Louisiana marshes to shell beds. (a) High marsh overlying extensive oyster shell pavement, Point au Fer. Marsh, dominated by *Distichlis spicata*, consists of extremely thin veneer; shell beds (*Crassostrea virginica* and other molluscs) are exposed in tidal creek. Willows in background. (b) Transgressive shell-deposit, Sable Island. Shell beds (*Crassostrea virginica*, mussel *Brachidontes* sp., derived from adjacent bay) are advancing over a small marsh consisting of *Spartina*, *Salicornia*, and "scrub" mangroves. Smaller embayment floored with shell debris and fine-grained inorganic sediments.

marshes may be composed entirely of carbonate mud, such as the "algal" marshes in Florida and the Bahamas; these are predominantly high marshes, containing sparse stands of *Distichlis* sp. and *Salicornia* spp.

Substrate Characteristics and Biogeochemistry

From a biological or ecological viewpoint, the most important biogeo-chemical aspects of marshes include nutrient and trace element cycling and

also plant–substrate–water interchanges during respiratory and other metabolic activities (e.g., Nixon and Oviatt, 1973; Valiela and Teal, 1974;
Wiegert *et al.*, 1975; Kraeuter, 1976; Pomeroy *et al.*, 1972, 1977; Pomeroy
and Wiegert, 1981). Geologists are concerned with many of the same cycles
or processes, especially as they relate to clays and various diagenetic
minerals. Relatively few of these relationships have been studied in detail,
however, and are mostly beyond the scope of this review.

The biogeochemistry of marshes is complicated by cycles or parts of
cycles related to, or interacting with, atmospheric conditions (during low
tide), hydrospheric conditions (during high tide, rainfall, or runoff), and
lithospheric or substrate conditions that fluctuate to some extent with both
tidal cycles and local edaphic factors. This brief discussion is restricted
mostly to the biogeochemistry of marsh substrates or pore spaces within
them. These, in turn, are related closely to the water table.

Marsh Water Table

In one way or another, levels of or hydraulic gradients within the marsh water
table are governed primarily by porosity and permeability of sediments. In
permeable marsh sediments the water table may rise to the substrate surface
during high tide, whereas in impervious sediments it commonly does not
(Clarke and Hannon, 1969). Porosity, permeability, aeration, water percolation, and chemical diffusion within the sediments may be enhanced by plant
roots and animal burrows (Basan and Frey, 1977), although the opposite
may be true where extensive bioturbation destroys stratification features
within the substrate (Edwards and Frey, 1977, Figure 6). The energy of
percolation and diffusion, or the hydraulic head, is determined primarily by
the overlying tidal prism; to our knowledge, however, the efficiency and exact
mechanism of this "tidal pump" have not been evaluated for marsh
substrates (cf. Chapman, 1976, Figure 5.5).

Numerous additional factors influence the water table, including height of
the preceding tide (neap, spring, storm), duration of tidal submersion,
distance of the local area from water source(s), elevation or microtopography
of the surface, abundance and kinds of plants present, and underlying
bedrock (Purer, 1942; Clarke and Hannon, 1969). Other factors are
mentioned below in discussions of salinity.

In many marshes the burrows of fiddler crabs extend down to the low-tide
water table, which provides a ready source of respiratory water (Miller,
1961; Basan and Frey, 1977). These burrows, where preserved, would be a
potential way of establishing water-table levels in ancient marshes. In fact,
artificial casts of the burrows (Frey *et al.*, 1973) could be used for the same
purpose in present-day marshes.

Salinities within marsh sediments are controlled largely by the frequency
and duration of tidal submersion, subaerial evaporation and transpiration,

Table 4-4. Geochemical aspects of salt marsh sediments near Savannah, Georgia (adapted from Barnes et al., 1973).

Samples		Core depth (cm)	pH	Eh (mv)	Total organics (%, by combustion)	Total iron (%)	Leachable iron (%)
Near mouth of Ogeechee River; annual salinity range 7–23‰	Core 1	0	6.1	130	24	1.8	0.24
		10	6.8	10	22		
		20	7.6	−185	24	1.3	0.32
		30	7.0	−320	22		
		40	7.0	−320	20	1.9	0.16
		50	7.0	−305	24		
		60	7.0	−300	23	1.7	0.12
		70	7.0	−270	24		
		80	7.1	−270	21	2.1	0.14
	Core 2	0	6.4	−30	20	1.9	0.4
		10	6.8	−10	23		
		20	6.9	−90	24	2.1	0.5
		30	6.8	−305	22		
		40	6.8	−240	23	1.8	0.3

Near Savannah River; annual salinity range 12–29‰

	Depth	pH	Eh			
	50	6.8	−130	22	1.9	0.2
	60	6.8	−245	23	1.0	0.2
	70	6.9	−285	21		
	80	7.3	−85	23		
Core 1	0	6.2	30	11	1.3	0.9
	10	6.7	−255	25	3.5	0.3
	20	7.1	−105	24	3.6	0.6
	30	7.0	−200	28	2.1	0.7
	40	6.8	−140	20		
	50	6.8	−135	25		
	60	6.9	−275	23		
Core 2	0	5.6	70	23	3.7	1.3
	10	6.4	0	30	3.8	1.6
	20	6.6	−40	30	2.6	0.6
	30	6.3	−240	25	2.7	0.9
	40	6.3	−185	21		
	50	6.4	−110	23		
	60	6.5	−90	20		

and leaching by fresh water. Latitudinal and seasonal gradients in air temperature and moisture greatly influence rates of evaporation and rainfall or runoff. Other modifications of salinity regimes include the proximity and complexity of tidal and freshwater drainages, porosity and permeability of sediments, aeration or reduction of organic matter and its abundance, and differences in microtopography or microhabitat (Redfield, 1965; Clarke and Hannon, 1969). In a given marsh, however, interstitial salinities tend to increase from that of adjacent lagoonal or estuarine conditions to a maximum at about the low-marsh/high-marsh transition and then to decrease landward (Macdonald, 1977b, and references cited therein); salinities decline abruptly beyond mean high-water spring, even where no freshwater runoff occurs.

Salinity levels also vary with depth in the substrate, especially during seasonal cycles. In a sense, the subsurface may be considered a reservoir for salts; salt concentrations increase downward during the wet season but decrease downward during the dry season (Beeftink, 1966; Clarke and Hannon, 1969; Macdonald, 1977b). Exceptions to this trend, as well as that of landward decreases in salinity, noted above, occur in marshes associated with hypersaline lagoons (Phleger, 1969b).

In porous sand barrens within Georia high marshes, interstitial salinities may exceed 100‰ (Basan and Frey, 1977; cf. Kurz and Wagner, 1957). Comparable hypersaline sand barrens occur in southern California marshes (Warme, 1971), as do scattered salt crusts (Purer, 1942). Although the biochemistry of plant–salt metabolism has received considerable attention, we are not aware of any extensive investigations of salt marsh evaporites, capillary actions, or reflux.

On a larger scale, brine pans and salt flats lie adjacent to marshes in Ojo de Liebre lagoon, Baja California (Phleger, 1969b). Evaporation of lagoonal water has led to precipitation of aragonite, halite, and gypsum. To some extent, this regime of circulation and evaporation influences interstitial salinities in adjacent marshes (see also Horodyski et al., 1977).

Salt marsh paleosalinities have been determined with some success by the sedimentary phosphate method (Meyerson, 1972), which utilizes the relative proportions of iron and natural calcium phosphate in argillaceous sediments. Problems with this method include the possible presence of detrital apatite, bone, or other preexisting calcium phosphates in the sediment.

Early Diagenetic Effects

Most low-marsh substrates are rich in H_2S and organic matter, have high potential for oxygen demand, have low pH but variable Eh, and contain abundant aerobic and anaerobic bacteria (Tables 4-4 and 4-5; Kurz and Wagner, 1957; Stevenson and Emery, 1958; Pomeroy, 1959; Oppenheimer, 1960; Teal and Kanwisher, 1961; Bradshaw, 1968; Friedman and Gavish, 1970; Wiedemann, 1972b; Day et al., 1973; Pomeroy and Wiegert, 1981).

Table 4-5. Mean oxygen demand and depletion capacity of estuarine–marsh sediments, Wassaw Sound, Georgia (adapted from Frankenberg and Westerfield, 1968; cf. Teal and Kanwisher, 1961; Mackin, 1962).

Environment	Oxygen demand[a]	Oxygen depletion capacity[b]
Salt marsh	3.25	766
Tidal creek bank	2.10	515
Tidal creek bottom	3.73	987
Tidal river bottom	2.08	477
Sound bottom	0.89	154

[a] mg O_2/cc sediment/day.

[b] cc water depleted/cc sediment suspended/day.

The sediments therefore tend to be highly reactive, and diagenesis begins contemporaneously or penecontemporaneously with deposition.

Reactivity of marsh sediments perhaps is most strikingly illustrated in areas that have been manipulated. In Georgia marshes impounded for experimental aquaculture, the surficial clays became oxidized (producing "cat clays"), the abundant sulfides were transformed into sulfates and sulfuric acid, and the pH fell to 2 (Linton, 1969). Impervious sediments prevented freshwater flushing and, because few carbonates were available in the sediment, the original buffering effect of sea water was quickly overcome. Although these seem to be merely "artificial peculiarities," such results may be somewhat analogous to those accompanying abrupt coastal emergence or rapid lowering of sea level.

Significantly, for calcium carbonate in coastal waters, the stable pH is closer to 8 than to 7 (Wiedemann, 1972a,b). In the marshes of Newport Bay, California, carbonate begins to dissolve at a pH of 7.8, below 25°C (Stevenson and Emery, 1958). In Dutch marshes, Beeftink (1966) noted that the carbonate content of sediments increases with clay content along tidal creek levees but that the opposite is true within marsh "basins" between creeks.

On a more general theme, Berner (1969) noted that the inorganic chemical composition of interstitial water in recent sediments may be affected to a considerable degree by bacterial decomposition of organic matter. This process may lower the pH, because of the release of biogenic CO_2 to solution; thus, the pH of sediments normally is lower than that of the overlying water. Where detrital calcium carbonate is abundant, increased concentrations of dissolved calcium in interstitial waters is caused mostly by these higher partial pressures of CO_2 and to the lower pH. Other bacterial processes, however, may cause a rise in pH, which, in turn, may initiate the precipitation of calcium, e.g., during the reduction of dissolved sulfate to form H_2S. Calcite concretions have been observed in Louisiana and Georgia marshes (Coleman and Gagliano, 1965; Wiedemann, 1972a) and siderite nodules in California marshes (Pestrong, 1972b), although the history of their development remains obscure.

In experiments, Berner (1969) found that the type of chemical reaction and consequent fluctuations in pH depend upon the kind of organic matter being decomposed by the bacteria. His experiments involved only animal matter.

Iron is abundant in numerous marshes (Table 4-4) and reacts with various organic and inorganic compounds (cf. Rashid, 1974; Berner et al., 1979). Ferric hydroxide, either in nodules (Bouma, 1963) or as linings of burrows or decaying roots, is perhaps most obvious. Other substances include ferrous monosulfide, ferrous manganese phosphate, jarosite, and iron sulfides (Wiedemann, 1972b).

In the old salt marsh muds of Cabretta Island, Georgia, Frey and Basan (1981) observed numerous filled-in crab burrows now cemented by iron-rich compounds. The nodules remain soft but nevertheless are much firmer than the host sediments. With continued diagenesis they presumably would become comparable to the hard ironstone nodules observed in various shales. In many instances these marsh nodules are now larger or more irregular than the original burrow systems, thus representing "overmineralization" or "overgrowth."

Processes Affecting Sedimentation in Salt Marshes

Physical processes, controlled mainly by tides, are extremely important for the establishment, development, and maintenance of marshes. In the sense of Odum (1973), these coastal environments are in fact energy-absorbing systems naturally subsidized by the impact of tides. The latter provide energy for introduction of sediments into the system, dispersal of sediment over the marsh surface, and maintenance of the marsh drainage network.

Chemical and biologial factors also are important in the establishment of a marsh, in governing the succession of plants and animals, and in affecting marsh maturation (Redfield, 1972; Ranwell, 1972; Reimold and Queen, 1974; Chapman, 1974, 1977; Macdonald, 1977a; Pomeroy and Wiegert, 1981). However, the emphasis here is mainly upon the sedimentary aspects of these factors, including the inseparable processes of plant colonization, sediment trapping, and marsh growth.

In most marshes, sediment accumulations amount to only a few millimeters per year (Letzsch and Frey, 1980a, Table 1; Pethick, 1981). Depositional rates may be somewhat greater along natural levees and are much greater during periods of barrier washover (Deery and Howard, 1977).

Sources of Sediment

The most abundant natural sediment that enters the coastal system is fine material composed of free to aggregated inorganic or organic particles. The sediment budget and specific mineralogic components of marsh ecosystems vary according to local biologic, geologic, geographic, climatologic, and oceanographic regimes. However, most marsh systems depend directly or indirectly upon one or more of the following general mechanisms for sediment nourishment: (1) riverine sources (Windom et al., 1971; Bartberger, 1976); (2) landward migration of fine-grained sediment from alongshore drift or the inner continental shelf (Meade, 1969; Windom et al., 1971); (3) barrier washover (Godfrey and Godfrey, 1974; Bartberger, 1976; Deery and Howard, 1977; cf. Cleary et al., 1979); (4) erosion of cliffed headlands (Kraft, 1971; Redfield, 1972) or other subaerial and subaqueous outcrops of older sediment (Howard and Frey, 1973, 1975b), which includes considerable recycling of marsh silt and clay within the estuary–marsh system (Pinet and Morgan, 1979); (5) wind-blown sediments (Phleger, 1977); and (6) organic aggregates (Riley, 1963) and in situ production of biogenic sediment, such as plant detritus, peat, and feces or pseudofeces (Kuenzler, 1961; Odum and de la Cruz, 1967; Haven and Morales-Alamo, 1968; Kraeuter and Haven, 1970; Redfield, 1972; Pryor, 1975). Lesser amounts of sediment may be derived from adjacent terrestrial sources (Stevenson and Emery, 1958; Edwards and Frey, 1977).

Implications of Water Movement in Salt Marshes

Salt marshes are intimately related to adjacent bodies of tidal water. Nutrients important to the sustenance of marsh plants and animals are exchanged between the marsh and estuary (Teal, 1962; Day et al., 1973; Odum, 1974; Wiegert et al., 1975; Hackney et al., 1976; Pomeroy et al., 1972, 1977; Pomeroy and Wiegert, 1981). In return, biologic components from the marsh enter the estuary in the form of organic detritus (Reimold et al., 1975) or soluble parts extracted from the plant matter prior to its export from the marsh (Kirby and Gosselink, 1976; Pinet and Frey, 1977). Physical aspects of salt marsh hydrography are equally related to those of nearby waters. The intimacy between these two aquatic systems thus requires that estuarine (or deltaic, or lagoonal) processes be investigated simultaneously with marsh processes. In fact, many of these processes are the same; only their scale or intensity differs. However, few detailed studies of marsh hydrography have been made, and equally few studies of lagoonal or estuarine hydrography have been concerned with the lagoon– or estuary–marsh transition (Lauff, 1967; Castañares and Phleger, 1969; Nelson, 1972; Dyer, 1973). In the following discussions, therefore, we have tried merely to point out some important generalities.

Hydrographic Setting

Many tidal channels and terrestrial streams have similar sedimentary deposits and respond to their respective processes in a similar manner (van Straaten, 1954; Pestrong, 1972b; Howard and Frey, 1980; Frey and Howard, 1980). These similarities may be attributed in part to the increased strength and duration of either ebb or flood tide over that of the other (van Straaten, 1961), thus producing a resultant "unidirectional" current dominance on patterns of erosion and deposition. Ebb currents dominate over flood currents in numerous southeast Atlantic estuaries (Howard and Frey, 1975b; cf. Pestrong, 1972b), whereas flood currents dominate many southern California bays and lagoons (Warme, 1971). During storms, however, flood tide is generally stronger than the ebb (Ranwell, 1972).

Currents reach their highest velocity in some tidal creeks during late ebb or early flood; at this time, creeks have a small "wet cross-section" that constricts flow and stimulates higher current velocities (Postma, 1967). In Georgia marsh creeks, however, the highest velocities are found at early to midebb and early to midflood (Letzsch and Frey, 1980a). The force of flood tide wanes progressively as water enters the marsh at creek heads or spills over natural levees, thence to move as a "sheet" over the marsh surface. The ebb-tide water mass is sluggish as it moves across the marsh flat or as it empties out from tidal creeks into broad, unvegetated tidal flats (Pestrong, 1972b); the later ebb water mass increases in velocity as it becomes progressively more confined to channels. Although ebb tides are dominant in Georgia tidal creeks, flood tides are dominant on the vegetated marsh surface (Letzsch and Frey, 1980a).

Wind friction may generate waves, drag water, and rework sediments on the flooded marsh surface (Edwards and Frey, 1977). However, most marshes are situated in a protected part of the coastal complex, behind various kinds of barriers; here wave force is minimal (Figure 4-11), and fetch is small. Grasses, in addition, tend to dampen the effect of wind-generated waves (Wayne, 1976). Waves probably are most effective as geologic agents in marshes when onshore storm winds coincide with high tide. Because of this storm surge, water masses are stacked in the higher parts of the marsh. At Sapelo Island, for instance, the normal tide range exceeds 2.0 m; but during northeast storms, tides may exceed 3.4 m. In extreme cases, ebb tide is almost negligible and water remains stacked in the marshes. Waves also are more important in open lagoons and bays or where marshes fringe the open coast, as along the panhandle of Florida.

On unvegetated tidal flats, flood tide seems to be the more important agent for erosion and transport; these processes are retarded on the ebb (cf. Figure 4-12). Accretion presumably is most active during the early stages of ebb tide (van Straaten and Kuenen, 1957; Evans, 1965; Pestrong, 1972b; Anderson, 1973). Although little is known about these processes in marshes, the effect of baffling, or damping, by vegetation undoubtedly alters the

Figure 4-11. Small wind waves on a low-marsh mud barren at high tide, Blackbeard Marsh, Georgia. Slightly larger waves rework surficial sediments in the barren but are effectively damped by grass (*Spartina alterniflora*) in adjacent marsh.

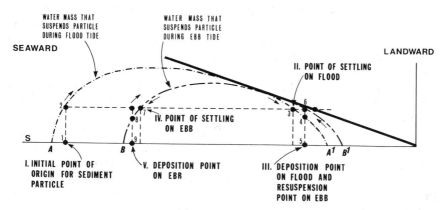

Figure 4-12. Hypothetical effect of settling lag (flood tide) and scour lag (ebb tide) on a sediment particle. S, substrate surface; Roman and Arabic numerals indicate presumed hydraulic and sedimentologic events, and letters indicate particle trajectory. During repeated tidal cycles, the net movement of an ideal grain should be successively more landward (adapted from Postma, 1967).

hydrologic and sedimentary patterns. Tidal flat processes tend to dominate in isolated small marshes, such as "marsh islands" (= "thatch islands;" Johannessen, 1964; Redfield, 1972, Figure 21), or along the fringe of low mashes that grade directly into unvegetated tidal flats without a conspicuous change in slope (Pestrong, 1972a, Figure 3). Nevertheless, any nonstorm water mass moving over the general surface of large marshes has extremely low velocity, not only because of plants but also because of gentle slopes and lack of channelization. The potential for erosion is much greater in drainage channels, which implies that lateral erosion through stream meandering is substantially more important than "sheet" erosion of the marsh surface (Letzsch and Frey, 1980a,b).

Tidal Hydraulics and Sedimentation

Most of the published information on hydraulic controls of grain-size distribution in intertidal environments comes from literature on bare tidal flats (Reineck, 1973). These, on the whole, are considerably simpler systems than vegetated tidal flats. We know of no investigations that directly confront hydraulic behavior of sediment in a vegetated system; remarks that follow are therefore speculative and conceptual.

Trends in sizes of sediment established initially, during the tidal flat stage, may be maintained partly by comparable processes that act on sparsely vegetated tidal flats. After fledgling marshes are well vegetated, however, current interference by grasses helps "trap" further sediment (Wayne, 1976), in addition to numerous other mechanisms that enhance deposition (Warme, 1971, pp. 25–26). Damping by grasses probably helps reduce wind velocities as well.

The gradual differentiation of sediment load in bare tidal flats results in a series of grain sizes that decrease from the lower to the upper part of the flat (Evans, 1965; Pestrong, 1972b). This pattern of sediment distribution is caused in large part by the decrease in tidal and wind energy across the tidal flat (van Straaten, 1954; Postma, 1967), as modified by tidal or estuarine circulation patterns (Dyer, 1973). Sediment lag and scour lag are conceptually important processes that result from this progressive decrease in physical energy (Evans, 1965; Postma, 1967), although few finite data are available with which to test the model rigorously. The net result of this inequality of tidal energy, at least ideally, is that a particle ultimately comes to rest landward from the point at which it first entered the system (Figure 4-12). Sediment probably is moved toward the upper reaches of the tidal flat and salt marsh by the process of settling lag and may be trapped in the marsh because of current baffling and scour lag—assuming that the water remains upon the marsh surface long enough for fine particles actually to settle from suspension (Edwards and Frey, 1977; cf. McCave, 1970; Pryor, 1975).

The discussion to this point has been concerned mostly with systems in which bare tidal flats grade continuously into vegetated ones, such as those found along many parts of the North Sea coast (van Straaten, 1954; Weimer et al., 1982). These contrast strikingly with marshes in which the lower margin is marked by steep banks or a conspicuous break in slope (Figures 4-7 and 4-9b); here, water and sediments enter the marsh mostly or totally through drainage creeks (Figure 4-6) rather than from gently sloping tidal flats.

The extended pathway necessary to pass the water mass first through lagoons or estuaries and subsequently into tidal creeks and out creek headwaters onto marsh surfaces seemingly would dissipate the energy of both wind and tidal currents and hence act as a mechanism for sediment trapping. Perhaps this is a special case of the model (Figure 4-12). However, the requisite processes and conditions remain virtually unknown: upper or lower flow regimes, water density stratification, tidal or nontidal circulation, and others (Dyer, 1973), acting first in successively smaller channels and then in "sheets" of water on the marsh. Also poorly understood is the possible effect of the estuarine "turbidity maximum" (Postma, 1967; Schubel, 1968) on adjacent marshes. Presumably, all of these factors play some role in trapping, eroding, suspending, or transporting sediment and in modifying the hydraulic behavior of individual particles. In spite of such uncertainty, however, numerous empirical observations are available. These, summarized below, constitute a useful point of departure for future research.

Sediments carried by tidal currents tend to be graded vertically within the water column (Passega and Byramjee, 1969). Actual transport is accomplished not only by traction, saltation, and suspension, depending upon grain size and current competence, but also by flotation of organic detritus. The heavy or dense components of the bed load are deposited mostly (1) within large tidal channels, as longitudinal bars and point bars (Schou, 1967; Howard et al., 1975); (2) along vegetated creek banks or natural levees (Stevenson and Emery, 1958); or (3) in the marsh itself, via smaller tidal channels (Pestrong, 1972b); deposition here is either in ponded headwaters or, less commonly, as crevasse splays. Silt, clay, and floating detritus tend to be carried correspondingly farther into the marsh interior, although appreciable quantities of these, too, may be trapped along banks and levees. The part of the suspended sediment load that can settle out presumably does so mostly during the time lag between tide-turns, when the competency of the current is lowest (van Straaten and Kuenen, 1957; Pestrong, 1972b; Anderson, 1973; Letzsch and Frey, 1980a).

Along deltaic coastlines, in contrast to tide-dominated ones, river flooding is an important aspect of sediment transport for marshes and coastal swamps. Marshes along such coastlines, especially ones that have low tidal ranges, receive new sediment mostly during river floods (J.M. Coleman, 1976, personal communication). Other marshes that are not associated with deltaic coastlines but that are associated with major river systems (e.g., marshes in the mid-Atlantic states) probably also recieve some sediment during river floods. In terms of hydraulics, the main difference between river flood and tidal flood is the duration; river slack allows suspended sediments much more time to settle out than does tidal slack.

The foregoing examples illustrate the importance of the actual path taken by sediment-laden currents as they enter the marsh (Beeftink, 1966). Most such sediments probably enter the marsh through drainage headwaters (apical deposition), whereas others—at high tide—may be transported over the tops of banks or levees, or through breaches within them (lateral deposition). Waters overflowing long gentle banks or low levees might not behave much differently than those emerging from drainage headwaters, yet the abrupt crevassing of a major levee might result in greatly accelerated currents, increased sediment transport, or even erosion of the marsh surface.

Nevertheless, in spite of models (Figure 4-12) and all the generalities mentioned above, many salt marshes are finer grained in their lower parts than in their higher parts (Figure 4-8). High marshes may contain considerably more sand because of the proximity of cliffed headlands or barrier islands. A fining-upward sequence (Evans, 1965) simply may be caused by a lack of coarse material in the marsh sediment budget; coarser sediment may be selectively trapped in the estuary, in adjacent tidal flats, or at the mouths of tidal creeks, or it may be restricted to parts of marshes that have immediate creek connections.

Furthermore, the settling lag–scour lag model cannot be divorced from the reality of absolute settling rates, extremely small particle sizes, and the actual frequency and duration of slack water. For example, depositional rates are very low in most parts of undisturbed Georgia marshes (Letzsch and Frey, 1980a). However, where water is somehow impounded—by dams (Linton, 1969), in abandoned creeks and headwaters, or simply in holes left where sediment cores have been taken (Edwards and Frey, 1977)—depositional rates are significantly higher. Such observations suggest that suspended sediments ordinarily do not have sufficient time to settle out before tide-turn but would settle if the duration of slack or sluggish water were somehow increased. Instead, the preponderance of fine material remains in the water column unless "trapped" in the lee of obstacles or by biogenic processes. Even in the most densely vegetated Danish marshes, no more than half the matter carried in suspension actually gets deposited (Schou, 1967).

Sediment Trapping

Settlement from suspension, alone, is inadequate to explain the thick accumulations of many kinds of muds (Pryor, 1975). Hence, other mechanisms must be sought to reconcile the large quantities of mud found in most marshes. Physiochemical flocculation of clays, formation of organoclays, and biogenic sediment trapping are obvious alternatives.

Animals and plants enhance sediment accumulation by numerous means (Miller and Egler, 1950; Warme, 1971; Pestrong, 1972a,b; Pryor, 1975; Howard and Frey, 1975b; Kraeuter, 1976; Edwards and Frey, 1977):

1. Grass everywhere in the marsh has a damping effect on wind-generated waves. Depending upon stages of the tide, *Spartina alterniflora* may reduce wave heights by as much as 71% and wave energy by 92% (Wayne, 1976).

2. Stems and leaves impede current flow at and above the sediment–water interface, which reduces current competence and helps trap suspended sediment. (A direct analogy may be made between marsh grasses and wind-blown sediments near dry washover fans, beach ridges, or terrestrial fringes; Deery and Howard, 1977.)

3. A seemingly important but unstudied aspect of plant obstruction is the complex eddy and turbulent current system created by stalks as the water mass moves over the marsh, at least in "marsh islands" and along low-marsh fringes. Such interference might enhance deposition in the lee areas while maintaining suspension or creating scour in turbulent areas (cf. Christensen, 1976).

4. Some plants may create chemical microenvironments that promote clay flocculation. Salt, which is extruded from plants in order to maintain osmotic balance, may increase salinity within the vicinity of the plants. Increased salinity, in turn, may promote clay flocculation, thereby increasing depositional rates of clays (Pestrong, 1972b). Nevertheless, simple increases in salinity alone are not necessarily sufficient to initiate clay flocculation.

5. Roots help secure the substrate beneath the sediment–water interface and thus promote accumulation of clayey, cohesive sediments. Plant roots may extend more than a meter in depth along Georgia streamside marshes and up to 50 cm in some adjacent habitats (Edwards and Frey, 1977; cf. Purer, 1942, Figures 8–16).

6. Algal, bacterial, and diatom films also help trap sediments and stabilize the substrate (Blum, 1968; Warme, 1971; Day *et al.*, 1973; Dale, 1974; Gouleau, 1976; Ralph, 1977), although most of these organisms remain dormant in winter.

7. Gregarious or colonial animals, either by their body parts or by their dwelling tubes (Figures 4-7 and 4-9), may influence deposition in ways comparable to grasses (items 1 through 3 and 5); substrate coherence is enhanced by the byssal anchorage of mussels (Pestrong, 1972a, Figure 12; 1972b, Figure 6) as well as by the rigid framework structures of oysters.

8. Macroinvertebrates also trap tremendous quantities of suspended organic and inorganic detritus, depositing it as feces or pseudofeces (Kuenzler, 1961; Haven and Morales-Alamo, 1970; Kraeuter, 1976). This, and "bioflocculation," probably are substantially more important than supposed physiochemical flocculations resulting from salinity changes (Pryor, 1975).

The importance of halophytes and various benthic microbes in sediment trapping and substrate stabilization has been stressed by previous workers, but the sedimentologic significance of benthic animals and planktic or pseudoplanktic microbes has been largely overlooked. Organic aggregations, or bioflocculation, may cause clay-size minerals to settle more rapidly than individual particles (Biddle and Miles, 1972; Ernissee *et al.*, 1974).

An equally important biologic process is the pelletization of clay and silt by suspension feeders (Pryor, 1975). Organic and inorganic material is extracted from the water column by such marsh-dwelling animals as shrimp, polychaetes, mussels, and oysters. Much of this organic material is consumed but the remainder, together with residual mineral matter, is compacted in the animal's gut and excreted.

Although fecal material is difficult to recognize in marsh sediments (because it is similar in color and texture to other sediment), distinct fecal pellets are seen in abundance on banks and bars in tidal creeks or along bayside (or lagoonside) beaches that adjoin marshes. Fecal materials produced by the oyster *Crassostrea virginica* constitute 40% of a mud deposit 1.5 m thick that covers an area of 15,000 m^2 near Dauphine Island, Alabama, including *Spartina* marsh (Pryor, 1975). The importance of pelletization and sediment reworking in Georgia salt marshes is indicated in Table 4-6.

Table 4-6. Average rates of biodeposition by Georgia salt marsh invertebrates (adapted from Kraeuter, 1976).

| Species[a] | Rates of deposition (per m^2) | | | |
| | Dry weight | | Ash-free dry weight | |
	g feces/day	g feces/year	g/day	g/year
Littorina irrorata	0.29	107	0.12	43
Geukensia demissa	1.50	549	0.34	125
Polymesoda caroliniana	0.003	1	0.001	0.3
Uca pugnax	2.25	821	0.66	239
Uca pugilator	0.63	231	0.13	47
Yearly deposition	—	1709	—	455

[a]The gastropod *L. irrorata* grazes upon epiphytic algae; the fiddler crabs *U. pugnax* and *U. pugilator* deposit feed at the marsh surface; the clam *P. caroliniana* and the mussel *G. demissa* suspension feed from the water column.

An aspect of biologic pelletization that needs further study is the hydraulic behavior of pellets in tidal drainages and on the marsh surface (Haven and Morales-Alamo, 1968). Do most pellets deposited on the marsh surface accumulate there, as is true for those defecated by *Geukensia demissa* in Georgia (Kuenzler, 1961)? Or, are most transported to or from creeks because they can be moved in suspension or traction? Do the pellets disaggregate or deflocculate when fresh water from rainfall or river flood enters the marsh?

Flocculation and the Distribution of Clays

As noted by many authors, individual particles of clay are so small or have such low densities that they stand little chance of settling out of suspension. However, numerous authors also have recognized the importance of electrolytic flocculation of freshwater clays when exposed to salt water, which produces clay aggregates of larger size and greater density and settling velocity (Whitehouse et al., 1960). The result of such studies is a large body of literature concerned with clay provenance, floccule sizes, modes and rates of transport and deposition, the environmental significance of differential flocculation and the consequent distribution patterns of characteristic clays or clay mineral suites, and their chemical alterations.

Selected examples are cited elsewhere in this chapter, but we make no attempt to synthesize this vast literature because very little of it pertains to marshes *per se*. Instead, we have turned to recent review articles and to discussions with various colleagues in an attempt to stress some of the important conclusions, variant ideas, and thorny problems remaining:

1. Evaporation or downward drainage of water from ponds, lakes, and playas may result in the deposition of suspended clays, but the simple settling out of suspended particles cannot account for widespread deposits of marine argillaceous sediments (Pryor, 1975).

2. Physiochemical flocculation of clays is an effective, important process in the deposition of argillaceous sediments where waters of extremely different salinities intermingle, as in deltas and estuaries (Pryor, 1975). Large floccules may even move in the traction load and form cross-bedded granular shoals. Increased salinity alone is not the full explanation, however. For some polyelectrolytes, a slight increase in salinity increases the sensitivity of hydrophobic sols to flocculation by salt whereas continued increases in salinity may render them stable (dispersed), even in concentrated solutions of salt (Whitehouse et al., 1960).

3. In broad interdeltaic or interestuarine environments (or bays, lagoons, and salt marsh estuaries), suspended argillaceous sediments are more or less in chemical equilibrium with salinity of the water medium; therefore,

physiochemical flocculation by contact with salts cannot account for significant deposits of mud found there (Pryor, 1975).

4. In any coastal or marine environment where both suspended matter and suspension-feeding animals are common, the biogenic pelletization of particulate material can account for large quantities of deposited mud (Pryor, 1975), whether or not flocculation also occurs. Indeed, the floccules themselves may be deposited biogenically instead of physically. The process of biogenic pelletization probably is rivaled in importance only by the formation of organoclays (items 5 through 7).

5. Even where salinity conditions are favorable, differential flocculation of clay minerals may be negated by natural organic and metallic coatings that can impart similar properties to different clay mineral species (Gibbs, 1977).

6. Volumetrically, however, such coatings may be much less important than actual clay–organic aggregates (Narkis et al., 1968; Greenland, 1971). Humic and fulvic acids strongly influence the stability of suspensions. In some cases these organic compounds increase colloidal stability to the extent that polycations or other flocculants must first react with and neutralize acids in solution or adsorbed on clay surfaces before conditions become suitable for bridging and flocculation. Polyvalent metals also may be involved in the bonding (Edwards and Bremner, 1967). The result may be "flocs" of sufficient size and density to settle out of suspension. Organoclays thus produced probably are much more abundant in natural systems than are salt–clay floccules.

7. Organic compounds in sea water, whether dissolved or sorbed in or on clays, rarely are free of bacteria or other microbes, which add their own bulk and mass to the aggregates and help transform them biogeochemically (Bader et al., 1960; Riley, 1963; Hanson and Wiebe, 1977).

8. Little direct evidence exists to indicate that salt–clay flocculation has much influence upon how and where fine-grained sediments ultimately are deposited in estuaries (Gibbs, 1977). The areal distribution of clays such as those in items 5 through 7 may be more a function of overall sediment size and physical sorting than of salinity gradients.

9. Similarly, the fact that clay mineral suites in the lower reaches of an estuary may resemble offshore suites more closely than those in the upper reaches of the same estuary does not, in itself, constitute evidence for the landward transport of clays from the inner shelf into the estuary. Instead, erosion of older deposits of mud in the lower reaches of the estuary may have released relicts of a different suite; these "contaminants" may in fact be moving seaward (Pinet and Morgan, 1979).

10. Chemical alteration of clays (including degradation and rejuvenation) may be more a biochemical process than a geochemical one. The

organic constituents mentioned in items 5 through 7, both surficial and interlayered, are an important but little studied example. All fecal matter represents a chemical change, because of the effects of the animal's digestive processes (Pryor, 1975), whether the clays are derived directly from the water column by suspension feeders or from the substrate by deposit feeders (Table 4-6). A delicate balance may in fact exist among suspension feeders, deposit feeders, and the physical and biogenic recycling of argillaceous sediments (Rhoads and Boyer, 1982).

Organic components of clay mineral agglomerates, including the live microbes mentioned above, probably are the single most neglected aspect of clays and clay flocculation. Nevertheless, we suspect that few clays in aquatic systems are entirely free of these organics and their influences. To what extent, then is "electrolic" flocculation truly an inorganic process? As a corollary, to what degree do "flocculated" clays "deflocculate" when suddenly exposed to fresh water? Deflocculation of clays during rains has been suggested as a cause for surficial erosion in marshes; in Georgia marshes, however, periods of intense rainfall tend to coincide with periods of increased surficial deposition (Letzsch and Frey, 1980a).

The major problem, of course, is that neither clays nor the organic matter associated with them in natural environments may be studied directly, whether in transit or in place; and because both the clays and the organics are highly reactive, they cannot be collected and subjected to "standard" laboratory preparations without being altered to some degree. In many cases the alteration probably is so drastic that the laboratory specimen bears little resemblance to the natural one. Unfortunately, this situation is not likely to improve without a corresponding improvement in our analytical techniques.

Seasonality of the Sediment Supply

Marshes, like most other coastal environments, are subject to natural periodicities in the intensity of both physical and biological processes. These, in turn, are related directly to seasonal changes in climate, including such factors as increased river flow and sediment transport during the rainy season, or reduced biologic activity and increased storm incidence during winter. In Bridgewater Bay, England, for instance, warm weather during spring and summer may cause an increase in both the biologic agglomeration of clay particles and the "flocculation" of particles because of increased salinity (Ranwell, 1972). Density stratification of estuarine waters also is greatest during this time. As a result, the sediments are deposited at seaward edges of sheltered marshes. Cooler autumn weather evidently reverses these effects. Sedimentation is correspondingly less, yet the sediments may be dispersed more widely. Some sediments are mobilized by storms, which help move them into more interior parts of the marsh. Seasonal deposition in Georgia marshes ranges from a minimum in autumn, through winter and spring, to a maximum in summer (Letzsch and Frey, 1980a); these differences correlate more or less with the seasonality of storm-wind incidence. Seasonal effects, on the whole, probably are more pronounced in fluviomarine systems than in bay or lagoonal marshes, however.

Physiochemical Conditions

In most parts of North America, spring rains enhance erosion of uplands and cause river flooding, which eventually brings an increased sediment load to the delta or estuary (Holeman, 1968). As summer progresses, dry months prevail and the sediment load decreases. During late fall and winter, sediment supply increases somewhat but is not as great as in spring. At times of low discharge, tidal influences are stronger than fluvial ones, which causes a net landward movement of the salt wedge (Dyer, 1973). In wet months the reverse situation prevails. Seasonality of sediment influx into deltas or estuaries, therefore, is a function not only of climate but especially of the local seasonal balance between marine and fluvial processes (McCulloch et al., 1970; Chakrabarti, 1971).

During certain times of year, density stratification of the water mass in an estuary may alternately trap, release, and circulate sediments (Meade, 1969; Visher and Howard, 1974; Rumboltz et al., 1976). Estuaries in temperate climates generally are best stratified during summer months, and at this time little sediment crosses the pycnocline (Gross, 1972). Periodic "flocculation" of clay particles also has been suggested as a cause for sediment seasonality in several estuaries (Neiheisel and Weaver, 1967; Pryor, 1975; Rumboltz et al., 1976). Increased salinity, which occurs during dry summer months because of the invasion of the salt wedge, may promote flocculation and cause deposition of material farther upstream, for example. Some clays tend to deflocculate with increased salinity, however.

Increased turbulence in winter, which results from more frequent and more forceful storm activity, may cause coastal erosion. Even moderate storms may hold clay and silt in suspension, which decreases rates of deposition but may increase rates of dispersal. These periods of increased wave or current energy are especially important in the erosion and redistribution of local sediments, some of which may be mistaken for sediments newly introduced into the system (Pinet and Morgan, 1979).

In northern climates ice may cover tidal flats and marshes throughout winter (Dionne, 1972; Knight and Dalrymple, 1976). Ice may obstruct or modify the substrate surface (e.g., ramparts, pans), or erode it (by plucking and rafting); yet a stable sheet of ice also may protect the underlying substrate and thus inhibit erosion or resuspension of fine-grained sediments. Even varvelike seasonal deposits have been reported from northern marshes (see Figure 4-18; Redfield, 1972).

Biologic Conditions

The annual growth of marsh plants is perhaps the most obvious example of biologic periodicity in the production and distribution of sediments, especially peat. Other biologic processes also are significant, however. During spring and summer months, when biologic activity is high, for instance, the production of feces and pseudofeces by suspension feeders may supply large quantities of sediment to the marsh system. During winter months most northern marsh–estuarine animals are inactive; hence this source of sediment essentially ceases. Southern marsh animals are less active in winter but ordinarily are not dormant (Day *et al.*, 1973; Kraeuter, 1976).

Similarly, many kinds of diatoms, bacteria, and algae, which secrete mucoidal substances that not only help trap newly settled particles but also increase the resistance of fine-grained sediments to erosion (Blum, 1968; Gouleau, 1976), are less active in winter than in other seasons. The same probably is true of other microbes in the water column, many of which enhance sedimentation rates through aggregation or bioflocculation (Riley, 1963; Biddle and Miles, 1972; Ernissee *et al.*, 1974; Hanson and Wiebe, 1977).

Tidal Drainage Systems

Striking dendritic patterns are typical of most marsh drainage systems (Figure 4-6; Steers, 1959, Figure 13; Redfield, 1972, Figure 13). Nevertheless, the density and complexity of drainages in a given marsh vary according to materials comprising the substrate and to the stage in marsh maturation (Table 4-1; Chapman, 1974). In general, densely vegetated marsh substrates tend to favor a broader spectrum of channel meanders than do sparsely vegetated ones, and thick peat deposits are more stable than are substrates of mud or sand (Pestrong, 1972b; Redfield, 1972).

To some extent, the initial pattern of marsh drainage is inherited from that of the ancestral tidal flat (Beeftink, 1966, Figure 3; Pestrong, 1972b). In turn, such late marsh features as ponds, pannes, and other incompletely filled depressions (Miller and Egler, 1950; Steers, 1959, 1977) may be relict features of earlier stages in the marsh drainage network (Evans, 1965; Redfield, 1972). The more common origins for such depressions include degenerate drainage headwaters, blockage of small channels by slumps or splays, decay of underlying peat (rotten spots), or simply insufficient sediment supply in the marsh system. In some Louisiana marshes, ponds form and marsh areas are lost because rates of substrate compaction and subsidence exceed rates of vertical accretion of sediments (Mackin, 1962); the plants tend to be killed off by abnormally long periods of submersion in highly saline waters during summer. Another peculiarity of marsh drainages is the occasional "hanging valley" confluence of tributaries (Figure 4-9b;

Redfield, 1972, Figure 48), which is contrary to the concept of a base level of erosion in fluviatile systems.

Equilibrium between Drainages and Marsh Growth

Pestrong (1965, 1972b) suggested that in southern San Francisco Bay, the lateral and vertical growth of marshes is reflected by commensurate growth of the marsh drainage system, particularly in the headward (apical) erosion of tidal channels (cf. Schou, 1967; Steers, 1977, Figure 2.7). This, in turn, leads to increased ebb-tide deposition. Large marsh tracts therefore expand more rapidly than small ones. At some equilibrium point, however, the expanding marsh so decreases the volume of the tidal prism that flood-tide deposition is reduced significantly. Ebb deposition continues, although this accumulation tends to be offset by tidal scour as the growing marshes prograde upon the main bay channels.

The degree to which this model applies to other marshes remains largely untested, although conspicuous exceptions are apparent. In the peat marshes of Barnstable Harbor, Massachusetts, for example, drainage channels have been remarkably stable since at least 1859 (date of publication of a detailed chart of the area); although substantial new marsh areas have been added incrementally, the older channels have not been extended appreciably by apical erosion (Redfield, 1972). Instead, they are relicts of drainage channels formed at an earlier time. In spite of much evidence for creek bank (lateral) erosion, losses in one place have been compensated by reconstruction in another; the basic positions of the channels remain essentially unchanged.

Southeastern Atlantic marsh drainages seem to be somewhat intermediate between those of San Francisco Bay and Barnstable Harbor. Redfield (1972, p. 235) suggested that drainage channels at Southport, North Carolina, and Sapelo Island, Georgia, show evidence of considerable lateral migration. However, an old map of the Blackbeard marsh (Figure 4-6) reveals more similarities than differences, especially with regard to apical erosion. Lateral erosion is obvious, as discussed below; but nearly all major drainages remain identifiable (Letzsch and Frey, 1980a).

Nevertheless, these marshes essentially already have filled the area available to them (Figure 4-6) and thus seem to have been more or less in equilibrium with overall coastal dynamics for a considerable period (Table 4-1); additional evidence for this stability includes thorough bioturbation of sediments by plants and animals (Edwards and Frey, 1977), implying that the substrates have been in continuous occupation for a long time. Hence, the "apical erosion" stage in the growth and development of these marshes (Ragotzkie, 1959) may have been surpassed long ago, before the presumed lagoonal waters were filled with sediment. Sediment cores are only now becoming available for the study of this succession.

Even after an estuary or lagoon is completely filled with marsh deposits, old tidal creek patterns may remain discernible on aerial photographs (Stevenson and Emery, 1958). These vestigial patterns are a potential aid in interpreting coastal sequences such as those proposed by Hoyt and Hails (1967), although the method has scarcely been used.

Deposition and Erosion along Channels

Formation of channels is an important process on tidal flats. Growth of drainage networks, in turn, is important for initial establishment of grasses; marsh plant roots require some degree of emergence in order to carry out photosynthesis (Chapman, 1974). High elevations on the intertidal surface, such as levees or other topographic features that develop from drainages, are early sites of plant colonization.

The establishment of grasses on previously unvegetated flats influences the subsequent character of drainages. Channels incised into vegetated substrates tend to be more sinuous and to have more pronounced meander curvatures than do channels cut in unvegetated flats (Pestrong, 1972b); rooted substrates offer more resistance to current or turbulent flow. Accretion of levees along the vegetated streamside also plays a role in channel formation; as scour deepens the channel, levees heighten the sides of the creek (van Straaten, 1954; Stevenson and Emery, 1958). After the channel network is formed, mussel or oyster beds established along the creek banks (Figure 4-9a) act as stabilizers to reduce erosion (Pestrong, 1972a,b). The amount and frequency of channel displacement, however, may be controlled partly by the kind of sediment that underlies the marsh, as mentioned previously. Another varable is the possible separation of flood and ebb channels (Schou, 1967).

In most marshes, channel processes continue to deposit and erode sediments throughout the channel's existence, although much of this activity may be on a comparatively minor scale. Processes include dominantly physical or dominantly biological ones or a combination of the two (Edwards and Frey, 1977; Letzsch and Frey, 1980b).

Channel erosion on bare tidal flats essentially takes place in response to three mechanisms: (1) direct, fluid shear stress against the substrate; (2) action of small-scale waves (wavelets) against channel banks; and (3) rotational slides (Reineck, 1970; Laury, 1971; Bridges and Leeder, 1976). On vegetated flats, similar processes act on channels; but bank undercutting and subsequent slumping and rotational sliding of coherent (root-bound) blocks may become a relatively more important factor than on some unvegetated flats.

The shear strength of sediments, which is determined by both sediment type and vegetation, also affects erosion and deposition in the marsh and its drainages (see Table 4-7). Channels incised in high, *Salicornia* marshes in California have steep slopes, whereas gentle creek bank slopes are more common in low, *Spartina* marshes (Pestrong, 1965).

Slump blocks and sod clasts are perhaps the most obvious indications of lateral erosion. In marshes near Sapelo Island, rotational slump blocks that have separated from the marsh creep slowly down the sloping banks, instead of instantaneously caving into the creek (Figure 4-13a). Similar blocks occur in New England peats, except that the fracture may gape open (Redfield, 1972, Figure 46). Slump blocks sometimes are degraded into large clasts of mud, but more commonly such clasts are torn directly from marsh sod (Figure 4-13b). These sod clasts may become incorporated into contemporaneous sediments of the lower creek bank or may be transported by tidal currents into deeper creek or estuary channels (Howard and Frey, 1973, 1975b; Basan and Frey, 1977; Edwards and Frey, 1977).

Comparable blocks and clasts have been observed in numerous other marshes (Pestrong, 1972a, Figure 11; Greensmith and Tucker, 1973). In the Barnstable Harbor marsh such blocks congest channels (Redfield, 1972, Figure 47), become incorporated into creek floors, and thereby raise drainage channels as the marsh surface itself is being raised; the size of the subsequent channel may change slightly as it readjusts itself to the volume of tidal water demanded of it.

Animals are responsible, directly or indirectly, for much of the lateral erosion in marshes. In poorly vegetated substrates or in adjacent tidal flats, snail trails on thin mud layers may exert some control on mud crack patterns by preconditioning the surface so that tensional stress, which occurs during desiccation, is relieved along these structures (Baldwin, 1974). Biogenic control of this sort, however, affects only the upper surface. Burrows of bivalves, shrimp, or crabs, which extend several centimeters into the substrate, may promote deep, penetrating cracks in coherent sediment (Figure 4-14) (Kues and Siemers, 1977). During desiccation, stress may be

Table 4-7. Shear strength of selected tidal sediments, California (adapted from Pestrong, 1969; cf. Pestrong, 1972b, Table 1)

Location	Sediment	Shear strength (psf)
San Francisco Bay		
Channel	Clay	72
Tidal flat	Silty clay	107
Spartina marsh	Clay	156
Salicornia marsh	Clay	299
Tomales Bay		
Channel	Clayey silt	84
Salicornia marsh	Coarse silt	262

(a)

(b)

Figure 4-13. Mass wasting along Blackbeard Creek banks, Georgia. (a) Successive rotational slump blocks. (b) Sod clasts ripped from upper marsh surface; the process is enhanced by abundant crab burrows, which abet tidal current plucking. Low tide.

relieved by fracturing along lines of inherent weaknesses produced by the burrows (Basan and Frey, 1977, Plate 1B). These fractures enhance large-scale mass wasting of the streamside marsh. Where burrows are extremely abundant (Figure 4-15), the process is accelerated (Letzsch and Frey, 1980b). Water circulating through these burrows, as well as turbulent flow over the roughened surface, further weakens the substrate and causes small-scale plucking and slumping along creek banks and the streamside marsh.

All of the above processes help steepen the slopes of creek banks, as well as aid slope retreat. Continued erosion, in turn, promotes channel meandering, sometimes to the point at which adjacent drainages intersect and "stream

Figure 4-14. Effects of decapod burrows upon creek bank erosion, Blackbeard Marsh. At top, crab burrows enhance tidal plucking of sod clasts (Figure 4-13b). At bottom, crab and shrimp burrows instigate and control pattern of desiccation cracks; continued desiccation and gaping cracks also enhance tidal scour or plucking.

piracy" occurs. On the whole, however, erosion in one place is compensated by deposition in another, whether physically (slipoff slopes, etc.) or biogenically (Figures 4-7 and 4-9).

Dynamics of the General Salt Marsh Surface

As stressed previously, deposition and erosion of marshes may occur surficially, apically, or laterally, and the processes are related intimately to tidal hydraulics, sediment supply, plant colonization, animal activities, and marsh growth. In southwestern Dutch marshes, for example, apical sediments are mainly clay and silt, whereas lateral sediments consist mainly of sand, at least during earlier stages of marsh maturation (Beeftink, 1966, Figure 3); the rate of creek bank and levee growth diminishes gradually, with respect to that of the marsh interior, provided that the tidal supply of clay and silt is sufficient. By this time the volume of the tidal prism has been reduced and currents are less competent to transport sand, even along creeks. The result is a vertical and temporal succession from coarser grained, mechani-

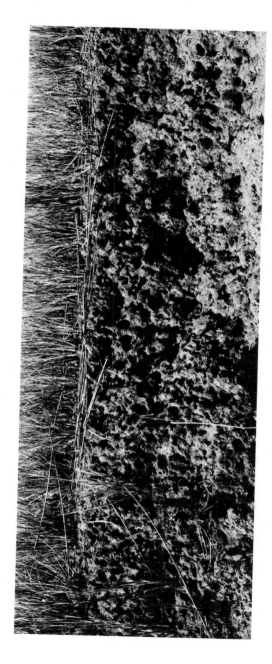

Figure 4-15. Bioerosion of salt marsh tidal creek bank by crabs, especially *Panopeus herbsti*. Approximately 45% of the mean creek bank surface has been evacuated of sediment, which is then recycled within the depositional system. Sapelo Island, Georgia (adapted from Letzsch and Frey, 1980b.)

cally unstable deposits to a finer grained, stable marsh (cf. Stevenson and Emery, 1958; Evans, 1965; Pestrong, 1965, 1972b).

In spite of the maturity and stability in such a marsh, however, one wonders whether tidal sedimentation can ever bring the marsh surface to grade—a perfect plane. Even where peat compaction does not occur in the Barnstable marsh, for example, careful analyses reveal faint microtopography on the surface that seems to be related to differential levels of the tide, as controlled, in turn, by the peculiarities of local drainages (Ayres, 1959; Redfield, 1972). What role is played by tidal harmonics, or nodes? And what may be the magnitude of their effect under a given tidal range?

Erosion, to some degree, also may continue throughout the history of such a marsh. In Georgia, surficial sediments may be flushed out of the marsh during autumn, the period of greatest storm incidence (Letzsch and Frey, 1980a). However, it is doubtful that surficial erosion, other than that occurring during storm conditions, is ever as important as lateral erosion, or even apical erosion in the early stages of marsh development. The force of everyday tides or wind-generated waves, which is damped by such surface obstructions as plants, probably is insufficient to lift and resuspend as much sediment as is deposited on the marsh surface during slack water (Figure 4-12; Warme, 1971). Small marsh islands (Johannessen, 1964) and the seaward or bayward fringe of larger marshes may be exceptions; but even here, wholesale erosion occurs most obviously by "cliff retreat" (Reineck, 1970, Figure 21; Redfield, 1972, Figures 43 and 44).

Of fundamental importance in the stability or susceptibility of marshes to erosion is the shear strength of sediments and the prowess of rooted plants (Purer, 1942, Figures 8 to 16). Some California marshes, for example, exhibit a trend toward increasing grain size and shear strength of sediments with increased marsh elevation; the highest values of shear strength (Table 4-7) are associated consistently with the organically richer, denser, more root-bound marsh deposits (Pestrong, 1969; 1972b, Table 1).

In contrast, increased physical stability of the substrate may enhance plant colonization. Marshes generally do not thrive where exposed to high-energy waves, yet the physical limits of colonization by *Spartina alterniflora* seem to be related more to unstable substrates than to mechanical damage to the plant (Redfield, 1972). Where rooted in firm peat, *S. alterniflora* marshes fringe the fully exposed shores of Cape Cod Bay and thus are subject to all the vagaries of wave action.

Root density alone probably is not as important in sediment binding as the actual distribution or configuration of the roots. Although Georgia low-marsh substrates, particularly those of the streamside and levees (Figure 4-3), contain greater amounts of undifferentiated root material than those of the high marsh, the deeper, predominantly vertical penetration of the main low-marsh roots leaves the upper substrate bound only by a loose network of small roots (Edwards and Frey, 1977, Plate 2, Figure 8). In contrast, the shallower, horizontal root networks of the high marsh seem to bind the upper sediments more tightly. Perhaps this difference also is compensated by the greater baffling effect of tall, luxuriant grasses along levees, where adjacent channel currents are greatest; currents ordinarily are negligible in the high marsh, despite shorter grasses.

Marsh erosion may be enhanced where such substrate stabilizers as halophytes will not grow (e.g., salt pans), or where adverse ecologic conditions cause temporary denudations of the marsh surface (e.g., barrens). Pans and pond holes (Miller and Egler, 1950; Redfield, 1972) are not ordinarily present in Georgia marshes (even the ponded water marsh, Figure 4-3, is exceedingly shallow and is vegetated); but "barrens" are common locally. These may form upon either low-marsh mud (Figure 4-16) or high-marsh sand. The origin of some barrens remains unclear, but numerous ones on high marshes may be attributed to the smothering of original grasses underneath tidal wracks, or rafts, of dead *Spartina* (Basan and Frey, 1977, Plate 1D; Edwards and Frey, 1977, Plate 9, Figure 45).

Grass wracks, which may be several meters wide and several tens of meters long, cause severe changes to local plant habitats by (1) blocking light for photosynthesis; (2) increasing substrate temperature underneath wracks, as waterlogged debris dries and decays, thereby (3) increasing the evaporation rate, and thus residual salinity (Kurz and Wagner, 1957). As a result of these changes in habitat, the marsh plants die and no longer are effective current bafflers or sediment binders. Physical processes become correspondingly more effective, even in these comparatively protected parts of the coastal system (Figure 4-11).

In well-vegetated marshes, bioturbation by animals, especially crabs, probably is more conducive to erosion than are physical processes. Compact sediment is loosened or reworked during feeding and burrowing activities

Figure 4-16. Low-marsh mud barren (Figure 4-11) during ebb tide. Plant stubble remains discernible on freshly denuded surface. The reason for this defoliation is not known.

(Figure 4-17; Edwards and Frey, 1977, Plate 5); this presumably makes the sediments more readily erodible, even in weak currents, than are clays deposited by a water medium (Rhoads and Boyer, 1982). Bioturbation also affects the chemistry and integrity of sediments and is an intense, ubiquitous process in many American marshes. The effects probably are much more profound than has been recognized to date. For example, virtually every particle of sediment on Georgia marsh surfaces, at one time or another, must have been manipulated or ingested by a fiddler crab (Table 4-6; Miller, 1961; Edwards and Frey, 1977). Certain bottom-feeding fish also may be important in the biogenic reworking of marsh substrates (Odum, 1968).

Desiccation, although important on some creek banks (Figure 4-14), is not a significant instigator of erosional processes in regularly inundated marshes; the substrate tends to remain moist and coherent. Many lower high-marsh areas dry out during part of each tidal cycle, however, and still higher areas may remain emergent for several days. Even some parts of the low marsh, such as levees, may be exposed for unusually long periods during spring low tide. Prolonged exposure of marsh mud causes dehydration and cracking, which enhances bed roughness and makes the substrate less coherent. The surface therefore may be more easily eroded by the incoming tide, given currents of sufficient strength. Some of the smaller desiccation crusts may be

Figure 4-17. Low, sandy, streamside marsh surface thoroughly reworked by feeding activities of fiddler crab *Uca pugilator*. Small isolated clumps of *Crassostrea virginica* are found more typically in ponded marshes or in low marshes where levees are absent. Blackbeard Creek, Georgia.

transported by flotation (Fagerstrom, 1967) or by wind traction (Warme, 1971).

Other processes that affect the general marsh surface, such as river flooding, rainfall and runoff, ice covers, and barrier washovers, were mentioned previously. A final, minor example is the transport and mixing of marsh mollusc shells by birds, raccoons, hermit crabs (Frey et al., 1975), and even tidal flotation (Warme, 1971).

Sedimentary Structures

Because marshes may intergrade with numerous other kinds of coastal environments, marsh peripheries typically display a diversity of physical and biogenic sedimentary structures that are more typical of other environments. The importance of such structures decreases progressively or abruptly toward the marsh interior. In Georgia, the most closely related environments include riverine and salt marsh estuaries (Howard and Frey, 1973, 1975b, 1980), tidal stream bars (Land and Hoyt, 1966; Howard et al., 1975), creek or estuary banks (Edwards and Frey, 1977; Frey and Howard, 1980), and washover fans (Deery and Howard, 1977). Elsewhere on American coasts, such environments as deltas, bays, and lagoons may be equally important. In most of these, however, associated suites of physical and biogenic sedimentary structures have not been studied in detail.

Physical Structures

Discontinuous wavy to parallel laminations or continuous laminations and thin bedding are the most prominent physical structures in many marshes (Table 4-8; van Straaten, 1954, 1961; Bouma, 1963; Evans, 1965; Coleman and Gagliano, 1965, Figure 2; Reineck, 1970, Figure 21; Pestrong, 1972b, Figure 14). Contacts between laminations or bedding usually are distinct, and individual layers exhibit little or no grading of sediments. Current regimes ordinarily have insufficient energy to generate bedforms on the coherent, plant-baffled substrate; therefore, lenticular laminations have been attributed mainly to deposition on a dry, uneven surface (van Straaten, 1961; Bouma, 1963). Some of the high-marsh sand barrens in Georgia are covered with a continuous but "crinkled" mat of blue-green algae (see Figure 4-21), somewhat like the "supratidal crusts" in carbonate terranes of the Florida Keys and the Bahamas. Such crusts also give rise to discontinuous or wavy laminations.

Parallel laminations in other environments are attributable primarily to wave or current reworking, algal trapping of sediments, or to settlement from suspension. Reworking is important locally in exposed or poorly vegetated marshes, but not on the general surface of most marshes. Similarly, algal or bacterial mats are present locally (Table 4-8) (Ralph, 1977), but not thick stromatolitic deposits. Therefore, settlement from suspension, whether by gravity or by such processes as biogenic pelletization, must be an important mechanism in these marshes. Bouma (1963) suggested that the lack of graded bedding within laminations may indicate deposition of the clay as flakes or granules of larger size. According to Pestrong (1972b), parallel laminations in San Francisco Bay marshes result from (1) sediment-size variations, (2) mineral and colloid concentrations within certain layers, and (3) flocculated and nonflocculated zones. He considered differential floc-

Table 4-8. Characteristic sedimentary structures in Mugu Lagoon Marsh, California (adapted from Warme, 1971)

Environment[a]	Physical structures	Biogenic structures
Low marsh (< 0.5 m)	Common, discontinuous thin beds and laminae, some being wavy; discontinuity caused by plant obstructions, burrows, and roots; shallow desiccation cracks	Thin algal mats; profuse crab burrows and root structures
Middle marsh (< 0.3 m)	Discontinuous thin laminae; desiccation cracks; small ?gas pockets between intercalated laminae of mud and organic matter	Algal films; deep vertical root structures; abundant burrows of worms, insects, and other arthropods, including larvae
Upper marsh (< 0.2 m)	Rare, very thin, discontinuous laminae; thin layers of detrital plant fibers	Root structures; arthropod burrows
Salt pans (?)	Well-bedded thin laminae to thin beds; bedding planes become discontinuous, planar to curved plates with development of desiccation cracks; gas bubble structures, both surficial and under desiccation cracks; salt crystal molds; thin layers of mica; silt pellet dunes formed of wind-blown desiccation crusts	Various arthropod burrows, mostly of larvae; mammal tracks; rare root structures

[a]Numbers in parentheses are estimated rates of deposition per 100 yr.

culation to be more important than the other two mechanisms. As stressed on previous pages, however, it is doubtful that flocculation *per se* plays much of a role in marsh sedimentation. The formation of organoclays and biogenic sediment trapping by suspension feeders probably are much more important than has been recognized previously.

McCave (1970), noting that slack periods in the tide ordinarily are too short for fine-grained material to settle, concluded that some other periodic force is responsible for parallel laminations; he suggested storm conditions followed by periods of calm. Periodic sedimentation of another kind has been invoked to explain well-stratified deposits in the Barnstable marsh (Figure 4-18). These are interpreted to represent annual layers, in the manner of varves (Redfield, 1972). Thin layers of nearly pure sand and silt, thought to reflect winter deposition, alternate with thin layers of fibrous organic matter, thought to represent summer deposition. The layers become progressively thinner toward the upper part of the sequence, evidently indicating an approach to equilibrium with the high water level; the layers are virtually indistinguishable at the top. The deposits studied by Bouma (1963) and others exhibit the same thinning-upward trend and conceivably have the same depositional significance, at least in terms of the local tidal range. This

(a)   (b)

Figure 4-18. Well-stratified marsh deposits, Barnstable, Massachusetts. (a) Undermining of slope by tidal creek causes blocks of sediment to fall away, leaving vertical cliffs. (b) Sediments consist of interlayered peaty materials, silt, and nearly pure sand; layers become progressively thinner toward top of exposure (photos courtesy A. C. Redfield).

relationship would be especially valuable in the study of possible ancient analogs.

Many of the discontinuous or wavy laminations in salt marshes may be attributed to desiccation cracking or to plant or animal bioturbation (Table 4-8; Horodyski et al., 1977), or even the hooves of mammals (Deery and Howard, 1977). Thicker bedding has been noted in peat (Redfield, 1972) and in Mississippi Delta marshes (Coleman and Gagliano, 1965, Table 1); the latter structures result from periodic introduction of fine clastics by river flooding. On the whole, however, Mississippi Delta marshes are noted for the lack of diversity among sedimentary structures.

A few other kinds of sedimentary structures have been described from marshes, including various ripple and drag marks (Evans, 1965). Most of these occur in bare or poorly vegetated areas, however, which are more closely allied to tidal flats (van Straaten, 1954). Ripple marks preserved in salt have been observed within salt-evaporation pans alongside California marshes (R. Pestrong, 1977, personal communication); apparently the salt crystals behaved as discrete clastic grains.

All of the marsh substrates and stratification features mentioned above are strikingly different from those in Georgia, however. Here the sediments are essentially homogeneous (Figures 4-19 and 4-20), having been thoroughly reworked by animals and plant roots (Edwards and Frey, 1977). The only notable exceptions are rare vestigial laminations from relict substrates (Frey and Basan, 1981) and equally rare, thin, discontinuous laminations in sand barrens and very sparsely vegetated high marshes; the latter evidently result from surficial reworking by small wind waves (Figure 4-11) but are quickly reworked, in turn, by the feeding and burrowing of fiddler crabs (Figure 4-17). Such thorough biogenic reworking of sediments surely is more common among other American marshes than current literature would imply.

Biogenic Structures

Although the literature on salt marshes contains scattered mentions of such features as crab or worm burrows, snail trails, or root mottles, exceedingly few of these structures have been studied in any detail. Studies dealing with natural assemblages of biogenic sedimentary structures in marshes are even rarer. To date, only three sites have been investigated to an appreciable degree: Mugu Lagoon, California (Table 4-8), Beaufort, North Carolina (Frey, 1970; Allen and Curran, 1974), and Sapelo Island, Georgia (Frey and Howard, 1969; Basan and Frey, 1977; Edwards and Frey, 1977; Frey and Howard, 1980).

Algal laminations ordinarily are not abundant in Georgia marshes, although blue-green algae make thin surficial mats on some high marsh sand barrens (Figure 4-21) and locally on meadow-marsh muds (Edwards and Frey, 1977, Plate 9, Figure 43). The structures described below are considerably most important overall.

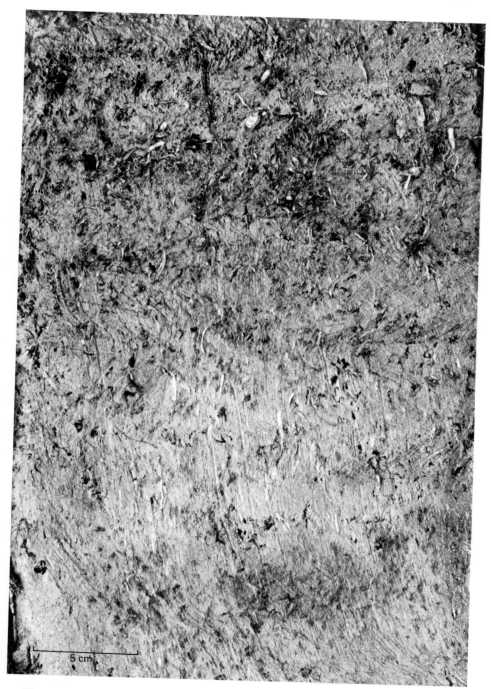

Figure 4-19. Trimmed sediment core from meadow marsh, illustrating typical lack of physical stratification (see Figure 4-20). Blackbeard Marsh, Georgia (from Edwards and Frey, 1977).

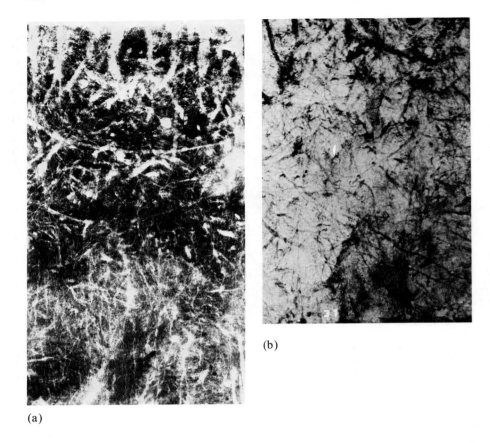

(b)

(a)

Figure 4-20. X-ray radiographic prints of sediment peels from can cores, salt marsh near Sapelo Island, Georgia. Both cores approximately 20 cm across. Physical sedimentary structures completely destroyed by plant-root penetrations and animal burrows. (a) Dense, relatively coarse root mats; muddy low marsh. (b) Sparse, relatively fine root mats; slightly sandy substrate, short-*Spartina* high marsh (see Figure 4-3).

Burrows and Root Mottlings

Snails, crabs, and a few other organisms leave various trails, grazing traces, or deposits of pelleted sediment upon marsh surfaces. Many of these activities are important in terms of surficial reworking of sediments (Figure 4-17), but few of the resulting structures have much potential for preservation as diagnostic trace fossils. Burrows and root traces are more important in both respects; the biogenic reworking may penetrate to considerable depth in the substrate, in some cases several decimeters (Figure 4-20; Warme, 1971) and the structures themselves have relatively good prospects for preservation, as evidenced by their common occurrence in 500-

(a)

(b)

Figure 4-21. Blue-green algae on surface of high-marsh sand barren, Sapelo Island, Georgia. (a) Mat forms solid cover over barren except for its periphery. At this location, several such barrens are strung together in pater noster fashion, parallel to terrestrial border of marsh. (b) Crinkled surface apparently caused by combination of desiccation, differential growth of algae, and disturbances by deer hooves. Wisps of fresh sand attest to sediment movement upon these barrens.

to 1000-yr-old marsh muds exposed on the coasts of Georgia (Frey and Basan, 1981) and South Carolina.

The main burrowing animals in Georgia marshes include the polycheate *Nereis succinea*; the mud crabs *Eurytium limosum, Panopeus herbsti*, and *Sesarma reticulatum*; and the fiddler crabs *Uca minax, U. pugilator*, and *U. pugnax*. Most of these animals exhibit characteristic habitat distributions within the marshes, which are reflected in the distribution and abundance of burrows (Figure 4-22). The polychaete burrows consist of smooth-walled, sinuous tunnels or simple to complex tunnel systems that ramify through the substrate in the manner of the trace fossil *Palaeophycus* (Frey *et al.*, 1973). Burrows of the different species of crabs tend to be distinctive in the main, although their morphologies overlap somewhat (Basan and Frey, 1977, Table 3). In the rock record, most crab burrows—depending on their extent and configuration—would appear as the trace fossils *Psilonichnus, Skolithos*, or *Thalassinoides*. *Uca minax* and *U. pugilator* excavate vertical or inclined shafts housing a single animal, whereas *U. pugnax* may construct a single burrow or a simple or complex burrow system, according to habitat conditions. The mud crabs typically construct only burrow systems; those of *Sesarma reticulatum* tend to penetrate to greater depths, and those of *Panopeus herbsti* to shallower depths, than those of *Eurytium limosum*. All burrow systems are communal domiciles, and the burrow systems of different species commonly intersect one another (marsh researchers therefore cannot reliably determine crab population densities merely by counting the burrow apertures; Frey *et al.*, 1973; Wolf *et al.*, 1975).

The roots of Georgia marsh halophytes tend to become progressively smaller and shallower but more dense from the upper creek bank to the high marsh (Figures 4-3 and 4-22), although those of *Juncus roemerianus* are

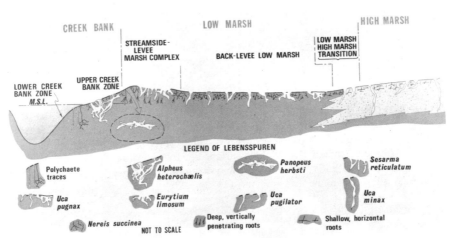

Figure 4-22. Characteristic distribution of burrows and root systems in salt marshes, Sapelo Island, Georgia (from Basan and Frey, 1977).

larger, deeper, and less dense than those of other high-marsh plants (Edwards and Frey, 1977). The basic trend holds true even for the different growth forms of *Spartina alterniflora*. Additional work is needed on the specific characteristics of roots and root pattern trends in Georgia marshes, however (Gallagher, 1974; cf. Purer, 1942, Figures 8 through 16).

Bioturbation

Significantly, of the various "fundamental" characteristics of salt marshes specified in the Bouma model (1963), three are decidedly atypical of Georgia and other southeastern marshes. According to the model, most marshes are typified by (1) well-stratified sediments, (2) scarcity or absence of burrowing animals, and (3) the insignificance of plant root disturbances relative to the more intense bioturbation by animals in other kinds of marine environments. As stressed in previous sections, the very abundance and effectiveness of burrowers and rooting plants in these marshes (Figures 4-17, 4-20, and 4-22) are the main reason for the dearth of well-stratified sediments. In fact, total biogenic reworking of sediments is more intense here than in any adjacent facies (Edwards and Frey, 1977; Frey and Howard, 1980). The same is true of numerous other North American marshes (Phleger, 1977).

Equally important, the intensity of bioturbation in a given coastal or marine environment may be a function of (1) population densities of plants and animals, (2) rates of deposition, or (3) a combination of the two (Howard and Frey, 1975b). Thorough biotic reworking of sediments in Georgia marshes therefore implies not only an abundance of organisms but also rates of deposition sufficiently slow to permit this reworking. In contrast, the paucity or absence of burrowing animals and densely rooted plants in such marshes as those studied by Bouma does not, in itself, reveal much about rates of deposition; in the rock record, however, the dominance of physical structures over biogenic ones probably would be interpreted as an indication either of rapid sedimentation or of frequent reworking by physical processes. The moral is clear; more criteria are needed from low-energy, back-barrier facies.

References

Adams, D. A., 1963. Factors influencing vascular plant zonation in North Carolina salt marshes. *Ecology*, **44**, 445–456.

Allen, E. A., and Cohen, A. D., 1977. Thin section analysis of coastal-marsh sediments and its use in paleoenvironmental reconstruction (Abs.). *Amer. Assoc. Petrol. Geol. Bull.*, **61**, 759.

Allen, E. A., and Curran, H. A., 1974. Biogenic sedimentary structures produced by crabs in lagoon margin and salt marsh environments near Beaufort, North Carolina. *J. Sed. Petrol.*, **44**, 538–548.

Amos, C. L., 1980. Physical processes and sedimentation in the Bay of Fundy. *In*: McCann, S. B. (ed.), *Sedimentary Processes and Animal-Sediment Relationships in Tidal Environments*. Geol. Assoc. Canada, Short Course Notes, **1**, 95–132.

Anderson, F. E., 1973. Observations on some sedimentary processes acting on a tidal flat. *Mar. Geol.*, **14**, 101–116.

Ayres, J. C., 1959. The hydrography of Barnstable Harbor, Massachusetts. *Limnol. Oceanogr.*, **4**, 448–462.

Bader, R. G., Hood, D. W., and Smith, J. B., 1960. Recovery of dissolved organic matter in seawater and organic sorption by particulate material. *Geochim. Cosmochim. Acta*, **19**, 236–243.

Baldwin, C. T., 1974. The control of mud crack patterns by small gastropod trails. *J. Sed. Petrol.*, **44**, 695–697.

Barnes, S. S., Craft, T. F., and Windom, H. L., 1973. Iron-scandium budget in sediments of two Georgia salt marshes. *Georgia Acad. Sci. Bull.*, **31**, 23–30.

Bartberger, C. E., 1976. Sediment sources and sedimentation rates, Chincoteague Bay, Maryland and Virginia. *J. Sed. Petrol.*, **46**, 326–336.

Basan, P. B., 1979. Classification of low marsh habitats in a Georgia marsh. *Georgia J. Sci.*, **37**, 139–154.

Basan, P. B., and Frey, R. W., 1977. Actual-palaeontology and neoichnology of salt marshes near Sapelo Island, Georgia. *In*: Crimes, T. P., and Harper, J. C. (eds.), Trace fossils 2. *Geol. J. Spec. Issue 9*. Seel House Press, Liverpool, pp. 41–70.

Beeftink, W. G., 1966. Vegetation and habitat of the salt marshes and beach plains in the south-western part of The Netherlands. *Wentia*, **15**, 83–108.

Belknap, D. F., and Kraft, J. C., 1977. Holocene relative sea-level changes and coastal stratigraphic units on the northwest flank of the Baltimore canyon trough geosyncline. *J. Sed. Petrol.*, **47**, 610–629.

Bernard, H. A., LeBlanc, R. J., and Major, C. F., 1962. Recent and Pleistocene geology of southeast Texas. *In*: Rainwater, E. H., and Zingula, R. P. (eds.), *Geology of the Gulf Coast and Central Texas, and Guidebook of Excursions*. Houston Geol. Soc., Houston, TX, pp. 175–224.

Berner, R. A., 1969. Chemical changes affecting dissolved calcium during the bacterial decomposition of fish and clams in sea water. *Mar. Geol.*, **7**, 253–274.

Berner, R. A., Baldwin, T., and Holdren, Jr., G. R., 1979. Authigenic iron sulfides as paleosalinity indicators. *J. Sed. Petrol.*, **49**, 1345–1350.

Biddle, P., and Miles, J. H., 1972. The nature of contemporary silts in British estuaries. *Sed. Geol.*, **7**, 23–33.

Bloom, A. L., 1964. Peat accumulation and compaction in a Connecticut coastal marsh. *J. Sed. Petrol.*, **34**, 599–603.

Blum, J. L., 1968. Salt marsh *Spartinas* and associated algae. *Ecol. Monogr.*, **38**, 199–221.

Bouma, A. H., 1963. A graphic presentation of the facies model of salt marsh deposits. *Sedimentology*, **2**, 122–129.

Bradshaw, J. S., 1968. Environmental parameters and marsh Foraminifera. *Limnol. Oceanogr.*, **13**, 26–38.

Bridges, P. H., and Leeder, M. R., 1976. Sedimentary model for intertidal mudflat channels, with examples from the Solway Firth, Scotland. *Sedimentology*, **23**, 533–552.

Broussard, M. L. (ed.), 1975. *Deltas: Models for Exploration*. Houston Geol. Soc., Houston, TX, 555 pp.

Castañares, A. A., and Phleger, F. B (eds.), 1969. *Coastal lagoons*. Ciudad Universitaria, Univ. Nacion. Autón., México, 686 pp.

Chakrabarti, A. K., 1971. Studies on sediment movement at the entrance of a tidal river. *Sed. Geol.*, **6**, 111–127.

Chapman, V. J., 1974. *Salt Marshes and Salt Deserts of the World*. J. Cramer, Lehre, preface + 392 pp.

Chapman, V. J., 1976. *Coastal Vegetation*. Pergamon Press, New York, 292 pp.

Chapman, V. J. (ed.), 1977. *Wet Coastal Ecosystems*. Elsevier, Amsterdam, 428 pp.

Christensen, B. A., 1976. Hydraulics of sheet flow in wetlands. *In: Inland Waters for Navigation, Flood Control and Water Diversions*. Amer. Soc. Civil Engrs., Symp. Proc., Colorado State Univ., Boulder, CO, pp. 745–759.

Clarke, L. D., and Hannon, N. J., 1969. The mangrove swamp and salt marsh communities of the Sydney district. II. The holocoenotic complex with particular reference to physiography. *J. Ecol.*, **57**, 213–234.

Cleary, W. J., Hosier, P. E., and Wells, G. R., 1979. Genesis and significance of marsh islands within southeastern North Carolina lagoons. *J. Sed. Petrol.*, **49**, 703–710.

Cohen, A. D., and Spackman, W., 1972. Methods in peat petrology and their application to reconstruction of paleoenvironments. *Geol. Soc. Amer. Bull.*, **83**, 129–141.

Coleman, J. M., 1966. Ecological changes in a massive fresh-water clay sequence. *Gulf Coast Assoc. Geol. Socs. Trans.*, **16**, 159–174.

Coleman, J. M., 1976. *Deltas: Processes of Deposition and Models for Exploration*. Continuing Educ. Publ. Co., Champaign, IL, 102 pp.

Coleman, J. M., and Gagliano, S. M., 1965. Sedimentary structures: Mississippi River deltaic plain. *In:* Middleton, G. V. (ed.), *Primary Sedimentary Structures and their Hydrodynamic Interpretation*. Soc. Econ. Paleont. Mineral. Spec. Publ. 12, Tulsa, OK, pp. 133–148.

Coleman, J. M., and Wright, L. D., 1975. Modern river deltas: variability of processes and sand bodies. *In:* Broussard, M. L. (ed.), *Deltas: Models for Exploration*. Houston Geol. Soc., Houston, TX, pp. 99–149.

Cooper, A. W., 1974. Salt marshes. *In:* Odum, H. T., Copeland, B. J., and McMahan, E. A. (eds.), *Coastal Ecological Systems of the United States*, Vol. II. Conservation Foundation, Washington, DC, pp. 55–99.

Cotnoir, L. J., 1974. Marsh soils of the Atlantic coast. *In:* Reimold, R. J., and Queen, W. H. (eds.), *Ecology of Halophytes*. Academic Press, New York, pp. 441–447.

Daiber, F. C., 1977. Salt-marsh animals: distributions related to tidal flooding, salinity and vegetation. *In*: Chapman, V. J. (ed.), *Wet Coastal Ecosystems*. Elsevier, Amsterdam, pp. 79–108.

Dale, N. G., 1974. Bacteria in intertidal sediments: factors related to their distribution. *Limnol. Oceanogr.*, **19**, 509–518.

Day, Jr., J. W., Smith, W. G., Wagner, P. R., and Stowe, W. C., 1973. *Community Structure and Carbon Budget of a Salt Marsh and Shallow Bay Estuarine System in Louisiana*. Publ. LSU-SG-72-04, Louisiana State Univ., Center Wetland Res., Baton Rouge, LA, 79 pp.

Deery, J. R., and Howard, J. D., 1977. Origin and character of washover fans on the Georgia coast, U.S.A. *Gulf Coast Assoc. Geol. Socs. Trans.*, **27**, 259–271.

DePratter, C. B., and Howard, J. D., 1977. History of shoreline changes determined by archaeological dating: Georgia coast, U.S.A. *Gulf Coast Assoc. Geol. Socs. Trans.*, **27**, 252–258.

Dionne, J. C., 1972. Charactéristiques des schorres des régions froides, en particulier de l'estuaire du Saint-Laurent. *Z. Geomorph. N. F.*, **13**, 131–162.

Dörjes, J., and Howard, J. D., 1975. Estuaries of the Georgia coast, U.S.A.: sedimentology and biology. IV. Fluvial-marine transition indicators in an estuarine environment, Ogeechee River-Ossabaw Sound. *Senckenberg. Marit.*, **7**, 137–179.

Duncan, W. H., 1974. Vascular halophytes of the Atlantic and Gulf coasts of North America north of Mexico. *In*: Reimold, R. J., and Queen, W. H. (eds.), *Ecology of Halophytes*. Academic Press, New York, pp. 23–50.

Dyer, K. R., 1973. *Estuaries: A Physical Classification*. John Wiley, New York, 140 pp.

Edelman, C. H., and van Staveren, J. M., 1958. Marsh soils in the United States and in the Netherlands. *J. Soil Water Cons.*, **13**, 5–17.

Edwards, A. P., and Bremner, J. M., 1967. Microaggregates in soils. *J. Soil Sci.*, **18**, 64–73.

Edwards, J. M., and Frey, R. W., 1977. Substrate characteristics within a Holocene salt marsh, Sapelo Island, Georgia. *Senckenberg. Marit.*, **9**, 215–259.

Egler, F. E., 1952. Southeast saline everglades vegetation, Florida, and its management. *Vegetatio*, **3**, 213–265.

Eleuterius, L. N., 1972. The marshes of Mississippi. *Castanea*, **37**, 153–168.

Ernissee, J. J., Abbott, W. H., Colquhoun, D. J., and Pierce, J. W., 1974. Bioflocculation: a significant process in estuarine sedimentation (Ab.). *Geol. Soc. Amer., Abs. Prog.*, **6**, 352–353.

Evans, G., 1965. Intertidal flat sediments and their environments of deposition in the Wash. *Geol. Soc. London Quart. J.*, **121**, 209–245.

Fagerstrom, J. A., 1967. Development, flotation, and transportation of mud crusts—neglected factors in sedimentology. *J. Sed. Petrol.*, **37**, 73–79.

Fischer, A. G., 1961. Stratigraphic record of transgressing seas in light of sedimentation on Atlantic coast of New Jersey. *Amer. Assoc. Petrol. Geol. Bull.*, **45**, 1656–1666.

Fornes, A. O., and Reimold, R. J., 1973. *The Estuarine Environment: Location of Mean High Water—Its Engineering, Economic and Ecological Potential.* Amer. Soc. Photogram., Proc., Fall Convention, Walt Disney World, Lake Buena Vista, FL, pp. 938–978.

Frankenberg, D., and Westerfield, Jr., C. W., 1968. Oxygen demand and oxygen depletion capacity of sediments from Wassaw Sound, Georgia. *Georgia Acad. Sci. Bull.*, **26**, 160–172.

Frey, R. W., 1970. Environmental significance of recent marine lebensspuren near Beaufort, North Carolina. *J. Paleontol.*, **44**, 507–519.

Frey, R. W., and Basan, P. B., 1981. Taphonomy of relict Holocene salt marsh deposits, Cabretta Island, Georgia. *Senckenberg. Marit.*, **13**, 111–155.

Frey, R. W., Basan, P. B., and Scott, R. M., 1973. Techniques for sampling salt marsh benthos and burrows. *Amer. Midland Naturalist*, **89**, 228–234.

Frey, R. W., and Howard, J. D., 1969. A profile of biogenic sedimentary structures in a Holocene barrier island-salt marsh complex, Georgia. *Gulf Coast Assoc. Geol. Socs. Trans.*, **19**, 427–444.

Frey, R. W., and Howard, J. D., 1980. Physical and biogenic processes in Georgia estuaries. II. Intertidal facies. *In*: McCann, S. B. (ed.), *Sedimentary Processes and Animal-Sediment Relationships in Tidal Environments.* Geol. Assoc. Canada, Short Course Notes, **1**, 183–220.

Frey, R. W., Voorhies, M. R., and Howard, J. D., 1975. Estuaries of the Georgia coast, U.S.A.: sedimentology and biology. VIII. Fossil and recent skeletal remains in Georgia estuaries. *Senckenberg. Marit.*, **7**, 257–295.

Friedman, G. M., and Gavish, E., 1970. Chemical changes in interstitial waters from sediments of lagoonal, deltaic, river, estuarine, and salt water marsh and cove environments. *J. Sed. Petrol.*, **40**, 930–953.

Gabriel, B. C., and de la Cruz, A. A., 1974. Species composition, standing stock, and net primary production of a salt marsh community in Mississippi. *Chesapeake Sci.*, **15**, 72–77.

Gallagher, J. L., 1974. Sampling macro-organic matter profiles in salt marsh plant root zones. *Soil Sci. Soc. Amer. Proc.*, **38**, 154–155.

Gallagher, J. L., and Reimold, R. J., 1973. *Tidal Marsh Plant Distribution and Productivity Patterns from the Sea to Fresh Water—A Challenge in Resolution and Discrimination.* Fourth Biennial Workshop Color Aerial Photography, Proc., pp. 166–183.

Gibbs, R. J., 1977. Clay mineral segregation in the marine environment. *J. Sed. Petrol.*, **47**, 237–243.

Godfrey, P. J., and Godfrey, M. M., 1974. The role of overwash and inlet dynamics in the formation of salt marshes on North Carolina barrier islands. *In*: Reimold, R. J., and Queen, W. H. (eds.), *Ecology of Halophytes.* Academic Press, New York, pp. 407–427.

Gosselink, J. G., and Kirby, C. J., 1974. Decomposition of salt marsh grass, *Spartina alterniflora* Loisel. *Limnol. Oceanogr.*, **19**, 825–832.

Gould, H. R., and Morgan, J. P., 1962. Coastal Louisiana swamps and marshlands.

In: Rainwater, E. H., and Zingula, R. P. (eds.), *Geology of the Gulf Coast and Central Texas, and Guidebook of Excursions*. Houston Geol. Soc., Houston, TX, pp. 286–341.

Gouleau, D., 1976. Le rôle des Diatomées benthiques dans l'engraissement rapide des vasières atlantiques découvrantes. *Compt. Rend., Ser. D.*, **283**, 21–23.

Greenland, D. J., 1971. Interactions between humic and fulvic acids and clays. *Soil Sci.*, **111**, 34–41.

Greensmith, J. T., and Tucker, E. V., 1973. Peat balls in late-Holocene sediments of Essex, England. *J. Sed. Petrol.*, **43**, 894–897.

Gross, M. G., 1972. *Oceanography, a View of the Earth*. Prentice-Hall, Englewood Cliffs, NJ, 581 pp.

Guilcher, A., 1963. Estuaries, deltas, shelf and slope. *In*: Hill, M. N. (ed.), *The Sea*, Vol. III. Interscience, New York, pp. 620–654.

Hackney, C. T., Burbanck, W. D., and Hackney, O. P., 1976. Biological and physical dynamics of a Georgia tidal creek. *Chesapeake Sci.*, **17**, 271–280.

Hanson, R. B., and Wiebe, W. J., 1977. Heterotrophic activity associated with particulate size fractions in a *Spartina alterniflora* salt-marsh estuary, Sapelo Island, Georgia, U.S.A., and the continental shelf waters. *Mar. Biol.*, **42**, 321–330.

Hathaway, J. C., 1972. Regional clay mineral facies in estuaries and continental margin of the United States and east coast. *In*: Nelson, B. W. (ed.), *Environmental Framework of Coastal Plain Estuaries*. Geol. Soc. Amer. Memoir 133, Boulder, CO, pp. 293–316.

Haven, D. S., and Morales-Alamo, R., 1968. Occurrence and transport of faecal pellets in suspension in a tidal estuary. *Sed. Geol.*, **2**, 141–151.

Haven, D. S., and Morales-Alamo, R., 1970. Filtration of particles from suspension by the American oyster *Crassostrea virginica*. *Biol. Bull.*, **139**, 248–264.

Hayden, B. P., and Dolan, R., 1979. Barrier islands, lagoons, and marshes. *J. Sed. Petrol.*, **49**, 1061–1072.

Hoese, H. D., 1967. Effect of higher than normal salinities on salt marshes. *Contr. Mar. Sci.*, **12**, 249–261.

Holeman, J. N., 1968. Sediment yield of major rivers of the world. *Water Resources Res.*, **4**, 737–747.

Horodyski, R. J., Bloeser, B., and Haar, S. V., 1977. Laminated algal mats from a coastal lagoon, Laguna Mormona, Baja California, Mexico. *J. Sed. Petrol.*, **47**, 680–696.

Howard, J. D., Elders, C. A., and Heinbokel, J. F., 1975. Estuaries of the Georgia coast, U.S.A.: sedimentology and biology. V. Animal-sediment relationships in estuarine point bar deposits, Ogeechee River-Ossabaw Sound, Georgia. *Senckenberg. Marit.*, **7**, 181–203.

Howard, J. D., and Frey, R. W., 1973. Characteristic physical and biogenic sedimentary structures in Georgia estuaries. *Amer. Assoc. Petrol. Geol. Bull.*, **57**, 1169–1184.

Howard, J. D., and Frey, R. W., 1975a. Estuaries of the Georgia coast, U.S.A.: sedimentology and biology. I. Introduction. *Senckenberg. Marit.*, **7**, 1–31.

Howard, J. D., and Frey, R. W., 1975b. Estuaries of the Georgia coast, U.S.A.: sedimentology and biology. II. Regional animal-sediment characteristics of Georgia estuaries. *Senckenberg. Marit.*, **7**, 33–103.

Howard, J. D., and Frey, R. W., 1980. Physical and biogenic processes in Georgia estuaries. I. Coastal setting and subtidal facies. *In*: McCann, S. B. (ed.), *Sedimentary Processes and Animal-Sediment Relationships in Tidal Environments*. Geol. Assoc. Canada, Short Course Notes, **1**, 153–182.

Hoyt, J. H., and Hails, J. R., 1967. Pleistocene shoreline sediments in coastal Georgia: deposition and modification. *Science*, **155**, 1541–1543.

Inman, D. L., and Nordstrom, C. E., 1971. On the tectonic and morphologic classification of coasts. *J. Geol.*, **79**, 1–21.

Johannessen, C. L., 1964. Marshes prograding in Oregon: Aerial photographs. *Science*, **146**, 1575–1578.

Johnson, D., 1925. *The New England-Acadian Shoreline*. John Wiley, New York, 608 pp.

Kaye, C. A., and Barghoorn, E. S., 1964. Late Quaternary sea-level change and crustal rise at Boston, Massachusetts, with notes on the autocompaction of peat. *Geol. Soc. Amer. Bull.*, **75**, 63–80.

Kerwin, J. A., and Pedigo, R. A., 1971. Synecology of a Virginia salt marsh. *Chesapeake Sci.*, **12**, 125–130.

Kirby, C. J., and Gosselink, J. G., 1976. Primary production in a Louisiana Gulf Coast *Spartina alterniflora* marsh. *Ecology*, **57**, 1052–1059.

Knebel, H. J., Conomos, T. J., and Commeau, J. A., 1977. Clay-mineral variability in the suspended sediments of the San Francisco Bay system, California. *J. Sed. Petrol.*, **47**, 229–236.

Knight, R. J., and Dalrymple, R. W., 1976. Winter conditions in a macrotidal environment, Cobequid Bay, Nova Scotia. *Rev. Géogr. Montréal*, **30**, 65–85.

Kolb, C. R., and van Lopik, J. R., 1966. Depositional environments of the Mississippi River deltaic plain—southeastern Louisiana. *In*: Shirley, M. L. (ed.), *Deltas in Their Geologic Framework*. Houston Geol. Soc., Houston, TX, pp. 17–61.

Kraeuter, J. N., 1976. Biodeposition by salt-marsh invertebrates. *Mar. Biol.*, **35**, 215–223.

Kraeuter, J., and Haven, D. S., 1970. Fecal pellets of common invertebrates of lower York River and lower Chesapeake Bay, Virginia. *Chesapeake Sci.*, **11**, 159–173.

Kraeuter, J. N., and Wolf, P. L., 1974. The relationship of marine macroinvertebrates to salt marsh plants. *In*: Reimold, R. J., and Queen, W. H. (eds.), *Ecology of Halophytes*. Academic Press, New York, pp. 449–462.

Kraft, J. C., 1971. Sedimentary facies patterns and geologic history of a Holocene marine transgression. *Geol. Soc. Amer. Bull.*, **82**, 2131–2158.

Kuenzler, E J., 1961. Structure and energy flow of a mussel population in a Georgia salt marsh. *Limnol. Oceanogr.*, **6**, 191–204.

Kues, B. S., and Siemers, C. T., 1977. Control of mudcrack patterns by the infaunal bivalve *Pseudocyrena*. *J. Sed. Petrol.*, **47**, 844–848.

Kurz, H., and Wagner, K., 1957. Tidal marshes of the Gulf and Atlantic coasts of northern Florida and Charleston, South Carolina. *Florida State Univ. Stud.*, **24**, 1–168.

Land, L. S., and Hoyt, J. H., 1966. Sedimentation in a meandering estuary. *Sedimentology*, **6**, 191–207.

Lauff, G. H. (ed.), 1967. *Estuaries*. Amer. Assoc. Advan. Sci. Publ. 83, Washington, DC, 757 pp.

Laury, R. L., 1971. Stream bank failure and rotational slumping: preservation and significance in the geologic record. *Geol. Soc. Amer. Bull.*, **82**, 1251–1266.

Leopold, L. B., and Maddock, T. Jr., 1953. *The Hydraulic Geometry of Stream Channels and Some Physiographic Implications*. Prof. Paper 252, U.S. Geol. Survey, Washington, D.C., 56 pp.

Letzsch, S. W., and Frey, R. W., 1980a. Deposition and erosion in a Holocene salt marsh, Sapelo Island, Georgia. *J. Sed. Petrol.*, **50**, 529–542.

Letzsch, S. W., and Frey, R. W., 1980b. Erosion of salt marsh tidal creek banks, Sapelo Island, Georgia. *Senckenberg. Marit.*, **12**, 201–212.

Linton, T. L., 1969. Physical and biological problems of impounding salt marshes. *In*: Castañares, A. A., and Phleger, F. B. (eds.), *Coastal Lagoons*. Ciudad Universitaria, Univ. Nacion. Autón., México, pp. 451–455.

Macdonald, K. B., 1969. Quantitative studies of salt marsh mollusc faunas from the North American Pacific Coast. *Ecol. Monogr.*, **39**, 33–60.

Macdonald, K. B., 1977a. Plant and animal communities of Pacific North American salt marshes. *In*: Chapman, V. J. (ed.), *Wet Coastal Ecosystems*. Elsevier, Amsterdam, pp. 167–191.

Macdonald, K. B., 1977b. Coastal salt marsh. *In*: Barbour, M. G., and Major, J. (eds.), *Terrestrial Vegetation of California*. John Wiley, New York, pp. 263–294.

Macdonald, K. B., and Barbour, M. G., 1974. Beach and salt marsh vegetation of the North American Pacific coast. *In*: Reimold, R. J., and Queen, W. H. (eds.), *Ecology of Halophytes*. Academic Press, New York, pp. 175–233.

Mackin, J. G., 1962. Canal dredging and silting in Louisiana bays. *In*: Mackin, J. G., and Hopkins, S. H. (eds.), *Studies on Oysters in Relation to the Oil Industry*, Vol. 7. Inst. Marine Sci., pp. 262–314.

Mackin, J. G., and Hopkins, S. H. (eds.), 1962. *Studies on Oysters in Relation to the Oil Industry*, Vol. 7. Inst. Marine Sci., 319 pp.

May, III, M. S., 1974. Probable agents for the formation of detritus from the halophyte, *Spartina alterniflora*. *In*: Reimold, R. J., and Queen, W. H. (eds.), *Ecology of Halophytes*. Academic Press, New York, pp. 429–440.

McCave, I. N., 1970. Deposition of fine-grained suspended sediment from tidal currents. *J. Geophys. Res.*, **75**, 4151–4159.

McCulloch, D. S., Peterson, D. H., Carlson, P. R., and Conomos, T. J., 1970. A preliminary study of the effects of water circulation in the San Francisco Bay estuary: Some effects of fresh-water inflow on the flushing of south San Francisco Bay. U.S. Geol. Survey Circ. 637A, Washington, DC, pp. A1–A27.

Meade, R. H., 1969. Landward transport of bottom sediments in estuaries of the Atlantic coastal plain. *J. Sed. Petrol.*, **39**, 222–234.

Meyerson, A. L., 1972. Pollen and paleosalinity analyses from a Holocene tidal marsh sequence, Cape May County, New Jersey. *Mar. Geol.*, **12**, 335–357.

Miller, D. C., 1961. The feeding mechanism of fiddler crabs, with ecological considerations of feeding adaptations. *Zoologia*, **46**, 89–100.

Miller, W. R., and Egler, F. E., 1950. Vegetation of the Wequetequock-Pawcatuck tide-marshes, Connecticut. *Ecol. Monogr.*, **20**, 143–172.

Narkis, N., Rebhun, M., and Sperber, H., 1968. Flocculation of clay suspensions in the presence of humic and fulvic acids. *Israel J. Chem.*, **6**, 295–305.

Neiheisel, J., and Weaver, C. E., 1967. Transport and deposition of clay minerals, southeastern United States. *J. Sed. Petrol.*, **37**, 1084–1116.

Nelson, B. W. (ed.), 1972. *Environmental Framework of Coastal Plain Estuaries.* Geol. Soc. Amer. Memoir 133, Boulder, CO, 619 pp.

Nixon, S. W., and Oviatt, C. A., 1973. Ecology of a New England salt marsh. *Ecol. Monogr.*, **43**, 463–498.

Odum, E. P., 1973. A description and value assessment of South Atlantic and Gulf Coast marshes and estuaries. *In: Fish and wildlife values of the estuarine habitat.* Proc. Bur. Sport Fish. Wildl., Atlanta, pp. 23–31.

Odum, E. P., 1974. Halophytes, energetics and ecosystems. *In*: Reimold, R. J., and Queen, W. H. (eds.), *Ecology of Halophytes.* Academic Press, New York, pp. 599–602.

Odum, E. P., and de la Cruz, A. A., 1967. Particulate organic detritus in a Georgia salt marsh-estuarine ecosystem. *In*: Lauff, G. H. (ed.), *Estuaries.* Amer. Assoc. Advan. Sci. Publ. 83, Washington, DC, pp. 383–388.

Odum, W. E., 1968. The ecological significance of fine particle selection by the striped mullet *Mugil cephalus. Limnol. Oceanogr.*, **13**, 92–98.

Oppenheimer, C. H., 1960. Bacterial activity in sediments of shallow marine bays. *Geochim. Cosmochim. Acta*, **19**, 244–260.

Passega, R., and Byramjee, R., 1969. Grain-size image of clastic deposits. *Sedimentology*, **13**, 233–252.

Penfound, W. T., and Hathaway, E. Ș., 1938. Plant communities in the marshlands of southeastern Louisiana. *Ecol. Monogr.*, **8**, 1–56.

Pestrong, R., 1965. The development of drainage patterns on tidal marshes. *Stanford Univ. Pubs. Geol. Sci.*, **10** (2), 1–87.

Pestrong, R., 1969. The shear strength of tidal marsh sediments. *J. Sed. Petrol.*, **39**, 322–326.

Pestrong, R., 1972a. San Francisco Bay tidelands. *California Geol.*, **25**, 27–40.

Pestrong, R., 1972b. Tidal-flat sedimentation at Cooley Landing, southwest San Francisco Bay. *Sed. Geol.*, **8**, 251–288.

Pethick, J. S., 1981. Long-term accretion rates on tidal salt marshes. *J. Sed. Petrol.*, **51**, 571–577.

Phleger, F. B, 1969a. Some general features of coastal lagoons. *In*: Castañares, A. A., and Phleger, F. B (eds.), *Coastal Lagoons.* Ciudad Universitaria, Univ. Nacion. Autón., México, pp. 5–25.

Phleger, F. B, 1969b. A modern evaporite deposit in Mexico. *Amer. Assoc. Petrol. Geol. Bull.*, **53**, 824–829.

Phleger, F. B, 1970. Foraminiferal populations and marine marsh processes. *Limnol. Oceanogr.*, **15**, 522–534.

Phleger, F. B, 1977. Soils of marine marshes. *In*: Chapman, V. J. (ed.), *Wet Coastal Ecosystems*. Elsevier, Amsterdam, pp. 69–77.

Pinet, P. R., and Frey, R. W., 1977. Organic carbon in surface sands seaward of Altamaha and Doboy Sounds, Georgia. *Geol. Soc. Amer. Bull.*, **88**, 1731–1739.

Pinet, P. R., and Morgan, Jr., W. P., 1979. Implications of clay-provenance studies in two Georgia estuaries. *J. Sed. Petrol.*, **49**, 575–580.

Pomeroy, L. R., 1959. Algal productivity in salt marshes of Georgia. *Limnol. Oceanogr.*, **4**, 386–397.

Pomeroy, L. R., Bancroft, K., Breed, J., Christian, R. R., Frankenberg, D., Hall, J. R., Maurer, L. G., Wiebe, W. J., Wiegert, R. G., and Wetzel, R. L., 1977. Flux of organic matter through a salt marsh. Circulation, sediments, and transfer of material in the estuary. *Estuarine Processes*, **2**, 270–279. Academic Press, New York.

Pomeroy, L. R., Shenton, L. R., Jones, R. D. H., and Reimold, R. J., 1972. Nutrient flux in estuaries. *In*: *Nutrients and Eutrophication*. Amer. Soc. Limnol. Oceanogr. Spec. Symp., **1**, 274–291.

Pomeroy, L. R., and Wiegert, R. G. (eds.), 1981. *The Ecology of a Salt Marsh*. Ecological Studies No. 38. Springer-Verlag, New York, 271 pp.

Postma, H., 1967. Sediment transport and sedimentation in the estuarine environment. *In*: Lauff, G. H. (ed.), *Estuaries*. Amer. Assoc. Advan. Sci. Publ. 83, Washington, DC, pp. 158–179.

Pryor, W. A., 1975. Biogenic sedimentation and alteration of argillaceous sediments in shallow marine environments. *Geol. Soc. Amer. Bull.*, **86**, 1244–1254.

Purer, E. A., 1942. Plant ecology of the coastal salt marshlands of San Diego County, California. *Ecol. Monogr.*, **12**, 81–111.

Ragotzkie, R. A., 1959. Drainage patterns in salt marshes. *In*: Ragotzkie, R. A., Pomeroy, L. R., Teal, J. M., and Scott, D. C. (eds.), *Proceedings of the Salt Marsh Conference*. Univ. Georgia Marine Inst., Sapelo Island, pp. 22–28.

Ragotzkie, R. A., and Bryson, R. A., 1955. Hydrography of the Duplin River, Sapelo Island, Georgia. *Bull. Mar. Sci.*, **5**, 297–314.

Ralph, R. D., 1977. The Myxophyceae of the marshes of southern Delaware. *Chesapeake Sci.*, **18**, 208–221.

Ranwell, D. S., 1972. *Ecology of Salt Marshes and Sand Dunes*. Chapman and Hall, London, 258 pp.

Rashid, M. A., 1974. Absorption of metals on sedimentary and peat humic acids. *Chem. Geol.*, **13**, 115–123.

Redfield, A. C., 1965. The thermal regime in salt marsh peat at Barnstable, Massachusetts. *Tellus*, **17**, 246–259.

Redfield, A. C., 1972. Development of a New England salt marsh. *Ecol. Monogr.*, **42**, 201–237.

Redfield, A. C., and Rubin, M., 1962. The age of salt marsh peat and its relation to recent changes in sea level at Barnstable, Massachusetts. *Proc. Natl. Acad. Sci.*, **48**, 1728–1735.

Reimold, R. J., 1977. Mangals and salt marshes of eastern United States. *In*: Chapman, V. J. (ed.), *West Coastal Ecosystems*. Elsevier, Amsterdam, pp. 157–166.

Reimold, R. J., Gallagher, J. L., Linthurst, R. A., and Pfeiffer, W. J., 1975. Detritus production in coastal Georgia salt marshes. *In*: Cronin, L. E. (ed.), *Estuarine Research*, Vol. 1. Academic Press, New York, pp. 217–228.

Reimold, R. J., and Queen, W. H. (eds.), 1974. *Ecology of Halophytes*. Academic Press, New York, 605 pp.

Reineck, H.-E. (ed.), 1970. *Das Watt*. Waldemar Kramer, Frankfurt am Main, 142 pp.

Reineck, H.-E., 1973. *Bibliographie geologischer Arbeiten über rezente und fossile Kalk und Silikatwatten*. Cour. Forsch.-Inst. Senckenberg, 57 pp.

Rhoads, D. C., and Boyer, L. F., 1982. The effects of marine benthos on physical properties of sediments: a successional perspective. *In*: McCall, P. L., and Tevesz, M. J. S. (eds.), *Animal-Sediment Relations: The Biogenic Alteration of Sediments*. Plenum Press, New York, pp. 3–52.

Riley, G. A., 1963. Organic aggregates in seawater and the dynamics of their formation and utilization. *Limnol. Oceanogr.*, **8**, 372–381.

Rumboltz, M., Arthur, J. F., and Ball, M. D., 1976. *Sediment Transport Characteristics of the Upper San Francisco Bay-Delta Estuary*. Proc. 3rd Fed. Interagency Sed. Conf., Denver, CO, pp. 4/12–4/25.

Rusnak, G. A., 1967. Rates of sediment accumulation in modern estuaries. *In*: Lauff, G. H. (ed.), *Estuaries*. Amer. Assoc. Advan. Sci., Washington, DC, pp. 180–184.

Rützler, K., 1969. The mangrove community, aspects of its structure, faunistics and ecology. *In*: Castañares, A. A., and Phleger, F. B (eds.), *Coastal Lagoons*. Ciudad Universitaria, Univ. Nacion. Autón., México, pp. 515–535.

Schou, A., 1967. Estuarine research in the Danish moraine archipelago. *In*: Lauff, G. H. (ed.), *Estuaries*. Amer. Assoc. Advan. Sci. Publ. 83, Washington, DC, pp. 129–145.

Schubel, J. R., 1968. Turbidity maximum of the northern Chesapeake Bay. *Science*, **161**, 1013–1015.

Sears, P. B., 1963. Vegetation, climate, and coastal submergence in Connecticut. *Science* **140**, 59–60.

Shanholtzer, G. F., 1974. Relationship of vertebrates to salt marsh plants. *In*: Reimold, R. J., and Queen, W. H. (eds.), *Ecology of Halophytes*. Academic Press, New York, pp. 463–474.

Smalley, A. E., 1959. The role of two invertebrate populations, *Littorina irrorata* and *Orchelimum fidicinium*, in the energy flow of a salt marsh ecosystem. Unpubl. Ph.D. Dissert., Univ. Georgia, Athens, GA.

Staub, J. R., and Cohen, A. D., 1979. The Snuggedy Swamp of South Carolina: a back-barrier estuarine coal-forming environment. *J. Sed. Petrol.*, **49**, 133–144.

Steers, J. A., 1959. Salt marshes. *Endeavour*, **18** (70), 75–82.

Steers, J. A., 1977. Physiography. *In*: Chapman, V. J. (ed.), *Wet Coastal Ecosystems*. Elsevier, Amsterdam, pp. 31–60.

Stevenson, R. E., and Emery, K. O., 1958. *Marshlands at Newport Bay, California*. Occas. Papers, Allen Hancock Foundation, **20**, 109 pp.

Swain, F. M., 1971. Biogeochemistry of sediment samples from Broadkill marsh, Delaware. *J. Sed. Petrol.*, **41**, 549–556.

Tanner, W. F., 1960. Florida coastal classification. *Gulf Coast Assoc. Geol. Socs. Trans.*, **10**, 259–266.

Teal, J. M., 1962. Energy flow in the salt marsh ecosystem of Georgia. *Ecology*, **43**, 614–624.

Teal, J. M., and Kanwisher, J., 1961. Gas exchange in a Georgia salt marsh. *Limnol. Oceanogr.*, **6**, 388–399.

Turner, R. E., 1976. Geographic variations in salt marsh macrophyte production: a review. *Contrib. Mar. Sci.*, **20**, 47–68.

Valiela, I., and Teal, J. M., 1974. Nutrient limitations in salt marsh vegetation. *In*: Reimold, R. J., and Queen, W. H. (eds.), *Ecology of Halophytes*. Academic Press, New York, pp. 547–563.

van Andel, T. H., 1960. Sources and dispersion of Holocene sediments, northern Gulf of Mexico. *In*: Shepard, F. P., Phleger, F. B, and van Andel, T. H. (eds.), *Recent Sediments, Northwest Gulf of Mexico*. Amer. Assoc. Petrol. Geol., Tulsa, OK, pp. 34–55.

van Straaten, L. M. J. U., 1954. Composition and structure of recent marine sediments in The Netherlands. *Leidse Geol. Mededelin.*, **19**, 1–110.

van Straaten, L. M. J. U., 1961. Sedimentation in tidal flat areas. *J. Alberta Soc. Petrol. Geol.*, **9**, 203–226.

van Straaten, L. M. J. U., and Kuenen, P. H., 1957. Accumulation of fine grained sediments in the Dutch Wadden Sea. *Geol. en Mijnbouw*, **19**, 329–354.

Visher, G. S., and Howard, J. D., 1974. Dynamic relationship between hydraulics and sedimentation in the Altamaha estuary. *J. Sed. Petrol.*, **44**, 502–521.

Warme, J. E., 1971. Paleoecological aspects of a modern coastal lagoon. *Univ. California Pubs. Geol. Sci.*, **87**, 1–131.

Wayne, C. J., 1976. The effects of sea and marsh grass on wave energy. *Coastal Res. Notes*, **4** (7), 6–8.

Weimer, R. J., Howard, J. D., and Lindsay, D. R., 1982. Tidal flats and associated tidal channels. *In*: Scholle, P. A., and Spearing, D. (eds.), *Sandstone depositional environments*. Amer. Assoc. Petrol. Geol. Memoir 31, pp. 191–245.

Whitehouse, U. G., Jeffrey, L. M., and Debbrecht, J. D., 1960. Differential settling tendencies of clay minerals in saline waters. *In*: Ingerson, E. (ed.), *Clays and Clay Minerals*. Proc. 7th Natl. Conf., Monograph 5, Pergamon Press, New York, pp. 1–79.

Wiedemann, H. U., 1972a. Shell deposits and shell preservation in Quaternary and Tertiary estuarine sediments in Georgia, U.S.A. *Sed. Geol.*, **7**, 103–125.

Wiedemann, H. U., 1972b. Application of red-lead to the detection of dissolved sulfide in waterlogged soils. *Z. Pflanzenernährung Bodenkunde*, **133**, 73–81.

Wiegert, R. G., Christian, R. R., Gallagher, J. L., Hall, J. R., Jones, R. D. H., and Wetzel, R. L., 1975. A preliminary ecosystem model of coastal Georgia *Spartina* marsh. Chemistry, biology and the estuarine system. *Estuarine Research*, Vol. 1. Academic Press, New York, pp. 583–601.

Windom, H. L., Neal, W. J., and Beck, K. C., 1971. Mineralogy of sediments in three Georgia estuaries. *J. Sed. Petrol.*, **41**, 497–504.

Winker, C. D., and Howard, J. D., 1977. Correlation of tectonically deformed shorelines on the southern coastal plain. *Geology*, **5**, 123–127.

Wolf, P. L., Shanholtzer, S. F., and Reimold, R. J., 1975. Population estimates for *Uca pugnax* (Smith, 1870) on the Duplin estuary marsh, Georgia, U.S.A. (Decapoda Brachyura, Ocypodidae). *Crustaceana*, **29**, 79–91.

Wood, E. J. F., Odum, W. E., and Zieman, J. C., 1969. Influence of sea grasses on the productivity of coastal lagoons. *In*: Castañares, A. A., and Phleger, F. B (eds.), *Coastal Lagoons*. Ciudad Universitaria, Univ. Nacion. Autón., México, pp. 495–502.

Yeo, R. K., and Risk, M. J., 1981. The sedimentology, stratigraphy, and preservation of intertidal deposits in the Minas Basin system, Bay of Fundy. *J. Sed. Petrol.*, **51**, 245–260.

5

Coastal Dunes

Victor Goldsmith

Introduction

The importance of eolian deposition in coastal areas is clearly demonstrated by the size and bulk of coastal sand dunes in many areas. The dune fields of Coos Bay, Oregon (Figure 5-1), the southern end of Lake Michigan (Figure 5-2), and the Cape Cod dune fields southeast of Provincetown (Figure 5-3) and on Sandy Neck (Figure 5-4) are just a few outstanding examples. Smaller sand dune accumulations are an integral part of almost all depositional coasts. More subtle forms of eolian deposition on beaches, marshes, and intertidal sand beaches, and in shallow bays and estuaries, may not be as noticeable but are also of importance. Eolian deposition on beaches, together with storm washover deposition in dune areas, may account in part for the lack of environmentally definitive grain size criteria in many areas (Leatherman, 1975). Clearly, eolian deposits form an imposing percentage of total sediment accumulations on depositional coasts.

Sand dunes may occur where there is a large supply of sand, a wind to move it, and a place in which it can accumulate. The coastline is an ideal location for these criteria to be met. Longshore transport supplies the sand, and waves accumulate the sand on a beach. Differential cooling and warming between land and sea assures an onshore wind at least some of the time, regardless of the general wind circulation pattern. Climate is definitely not a criterion for accumulations of eolian deposits along the coast because large dunes form in arid climates (e.g., Guerrero Negro, Mexico; Inman *et al.*, 1966), temperate climates (Provincetown dunes; Messenger, 1958), and in humid areas with up to 300 cm of annual rainfall (Oregon, Washington, and southeast Alaskan coasts; Cooper, 1958). The only climatic condition lacking extensive coastal dunes appears to be humid-tropical. This may be related to a lack of sand supply because of intense chemical weathering, higher shear stress required to move almost continuously wet sand, and dense vegetation adjacent to the beach (Jennings, 1964; Nossin, 1962; Davies, 1973; Swan, 1979). Large-scale migrating dunes or transverse dune ridges (Cooper, 1958, 1967), which form through high-angle slipface deposition, appear to be the most common types of eolian deposition on arid coasts, whereas in deserts the most common type of dune is the compound longitudinal (McKee, 1979). However, such dunes are often present on humid coasts as well (e.g., the medaños on the Outer Banks of North Carolina, such

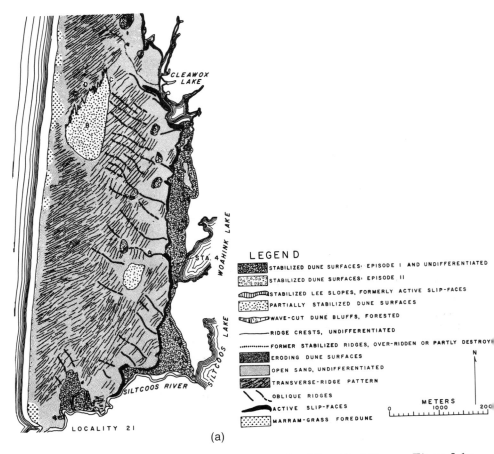

LEGEND

▨ STABILIZED DUNE SURFACES; EPISODE I AND UNDIFFERENTIATED

▨ STABILIZED DUNE SURFACES; EPISODE II

▨ STABILIZED LEE SLOPES, FORMERLY ACTIVE SLIP-FACES

▨ PARTIALLY STABILIZED DUNE SURFACES

▨ WAVE-CUT DUNE BLUFFS, FORESTED

— RIDGE CRESTS, UNDIFFERENTIATED

⋯⋯ FORMER STABILIZED RIDGES, OVER-RIDDEN OR PARTLY DESTROYE

▨ ERODING DUNE SURFACES

▨ OPEN SAND, UNDIFFERENTIATED

▨ TRANSVERSE-RIDGE PATTERN

▨ OBLIQUE RIDGES

▨ ACTIVE SLIP-FACES

▨ MARRAM-GRASS FOREDUNE

METERS
0   1000   200

(a)

Figure 5-1. A portion of the extensive dune fields of Coos Bay, Oregon. Figure 5-1a is from Cooper (1958) and 5-1b and 5-1c are photos taken by the author in 1972 illustrating the "precipitation" of sand along the active slipface of a transverse dune at the east margin of the dune field shown in a view looking west. The dune field here is approximately 300 km² in extent, with dunes up to 110 m high.

(b)

(c)

Figure 5-1. *(Continued.)*

Victor Goldsmith

(a)

(b)

as Jockey Ridge at Kitty Hawk, Cape Hatteras, and the Cape Cod and Coos Bay dune fields). Coasts with abundant rainfall most commonly contain eolian deposits largely fixed in place from the beginning by the abundant and well-adapted coastal vegetation. Such eolian deposits take the form of interconnected dune ridges with gently undulating upper surfaces and of distinct parabolic dunes (Figure 5-5). These vegetated dunes grow upward largely in place and are characterized by low-angle dipping beds or a combination of low-angle and high-angle beds (Goldsmith, 1971, 1973).

This chapter first examines the external forms and internal geometry of eolian sedimentation in coastal areas; second, the physical processes and materials of sediment accumulation; and finally, the controversial role of humans in both decolonizing dunes and in promoting dune growth via artificially inseminated vegetated coastal dunes.

Coastal Sand Dunes

World Survey

Regional geomorphic studies of coastal sand dunes are quite abundant. In the United States such investigations have been conducted on the coastal dunes of Cape Cod, Massachusetts (Messenger, 1958; Smith and Messenger, 1959), Plum Island, Massachusetts (McIntire and Morgan, 1963; Larsen, 1969), California (Zeller, 1962; Johnson, 1965; Cooper 1967; Hunter et al., 1983), Washington and Oregon (Cooper, 1958), and the Lake Michigan shoreline (Cressey, 1928; Olson, 1958). In Mexico, studies have been conducted on both the east (Marta, 1958) and west coasts (Inman et al., 1966), and along the New Brunswick coast of Canada by Rosen (1979). The Peruvian coastal dunes have also been studied (Finkel, 1959), as well as the dunes of Bermuda (MacKenzie, 1964a,b; Vacher, 1973) and of Barbados, British West Indies (Gooding, 1947).

European studies were made in England (Landsberg, 1956; Barnes and King, 1957; Kidson and Carr, 1960; King, 1972), Scotland (Mather, 1979), France (Buffault, 1942; Hallegouet, 1978), The Netherlands (van Straaten, 1965; Jelgersma and Van Regteren Altena, 1969), Germany (Hempel,

Figure 5-2. Two views of a large transverse dune at the southeast end of Lake Michigan at Tunnel Park, Michigan, illustrating the landward migration of the dune. View (a) is looking "seaward," and view (b) is along the shoreline, to the southwest.

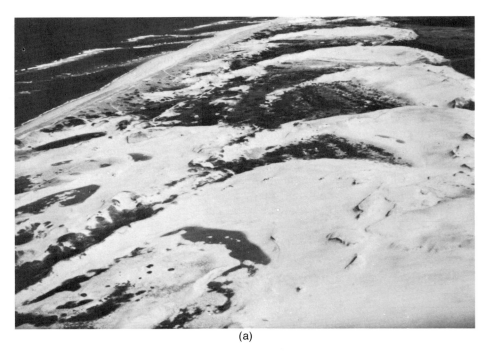

(a)

Figure 5-3. Two aerial photographs of the migrating, unvegetated parabolic dunes south of Provincetown, Cape Cod, Massachusetts, in March 1972. View (a), looking southeast, shows the importance of the prevailing northwest winds. The Atlantic Ocean is to the east (left) and Pilgrim Lake to the west (right). View (b) (looking west) shows the "precipitation" of sand inundating a pine forest.

1980), Poland (Borowka, 1980), Norway (Klemsdal, 1969), Finland (Alestalo, 1979), and the USSR (Zenkovich, 1967). African studies are represented by Kaiser (1926), Guilcher (1959), Begnold (1941), Walsh (1968), and McKee (1982); and Israeli studies by Streim (1954), Yaalon (1967), Karmeli et al., (1968), and Tsoar (1974, 1983b).

Other areas of extensive coastal dune studies are Java (Verstappen, 1957), New Zealand (Brothers, 1954), Tasmania (Jennings, 1957a,b), Australia (Bird, 1964; Simonett, 1950; Chapman et al., 1982), and Japan (Mii, 1958; Tanaka, 1967). Studies on the Baltic, Black Sea, and Caspian coasts have been summarized by Zenkovich (1967). Coastal desert studies have been summarized by Meigs (1966).

Studies of eolian cross-bedding and other internal features have been made by Land (1964), McBride and Hayes (1962), Hayes (1967), Bigarella et al., (1969), Goldsmith (1971, 1972, 1973), Rosen et al., (1977), Yaalon and Laronne (1971), Yaalon (1975), and McKee and Bigarella (1972).

(b)

Figure 5-3. *(Continued.)*

Similar studies on interior dunes have been made by McKee and Tibbitts (1964), McKee (1966) and Tsoar (1983a). Characteristics of eolian environments in general have been described by Smith (1960, 1968), Screiber (1968), and Bigarella (1972).

Dune Classification

Coastal sand dunes may be described and classified on the basis of either (1) description (e.g., external form and internal geometry) or (2) genetics (e.g., mode of formation). In this chapter the latter type of clasification system is used, and the basis for its use is discussed in later sections. The two main

Figure 5-4. Air view of the extensive dunes on Sandy Neck, Cape Cod, Massachusetts, a 15 km long barrier spit containing dunes up to 20 m in height. Note the eolian sand encroaching on the marsh (photo taken at an altitude of 1300 m in 1969 by C. H. Hobbs, Ill).

types of coastal sand dunes are (1) vegetated dunes (i.e., fixed) and (2) transverse dune ridges (i.e., generally migrating and bare of vegetation) (Figures 5-1, 5-5).

Goldsmith *et al.* (1977) have refined this basic coastal dune classification into four basic dune types: vegetated, artificially induced, medaños, and parabolic dunes; the first letter of each type conveniently form the acronym VAMP. The parabolic and artificially induced dunes are transitional types between the two extremes, completely vegetated and the unvegetated medaño. A fifth group contains less common dune types, such as lunettes (made of silt-clay and formed at the downwind end of dry lakes). The geomorphology, internal geometry, and origin of the four major dune types forming the VAMP classification are discussed throughout this chapter.

The occurrence of dunes on the coast appears to be unrelated to present climate but is directly related to sand supply and a favorable wind regime. The original sand source could be glaciofluvial sediments, as on the Baltic (Zenkovich, 1967), Cape Cod (Messenger, 1958; Goldsmith, 1972), and Lake Michigan (Cressey, 1928), or abundant fluvial sediments, as on the west coasts of the United States (Zeller, 1962; Cooper, 1967) and Australia (Bird, 1964). In any case, these sediments are moved to the shoreline and

(a)

(b)

Figure 5-5. Photographs (a) of the upper dune surfaces of low relief (3–5 m) interior dunes and (b) of the highest dunes (12 m) on Monomoy Island, Cape Cod, Massachusetts, in July 1969. Both the upper surfaces are relatively flat to gently undulatory. The dunes grow upward via vertical accretion with the dune grass acting as a baffle, trapping the sand, resulting in dune cross-bed sets dipping at low angles ($\bar{X}_D = 11.2°$).

deposited on the beach by the longshore currents and waves. The sand grains are then picked up and moved by the winds at low tide.* Leatherman (1975) has suggested that an additional major source of dune sand is storm overwash deposition.

Wind Regime

There is no simple relationship between the directional distribution of wind velocity and sand dune development; very low velocity winds are unable to move the sand, whereas extreme velocity winds may tend to destroy dunes. The higher velocity storm winds may orient the dunes (as on the Sinai and Oregon coasts) and even when accompanied by rain (as the Oregon coast) may move more sand per unit time than the lower velocity, more frequent prevailing winds. However, because of their much lower frequency, the high velocity winds may not be as important in dune development, in some areas, as the prevailing winds (Tsoar, 1974; Hunter *et al.*, 1983). Also, the orientation of the shoreline with respect to both dominant and prevailing winds is critical. The best way of ascertaining the characteristics of wind regime is to draw vector diagrams (Bird, 1964; Jennings, 1964; Landsberg, 1956) that take into account the relative frequency of occurrence of winds of different velocities (Figure 5-6). Landsberg (1956) has suggested the following formula for constructing a vector diagram that also takes into account the sand-moving power of the wind. The latter depends on the cube of the wind velocity above 16 km/hr (King, 1972, pp. 181–182):

$$b = s \ \Sigma \ n_j(v_j - v_t)^3, \qquad\qquad (5\text{-}1)$$

where b = resultant vector; s = scaling factor of 10^{-3}; v_j = Beaufort Force (km/hr); $v_t = 16.0$ km/hr; and n_j = frequency. However, this type of analysis has been applied in only a few dune studies (e.g., Gutman, 1977a; Rosen, 1979).

Maximum efficiency in dune growth occurs when the resultant wind vector is oriented normal to the coast. When the resultant wind vector is not normal to the coast, dunes with crests oblique to the adjacent shoreline may form, as on the Baltic coast (Figure 5-7), the Sinai coast of Egypt (Streim, 1954; Tsoar, 1983a), and the central Oregon coast (Hunter *et al.*, 1983).

Note that the equation above only takes into account onshore winds. On lee shores, the role of interior vegetation in decreasing sand transport must be taken into account (e.g., Gutman, 1977b). The relationship of wind direction to dune growth and internal dune geometry is discussed in the next section.

---

*The exact mechanisms of sand movement are described in detail in a later section.

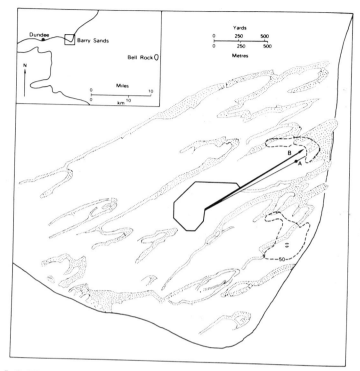

Figure 5-6. The relationship between dune orientation and wind direction using a vector diagram, Barry, England (after Landsberg, 1956; from King, 1972, Figure 6-11, p. 182).

Vegetated Dunes

The most common type of coastal dunes is the vegetated dune. Vegetated dunes are generally in the form of ridges with flat to undulating upper surfaces and continuous but irregular crests, often punctuated by blowouts (i.e., low places in the dune crest through which eolian transport occurs) and washover sluice channels (i.e., low places in the dunes through which water transport occurs during storms). Vegetated dune ridges are commonly made up of stabilized parabolic dunes (Smith, 1941), with the ends anchored by vegetation and the centers recessed back from the beach (Bird, 1964, Figure 42, p. 101).

Commonly there occurs a series of dune ridges, usually but not always parallel to each other and the coastline, reflecting the accretional history of the coastline much as growth rings on a tree reflect its history (Figure 5-8). Elsewhere they tend to reflect the wind regime instead of the coastal outline.

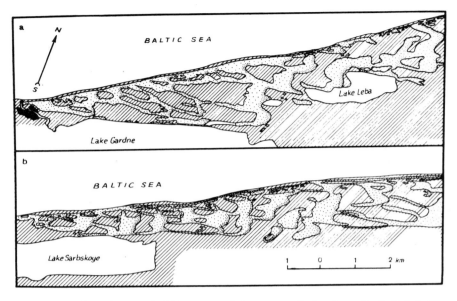

Figure 5-7. Sketch map of oblique coastal dunes near Leba along the Polish Baltic coast. The dune orientation is a result of the prevailing westerly winds blowing almost parallel to the coastline (from Zenkovich, 1967, Figure 294, p. 603).

Incipient Vegetated Dunes

Coastal sand dunes are initiated on accretional coasts above the spring high-tide line. Sand accumulation usually begins behind some obstacle or roughness element on the beach, such as flotsam washed up on the beach during storm conditions, or behind vegetation. King (1972) notes that the growth of incipient dunes on the Lincolnshire coast is promoted by the collection of flustra at the high-tide level.

This roughness element, whatever it may be, deflects the air stream (i.e., wind) around it. In the extreme case of a continuous impermeable obstacle (Figure 5-9a), such as wind-facing cliffs, "echo dunes" form at a distance slightly less than the cliff height because of the formation of a reverse eddy

Figure 5-8. Two aerial views, looking southwest, of the vegetated dune ridges at the south end of Monomoy Island, Cape Cod, Massachusetts, in January 1969. View (b) is a closeup taken just to the right of the view (a) photograph. These dune ridges, which have formed in the last 100 years, are up to 11 m in elevation, have a gently undulatory upper surface, and are for the most part anchored by vegetation. However, note the large blowouts of sand onto the beach, emphasizing the importance of eolian sedimentation in all of the coastal environment.

(a)

(b)

(Tsoar, 1983b). In the "shadow zone" downwind of a permeable obstacle, the air flow is in the form of swirls and vortices and the net forward velocity of this air is much less than the airstream outside the shadow zone (Figure 5-9b). Sand grains moved by the wind do not follow the exact flow lines of the wind, for the bulk of the sand is moved by creep and saltation. Eventually, many grains come to rest inside this relatively stagnant shadow zone, accumulating in a growing heap with the slopes standing at the angle of repose.*

An excellent example of incipient dune growth aided by vegetation is described for the Texas coast by McBride and Hayes (1962) and Hayes (1967). They describe the formation of "pyramidal" dune wind shadows behind large isolated clumps of grass on the supratidal beach (Figure 5-10b). These beds then dip away from the dune crest in two oblique directions, with the dune crest being parallel to the prevailing wind direction. The direction of dip of the resulting beds, the azimuth, is bimodal and the two dip directions of the beds are bisected by the prevailing wind direction (Figure 5-10a).

Incipient dune growth was monitored for 3 yr on the accretional portions of Monomoy Island and Nauset Spit at the southeastern corner of Cape Cod, Massachusetts (Goldsmith, 1972). These observations indicate that the dunes undergo a growth of 0.3 to 0.5 m/yr. The dunes begin around small isolated hummocks of dune vegetation. Wind shadows also form behind logs, peat blocks, and other material deposited by storms at the former storm high-tide line.

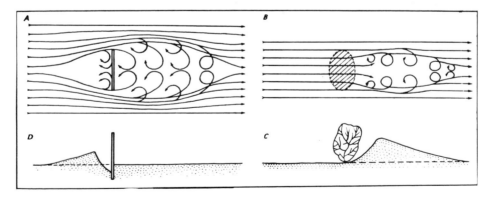

Figure 5-9. Plane sections and profiles illustrating the formation of "echo dunes" in front of a continuous obstacle and wind shadows and incipient vegetated dunes behind a permeable obstacle. For discussion, see text (from Zenkovich, 1967, Figure 282, p. 587).

---

*The angle of repose for dry sand varies between 32 and 34°, depending on the grain size, with coarser sand lying at steeper angles Bagnold, 1954).

DUNE CROSS-BEDS MUSTANG ISLAND, TEXAS

(After McBride and Hayes, 1963)

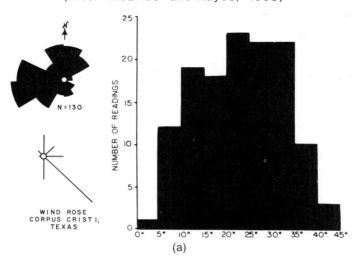

(a)

(b)

Figure 5-10. (a) Dune cross-bed azimuths, dip distribution, and winds at Texas (after McBride and Hayes, 1962). Note the bimodal azimuth distribution, with most of the cross-beds forming high-angle (> 20°) dips. (b) Pyramidal wind shadow dune, approximately 1.5 m in height, on Mustang Island, Texas. View is looking southeast, toward the prevailing wind. This type of dune forms via slipface accretion in the lee of vegetation hummocks. Note that the sand surface slopes away from the crest of the dune shadow in two directions. This accounts for the bimodal azimuth distribution and the abundance of high-angle dipping beds in the Texas dunes (photographed by Miles O. Hayes).

Dune Vegetation

This incipient dune area, generally in the supratidal zone, is rapidly colonized by plants. On Cape Cod the first plants are usually marram grass (*Amophila arenaria*) and saltwort (*Salsola kali*), and on Cape Hatteras, where artificial plantings are common, they are American beach grass (*Amophila breviligulata*) and sea oats (*Uniola panicilata*) (documented with aerial infrared photography by Stembridge, 1978). In Britain the first colonizers are marram grass, sand couch grass (*Agropyron junciforme*), saltwort (*Salsola kali*), and sea rocket (*Cakile maritima*). On the Baltic Coast the first plants to arrive are Lyme grass (*Elymus arenarius*), sea sandwort (*Honkenya peploides*), and sea rocket, whereas on the coast of the North Sea and the Atlantic coast of Europe the first plants are usually sand couch grass and marram grass (Zenkovich, 1967, p. 594). On the northwest coast of the United States, dune vegetation is not widespread, despite the high rainfall. This may be in part because the usually abundant marram grass is not native to this coast and was only introduced to San Francisco in 1869 and to Coos Bay, Oregon, in 1910 (Cooper, 1958).

All these plants are characterized by high salt tolerance and long, elaborate root systems, which reach down to the freshwater table and have additional rhizomes that grow parallel to the upper dune surface. Maximum growth and branching of the root system is achieved when the marram grass is covered over with abundant sand (Chapman, 1964). Thus the effect of these plants on stabilization of the dune is immense. Within 3 yr the areas between the many isolated wind shadows fill in as the grass spreads along the rhizomes that grow out from the initial plant. After small heaps of sand reach a height between 2 and 3 m, the rate of growth decreases because sand is not as easily transported to the upper dune surface. With continued growth of the dunes, the roots continue to reach toward the water table (Figure 5-11). Low angle dipping beds are therefore quite abundant. At the top of the highest dunes on Monomoy Island, the surface is gently undulatory (Figure 5-5b), as a result of the upward growth of the grass coincident with the vertical growth of the dunes. An average of 1.5 m of sand accumulated in 6 yr on the foredunes at Gibralter Point (Lincolnshire coast, Britain), at dune elevations between 4 and 7 m (Figure 5-12).

Because dune vegetation is quite sensitive to salt spray and elevation above the water table, its distribution is quite distinctive and changes in a regular floral succession with distance inland from the beach. The second

Figure 5-11. These two photographs show that the roots of the marram grass maintain contact with the water table, with vertical growth of the dune.

and third lines of dunes have a greater variety of plants than the foredunes, including sea rocket (*Eatule adentula*), bayberry bushes (*Myrica pensylvanica*), goldenrod, salt spray rose (*Rosa virginiana*), bayonet grass (*Scirpus paludosus*), and in England, sea buckthorn (*Hippophae rhamnoides*). The presence of these latter plants in the first dune line is indicative of severe beach erosion, with the original foredune ridge having been removed. This vegetation succession in dunes is described in more detail in Chapman (1964) and Hoiland (1978).

Godfrey *et al.* (1979) have shown the importance of regional variations in dune vegetation, on the U. S. east coast, in determining barrier beach topography. Further, despite their high salt tolerance and growth in the highly exposed barrier island, many of these plants are quite fragile if walked or driven on. A path 5 m long and 0.6 m wide was worn through the 1.2 m high dune grass at a beach profile locality by walking over the dune only once every 2 weeks. The most important contribution that humans can make toward preservation of barrier islands is to prevent damage to the dune vegetation.

The effect of plants and their root systems on stabilization and growth of vegetated (i.e., fixed) dunes is immense, therefore, and just as important as the steel girder framework of tall buildings. The vertical growth of these dunes, coincident with very little horizontal migration, results in a distinctive internal dune geometry.

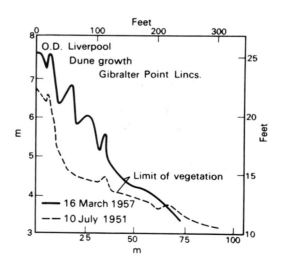

Figure 5-12. Dune growth at Gibralter Point, Lincolnshire, England (after King, 1972, Figure 6-9, p. 178).

Internal Geometry

In order to gain insight into the processes of vegetated coastal dune formation, measurements of 301 eolian cross-bed sets were made on Monomoy Island, Cape Cod, Massachusetts (Figure 5-13) and compared with similarly obtained data from Brazil (Bigarella *et al.*, 1969), Texas (McBride and Hayes, 1962), Georgia (Land, 1964), and Israel (Yaalon and Laronne, 1971).

Monomoy Island. Cape Cod, Massachusetts. Measurements consisting of the amount of dip, the azimuth, and the elevation of the cross-bed were made on fresh erosional scarps in the foredune ridge and on scarps in the highly vegetated interior dunes. With Monomoy Island undergoing an average permanent beach retreat of 13 m/yr, scarps were quite abundant in the foredune ridge. In addition, many of the interior dunes are cut by large blowouts so that the dune bedding is exposed (Figure 5-8).

Data from grain-size distributions indicate that the sand on the foredunes is in general slightly coarser and better sorted than the berm sand. This is because of the formation of lag deposits of heavy minerals, coarse sand, and gravel, caused by the wind removing the finer quartz sand from the storm and overwash deposits. These coarse layers make good marker beds for measurements of dune bedding. All measured dune cross-beds were either planar or exhibited only a very slight curvature. When visible on large scarps, the beds could generally be followed for 5 to 10 m while remaining nearly parallel to each other. In a few instances the same beds were visible in scarps on three sides of a 5 m high dune and displayed the planar and nearly parallel nature of most of the dune cross-beds. Unfortunately, not enough of these large exposures were available to measure bed thickness.

Surprisingly, little disturbance of the beds resulted from the roots of the vegetation, which further substantiates the conclusions reached in the study. Bed orientations were not measured in the vicinity of such disturbances.

Techniques used in the measurements were very simple (Colonell and Goldsmith, 1971). Small trenches were dug in the scarps in both the interior and foredune ridges. An 18-cm disk with a center-mounted level was inserted in the trench parallel to the beds and rotated until the bubble in the level was centered. The azimuth and the angle of dip of the cross-beds was then measured with a Brunton compass.

The statistical distribution of dip angles was determined by standard linear statistics (Research Computing Center University of Massachusetts, 1971). Azimuth distributions involve directional properties that require the use of circular distributions. Statistical parameters for the azimuth distributions were calculated using a vector summation technique (Curray, 1956, pp. 118–120), with each observation considered to be a vector with direction and magnitude. This method, also recommended by Pincus (1956, pp. 544–546), involves the calculation of the azimuth of the resultant vector, which

DUNE CROSS-BEDS, MONOMOY ISLAND, MASS.
AZIMUTHS AND AMOUNT OF DIP

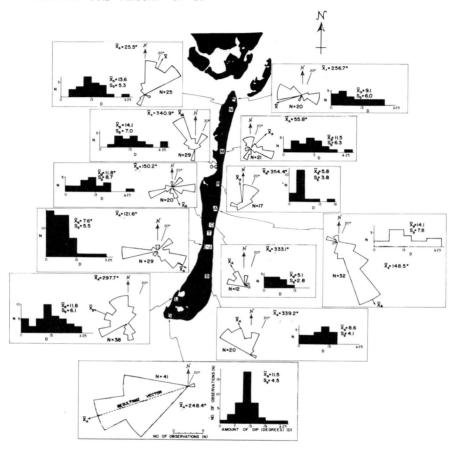

Figure 5-13. Dune cross-bed dips and azimuths at each of 15 sample locations on Monomoy Island. Note the low dips and the large variation in azimuths, with the beds tending to dip seaward around the island. The symbols employed in this and the following figures are listed below:

$\bar{X}_D$ = mean dip
S_D = standard deviation of the dip
$\bar{X}_A$ = azimuth of the resultant vector

indicates the preferred orientation direction, and the vector magnitude, which is a measure of dispersion.

The combined azimuth distribution shows a statistically significant correlation with three prevailing wind directions, northwest, southwest, and southeast, instead of with the dominant northeast storm winds (Figure 5-14). The innermost winds in the roses ($<$ 16 km/hr) can be ignored because these winds are generally incapable of moving sand. As will be shown later, this high variability and polymodal nature of the Monomoy dune azimuth distribution is anomalous with respect to coastal dunes in other areas and is caused by the large wind variability because of the exposure of Monomoy to the sea on three sides. This azimuth distribution is valid for cross-bed sets at any dune elevation (Goldsmith, 1973). There was, however, a wide variation in the azimuth distribution with the side of the dune on which the beds were measured.

Mustang Island, Texas. The azimuth distribution of the dune cross-bed sets here is bimodal, and these two modes are bisected by winds from the southeast (Figure 5-10). The high-angle dips that results from the formation of pyramidal dune wind shadows are shown clearly in the dip distribution of cross-beds (Figure 5-10). Here a high mode, containing most of the 130 dip measurements, is located approximately between 20 and 35°. A low mode, between 10 and 15°, is also present. In this case, the reported mean of the distribution, 24°, does not give an adequate indication of the true nature of the cross-bed distribution.

The characteristics of this distribution become more apparent when the distributions for five different areas are compared. For example, on Mustang Island the dip distribution is bimodal, with the lower mode similar to the dip mode of the Monomoy dune beds. The absence of a higher dip mode on Monomoy may perhaps be explained by the rarity of pyramidal dune wind shadows on Monomoy. This may in part be a result of the rapid dune retreat on Monomoy. Along the Texas coast the pyramidal dunes form primarily during the extensive accretional stage that occurs after severe hurricane erosion.

Sapelo Island, Georgia. The azimuth distributions from two locations on Sapelo Island, Georgia, a presently active foredune ridge and a dissected beach ridge approximately 200 yr old (Land, 1964, p. 390), are illustrated in Figure 5-15. There is very good correlation of the west-northwest dips of the beds in the foredune ridge and dissected ridge with the prevailing southeast winds. However, the second mode of the dip distribution of the cross-beds in the old dissected beach ridge is unaccounted for by the present wind distribution. It has been postulated that this is because of a change in wind direction (Land, 1964, p. 390). However, the Monomoy Island data suggest that this could also be caused either by the relict geometry of Sapelo Island and the location on the island where these beds have been measured, or by the side of the dune on which these beds have been measured. There is a wide

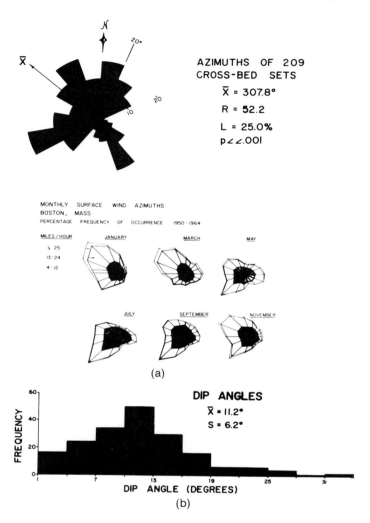

Figure 5-14. Summary of azimuths (top), surface wind azimuths for Boston, Massachusetts (1950–1964) (bottom), and dips of 209 dune cross-beds from 11 sample locations on Monomoy Island, Cape Cod, Massachusetts. Note the comparison between the wind rose, with the dominant northeast winds and prevailing southwest and northwest winds, and the quadrimodal azimuth distribution. The probability that the azimuth distribution is caused by chance is quite low (0.1%). Note also the consistently low cross-bed dip distribution (after Goldsmith, 1972).

variation in amounts of dip of the cross-bed sets (Figure 5-11). Notice that Land deleted all beds with dips less than 10°, which constitute about one-half of his data (Land, 1964, pp. 390–391, 393). Therefore, the dunes on Sapelo Island contain cross-beds with both high-angle dips and, most noticeably, an abundance of low-angle dips (< 20°). Dune studies by Oertel and Larsen (1976) along three other Georgia barrier islands to the north of Sapelo (Tybee, Wassaw, and Ossabaw) revealed mean dip angles of 9.2° for the dune cross-beds.

Praia de Leste, Brazil. Bigarella et al. (1969) and Bigarella (1972) furnished detailed sketches and measurements of the dune bedding at two locations in Brazil (Bigarella et al., 1969; Figure 5-16). It therefore was possible to use the same statistical techniques and computer programs developed for the Monomoy data for the original Brazilian dip and azimuth measurements. Data from two locations along the Brazil coast are summarized in Figure 5-17. The azimuth distribution of Jadim Sao Pedro is bimodal, and these two modes are cut by the mean vector of the summer prevailing winds. This resembles the Texas case. The other areas, Porto Nova and Guairamar, show a direct correlation between the largely unimodal dip distribution and the mean vector of the prevailing southeast wind. This resembles the Monomoy case. It is interesting to note that the mean vectors of the cross-bed azimuths $(\overline{\chi_A})$ for both areas in Brazil are within about 20° of each other and that they also conform closely with the prevailing wind vector. However, as pointed out earlier, there is a danger in using mean vectors in a bimodal distribution.

Both these Brazilian areas have similar cross-bed dip distributions and both appear to be bimodal (Figure 5-17). The higher modes are between 25 and 37°. It has been suggested (Goldsmith, 1973) that the higher dip modes may result from dune wind shadows that form via slipface deposition. The lower dip mode, which is about 12°, conforms closely with the mean dip of 11.2° for Monomoy Island.

Mediterranean Coast of Israel. The dip and azimuth distributions of 609 cross-beds of Holocene to Pleistocene eolianites, composed of carbonate fragments and quartz grains, from seven locations along the Mediterranean coast of Isreal were measured by Yaalon and Laronne (1971). The azimuth distribution was unimodal in the low-angle beds (< 20°) and bimodal in the high-angle beds (> 20°). This is what would be expected if the cross-beds represented by the lower dip mode formed through vertical accretion concomitantly with the growth of the vegetation, and if cross-beds of the higher dip mode formed as pyramidal wind shadow dunes. The mean vectors of the two azimuth distributions correlated directly with the prevailing winds (Figure 5-18a).

The dip-angle distribution was bimodal, with modes at approximately 10 to 15° and 25 to 30° (Figure 5-18a). There was no correlation of the two dip modes with foreset and backset deposition (Yaalon and Laronne, 1971). The

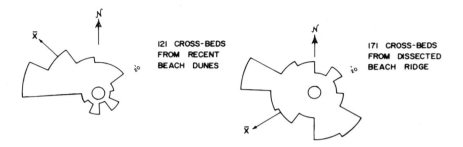

DUNE CROSS-BED AZIMUTHS
SAPELO ISLAND, GEORGIA
(AFTER LAND, 1964)

121 CROSS-BEDS
FROM RECENT
BEACH DUNES

171 CROSS-BEDS
FROM DISSECTED
BEACH RIDGE

SURFACE WIND AZIMUTHS
SAPELO ISLAND, GEORGIA
(AFTER LAND, 1964)

WINDS OF
8-18 MILES/HOUR
35% OF TOTAL

WINDS
⅃ 18 MILES/HOUR
4% OF TOTAL

(a)

DUNE CROSS-BED DIP ANGLES
SAPELO ISLAND, GEORGIA
(AFTER LAND, 1964)

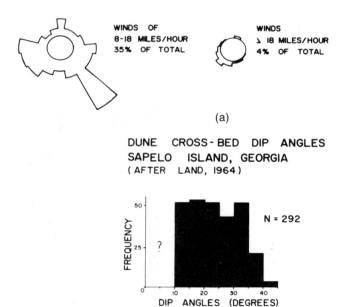

N = 292

(b)

Figure 5-15. (a) Dune cross-bed azimuths (top), surface wind azimuths (bottom). (b) Dip angles from two sample locations on Sapelo Island, Georgia (after Land, 1964). The southeast-dipping cross-beds from the dissected beach ridge do not appear to be directly related to dominant or prevailing wind directions. Note the wide dip-angle distribution. Land discarded all dip angles less than 10°, which he admitted were relatively abundant.

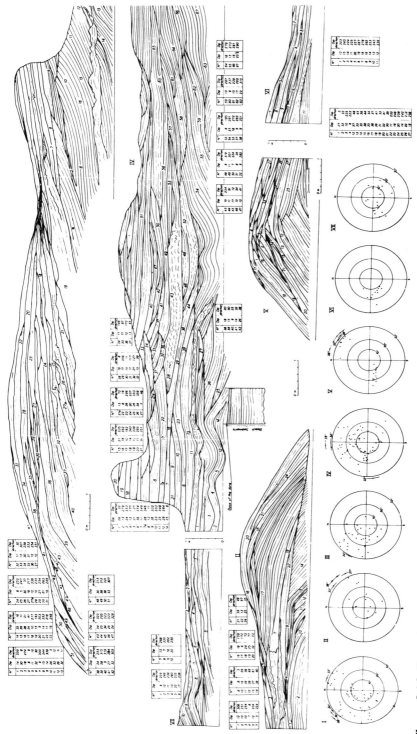

Figure 5-16. Coastal and sand dune ridge cross-sections showing the internal dune structures at Jardim Sao Pedro dunes in coastal Brazil (after Bigarella et al., 1969).

very striking bimodal dip distribution, and the presence of a bimodal azimuth distribution only in the high-angle dips, suggests quite strongly the similarity with the Texas case and that these beds may have formed as pyramidal wind shadow dunes. Yaalon (1975) suggests that the bimodal azimuth distribution of the Israeli high-angle cross-beds may also be a result of the fact that

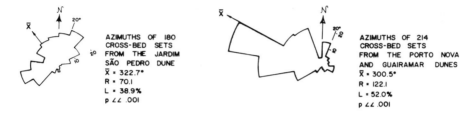

DUNE   CROSS-BED   AZIMUTHS – PRAIA DE LESTE,   BRAZIL
(AFTER   BIGERELLA, ET AL.,   1969)

AZIMUTHS OF 180
CROSS-BED SETS
FROM THE JARDIM
SÃO PEDRO DUNE
X̄ = 322.7°
R = 70.1
L = 38.9%
p ∠∠ .001

AZIMUTHS OF 214
CROSS-BED SETS
FROM THE PORTO NOVA
AND GUAIRAMAR DUNES
X̄ = 300.5°
R = 122.1
L = 52.0%
p ∠∠ .001

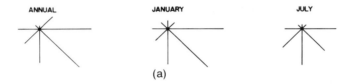

SURFACE   WIND   AZIMUTHS – PARANA,   BRAZIL
(AFTER   BIGERELLA, ET AL., 1969)

ANNUAL   JANUARY   JULY

(a)

DUNE   CROSS-BED   DIP   ANGLES – PRAIA DE LESTE,   BRAZIL
(AFTER BIGERELLA, ET AL., 1969)

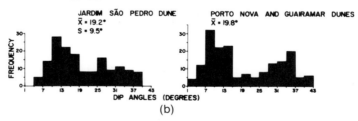

JARDIM SÃO PEDRO DUNE
X̄ = 19.2°
S = 9.5°

PORTO NOVA AND GUAIRAMAR DUNES
X̄ = 19.8°

FREQUENCY

DIP ANGLES (DEGREES)

(b)

Figure 5-17. Dune cross-bed azimuths (top), surface wind azimuths (bottom), and cross-bed dip angles summarized from three locations at Praia de Leste, Brazil (after Bigarella *et al.*, 1969.) The Jardim Sao Pedro dunes have a cross-bed azimuth distribution similar to that for Mustang Island, Texas; the other dune cross-bed azimuths correlate directly with the annual prevailing winds. The cross-bed dip angles are bimodal at all locations. The lower mode is about 12°, which is quite similar to the 11.2° mode for the Monomoy dunes.

distinct segments of the dune ridge resemble and advance as transverse dunes, as revealed in aerial photographs. The low-angle cross-beds probably formed as a result of the influence of vegetation as described below.

Origin of Vegetated Coastal Sand Dunes. Data from the above widely scattered localities show that vegetated coastal sand dunes have a distinctive

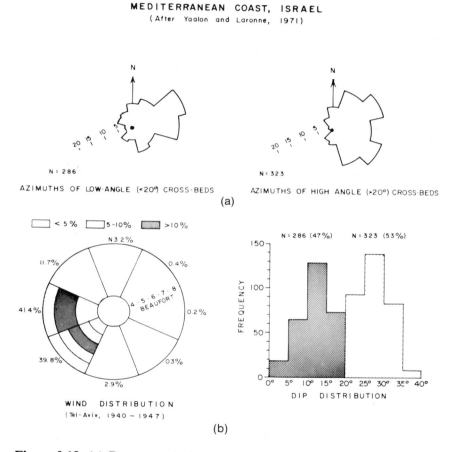

DUNE CROSS-BEDS
MEDITERRANEAN COAST, ISRAEL
(After Yaalon and Laronne, 1971)

N = 286

AZIMUTHS OF LOW-ANGLE (<20°) CROSS-BEDS

N = 323

AZIMUTHS OF HIGH ANGLE (>20°) CROSS-BEDS

(a)

WIND DISTRIBUTION
(Tel-Aviv, 1940 – 1947)

DIP DISTRIBUTION

(b)

Figure 5-18. (a) Dune cross-bed azimuths (low angle and high angle), (b) wind azimuths, and cross-bed dip angles from seven locations of Israeli coastal dunes (after Yaalon and Laronne, 1971.) The low-angle dips (dark shading) have a unimodal azimuth distribution that correlates directly with the prevailing winds, whereas the high-angle dips (> 20°) (light shading) have a bimodal dip distribution, similar to the pyramidal wind shadow dune of McBride and Hayes (1962).

internal geometry. The azimuth distributions correlate closely with both the medium-velocity prevailing wind and maximum wind direction. The cross-beds may dip with the same orientation as the prevailing mean wind vectors (e.g., Monomoy Island; Sapelo Island; Israel; and both Porto Nova and Guairamar, Brazil) or may dip in two directions. The bimodal azimuth distributions are bisected by the mean wind vector (e.g., Mustang Island, Texas; Jardim Sao Pedro, Brazil; and Israel).

The dip distributions from all five coastal dune areas contain an abundance of low-angle cross-beds. The low-angle cross-beds form as sand accumulates around the dune vegetation, which acts as a baffle, trapping the wind-blown sand, as observed in the Monomoy coastal dunes. The vegetation therefore anchors and stabilizes the dunes, preventing dune migration and encouraging the formation of low-angle cross-beds. It follows then that there should be a relation between vegetation density and proportion of low-angle beds (Yaalon, 1975). A relationship between the type of plant and the dip of the dune cross-beds has been suggested for the Georgia coast by Oertel and Larsen (1976), with low dip angles ($< 10°$) associated with short grasses, moderate dip angles (10 to 25°) associated with bushy plants, and high dip angles ($> 25°$) associated with the slipfaces of dune ridges vegetated with tall grasses.

The higher angle dipping dune beds probably form as pyramidal wind shadow dunes (i.e., lee dunes) with sand accumulating as slipface deposition. A bimodal azimuth distribution (e.g., Texas, Brazil, and Israel) is highly suggestive of a pyramidal wind shadow dune origin as postulated for the Texas coast by McBride and Hayes (1962).

Transverse Dune Ridges

Transverse dune ridges (Cooper, 1958, 1967), or migrating dunes, are characterized by a lack of anchoring vegetation, move generally landward in response to the prevailing winds, contain a well-defined slipface at or close to the angle of repose, and stand as a single large distinct feature. The large transverse ridges either have a single straight to sinuously shaped crest up to 1 km long (Figure 5-19) or take the form of isolated barchan dunes (Figure 5-20) that migrate inland at rates of 10 to 30 m annually, as in Mexico and Peru (Inman et al., 1966; Finkel, 1959). They are as much as 30 to 50 m in height and may or may not be associated with vegetated dunes. (The "walking dune of Arcachon," on the southwest coast of France, has an elevation of over 100 m.) The transverse dune ridges, therefore, fit better than vegetated dunes into Bagnold's definition of a true dune, which is " . . . a mound or hill of sand which rises to a single summit" (Bagnold, 1954, p. 188).

Precipitation dunes, a form of transverse ridges, deposit sand in front of and often migrate over, houses, roads, streams, forests, and even the marshes and estuaries behind the barrens (Figures 5-1, 5-2, and 5-3). The origin of the name is derived from the process whereby the sand precipitates down on the forests, houses, or anything else in its way. The precipitation ridges of the west coast of the United States are spectacular examples of this process (Cooper, 1958, 1967). Uninterrupted growth and migration of dunes may cut off streams flowing into the sea and may cause accumulation of water in the troughs between dune ridges, especially if the groundwater table is high enough (Jennings, 1957a; Inman et al., 1966). This aspect is promoted by deep eolian deflation in the interdune lows. These immense mounds of sand, therefore, largely bare of vegetation, are often associated with small bodies of water.

A medaño (the Spanish word for coastal sand hill) is a distinctive type of unvegetated dune (Goldsmith et al., 1977). A medaño is a high, steep, isolated sand hill from tens to hundreds of meters in height, formed by a bimodal or polymodal wind regime moving sand upward toward the summit from several directions. Although it may change in profile shape, reflecting daily wind changes, there is very little net annual migration despite the lack of vegetation, because of equal sand transporting ability of opposing wind directions. It appears to be distinctive to the coastal zone, not having been observed in desert areas.* Examples are at Currituck Spit, Virginia–North Carolina (Gutman, 1977b); Coos Bay, Oregon; and the south end of Lake Michigan. Under high-velocity wind conditions, superimposed transverse bedforms, about 1 m in height and 20 m in wavelength, form with different orientations in response to different wind directions. The internal geometry of these features therefore would be expected to consist of high-angle cross-beds with bimodal dip directions. When new vegetation blocks the transport of sand from one direction, the height of the medaño decreases and the resulting dune better resembles a transverse dune, which then migrates over the blocking vegetation. Thus, planting vegetation on one side of a stable sand hill causes it to transition into a moving transverse dune (e.g., Outer Banks of North Carolina; Hennigar, 1977b).

Formation

A sand dune is an accumulation of loose sand that acts as an obstacle to the wind and is also subject to deformation by the wind. Figure 5-21 shows an

---

*The closest desert example is the star dune found in the Namib and Sahara deserts (McKee, 1979).

(a)

Figure 5-19. Two photographs of Jockey's Ridge, Nags Head, Outer Banks, North Carolina (November, 1972). The first photo (a) was taken from the elongated sinuous crest, looking down on the southwest facing slipface, approximately 30 m in elevation. The second photo (b) shows the gentle northeast slope of the dune leading up to the crest (shown in the above photograph).

idealized schematic of flow past a streamlined obstacle. Sand particles start to move when the shear stress exerted by the wind exceeds a critical value. The value of the shear stress varies from point to point on the obstacle because the fluid flow is not uniform, as indicated by the streamlines. The air flow velocity, and therefore the shear stress, increases on the windward face (A–B) and decreases on the lee face (D–E). Where the air flow is accelerated, as indicated by the converging flow lines, the shear stress increases, and where the flow is being retarded by the flow separation on the lee side of the crest, the shear stress decreases. Because the rate of sand movement is directly related to the shear stress, which in turn is related to the air flow velocity, the result is that sand is removed and transported from the windward face to the lee face, where it is deposited (Bagnold, 1954, p. 197). Sand can only leave the dune in the absence of a slipface (King, 1973). With

(b)

Figure 5-19. *(Continued.)*

increasing sand transport, a slipface develops on the lee side (Figure 5-22b). In the zone of flow separation (C–D) reverse flow occurs, enhancing the dune height and steepness of the slipface.

The streamlines, and hence the dune shape, also vary with the velocity of the wind. With low-velocity winds the shear stress does not reach the critical value for sand removal and transport except near the crest of the dune. Sand is then removed from the windward side of the crest and deposited in the lee of this crest where the streamlines diverge. Therefore, low-velocity winds have a tendency to flatten and lengthen dune profiles. During increasing wind

Figure 5-20. Aerial photograph of the barchan dune fields along the coast of Baja California, Mexico, in the vicinity of the area studied by Inman *et al.* (1966). Pacific Ocean is to the right (east). This dune field is inundated during spring and storm tides (photographed by Miles O. Hayes in August 1972).

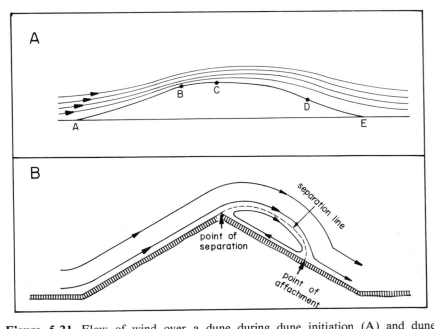

Figure 5-21. Flow of wind over a dune during dune initiation (A) and dune maintenance (B) (5-21A after Bagnold, 1954, p. 198 and B after H. Tsoar). See text for explanation.

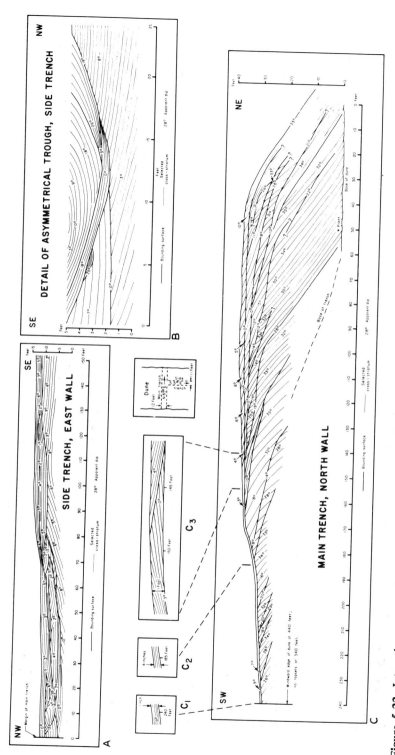

Figure 5-22. Internal geometry of a transverse dune from White Sands National Monument, New Mexico (after McKee, 1966, Figure 7, pp. 22–24).

velocity the location of critical value of shear stress (i.e., the convergence of flow lines) either moves down the windward dune slope toward the toe or, under some conditions, stays on the crest. High-velocity winds will thus steepen and increase the height of the dune. During extremely high wind velocities the wind may arrive at the dune already charged with sand, much of which may be deposited on the lee slope. When this occurs the dune grows wider and higher as it advances.

Deposition of sand on the lee slope occurs a certain distance downwind of the air flow divergence because of a lag effect. The location of the flowline divergence on the lee slope exhibits less variation in location with variation in wind velocities than the location of the flowline convergence on the windward slope. Therefore, maximum deposition occurs more or less a set distance downwind from the dune crest, for a large range of wind velocities. As the size of the dune increases, the location of maximum deposition increases in elevation and moves closer to the dune crest. This results in the upper part of the lee face receiving more deposition than the lower part, eventually causing oversteeping of the lee face and the formation of a slipface with sand grains resting at the angle of repose and avalanching down the slipface when the angle of repose is exceeded. Once formed, this slipface acts as a sediment trap for all the sand moved over the crest by the wind. The rate of advance of such a dune is therefore directly proportional to the quantity of sand moved over the crest and inversely proportional to the height of the slipface (Bagnold, 1954, p. 204).

A blowout will form where there is a lowering of the dune crest. Wind will be "funneled" across the lowered crest, resulting in an increased velocity and sand removal from the crest, which in turn increases the size of the gap at the dune crest. Eventually this sand is transported beyond the base of the original dune and may accumulate at a new location resulting in the formation of a new dune. Blowouts also occur on vegetated foredune ridges and are commonly associated with parabolic dunes. Blowouts result in large amounts of windblown sand transported onto the beach (Figure 5-8). A layer of sand over 2 cm thick and spread out over a beach 50 m wide was observed to form by an offshore wind in one tidal cycle.

Internal Geometry of Transverse Dunes

Local exposures of portions of these dunes have indicated the dominance of high-angle bedding. The first detailed analysis of the internal geometry of a transverse dune ridge was made by McKee (1966) on a dune in White Sands National Monument, New Mexico, which appears to be similar to those formed along the coast. This particular dune, 130 m across in the direction of wind movement, 13 m in height, and with a crest 270 m in length, was first saturated with water and then trenched with bulldozers both perpendicular to and parallel with the one effective prevailing wind direction (Figure 5-22).

In the main trench, oriented parallel to the general wind direction, steeply dipping laminae that formed the lower two-thirds of the dune closely approached the angles of true dip, because they ranged from about 30 to 34° In this part of the dune, foresets were mostly large scale, some laminae extending for as much as 17 m. Bounding surfaces of sets had moderately high dip, mostly from 20 to 26°, but some had even greater dips. Later deposits, in the upper third of the dune, consisted of relatively thin, gently dipping to horizontal sets, mostly 60–90 cm thick, which contained cross-strata of moderate dip (11–15°). Finally, in a very late stage, strata with reverse or windward dips of 5 to 10° were deposited and preserved. The low angles suggest that these are upwind deposits, rather than foresets formed by wind from another direction. In a detailed analysis of the cross-strata are examples of wavy and contorted laminae among the steeply dipping foresets near the dune center, probably the result of irregular slumping or avalanching (McKee, 1966, pp. 31 and 39 and Figure 7). A more complicated internal structure occurs in longitudinal (i.e., also called seif, or linear) dunes (Tsoar, 1983a). Such dunes are common on many desert coasts including the western Sahara and Oman (Breed and Grow, 1979), the Sinai coast, and some nondesert coasts as well. The mutual relation between the surface geometry and internal structure is related to the deflection of oblique wind flow into a longitudinal flow on the lee flanks. Thus, two kinds of depositional units result, involving a wide range of dips (10–30°) (Tsoar, 1983a).

Transverse Dunes Associated with Vegetated Dunes

Although most spectacular on the U.S. west coast, transverse dunes form on coasts all over the world and are usually associated with vegetated dune fields. Some of these areas are discused below.

Coos Bay, Oregon. At Coos Bay, Oregon, these dunes occur along a 60-km stretch of coast and can be seen inundating forests and filling in the lagoon behind the barrier (Figure 5-1). Here, as at other areas on the west coast, the source of sediment appears to be the numerous rivers, which transport large quantities of sediment during the rainy season. Indeed, most of the large west coast dune fields of the United States form adjacent to large rivers. Hunter *et al.* (1983) studied the oblique dunes of the central Oregon coast (including Coos Bay), which they define as " . . . a relatively straight crested dune whose average trend is markedly oblique (15°–75°) to the resultant sand transport direction; . . . intermediate between transverse dunes and longitudinal dunes." They found that " . . . the dune migration takes place largely during fall, winter and spring storms." Actual sand transport by these wet south-southwesterly winds was only 39% of the amount calculated assuming dry conditions. This resulted in dune migration of 3.8 m/yr to the north-northeast, at an angle of 28° with the calculated resultant transport direction. Thus, these dune crests are oriented more or less perpendicular to the shoreline. The dry, moderate, north-northwesterly summer winds modified the dune form but not the dune trend.

Provincetown–Truro Dune Field, Cape Cod, Massachusetts. These dune fields are probably the most extensive and well-known dunes of the east coast of the United States (Figure 5-23). They cover an area of approximately 26 km² and are up to 30 m in height. They have been studied by Messenger (1958), Smith and Messenger (1959), and by this author. Smith and Messenger described four distinct lines of dunes subparallel to each

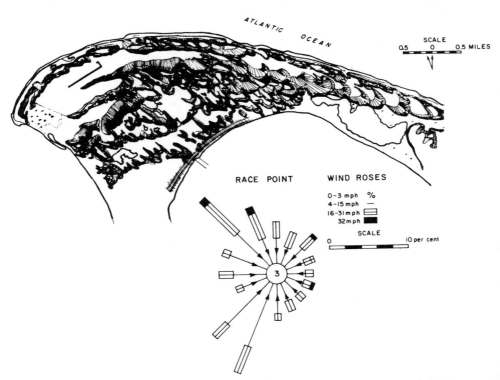

Figure 5-23. Shaded relief diagram of the Provincetown, Cape Cod, dune field (from Messenger, 1958). The aerial photographs in Figure 5-3 are of the parabolic dunes on the right side of the diagram. These dunes are oriented into the northwest winds, the most important wind direction for higher velocity winds (10 km/hr), as shown by wind rose for Race Point. Race Point is located at the extreme left side of the diagram.

other and to the shoreline, whose growth was related to the eustatic changes in sea level. Observed rates of growth of sand dunes of this coast suggest that, instead, the growth and formation of this dune field is related to the normal coastal processes of a highly accretional coast where there is an abundant supply of sediment available.* It therefore seems more plausible that this growth has occurred in the last 3000 to 4000 yr, since the end of the last rapid rise in sea level. The sediment supply originates in the glacial deposits composing the Highland Cliffs, which rise 50 m above sea level, and is transported northwest to the dune field by the dominant longshore drift system driven by the northeast storms. The Highland Cliffs have retreated at the rate of 1 m/yr in the last 100 yr (Zeigler *et al.*, 1964) or over 3 km since the end of the last rise in sea level. Prevailing and high velocity winds in the area are from the northwest. These winds pick up and transport the large amount of sediment supplied by the longshore drift back into the large Provincetown–Truro dune field area. Figure 5-3 shows several en echelon large parabolic shaped "precipitation" dunes migrating to the southeast and inundating forests, roads, and houses. Note that the western (right side of photograph) arms are larger in size, which is probably because of the larger eolian contribution of the northeasterly winds. The sand dune in the distance is being blown back into the ocean just north of and slightly updrift of the Highland Cliff source area. Therefore, much of this sand is being "recycled" with the aid of the northwesterly longshore transport system.

Westgate (1904) attempted to stabilize these dunes through the planting of vegetation and was very much aware of movement of the sand dunes. He suggested that the destruction of woods and pasturing of stock in the 1700s and 1800s caused destruction of the protective cover and resulted in the reactivation of the dunes. Westgate's attempts at reestablishing and stabilizing dune vegetation on the migrating very coarse sand dunes largely failed, as had most attempts at planting dune grass going back to 1826 (see concluding section of this chapter).

Whether the failure of attempts at artificial planting of dune vegetation at Cape Cod and elsewhere results from bad agricultural practices, the coarseness of the sand, or subtle climatic changes has been the subject of some controversy in recent years (Larsen, 1969).

Lake Michigan Dune Fields in Indiana and Michigan. Cressey (1928) and Olson (1958) studied the sand dunes along the southern and eastern shores of Lake Michigan in Indiana and Michigan. These compare favorably with coastal sand accumulations found anywhere in the world. One area in Indiana contains approximately 400 km² of dunes with many dunes between 50 and 70 m high. The fact that these dunes are formed adjacent to an inland water body (Lake Michigan) makes it easy to see how wind can dominate the origin and development of coastal sand dunes. The wind blowing across the lake controls the waves and currents of the water body. The waves erode the

---

*See earlier discussion of incipient vegetated dunes.

bluffs of glacial drift along the eastern shore of the lake and along the area just north of the dunes. This eroded material is the sediment source of the dunes. The longshore current resulting from the wind and waves then moves the material to the area of accumulation, where the wind picks up the sand and forms the dunes. In nearly all respects these dunes are similar to other vegetated and transverse coastal dunes (Figure 5-2). The moderate rainfall, 80 cm/yr, results in a luxuriant vegetation cover over most of the dunes, owing, in part, to a lack of salt in Lake Michigan. Dune movement and blowouts are also similar to the west coast dunes. During the winter months, however, the dunes tend to freeze and are completely immobilized. Most of the dunes are generally stable throughout the year because of the extensive vegetation cover.

Parabolic Dunes

Large fields of coastal parabolic dunes form behind foredunes. The dunes are in transition from medaños to transverse to parabolic dunes under the increasingly stabilizing effects of vegetation. Thus, these are different in origin from the small, isolated parabolic dunes that form from blowouts in vegetated dune ridges. They are characterized by their distinctive planimetric shape and are generally 5 to 10 m in height and 1 to 2 km in length. They are much lower in elevation than their predecessors. In their early stage (e.g., the Provincetown dunes, Figure 5-23), parabolic dunes have an upwind scour zone, deflated down to the water table within the two arms, and a slipface on the downwind convex side. In the later stage, because of the stabilizing effects of vegetation throughout the dune, the internal geometry closely resembles vegetated dunes (Goldsmith et al., 1977). Adjacent vegetation (i.e., maritime forests) will affect the orientation of these large parabolic dune fields by changing the effective wind direction (Gutman, 1977a).

Parabolic dunes differ from barchan dunes in that the former are open to the upwind direction and the latter are open in the downwind direction.

Transverse Dune Ridges on an Arid Coast

In arid climates, where there is an adequate supply of sand, lack of vegetation is the dominant factor in the geomorphology of coastal sand dunes. The sand forms large, distinct, geometrically regular dune masses that migrate inland under the influence of the prevailing wind (McKee, 1979). A common dune form under these conditions is the barchan dune. One of the classic areas for the study of barchans is along the coast of southern Peru.

Southwest Coast of Peru. Finkel (1959) made a detailed study of 43 distinct barchans located adjacent to the coast of southern Peru. Here a large source of sand is available from sand originally deposited on the coastal shelf before the Quaternary period. This shelf has since been uplifted as much as 2000 m, exposing these sands to the highly arid climate of the Pacific coastline of southern Peru. This coastline has essentially no rainfall, very low humidity, no vegetation, and a very small annual temperature fluctuation. This lack of vegetation means that there is little to stop the sand once it starts moving. The wind acting on this sand is of low velocity but unidirectional to the north. Calculations of Finkel (1959), based on Bagnold (1954), show that the effective wind has a velocity of 19.1 km/hr. A reversal of wind in the winter mornings does not seem to affect the dune movement. The average height (which showed a Gaussian distribution) was 3.5 m; the average length was 120 m; and the average width was 37.2 m. The lee slipface was at the angle of repose (32 to 33°), while the gentle windward slope was at an angle of 5 to 10°. The east horn was, on the average, 15% longer and was attributed to the westward slope of the land. The orientation of the horns, the most active part of the barchans, was between 330 and 341°, reflecting the prevailing southerly wind. A further indication of the northerly movement is found in the grain-size distribution, which shows a decrease in grain size in the direction of transport. Selective sorting by the wind is also indicated by the observation that the tails of the barchans (i.e., the toe of the windward slope) have a coarser grain size than the horns and crests. These dunes move forward by transportation of sand to the slipface, which becomes unstable. The sand then continues to slip downhill until the angle of repose is regained. Sometimes the smaller, faster barchans overtake the larger ones and blend with them. On the average, the dunes advance at a rate of 10 to 30 m annually. They do not appear to be stopped, or even slowed down, by hills, mounds, or higher sloping topography; the dunes even have a tendency to migrate uphill. This is probably because the wind strikes a surface tilted toward it harder than one tilted away from it. In general, one gets a feeling of great regularity in the form and processes of barchan dune development. This is confirmed by Finkel's observation that the total number of dunes on a line perpendicular to the wind direction remains constant with time, although the dunes tend to spread laterally (especially uphill to the east) as they move forward (Finkel, 1959).

Guerrero Negro, Baja California, Mexico. Coastal sand dunes similar in their form and in the processes acting on them to those of southern Peru have been studied by Inman et al. (1966) in a similar climatic situation at Guerrero Negro, Baja California, Mexico. Here the main difference is that the dunes are not as large or as distinct but occur as a field of uniformly spaced transverse ridges about 40 km^2 in area composed of en echelon barchan dunes approximately 6 m in height. The dune field has accumulated in the last 1800 yr (Figure 5-20). This difference in dune form may be caused by a larger and more accessible sand source in addition to a prevailing strongly unidirectional northwest onshore wind in the latter case, whereas in southern Peru the prevailing wind is overland from the south. These dunes, averaging 70 m in length, advanced at a mean rate of 18 m/yr during 3.5 yr of observations, giving a discharge rate of 23 m³/m of width/yr. The slope of the slipface was uniform at ±33°. Interestingly, the interdune lows of the entire dune field are flooded during spring high tides and the wave action during the time of flooding produces miniature beaches, erosional scarps, and swash laminations on the windward sides of the dunes.

The measured wind profile over one of these barchan dunes is shown in Figure 5-24. Note the reversal in wind direction at the toe of the slipface which, according to Inman et al. (1966), probably has the effect of encouraging deposition on the upper part of the slipface, preventing loss of sand from the dune mass, and conserving the size of the dune during its travel.

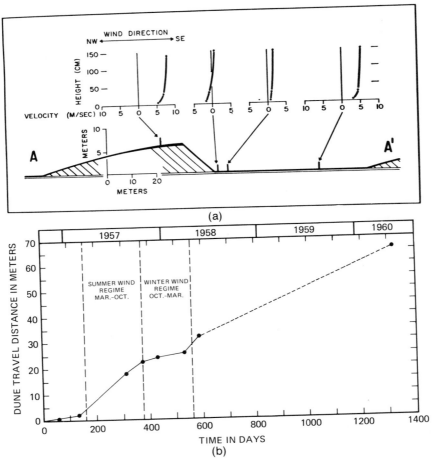

Figure 5-24. Measurements of a coastal barchan dune in Baja California, Mexico (after Inman et al., 1966, pp. 795–796.) (a) Wind velocity profile (top) over a dune. Note the reversal in wind direction at the toe of the slipface. (b) Travel distance of a dune; total travel distance of 65 m in 1322 days, or 18 m/yr.

Climbing Dunes and Cliff-Top Dunes

Climbing dunes are banked up against a cliff and are more or less currently active. Excellent examples are found in north Scotland and Iceland. Cliff-top dunes are generally relict features and have been described from several areas in Australia; on King Island, Tasmania (Jennings, 1967); and on the Auckland west coast of New Zealand (Brothers, 1954). The Auckland dunes will be described in some detail because this area demonstrates how all the types of dunes (vegetated dunes, transverse ridges, parabolic dunes, large wind shadows, cliff-top dunes, and climbing dunes) can coexist in the same climatic and geographic area.

Brothers (1954) has divided these dunes into three distinct groups, all of Recent age and all composed of black sands (Figure 5-25). These black sands were introduced into the longshore drift system by the Waikato River, which carried them from the central volcanic plateau region of rhyolites and andesites. The oldest group of sand dunes (group 1) has migrated inland and up the surface of the Pleistocene sandstones to a height of 170 m under the influence of the prevailing southwest winds. These dunes (Figure 5-25) have been further subdivided by Brothers into two smaller groups. The first of the groups contains the "primary forms" (i.e., transverse dunes) that were originally developed in free-moving sand, whereas the second of these groups, the "secondary forms" (i.e., vegetated dunes), was influenced by the stabilizing effects of vegetation. Together the group 2 dunes form parallel elongated dune ridges perpendicular to the southwest winds in the deep gullies of the Pleistocene sandstones. They also form long wind shadow dunes, tapering out to the northeast behind sandstone erosional remnants at the top of the sandstone cliffs. The latter are probably formed by the funneling of sand-bearing wind up the gullies to a height of 170 m. Sand accumulations in the intermediate area (e.g., higher elevation gullies) take

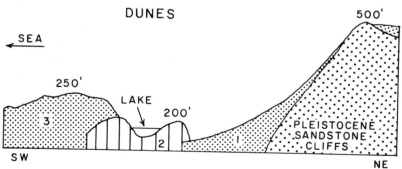

Figure 5-25. Diagrammatic cross-section of the dune formations along the Auckland west coast of New Zealand (after Brothers, 1954, Figure 3, p. 51). These dunes are divided into three distinct groups, 1, 2, and 3, in order of decreasing age and indicated on the sketch. See text for explanation.

" . . . a U-form like a drawn out hairpin with the prongs opening out into the wind" (Brothers, 1954, p. 51). The parabolic dunes have also developed on the tops of the cliffs, where the protective cover of vegetation is lacking, as secondary dunes from the loose sand of the primary dunes. Some of the transverse dunes have advanced inland at a recorded rate of up to 3 m/yr over a 44-yr period.

The younger group 2 dunes (Figure 5-25) form a series of three or four parallel ridges oriented perpendicular to the southwest winds along the base of the sandstone cliffs. These dunes contain both steep leeward and steep windward slopes fixed by vegetation. Lakes are located in the hollows between the ridges. These lakes are further confined by trailing lines of sand formed in the lee of the larger foredunes, which join the parallel ridges to each other.

The group 3 dunes (Figure 5-25), youngest of the three groups and closest to the coast, consist of thick sheets of loose sand in lobate ridges advancing over the group 2 dunes. These dunes are now forming directly from the wind picking up sand from the long, steep, wide beach, which is in turn supplied by the extensive longshore drift system. Despite the dependence of the dunes on the beach for their sand supply, it is interesting to note that the ridge crests form an angle of as much as 10 to 14° to the beach (i.e., oblique dunes). This is because the sand ridges are oriented perpendicular to the prevailing winds instead of following the curving beach outline. One of these foredune ridges, about 10 m high and 23 m wide, extends the length of the beach almost without interruption. Many large parabolic dunes are formed in these foredune ridges and are anchored on the tails by hummocks of marram grass. Some of these hummocks are being eroded by an encroaching sea to form scarps up to 3 m high.

Eolianites

Under tropical conditions a sand composed of calcium carbonate, instead of quartz, is common. Calcium carbonate sand, acted on by the wind, will produce a distinct type of coastal dune referred to as eolianite, or eolian calcarenite, commonly known as dune limestone. This type of dune forms where eolian accumulations of calcareous sand have become lithified. A second type of eolianite (called *quartz eolianite*) consists of quartz sand cemented with calcium carbonate. It commonly occurs in the Mediterranean climates of Israel, South Africa, and Australia. During the rainy season, calcium carbonate from the shell fragments, blown into the dunes, goes into solution (Yaalon, 1967). During the long dry season, the calcium carbonate precipitates, forming the cement. Thus, quartz eolianites require for formation: sufficient rainfall contained within a rainy season and a dry season accompanied by a warm climate (Yaalon, 1967). The processes of dune formation and growth for eolianite sand dunes are similar to the processes for quartz sand dunes. Whereas an eolianite dune is permanently lithified, a quartz sand dune fixed by vegetation may suddenly start to move again if the vegetation cover is destroyed. Eolianite may show most of the common sand dune forms. In the southeast of South Australia, for example, there are several ridges of dune limestone parallel to the shore, which were formed during emergence of the coast resulting from both eustatic lowering of sea level and uplift of the land. These parallel eolianite ridges are now separated by broad troughs containing lakes and swamps (Bird, 1964). Occasionally parabolic dune forms are found preserved. Elsewhere in South and Western Australia extensive eolianite has been noted along the coast where the dunes have been partially planed off by an advancing sea. They have even been found underwater along the Australian (Bird, 1964) and Israeli coasts, testifying to their ability to preserve their form intact and to a lower Pleistocene sea level at the time of their formation. Other studies of eolianites have been conducted by Yaalon and Laronne (1971) and Vacher (1973).

Age and Pulsating Dune Growth

Most examples of large-scale coastal sand dunes scattered throughout the world that have been discussed in some detail relate dune history to the effects of estuatic sea level changes, and to the Flandrian sea level rise in particular. A few authors (Brothers, 1954; Jennings, 1957a; Bird, 1964) cite a postglacial 3- to 4-m rise in sea level above the present level during the thermal maximum to account for certain dune features, especially the several parallel lines of dune ridges. It is interesting to note that all three of these authors are from the same region, Australia and New Zealand. Their evidence, if real, may not indicate a eustatic sea level change but may instead be related to regional tectonic activity. In either case, one or two relative fluctuations in sea level do not account for the three, four, or more parallel lines of dune ridges formed in most large areas of dune accumulations. Moreover, many of these large dune areas are on shorelines that are at present prograding. These areas could not have begun to prograde at the present rapid rate until the sea level rise slowed down about 3000 to 4000 yr b.p. The fact that this would allow more than enough time for four or more lines of dune ridges to form is substantiated wherever rates of dune formation have been monitored.

The very extensive dune system of the Dutch coast described in great detail by Van Straaten (1961, 1965) consists of both vegetated and transverse dunes aligned in several distinct patterns. One such pattern consists of lines of nearly parallel dunes up to 25 to 30 m in height, and another pattern contains three or four high, continuous transverse dune ridges up to 56 m high. These larger dunes are clearly post-Roman in age, as indicated by the pottery beneath them; most seem to have formed in the

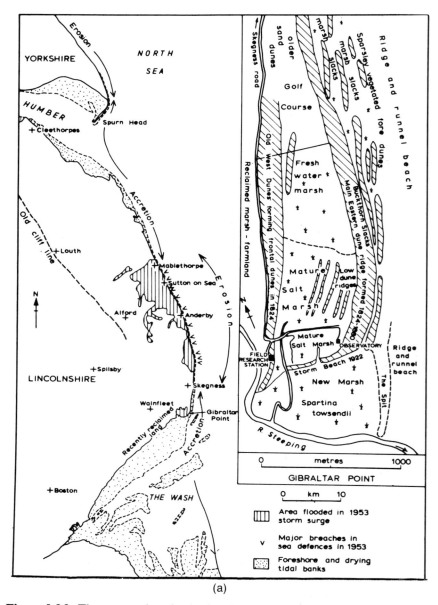

(a)

Figure 5-26. Three examples of pulsational spit growth. See text for discussion. (a) Gibralter Point, Lincolnshire, England (after King, 1972, Figure 6-8, p. 177). (b) Fishing Point, Assateague Island, Virginia (after Gawne, 1966, in Shepard and Wanless, 1971, Figure 419, p. 95.) (c) Monomoy Island, Cape Cod, Massachusetts (after Shephard and Wanless, 1971, Figure 317, p. 50; and Goldsmith, 1972.) (d) Aerial photograph of the Monomoy Spit taken in June 1971, showing the older dune ridges up to 12 m in height (in c above) and the new spit (lower left), approximately 1000 m long, which formed in less than 3 yr. North is toward the upper left.

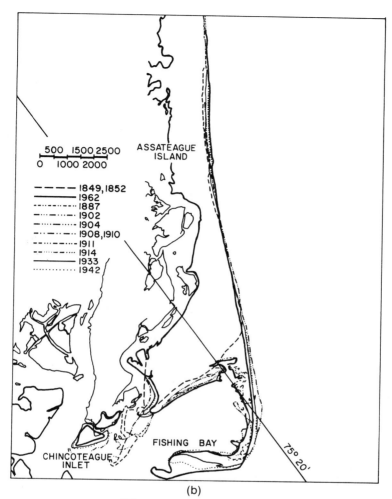

500 1500 2500
0   1000 2000

- - - - - 1849,1852
————— 1962
·········· 1887
-··-··- 1902
-···-···- 1904
-··-···- 1908,1910
--··-··- 1911
--··-··- 1914
————— 1933
·········· 1942

ASSATEAGUE
ISLAND

FISHING BAY

CHINCOTEAGUE
INLET

75° 20'

(b)

Figure 5-26. *(Continued.)*

Middle Ages, and some after 1610 a.d. A similar age is assigned to the dunes
at Plum Island, Massachusetts (McIntire and Morgan, 1963).

Examples of pulsating growth on prograding shoreline within historical
times are illustrated in Figure 5-26. In the first example, from Gibralter
Point, Lincolnshire coast, Britain, two distinct lines of dunes are visible. The
inner line of dunes at Gibralter Point was shown as the frontal dune system
on Armstrong's map of 1779. By 1824 the outer main dune ridge had not
formed, but it had by the end of the century; it therefore formed during the
nineteenth century, and the outer dune ridges, which are founded on beach

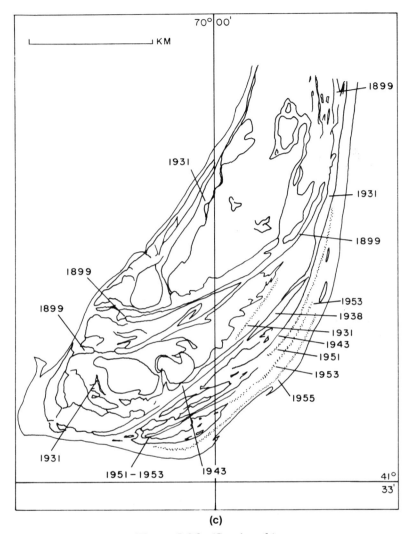

Figure 5-26. *(Continued.)*

ridges, formed during the twentieth century. The spit has formed since 1922, when a storm eroded the southern end of the eastern dune ridge. The dunes form arcuate ridges instead of long lines in most cases (King, 1972).

Another example is Chincoteague Spit on the Maryland–Virginia border of the Delmarva Peninsula on the Middle Atlantic Coast of the United States, which was studied by Gawne (1966). The southern end of Assateague Island, also on the Virginia–Maryland border, Delmarva Peninsula, Middle Atlantic coast of the United States, has displayed pulsating growth as well. Figure 5-27 shows that a period of spit growth, with numerous dune ridges, which occurred prior to 1849, was followed by a 50-

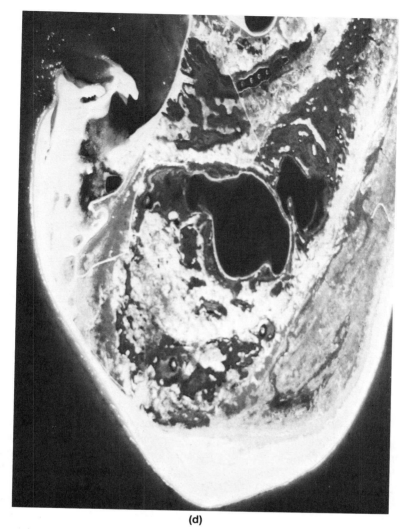

(d)

Figure 5-26. *(Continued.)*

yr hiatus, until 1904 when a new recurved spit, also containing dune ridges, began to form. Gawne (1966) was able to identify 39 successive dune ridges within six distinct growth periods, all of which formed on this new spit, called Fishing Point, since 1904.

The third example is Monomoy Island (Massachusetts) on the northeast coast of the United States. Here the earliest dune ridges, approximately 8 to 10 m in height, formed 100 to 125 yr ago (Shepard and Wanless, 1971, p. 50; Goldsmith, 1972). A second system of foredune ridges formed since the 1920s and a third set (Figures 5-26c and 5-26d) is presently forming on the extreme southwest corner of Monomoy Island.

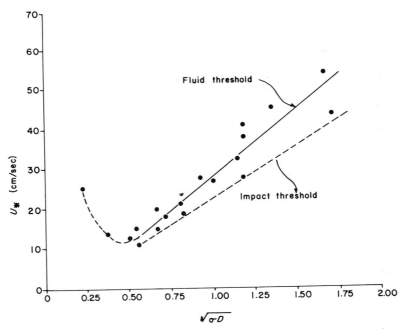

Figure 5-27. Experimental thresholds of movement for quartz-density sand under wind action (data from Horikawa and Shen, 1960; after Allen, 1970). U_* = shear velocity, σ = density of sand, D = mean grain diameter of sand.

To summarize the times of dune growth of these areas, there appears to be evidence for definite periods of relatively increased storm frequency brought upon by periodic shifts in the climatic belts (Goldsmith, 1972). Geomorphic evidence for storm frequency is also suggested by studies in other areas (Mather, 1965). The problem, however, is the lack of meterologic data during the 1800s. Cry (1965) has compiled a histogram of storm frequency going back to 1871. The trend is indeed suggestive of a storm frequency curve postulated from the historical study of these three areas. However, statistical analysis of the data does not give a sufficiently high confidence interval to support Cry's storm frequency hypothesis with a relative degree of certainty.

Namias (1970) discussed a climatic anomaly that occurred over the United States during the 1960s. These climatic anomalies may cause variations in storm frequency. Thus, such geomorphic features as coastal sand dunes, which are so directly dependent on the wind patterns, should certainly be expected to respond to subtle changes in the wind regime brought on by slight local climatic changes along the coast. These climatic changes may be reflected in the patterns of eolian sedimentation.

Physical Processes of Eolian Sedimentation

In the preceding sections the external and internal geometry of different types of coastal sand dunes from many areas around the world has been described. This section discusses the physical processes of transportation and deposition of individual sand grains, the collective movement of a mass of sand grains, and changes undergone by the individual grains within the mass of grains (i.e., rounding and sorting).

Entrainment and Transport of Sediment by Wind

Sand transport by wind occurs by saltation, creep, and suspension (in order of importance). Because of the large differences of density between sand grains and air, transportation by suspension is relatively unimportant in coastal dunes. This is one of the great differences between eolian and subaqueous transport. Knowledge of these processes comes from field observations, theoretical considerations, wind tunnel experiments, and sand trap studies (Bagnold, 1954; Horikawa and Shen, 1960; Belly, 1963; Williams, 1964; Johnson, 1965; Johnson and Kadib, 1965; Inman et al., 1966; Byrne, 1968; Allen, 1970).

Basic Theory

Basically, a stationary particle at rest will begin to move when the shear stress (τ_0) at the grain surface from the wind exceeds some critical value. The particle can be entrained in one of two ways, either directly by the wind or by being struck by another particle already in motion (i.e., impact). The critical shear velocity is dependent on the relative density of the particle and its size as follows (Bagnold, 1954, p. 101):

$$U_*^c = A \sqrt{\frac{\tau_{oc}}{\rho}} = A \sqrt{\frac{\sigma - \rho}{\rho} g D}, \qquad (5\text{-}2)$$

where U_*^c = critical shear velocity; A = constant; τ_{oc} = critical shear stress; ρ = fluid density (i.e., wind); and σ = grain density, g = gravity acceleration; D = grain diameter. Figure 5-27 shows these relationships for both fluid and impact entrainment. Notice that grains with a diameter of about 0.10 mm are the first to be entrained by a wind of increasing velocity; also note that the impact threshold velocity is about 80% of the fluid threshold. These relationships hold best for well-sorted grains.

Saltating grains (bounding motion initiated by impact) take a path with an initial, almost vertical, rise followed by a long, low trajectory almost parallel to the ground and a velocity close to that of the wind velocity (Figure 5-28).

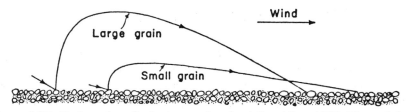

Figure 5-28. Trajectories of saltating sand grains (after Allen, 1970, Figure 3.4, p. 98). Note that the small grains do not bounce as high but move nearly as far as the larger grains. The same grains often alternate between saltation and creep.

The loss of momentum from the wind thus gained by this particle is transferred to one or more other particles upon impact or may result in the same original particle bouncing back up into the airstream.

Experimental data also show that during transport the coarsest grains are concentrated at two levels, at the bed and several centimeters or more upward. This complex vertical distribution is explained by the fact that when grains undergo saltation the bigger grains bounce higher (Allen, 1970, p. 101).

Coincidently with saltation, surface creep is occurring by other grains, or sometimes alternating with the saltating grains. Creeping particles travel slowly and irregularly along the surface under the influence of the bombarding grains or directly under the influence of the fluid.

The movement of these grains is dependent on the critical shear velocity (U_{*c}), which in turn is directly proportional to the rate of increase of the wind velocity with the log height; and using the Prandtl equation (Horikawa and Shen, 1960, p. 2):

$$U = \frac{U_*}{K} \; \log \frac{Z}{Z_0} , \qquad (5\text{-}3)$$

where U = wind velocity at elevation Z; K = Karman constant (≈ 0.40); and Z_0 = roughness at the surface.

Bagnold (1954) suggests that $Z_0 = D/30$ where D = grain diameter. Zingg (1953) suggests that $Z_0 = 0.081 \log (D/0.18)$. The log-height equation is of great advantage because it relates the fluid velocity at any height with both the surface roughness and critical shear stress at the surface. However, once U_{*c} is exceeded and sand particles are entrained, the wind velocity profile is affected by the sand movement. With increasing wind velocities, the layer affected increases in height, because of the increased height of sand trajectories.

For the same sized sand, though, all the wind velocity distributions plotted on semilog paper against height pass through a common point, called the focal point. This means that no matter how strongly the wind blows, no matter how great the velocity gradient, the wind velocity at a certain height (about 3 mm) remains almost the same (Bagnold, 1954, p. 58, Figure 17).

The reason for this phenomenon is unclear, but it allows for the simplification of Equation 5-3 to (Horikawa and Shen, 1960, p. 3):

$$U = 5.75U_* \log_{10} \frac{Z}{Z'} + U',$$

(5-4)

where according to Zingg (1953) $Z' = 10D$ (in mm) at the focal point; $U' = 20D$ (in mile/hour) at the focal point.

Of great interest is the total amount of sand transported. Two theoretical solutions of note are by Bagnold and by Kawamura (Horikawa and Shen, 1960).

According to Bagnold (1954), the transport of sand represents a continual loss of momentum by the wind. Bagnold further assumes that most transport occurs through saltation, about 25% as creep and a small remainder as suspension. Using these considerations he derived the following equation in which the total transport (q) is directly proportional to the cube of the shear velocity:

$$q = C \sqrt{\frac{d}{D}} \frac{\rho}{g} U_*^3,$$

(5-5)

where q = total sediment transport; D = grain diameter of a standard 0.25 mm sand; and C = sorting constant: $C = 1.8$ for naturally sorted sands, $C = 1.5$ for well sorted sands, and $C = 2.8$ for poorly sorted sands.

Kawamura stated that the shear stress (τ_0) is caused both by the impact of the grains and directly by the wind. Following these considerations he derived the equation:

$$q = C' \frac{\rho}{g} (U_* - U_{*_c})(U_* + U_{*_c})^2,$$

(5-6)

where C' = constant experimentally derived. Both equations are nearly identical for large transport rates but differ increasingly as U_* approaches U_{*_c}, because q should then equal zero, which is more closely approximated by Kawamura. However, there is experimental evidence in support of both (Allen, 1970), and such differences may be attributed to the derivation of the constants, differences in grain roundness, temperature, humidity, and other factors. Further studies on the relationship between sediment transport and fluid power have been made by Bagnold (1956, 1966) and Inman et al. (1966); laboratory studies and the problems of sand traps are discussed by Horikawa and Shen (1960). The interested reader is referred to these works.

Effect of Moisture

Johnson (1965), using experimental data, found that with a relatively high water content (in excess of 1%) the wind speed necessary to initiate sand movement becomes greatly affected (p. 754). More quantitatively, modifying Equation 5-2:

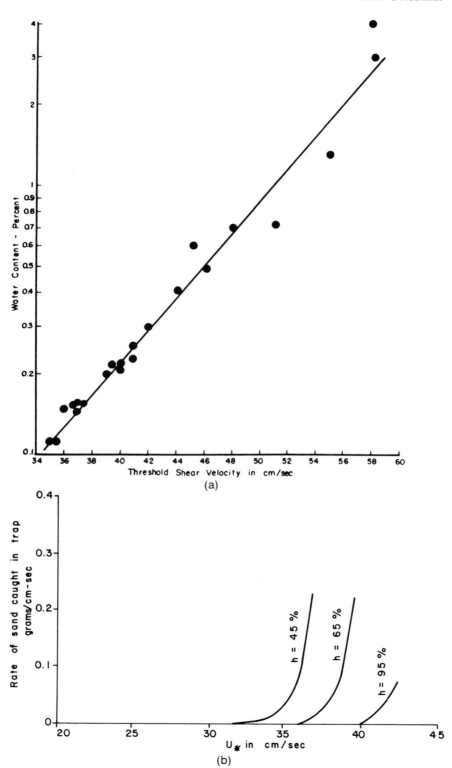

(a)

(b)

$$U = A \sqrt{\frac{\sigma - \rho}{\rho} gD} \, (1.8 + 0.6 \log_{10} W), \qquad (5\text{-}7)$$

where W = water content. These effects are illustrated in Figure 5-29. Both Tsoar (1974) and Hunter et al. (1983) noted large transport decreases during rainstorms. An equation for predicting U_{*c} on a wet sand surface was calculated by Hotta et al. (1985), based on field experiments.

Application of Theory

Effect of Vegetation on Surface Roughness. With increasing surface roughness, the effect of vegetation is felt not only close to the surface but at heights of tens of meters above the surface as well. This is a result of the turbulent exchange that occurs through eddies between layers to some height above the ground. Figure 5-30 (Olson, 1958) clearly shows the striking increase in the surface roughness with the planting of dune grass on the formerly bare slope of a Michigan dune. The thickness of the calm wind layer, Z_0, appears to have increased 30-fold from 0.04 to 1.0 cm. This effect of increased surface roughness was transferred upward, as indicated by the change in slope of the wind velocity profile.

These dramatic changes in the wind profile measurements were associated with changes in the eolian processes. Before vegetation was planted, erosion had occurred on the dune slope. After the planting, deposition occurred, and at the end of 2 yr up to 60 cm of sand had accumulated. Rapid deposition was continuing as long as 8 yr later, as the vegetation continued to thrive and spread out from the original area of planting.

Bressolier and Thomas (1977) extended these studies by calculating the surface roughness for various plant species, and their densities and heights, along the French Atlantic coast; Hesp (1979) studied this problem on the Australian coast.

Depositional Characteristics

Grain-size Parameters

During transportation by wind the entrained sediment is subject to sorting, rounding, and the etching of surface features, all of which impart distinctive characteristics on eolian sediments. A considerable effort has been undertaken to determine the grain-size characteristics of eolian sediments in order to detect eolian environments and to distinguish them from beach, fluvial, offshore, and other sedimentologic environments in the rock record. The standard textural parameters employed are mean grain diameter, standard

---

Figure 5-29. Experimental data illustrating the effects of (a) water content and (b) humidity on shear velocity (U_*) (after Johnson, 1965, Figures 17 and 14, pp. 755 and 754). Note how subtle changes in water content and humidity result in relatively large changes in U_*.

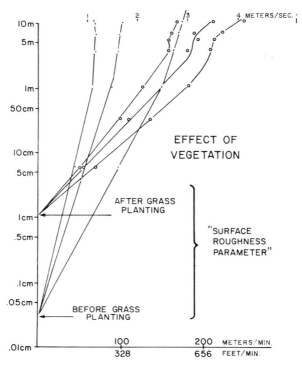

Figure 5-30. The change in surface roughness and dune wind profile before and after grass planting on an Indiana coastal sand dune (after Olson, 1958, Figure 2, p. 256.) After planting, Z_0 increased more than 30-fold and the dune surface changed from erosional to accretional. Within 2 yr after planting, 65 cm of sand had accumulated around the grass.

deviation, skewness, and kurtosis. These may be calculated either through the method of moments or by graphic measures. The use of grain-size parameters in distinguishing eolian sediments is thoroughly reviewed by Folk (1966, 1971), Anan (1971), King (1972, pp. 172–175), and others. The reader is referred to these sources for a thorough discussion. The results of these studies indicate characteristic tendencies, but no all-inclusive definitive statements can be made.

In general, dune sands are finer, better sorted, positively skewed, have a higher kurtosis, and are more rounded than the source sediments on the adjacent beaches. Although this is true for many areas, Figure 5-31 illustrates that there are exceptions and, moreover, that the differences between beach and dune sediments can vary widely along the same barrier island.

Anan (1971) suggests that these differences may be a function of sampling procedures. For example, samples collected on the upper third of the

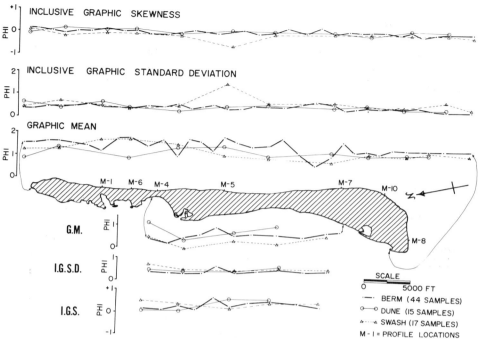

Figure 5-31. Variations in graphic mean grain size, inclusive graphic standard deviation, and inclusive graphic skewness (phi size) of the berm, swash, and dune samples. Note that the mean grain size of the berm (samples at 330 m intervals) is exactly the same at the north and south ends of the island, indicating that the beach sand is in equilibrium with the wave-energy conditions. Also, the dune sands are generally coarser than the berm sands on the east (ocean) side of Monomoy, but the reverse is true on the west (bay) side. Note the uniformity in sorting around the island and between environments. The samples on the west (low-energy) side were slightly better sorted than those on the east side.

vegetated dunes on the Massachusetts coast were finer and more positively skewed than samples collected from the base of the dunes. This is because the larger particles have more of a tendency to roll down, or less of a tendency to be moved up the dune by the wind. When only one sedimentologic unit* is sampled, the adjacent beach samples are calculated to be better sorted.

When several sedimentologic units are sampled, dune samples are often found to be bimodal. This is because of the formation of a wind lag deposit whereby the finer sediments may be differentially moved by the wind, leaving

---

*A sedimentologic unit on a beach or dune may be only millimeters thick.

a coarse lag deposit on the dune surface. Such a coarse surface layer may also inhibit further eolian transport. Gutman (1977c) has shown that the density of dune vegetation and the most recent winds (i.e., offshore versus onshore) significantly affects all the grain size parameters.

In general, the best textural characteristic for distinguishing eolian from subaqueous transport is probably a positive skewness, indicative of fines carried by the wind in suspension. Fines transported by water would be less likely to be deposited except under specific conditions (e.g., lagoons) that would leave other paleo-indicators.

Sorting

Bagnold (1954) found that the rate of sediment discharge increases as the sorting increased from one grain size to the "natural" distribution in dune areas. Moreover, this "natural grading" of the sand moved by the wind can be separated into three components that correspond to the three modes of transportation, saltation, creep, and suspension (Bagnold, 1954, pp. 138–139). With a constant wind, the surface creep component tends to advance the farthest, resulting in a coarsening and a shift in the grain-size curve toward negative skewness.

In an exhaustive experimental study of eolian saltation load, Williams (1964) found that there was a distinct size distribution in sand captured at different heights, and that sorting increased directly with the logarithm of the height. Surprisingly, this size distribution remained constant for a variety of wind speeds and grain shapes (p. 271). However, with a rough and irregular surface the saltation load can move ahead and be separated from the surface creep. Because of the strong interaction and exchange between these two components, any separation and coarsening is a gradual and subtle process inasmuch as the sand is initially well sorted. Recent work on eolian grain size distributions has been conducted by Barndorff-Nielsen et al. (1980).

A more important sorting process in coastal areas occurs on the faces of the dunes. Coarse sediment assimilates on the lower part of the dunes and in the interdune lows (previously discussed). Often the difference in mean grain size between these two dune environments is greater than between dune and adjacent beach (Goldsmith, 1972). Another important sorting mechanism is ripple formation and movement (discussed in a later section).

Particle Shape

Williams (1964) noted that opinion seems to vary as to whether a spherical particle moves more readily than an angular one. Also unknown is the effect of grain angularity and sphericity on height of saltation jump. This uncertainty may be caused in part by differences in grain behavior under different wind velocities, enhancing the possibility of shape selection as a criterion for sediment movement. Under low wind velocities the rate of sediment movement of particles of lower spherical grains displayed a greater tendency for saltation transport than angular grains and formed a greater percentage of the total saltation load with increasing saltation height (Williams, 1964, p. 286).

Surface Features

Frosting. Eolian abrasion of sand grains is evidenced by the greater rounding in eolian grains and in the abundance of frosted grains in dune environments. This appears to be related to the relatively high amount of transport by impact and saltation. Frosted grains apparently also occur in coastal dune environments. However, observations by the author indicate the frosted grains are just as prevalent in the beaches of Monomoy Island (Goldsmith, 1972) and also that polished grains, generally considered indicative of subaqueous action, such as waves, are quite prevalent in the dune sediments. This may indicate a large amount of environmental mixing, which may not be as prevalent in less dynamic coastal areas.

Kuenen and Perdok (1962), in an experimental study of frosting and defrosting of quartz grains, suggest that frosting is caused only to a minor degree by mechanical action by wind and that, instead, it is mainly caused by chemical action by corrosive solutions and by alternate solution and deposition of matter, especially in desert areas (p. 648). If this is true then desert coastal dunes, such as occur along the coast of Baja California, may contain a higher percentage of frosted grains in their sediments than do vegetated coastal dunes.

Additional observations of desert dunes indicate that mechanical frosting may be most common in larger size grains ($>$ 0.2–0.4 mm) and that frosting becomes less pronounced over a range of sizes (0.15–0.50 mm) (Kuenen and Perdok, 1962, p. 648). Comparative microscopic studies are needed of coastal and desert dune sediments.

Electron Microscopy. Several recent studies have suggested that the mechanical and chemical textures present on the surfaces of sand grains as revealed by electron microscopy are reflections of the processes of formation and therefore are definitive environmental indicators. This work is summarized by Krinsley and Donahue (1968) and Krinsley and Smalley (1972). These authors further distinguish between coastal dune sands and tropical desert dunes. Coastal dune sand grains have small meandering ridges that result from impact breakage, probably by other grains, and graded arcs that are probably percussion fractions. In addition to the above, desert tropical grains also contain flat pitted surfaces, which may obliterate the above two features (Krinsley and Donahue, 1968, p. 746). However, the pitted surfaces might also be expected to occur on tropical coastal dunes. The meander ridges and graded arcs were experimentally duplicated by Kuenen (1963) on grains in a wind device.

Eolian Ripples

Eolian ripples are quite common on all parts of the dunes and beach. The following commonly occurring characteristics make wind-formed ripples distinct from water-formed ripples:

1. Eolian ripples fall into two distinct types: (a) sand ripples, composed of well-sorted medium- to fine-grained sand; and (b) granule ripples, isolated ripple forms often parabolic shaped, composed entirely of coarse sand or granule-sized particles. (The next three characteristics refer primarily to the sand ripples.)
2. Sand ripples often occur on the sides of dunes, with their long axes parallel to the slope.
3. There is a tendency for the coarsest grains and/or heavy mineral grains to accumulate on the crests of the sand ripplies rather than the troughs of the ripples.
4. Eolian ripples tend to have large ripple indices ($>$ 17) and symmetry indices between 2.0 and 4.0 (Tanner, 1964, 1967), although there are many exceptions.

The sharp distinction between sand ripples and granule ripples is recognized not only for the coastal zone but for most desert environments as well (Bagnold, 1954; Sharp, 1963) (Figure 5-32). Granule ripples are usually asymmetrical whereas sand ripples are either symmetrical or asymmetrical.

Granule Ripples

Granule ripples form adjacent to areas of coarse lag deposits. In the coastal zone the two main areas are in the interdune lows and at the high-tide swash. They also appear to be most common on a hard sand "erosional"-type surface. Here the coarse-grained ripples are moved across the finer sand surface under the influence of the wind as isolated cuspate or linear ripple forms. Many of these ripples accumulate in en echelon or linear chains and display a striking resemblance to barchan and/or parabolic dunes in form and movement. They are usually more asymmetrical, are larger, and move more slowly than the sand ripples. Observations in the coastal zone indicate wavelengths of up to 2 m and heights up to 15 cm, although most are about half these dimensions. The desert granule ripples have been noted to be much larger. Studies of the internal structures of granule ripples in the Mojave Desert by Sharp (1963, p. 633) show the presence of large-scale foresets, with many ripples having foresets in two directions. This indicates how easily the ripples change their symmetry with a change in wind direction, a characteristic of all wind-formed ripples. The troughs are composed primarily of the finer sand substrate over which the granule ripples are

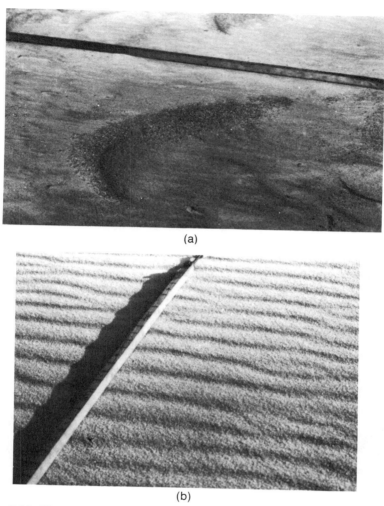

(a)

(b)

Figure 5-32. The two types of eolian ripples observed on Monomoy Island, Cape Cod, Massachusetts, are seen in (a) the barchan-type granule ripple, composed entirely of coarse sand and gravel; (b) the ripple with well-sorted sand in both crest and trough; (c) both types of ripples, superimposed; and (d) the well-sorted sand type of ripple, often found on dune faces and in wind shadows, with the crests parallel to the slope.

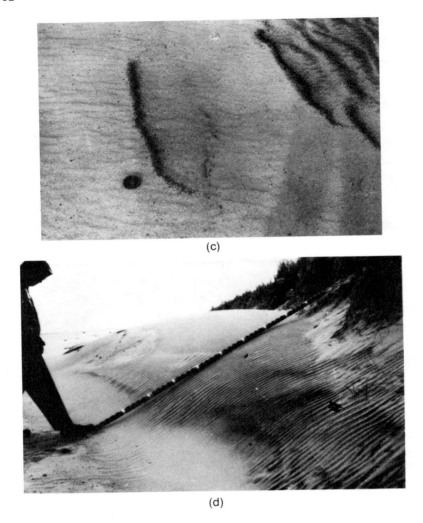

(c)

(d)

Figure 5-32. *(Continued.)*

moving, resulting in a strong grain-size bimodality for these ripples. Although they take longer to form than the sand ripples (hours versus minutes), granule ripples tend to last much longer after the wind forming them has decreased in velocity.

Granule ripples appear to form by surface creep driven by saltation impact under higher velocity wind conditions. The grains travel up the windward slope and cascade down the "slipface," accumulating at the angle of repose. The spacing (i.e., wavelength) is probably largely a function of available coarse material. These features therefore appear to resemble the larger eolian forms, dunes.

Sand Ripples

Sand ripples contain essentially the same grain size as the underlying substrate and often form in areas of sand accumulation, although they may be found associated with granule ripples. Sand ripples in the coastal zone provide a good indicator of the local wind directions and the complexity of the wind currents around the dunes. Their presence on the downwind side of a transverse dune crest, for example, may show wind transport parallel to the dune crest, resulting from local helicoidal wind flow (Tsoar, 1983a). Very often different sets of ripples oriented in various directions, reflecting the passage of a weather front, can be discerned in the dune areas. The sand ripples commonly form on wind shadows on the lee side of the dune, where wind and sand is funneled through a low area in the dune crest. Here their crests are almost always parallel to the slope and are often accentuated by heavy mineral accumulations (garnet, magnetite, etc.).

The tendency toward heavy mineral concentrations in the dune areas is especially noticeable on the surface of sand ripples. This may be caused in part because of the importance of grain roughness in eolian transport. A quartz grain having the same weight as a heavy mineral grain is larger, stands higher, and therefore provides a greater roughness factor. In areas of greater wind shear stress (i.e., the crests of the ripples) the quartz grains would be expected to be preferentially moved, resulting in a heavy mineral lag concentrate. This is discussed at length in Leatherman (1975).

Many observers (e.g., Sharp, 1963, p. 621) have suggested a tendency for the coarsest grains to accumulate on the ripple crests. However, this apparent phenomenon may be caused in part by the difference in rates of creep with different particle sizes. Of the many wind-formed sand ripples observed by the author in different coastal areas, approximately 75% had slightly coarser grains close to the crest than in the trough. However, in most cases the coarser zone, like the heavy mineral zone, was slightly offset from the crest toward the windward or leeward sides.

Observations of the ripples during formation and movement indicate that the sand accumulations via creep under the influence of saltation impact

(Sharp, 1963) and that ripple spacing (i.e., wavelength) is apparently not necessarily directly related to the length of the path of the saltating grain. Wind velocity and grain size also are critical in determining height and wavelength. Ripple height, for example, generally increases with grain size and decreases with increasing wind velocities, whereas ripple wavelength increases with both grain size and increasing wind velocities, although these relationships are not always linear. Ellwood *et al.* (1975) sampled grain sizes and ripple size in the Algerian Sahara and distinguished two ripple types: (1) small wave length (1–20 cm) formed in fine unimodal sand, and (2) larger bedforms (20 cm to 20 m) formed in coarse bimodal sand; transitional types were common. They attribute all of these to saltation, with the transitional types resulting from improved rebound of the fine fraction off the coarse fraction in the bimodal sand. With very high wind velocities the sand ripples disappear, and with a further increase in velocity, the granule ripples then also disappear. There also appears to be a fair amount of exchange of sand grains between the "saltating curtain" and the creeping grains. Essentially, though, a grain pushed over the ripple crest by a saltating grain reappears on the windward side of the ripple, to creep once again up the side of the ripple under the influence of the saltating grains. With increasing grain size, this rate of ripple migration decreases.

Even more important, Bagnold (1954) suggests that the sand ripples are initiated when small accumulations of larger grains, which travel more slowly, begin to form, creating a shadow zone effect. This saltation shadow grows as the smaller grains are "trapped" in the spaces between the larger grains. As the accumulation grows this effect as a sand trap increases, until the ripple form emerges. Experiments indicate that under winds of approximately 40 km/hr sand ripples can form in just minutes (Sharp, 1963, p. 628).

Ripple Index (RI)

The ripple index (wavelength/ripple height) has been the subject of much controversy. A RI > 17 is supposed to be an indicator of an eolian origin (Tanner, 1964, 1967). However, the ripple index of sand ripples varies inversely with grain size and directly with wind velocity. For example, the mean RI of 15 ripples observed forming and moving under a 35 km/hr wind and composed of medium- to coarse-grained sand was only 8.4 (Goldsmith, 1972, p. 364).

Tanner (1967) measured seven different dimensionless parameters on many different ripples and suggested that plots of a combination of two of these parameters, ripple index (RI) and ripple symmetry index (RSI), provided the best indication of an eolian origin (Tanner, 1967, Figure 1). The RSI is defined by Tanner (1967, p. 95) as the horizontal distance from the crest to the trough of the gentler face divided by the horizontal distance from the crest to the trough of the steeper face. Eolian ripples generally have a RSI between 2.0 and 4.0. Although far from foolproof, these indices do provide a solid basis for further investigations of additional parameters influencing eolian ripple formation and movement.

Preservation

A much greater problem, however, is the recognition of these ripples within modern coastal dunes. Of all those who studied the internal geometry of coastal sand dunes only Bigarella et al. (1969) noted these ripples preserved in the dunes. Yaalon (1967, p. 1196) suggested that the reason eolian ripples are rarely observed in eolian beds is the process of sand accumulation whereby " . . . each surface ripple is leveled out by the subsequent one." Eventual leveling out results in two distinct planar laminae representative of the original grain-size variation between the crest and the trough of the ripples. This same original grain-size variation between crest and trough has been noted in the eolian ripples on Monomoy Island (see also Hunter, 1977). Another possibility is that Bagnold's (1956) fluid dispersive mechanism is at work here. Kick the top of any dune and the coarsest particles will get up on top and outrun the fines in the resulting sand slide. In either case, the dune bedding observed on Monomoy appears to be similar to that observed by Yaalon (1967) and Yaalon and Laronne (1971) in the coastal dunes of Israel.

Artificially Inseminated Coastal Sand Dunes

Humans as Coastal Agents

The eolian history of many coastal areas in Europe, North America, Australia, and South Africa is quite similar in that once-stable forest-covered dunes have been deforested by human activities, e.g., as early as the fourteenth century in The Netherlands (Norrman, 1981). This has had a definite effect on dune processes and coastal geomorphology in many of these areas. On the northeast coast of the United States the first European settlers in the seventeenth century adversely affected dune vegetation almost immediately (Larsen, 1969; Messenger, 1958, p. 30). Similar intense farming practices along the Capetown area of the South African coast resulted in extensive sand drifting by the mid-1800s (Walsh, 1968). Australia has similar examples (Chapman et al., 1982, p. 140). Dune rebuilding and stabilization therefore has had a long history and is definitely not a new problem. The adverse effects of grazing on dunes have been documented for the present on Currituck Spit, Virginia—North Carolina, by Hennigar (1977a) and for Ossabaw and St. Catherines Islands, Georgia, by Oertel and Larsen (1976). Recently, the impact of off-road vehicles on the coastal ecosystem (Leatherman and Steiner, 1979) and the effect of man-made structures such as houses on dune dynamics (Nordstrom and McCluskey, 1982) have been studied. The destruction of the high sand dunes in Queensland, Australia, because of mining of rutile, zircon, ilmenite, and their subsequent reclamation, is well documented in Morley (1981).

The importance of dunes in stabilizing the shoreline and providing a sand reservoir for eroding beaches has been discussed by Oertel (1974) and Gares *et al*. (1979). Short and Hesp (1982) proposed a morphodynamic classification of surf zones, beaches, and dunes, based on examples from southeastern Australia. They suggest that eolian sand transport rates are potentially highest on low gradient dissipative beaches and lowest on steep reflective beaches, resulting in large-scale transgressive dune sheets associated with the former. However, there are problems with this proposed classification when applied to other coasts, e.g., Israel and the east coast of the United States. Clearly, the relationships between dunes and associated beaches are of utmost importance for dune stabilization programs.

Methods

The two main methods for creating and promoting stabilized coastal barrier dunes are through artificial plantings and sand fencing. Alternative methods, such as using junk cars on the beaches at Galveston, Texas (Gage, 1970), have been attempted without much success. Much of the literature on this subject is in United States Army Corps of Engineers publications (e.g., U.S. Army Corps of Engineers, 1977) or results from Corps-sponsored projects, inasmuch as the Corps has a vested interest in stable barrier islands (Davis, 1957; Savage, 1963; Woodhouse and Hanes, 1967; Woodhouse *et al*., 1968; Savage and Woodhouse, 1969; Campbell and Fuzy, 1972; Knutson, 1977). A review is given in Phillips and Willetts (1978) and a world survey of current management policies is presented in Norrman (1981).

Artificial Plantings

Much difficulty has been encountered in the past in achieving satisfactory plantings. This has led to the suggestion that the climate may be drier now than in the past (Larsen, 1969). More recent studies indicate that part of the problem may be the need for abundant fertilizer, for a more "irregular" spacing and density pattern of plantings, and to protect the young plants during the initial period of growth (1 to 2 yr). Also, such plants are often employed in areas only when the problem has reached critical proportions and the shoreline in undergoing rapid erosion. This problem has been discussed by Jagschitz and Wakefield (1971) and by Zak and Bredakis (1963).

Many of the studies of artificial plantings have taken place on the Cape Hatteras Barrier Island coast of North Carolina. Knowledge gained in these studies has been summarized by Savage and Woodhouse (1969), who suggest:

1. Widespread use of American beach grass over other plants, such as sea oats.
2. Plots with a width perpendicular to the beach of 10–15 m are sufficient to trap all the windblown sand effectively.
3. A nonuniform plant spacing pattern (approximately 0.5 m apart) with wide spacings at the outer edge and narrow spacings in the center for most effective sand trapping and long life for the plants.
4. Sand fencing should be employed to protect the plants the first year in areas subject to storm damage.

Tsuriell (1974) found that rows of plants (e.g., *Amophila arenaria, Agropyrum junceum, Lotus creticus*, and *Retuma roetam*) sown parallel to the prevailing winds along the Israeli coast " . . . have proved to be more effective than rows sown vertically. The intervals between the rows allow a free passage to the wind, hence causing the gradual unloading of the sand in the air." Vertical rows decrease the wind more abruptly, but blow out the seedlings and even the adult plants (Tsuriell, 1974).

The South Africans overcome this problem through extensive "brushing", i.e., emplacement of dead bushes, approximately 1 m high, over most of the area of drift sands. This brushing then protects the seeds planted concomitantly in the same area from the wind (Walsh, 1968).

Sand Fencing

These fences generally consist of slats about 1.0 to 1.5 m high, and having a porosity of 25 to 50%. Numerous tests indicate that single straight strands of fencing trap more sand per unit length of fence than multiple lines of fencing or more complex patterns. Too much roughness will act as an impermeable barrier to the wind rather than decrease the wind velocity and encourage sand deposition, the purpose of the fencing. Also, there may not be enough sand available to fill several lines of fencing. When the sand accumulation has reached to the top of the fence, a second fence can be added on top and seaward of the first fence, and so on (Figure 5-33).

Sand fencing has proved to be very effective in building incipient dunes in most coastal areas. The amount of sand trapped generally averages about 1 m^3/meter of fence width/year.

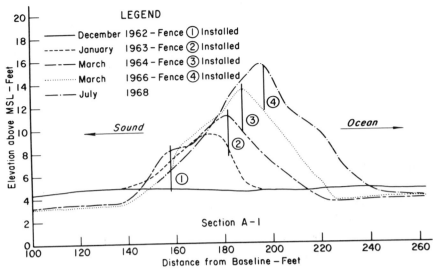

Figure 5-33. Sand accumulation using method of multiple fences on the Outer Banks of North Carolina. This generally successful method trapped 3.8 cubic yards per foot of fence and raised the dune height approximately 12 ft over a period of 6 yr (after Savage and Woodhouse, 1969, Figure 11, p. 698.)

In addition to planned dunes, artificially inseminated dunes often occur accidentally through natural sand accumulation around shipwrecks, abandoned cars and houses, and other debris.

Implications

The great success of the dune rebuilding program since the 1930s via sand fencing in the coastal areas of Virginia and North Carolina, between Chesapeake Bay and Ocracoke Inlet, has resulted in some changes that require a reassessment of the program. Studies of the massive dune program undertaken since the 1930s along the outer banks of North Carolina indicate that these artificially heightened foredunes have a distinctive internal geometry. Although the beds are low-angle dipping, typical of vegetated dunes, their direction of dip is bimodal and away from the long axis of the dune ridges (i.e., either offshore or onshore). Therefore, these beds do not dip in response to the important northwest, northeast, and southwest winds as do naturally vegetated dunes (Goldsmith et al., 1977; Rosen et al., 1977). Godfrey and Godfrey (1973) have discussed the large botanical changes to the dune ecosystem and their disturbing implications, resulting from the development of the "artificial" ecological system on Cape Hatteras. Dolan

(1972a,b) has suggested that a coarsening, narrowing, and steepening of the beach profiles in front of the stabilized Cape Hatteras dunes can be directly attributed to the success of this dune fencing program. Waves impinging on the coast expend much of their energy during shoaling and wave runup. During storms the foredune ridge is often eroded. However, much of this eroded sand is moved directly into the dune areas by wave overtopping and wave surge through overwash channels (i.e., low openings in the foredune ridge) and overwash deposits. However, Leatherman (1975) has suggested that much of the overwash sediments on Assateague Island, Maryland, are blown back onto the beach and into the dunes. Creating such a stable barrier may have two very important ramifications.

First, the wave energy is expended over a narrower area of the beach, resulting in changes in the geomorphology of the beach (Dolan, 1972a) that may have long-term implications. This results from the inability of the fixed dune line to respond to the natural processes of a dynamic system. Conversely, Leatherman (1979) studied the storm effects on two barrier islands containing these "artificial dunes" along the east coast of the United States and reports that the narrowing and steepening of the beach was only a post-storm effect. The profile reestablished its former equilibrium shape by the seaward-building offshore bar and landward erosion of the dune. Moreover, no significant difference was observed in the beach slope in front of a steep dune or without such a flanking dune. Further, the grain size present in the beach should be a function of sediment source and wave energy. Finally, Leatherman (1979) suggests that there is no evidence that wave reflection from dunes occurs in any significant manner, nor has it been shown to be an important process here.

Second, the sand eroded by waves from the foredune during storms that had formerly reached the backdune area was now prevented from doing so. Observations indicate that this storm wave deposition is at least as important a process in some coastal dunes as wind deposition (Goldsmith, 1972). Moreover, Godfrey (1970), Dolan (1972a), and Leatherman (1979) suggest that storm overwash and washovers onto the marshes and bays behind the barrier islands may be a critical process in maintaining the barrier's existence. Studies by Kraft et al. (1972) on the Delmarva barrier island chain further emphasize the importance of overwash deposits in terms of the high percentage of the total volume of barrier deposits. Historical studies of Monomoy Island, Cape Cod, Massachusetts (Goldsmith, 1972) indicate rates of ocean erosion averaging 20 m/yr, and that this island has migrated westward (i.e., retreated) a distance approximately 1.5 times its width since first discovered by the Pilgrims in 1620. Without overwash deposition this island, as well as many other barriers, would not still exist. However, Hennigar (1977b) has suggested that the buildup of the foredunes has resulted in stabilization of the interior by cutting off the sand supply and allowing the growth of interior vegetation.

Humans must carefully learn nature's ways before tampering with natural processes. Future studies of eolian sedimentation processes in the highly populated, well-used coastal zone should certainly be a high-priority item under the stimulus of our increasing concern for the preservation of natural coastal areas. Further, if the predictions of a dramatic increase in the rate of sea level rise in the coming decades (e.g., Emery, 1980; Etkins and Epstein, 1983) prove to be true, then an increased understanding of dune readjustment to a transgressive sea will be required for managing the coastal dunes, and indeed our whole coastal zone.

References

Alestalo, J., 1979. Land uplift and development of the littoral and eolian morphology on Hailuoto, Finland. *Acta Univ. Oul. A82 Geol.*, **3**, 109–120.

Allen, J. R. L., 1970. Wind and their deposits. *In: Physical Processes of Sedimentation*. Amer. Elsevier, New York, pp. 92–117.

Anan, F. S., 1971. *Provenance and Statistical Parameters of Sediments of the Merrimack Embayment, Gulf of Maine*. Ph.D. Dissert., Univ. of Massachusetts, Amherst, 377 pp.

Bagnold, R. A., 1941. *Libyan Sands*. Hodder and Stoughten, St. Paul's House, London, E. C. 4, 288 pp.

Bagnold, R. A., 1954. *The Physics of Blown Sand and Desert Dunes*. William Morrow and Co., New York, 265 pp.

Bagnold, R. A., 1956. Flow of cohesionless grains in fluids. *Roy. Soc. Phil. Trans. (London)*, **249**, 235–297.

Bagnold, R. A., 1966. *An Approach to the Sediment Transport Problem from General Physics*. U.S. Geol. Survey Prof. Paper 422-1.

Barndorff-Nielsen, O., Dalsgaad, K., Halgreen, C., Kuhlman, H., Moller, J. T., and Schou, G., 1980. *Variation in Particle Size Distribution Over a Small Dune*. Dept. Theor. Stat. Aarhus Univ. Res. Rept. 5.

Barnes, F. A., and King, C. A. M., 1957. The spit at Gibraltar Point, Lincolnshire. *East Midland Geogr.*, **8**, 22–23.

Belly, P. V., 1963. *Sand Movement by Wind*. Tech. Memo. No. 1, U.S. Army Coastal Engr. Res. Center, EC., Washington, DC.

Bigarella, J. J., 1972. Eolian environments—their characteristics, recognition and importance. *In:* Rigby, J. K., and Hamblin, W. K. (eds.), *Recognition of Ancient Sedimentary Environments*. Soc. Econ. Paleont. Mineral. Spec. Publ. 16, Tulsa, OK, pp. 12–62.

Bigarella, J. J., Becker, R. D., and Duarte, G. M., 1969. Coastal dune structures from Paran, Brazil. *Mar. Geol.*, **7**, 5–55.

Bird, E. C. F., 1964. *Coastal Landforms. An Introduction to Coastal Geomorphology with Australian Examples*. Australian National Univ. Press, Canberra, 193 pp.

Borowka, R. K., 1980. Present day dune processes and dune morphology on the Leba barrier, Polish coast of the Baltic. *Geogr. Ann. Ser. A*, **62**, 75–82.

Breed, C. S., and Grow, T., 1979. Morphology and distribution of dunes in sand seas observed by remote sensing. *In*: McKee, E. D. (ed.), *A Study of Global Sand Seas*. U.S. Geol. Survey Prof. Paper 1052, U.S. Govt. Printing Office, Washington, DC, pp. 253–304.

Bressolier, C., and Thomas, Y.-F., 1977. Studies on wind and plant interactions on French Atlantic coastal dunes. *J. Sed. Petrol.*, **47**, 331–338.

Brothers, R. N., 1954. A physiographic study of Recent sand dunes on the Auckland West Coast. *New Zealand Geogr.*, **10**, 47–59.

Buffault, P., 1942. *Histoire des Dunes Maritimes de la Gascogne*. Editions Delmas, Bordeau, 446 pp.

Byrne, R. J., 1968. Aerodynamic roughness criteria in aeolian sand transport: *J. Geophys. Res.*, **73** (2), 541–547.

Campbell, W. V., and Fuzy, E. A., 1972. Survey of the scale insect effect on American beach grass. *Shore and Beach*, **40**, 18–19.

Chapman, V. J., 1964. *Coastal Vegetation*. MacMillan Co., New York, 245 pp.

Chapman, D. M., Geary, M., Roy, P. S., and Thom, B. G., 1982. *Coastal Evolution and Coastal Erosion in New South Wales*. Coastal Council of N.S.W., Sydney, 341 pp.

Colonell, J., and Goldsmith, V., 1971. Computational methods for the analysis of beach and wave dynamics. *In*: Morisawa, M. (ed.), *Proc. Quantitative Geomorph. Symposium*, State Univ. of New York, Binghamton, NY, pp. 198–222.

Cooper, W. S., 1958. *Coastal Sand Dunes of Oregon and Washington*. Geol. Soc. Amer. Memoir 72, 169 pp.

Cooper, W. S., 1967. *Coastal Dunes of California*. Geol. Soc. Amer. Memoir 104, 131 pp.

Cressey, G. B., 1928. The Indiana sand dunes and shorelines of the Lake Michigan basin. *Geogr. Soc. Chicago Bull.*, No. 6, 80 pp.

Cry, G. W., 1965. *Tropical Cyclones of the North Atlantic Ocean: Tracks and Frequencies of Hurricanes and Tropical Storms 1871–1963*. U.S. Dept. Commerce Weather Bureau Tech. Paper No. 55, 148 pp.

Curray, J., 1956. The analysis of two-dimensional orientation data. *Jour. Geol.*, **64**, 117–131.

Davies, J. L., 1973. *Geographical Variation in Coastal Development*. Hafner Publ. Co., New York, 204 pp.

Davis, J. H., 1957. *Dune Formation and Stabilization by Vegetation and Plantings*. U.S. Army Coastal Research Center Tech. Memo. No. 101, 47 pp.

Dolan, R., 1972a. Barrier dune system along the Outer Banks of North Carolina: A reappraisal. *Science*, **176**, 286–288.

Dolan, R., 1972b. Barrier islands: natural and controlled. Paper Presented at *Coastal Geomorph. Symposium*, State Univ. of New York, Binghamton, NY.

Ellwood, J. M., Evans, P. D., and Wilson, I. G., 1975. Small scale aeolian bedforms. *J. Sed. Petrol.*, **45**, 554–561.

Emery, K. O., 1980. Relative sea levels from tide-gage records. *Proc. Natl. Acad. Sci.*, **77**, 6968–6972.

Etkins, R., and Epstein, E. F., 1983. Global mean sea level: indicator of climatic change? *Science*, **219**, 997–998.

Finkel, H. J., 1959. The barchans of southern Peru. *J. Geol.* **67**, 614–647.

Folk, R. L., 1966. A review of grain size parameters. *Sedimentology*, **6**, 73–93.

Folk, R. L., 1971. Longitudinal dunes of the northwestern edge of the Simpson Desert, Northern Territory, Australia, part 1: geomorphology and grain size relationships. *Sedimentology*, **16**, 5–54.

Gage, B. O., 1970. *Experimental Dunes of the Texas Coast*. U.S. Army Corps of Engineers Misc. Paper No. 1-70, 30 pp.

Gares, P. A., Nordstrom, K. F., and Psuty, N. P., 1979. *Coastal Dunes: Their Function, Delineation, and Management*. Center for Coastal and Environmental Studies, Rutgers Univ., New Brunswick, NJ, 112 pp.

Gawne, C. E., 1966. *Shore Changes on Fenwick and Assateague Islands, Maryland and Virginia*. Univ. of Illinois, B.S. Thesis.

Godfrey, P. J., 1970. *Oceanic Overwash and its Ecological Implications on the Outer Banks of North Carolina*. Office of Natural Science Studies, National Park Service, U.S. Dept. of Interior, Washington, DC, 37 pp.

Godfrey, P. J., and Godfrey, M. M., 1973. Comparison of ecological and geomorphic interactions between altered and unaltered barrier island systems in North Carolina. *In*: D. R. Coates (ed.), *Coastal Geomorphology*. Publ. in Geomorphology, State Univ. of New York, pp. 239–258.

Godfrey, P. J., Leatherman, S. P., and Zaremba, R., 1979. A geobotanical approach to classification of barrier beach systems. *In*: Leatherman, S. P. (ed.), *Barrier Islands*. Academic Press, New York, pp. 99–126.

Goldsmith, V., 1971. The formation and internal geometry of coastal sand dunes: an aid to paleographic reconstruction (Abs.): *Geol. Soc. Amer. Ann. Mtg.*, Washington, DC, Nov., 1971, 3, 582–583.

Goldsmith, V., 1972. *Coastal Processes of a Barrier Island Complex, and Adjacent Ocean Floor: Monomoy Island–Nauset Spit, Cape Cod, Massachusetts*. Ph.D. Dissert., Geology Dept., Univ. of Massachusetts, Amherst, 469 pp.

Goldsmith, V., 1973. Internal geometry and origin of vegetated coastal sand dunes. *J. Sed. Petrol.*, **43**, 1128–1143.

Goldsmith, V., Hennigar, H. F., and Gutman, A. L., 1977. The "VAMP" coastal dune classification. *In*: Goldsmith, V. (ed.), *Coastal Processes and Resulting Forms of Sediment Accumulation, Currituck Spit, Virginia/North Carolina*. SRAMSOE No. 143, Virginia Institute of Marine Science, Gloucester Point, VA, pp. 26-1–26-20.

Gooding, E. G. B., 1947. Observations on the sand dunes of Barbados, W. Indies. *J. Ecology*, **34**, 111–125.

Guilcher, A., 1959. Coastal sand ridges and marshes and their environment near Grand Popo and Ouidah, Dahomey. *Second Coastal Geography Conf., National Research Council, Committee on Geography*, Washington, DC, pp. 189–212.

Gutman, A. L., 1977a. Orientation of coastal parabolic dunes and relation to wind vector analysis. *In*: Goldsmith, V. (ed.), *Coastal Processes and Resulting Forms of Accumulation, Currituck Spit, Virginia/North Carolina*. SRAMSOE No. 143, Virginia Institute of Marine Science, Gloucester Point, VA, pp. 28-1–28-17.

Gutman, A. L., 1977b. Movement of large sand hills: Currituck Spit, Virginia/North Carolina. *In*: Goldsmith, V. (ed.), *Coastal Processes and Resulting Forms of Sediment Accumulation, Currituck Spit, Virginia/North Carolina*. SRAMSOE No. 143, Virginia Institute of Marine Science, Gloucester Point, VA, pp. 29-1–29-19.

Gutman, A. L., 1977c. Aeolian grading of sand across two barrier island transects, Currituck Spit, Virginia/North Carolina. *In*: Goldsmith, V. (ed.), *Coastal Processes and Resulting Forms of Sediment Accumulation, Currituck Spit, Virginia/North Carolina*. SRAMSOE No. 143, Virginia Institute of Marine Science, Gloucester Point, VA, pp. 35-1–35-16.

Hallegouet, B., 1978. L'evolution des massifs dunaive du Pays de Leon (Finistere). *Penn ar Bed Brest*, 11, 417–430.

Hayes, M. O., 1967. *Hurricanes as Geologic Agents: Case Studies of Hurricanes Carla, 1961, and Cindy, 1963*. Univ. of Texas Bureau Economic Geology, Rept. of Inv. 61, 54 pp.

Hempel, L., 1980. Zur genese von dunengenerationen an Flachkusten, Beobachtungen auf den Nordseeinsein Wangerooge und Spiekeroog. *Z. Geomorph.*, 24, 428–447.

Hennigar, H. F., 1977a. A brief history of Currituck Spit (1600–1945). *In*: Goldsmith, V. (ed.), *Coastal Processes and Resulting Forms of Sediment Accumulation, Currituck Spit, Virginia/North Carolina*. SRAMSOE No. 143, Virginia Institute of Marine Science, Gloucester Point, VA, pp. 3-1–3-21.

Hennigar, H. F., 1977b. Evolution of coastal sand dunes: Currituck Spit, Virginia/North Carolina. *In*: Goldsmith, V. (ed.), *Coastal Processes and Resulting Forms of Sediment Accumulation, Currituck Spit, Virginia/North Carolina*. SRAMSOE No. 143, Virginia Institute of Marine Science, Gloucester Point, VA, pp. 27-1–27-20.

Hesp, P., 1979. Sand trapping ability of culms of marram grass (*Amophila arenaria*). *J. Soil Cons. Serv., N.S.W.*, 35, 156–160.

Hoiland, K., 1978. Sand dune vegetation of Lista, SW Norway. *Norw. J. Bot.*, 25, 23–45.

Horikawa, K., and Shen, H. W., 1960. *Sand Movement by Wind Action (On the Characteristics of Sand Traps)*. U.S. Army Corps of Engineers, Tech. Memo. 119, 51 pp.

Hotta, S., Kubota, S., Katori, S., and Horikawa, A., 1985. Sand transport by wind on a wet sand surface. *Proc. 19th Inter. Coastal Eng. Conf.*, ASCE, New York, in press.

Hunter, R. E., 1977. Terminology of cross-stratified sedimentary layers and climbing ripple structures. *J. Sed. Petrol.*, 47, 697–706.

Hunter, R. E., Richmond, B. M., and Alpha, T. R., 1983. *Storm-Controlled Oblique Dunes of the Oregon Coast*. Geol. Soc. Amer. Bull., 94, 1450–1465.

Inman, D. L., Ewing, G. C., and Corliss, J. B., 1966. Coastal sand dunes of Guerrero Negro, Baja California, Mexico. *Geol. Soc. Amer. Bull.*, **77** (8), 787–802.

Jagschitz, J. A., and Wakefield, R. C., 1971. *How to Build and Save Beaches and Dunes*. Univ. of Rhode Island Agric. Exp. Station Bull. 408, Kingston, RI, 12 pp.

Jelgersma, S., and Van Regteren Altena, J. F., 1969. An outline of the geological history of the coastal dunes in the western Netherlands. *Geol. en Mijnbouw*, **48**, 335– 342.

Jennings, J. N., 1957a. Coastal dune lakes as exemplified from King Island, Tasmania. *Geogr. J.* **123**, 59–70.

Jennings, J. N., 1957b. On the orientation of parabolic or U-dunes. *Geogr. J.*, **123**, 474–480.

Jennings, J. N., 1964. The question of coastal dunes in tropical human climates. *Z. Geomorph.*, (Mortenson Birthday Volume), **8**, 150–154.

Jennings, J. N., 1967. Cliff-top dunes: *Australian Geogr. Stud.*, **5**(1), 40–49.

Johnson, J. W., 1965. Sand movement on coastal dunes. *In: Federal Inter-Agency Sedimentation Conference Proceedings*. U.S. Dept. of Agriculture, Misc. Publ. 970, pp. 747–755.

Johnson, J. W., and Kadib, A. A., 1965. Sand losses from a coast by wind action. *Ninth Conference on Coastal Engineering*, pp. 368–377.

Kaiser, E., 1926. Die diemantenwuste sud westafrikas. *In: 2 Banden*. D. Reimer, Verlag, Berlin.

Karmeli, D., Yaalon, D. H., and Ravina, I., 1968. Dune sand and soil strata in quaternary sedimentary cycles of the Sharon coastal plain. *Israel J. Earth Sci.*, **17**, 45–53.

Kidson, C., and Carr, A. P., 1960. *Dune Reclamation at Braunton Burrows*. Devonshire Chart Serv. (Dec.), pp. 3–8.

King, C. A. M., 1972. The effect of wind. *In: Beaches and Coasts*. Edward Arnold, London, Chapter 6, pp. 165–170.

King, C. A. M., 1973. Dynamics of beach accretion in south Linconshire, England. *In*: Coates, D. R. (ed.), *Coastal Geomorphology, Proc. of Third Annual Geomorph. Symposium*, Binghamton, NY, Publ. in Geomorph., State Univ. of New York, Binghamton, NY, pp. 73–98.

Klemsdal, T., 1969. Eolian forms in parts of Norway. *Norsk Geogr. Tidsskr.*, **23**, 49–66.

Knutson, P., 1977. *Planting Guidelines for Dune Creation and Stabilization*. CERC Tech. Aid No. 77-4, Coastal Engr. Res. Cent., Fort Belvoir, VA, 26 pp.

Kraft, J. C., Biggs, R. B., and Halsey, S. D., 1972. Morphology and vertical sedimentary sequence models in Holocene transgressive barrier systems. *Coastal Geomorph. Symposium*, State Univ. of New York, Binghamton, NY.

Krinsley, D. H., and Donahue, J., 1968. Environmental interpretation of sand grain surface textures by electron-microscopy. *Geol. Soc. Amer. Bull.*, **79**, 743–748.

Krinsley, D. H., and Smalley, I. J., 1972. Sand. *Amer. Sci.*, **60**, 286–291.

Kuenen, P. H., 1963. Pivotal studies of sand by a shape-sorter: *In: Deltaic and*

Shallow Marine Deposits, Developments in Sedimentology, Vol. 1. Elsevier, Amsterdam, pp. 207–215.

Kuenen, P. H., and Perdok, W. G., 1962. Experimental abrasion 5. Frosting and defrosting of quartz grains. *J. Geol.*, **70**, 648–658.

Land, L. S., 1964. Eolian cross-bedding in the beach dune environment, Sapelo Island, Georgia. *J. Sed. Petrol.*, **32**, 289–394.

Landsberg, S. Y., 1956. The orientation of dunes in Britain and Denmark in relation to the wind. *Geogr. J.*, **122**, 176–189.

Larsen, F. D., 1969. Eolian sand transport on Plum Island, Massachusetts. *In*: Hayes, M. O. (ed.), *Coastal Environments Northeast Massachusetts and New Hampshire*. C.R.G. #1, Geology Dept., Univ. of Massachusetts, Amherst, pp. 356–367.

Leatherman, S., 1975. *Quantification of Overwash Processes*. Ph.D. Dissert. Dept. of Environmental Sciences, Univ. of Virginia, 245 pp.

Leatherman, S. P., 1979. Beach and dune interactions during storm conditions. *Quart J. Engr. Geol.*, **12**, 281–290.

Leatherman, S. P., and Steiner, A. J., 1979. *Recreational Impacts on Foredunes: Assateague Island National Seashore*. Proc. 2nd Sci. Conf. on the Natl. Parks, San Francisco, CA, 16 pp.

Mackenzie, F. T., 1964a. Geometry of Bermuda calcareous dune cross-bedding. *Science*, **144**, 1449–1450.

Mackenzie, F. T., 1964b. Bermuda Pleistocene eolianites and paleowinds. *Sedimentology*, **3**, 51–64.

Marta, M. D., 1958. Coastal dunes: A study of the dunes at Vera Cruz. *6th Conf. on Coastal Engineering*, pp. 520–530.

Mather, A. S., 1979. Physiography and management of coastal dune systems in the Scottish Highlands and Islands. *In*: Guilcher, A. (ed.), *Les Cotes Atlantique d'Europe; Evolution, Amenagement, Protection*. Publ. CNEXO, Act. Coll., Vol. 9, pp. 251–260.

Mather, J. R., 1965. Climatology of damaging storms. *In*: Burton, I., Kates, R., Mather, J. R., and Snead, R. (eds.), *The Shores of Megalopolis: Coastal Occupance and Human Adjustment to Flood Hazard*. Thornwaite Assoc., Elmer, NJ, pp. 525–549.

McBride, E. F., and Hayes, M. O., 1962. Dune cross-bedding on Mustang Island, Texas. *Amer. Assoc. Petrol. Geol. Bull.*, **64** (4), 546–551.

McIntire, W. G., and Morgan, J. P., 1963. *Recent Geomorphic History of Plum Island, Massachusetts and Adjacent Coasts*. Louisiana State Univ., Coastal Studies Ser. 8, 44 pp.

McKee, E. D., 1966. Structure of dunes at White Sands National Monument, New Mexico (and a comparison with structures of dunes from other selected areas). *Sedimentology*, **3**, 1–69.

McKee, E. D., 1979. Introduction to a study of global sand seas. *In*: McKee, E. D. (ed.), *A Study of Global Sand Seas*. Geol. Survey Prof. Paper 1052. U.S. Govt. Printing Office, Washington, DC, pp. 1–21.

McKee, E. D., 1982. *Sedimentary Structures in Dunes of the Namib Desert, South West Africa*. Geol. Soc. Amer. Spec. Paper 188, Boulder, CO, 64 pp.

McKee, E. D., and Bigarella, J. J., 1972. Deformational structures in Brazilian coastal dunes. *J. Sed. Petrol.*, **42**, 670–681.

McKee, E. D., and Tibbitts, G. C., 1964. Primary structures of a seif dune and associated deposits in Libya. *J. Sed. Petrol.* **34**, 5–17.

Meigs, P., 1966. *Geography of Coastal Deserts*. UNESCO, 140 pp.

Messenger, C., 1958. *A Geomorphic Study of the Dunes of the Provincetown Peninsula, Cape Cod, Mass.* M.S. Thesis, Univ. of Massachusetts.

Mii, H., 1958. Coastal sand dune evolution of the Hachiro-Guta. *Saito Ho-on Kai Mus. Res. Bull. (Akita)*, **27**, 7–22.

Morley, I. W., 1981. *Black Sands: A History of the Mining Industry in Eastern Australia*. Univ. of Queensland Press, 278 pp.

Namias, J., 1970. Climatic anomaly over the United States during the 1960's. *Science*, **170**, 741–743.

Nordstrom, K. F., and McCluskey, J. M., 1982. *The Effects of Structures on Dune Dynamics at Fire Island National Seashore*. Final Rept., Center for Coastal and Environ. Studies, Rutgers Univ., New Brunswick, NJ, 70 pp.

Norrman, J. O., 1981. Coastal dune systems. *In*: Bird, E. C. F., and Koike, K. (eds.), *Coastal Dynamics and Scientific Sites*. Commission on the Coastal Environment, Intl. Geogr. Union, pp. 119–158.

Nossin, J. J., 1962. Coastal sedimentation in northeastern Johore (Malaya). *Z. Geomorph., Annales de Geomorphologie, N. S., N.F.G.*, **6**, 296–316.

Oertel, G. F., 1974. A review of the sedimentologic role of dunes in shoreline stability, Georgia coast. *Georgia Acad. Sci. Bull.*, **32**, 48–56.

Oertel, G. F., and Larsen, M., 1976. Developmental sequences in Georgia coastal dunes and distribution of dune plants. *Georgia Acad. Sci. Bull.*, **34**, 35–48.

Olson, J. S., 1958. Lake Michigan dune development, 1-2-3. *J. Geol.*, **56**, 254–263, 345–351, 413–483.

Phillips, C. J., and Willetts, B. B., 1978. A review of selected literature on sand stabilization. *Coastal Engr.*, **2**, 133–147.

Pincus, H. J., 1956. Some vector and arithmetic operations on two-dimensional orientation variates with application to geological data. *J. Geol.*, **64**, 533–557.

Research Computing Center University of Massachusetts, 1971. System library files: statistics. *Bits & Bytes*, April, 1971, 7–8.

Rosen, P. S., 1979. Aeolian dynamics of a barrier island system. *In*: Leatherman, S. P. (ed.), *Barrier Islands*. Academic Press, New York, pp. 81–98.

Rosen, P. S., Barnett, E. S., Goldsmith, V., Shideler, G. L., Boule, M., and Goldsmith, Y. E., 1977. Internal geometry of foredune ridges, Currituck Spit area, Virginia/North Carolina. *In*: Goldsmith, V. (ed.), *Cosatal Processes and Resulting Forms of Sediment Accumulation, Currituck Spit, Virginia/North Carolina*. SRAMSOE No. 143, Virginia Institute of Marine Science, Gloucester Point, VA, pp. 30-1–30-16.

Savage, R. P., 1963. Expeirmental study of dune building with sand fences. *Proc. 8th Cont. on Coastal Engineering*, Mexico City, pp. 380–396.

Savage, R. P., and Woodhouse, Jr., W. W., 1969. Creation and stabilization of coastal barrier dunes. *Proc. 11th Conf. on Coastal Engineering*, Amer. Soc. Civil Engr., pp. 671–700.

Screiber, Jr., J. F., 1968. Desert coastal zones. *In*: McGinnies, W. G., Goldman, B. J., and Paylore, P. (eds.), *Deserts of the World*. Univ. of Arizona Press, pp. 647–724.

Sharp, R. P., 1963. Wind ripples. *J. Geol.*, **71**, 617–636.

Shepard, F., and Wanless, H., 1971. *Our Changing Coastlines*. McGraw-Hill, New York, 579 pp.

Short, A. D., and Hesp, P. A., 1982. Wave, beach and dune interactions in southeastern Australia. *Mar. Geol.*, **48**, 259–284.

Simonett, D. S., 1950. Sand dunes near Castelreagh, New South Wales. *Australian Geogr.*, **5**, 3–10.

Smith, H. T. U., 1941. Review of Hack, J. T., 1941. Dunes of the western Navajo country. *Geogr. Rev.*, **31**, 240–263; *In: J. Geomorph.*, **4**, 250–252.

Smith, H. T. U., 1960. *Physiography and Photo-interpretation of Coastal Sand Dunes*. Final report, Contract Nonn-2242(00) ONR, 26 pp.

Smith, H. T. U., 1968. Geologic and geomorphic aspects of deserts. *In*: Brown, G. W. (ed.), *Desert Biology*. Academic Press, New York, pp. 51–100.

Smith, H. T. U., and Messenger, C., 1959. *Geomorphic Studies of the Provincetown Dunes, Cape Cod, Mass.* Tech. Report No. 1, Contract Nonr-2242(00), Office of Naval Research Geography Branch, Univ. of Massachusetts, Amherst.

Stembridge, Jr., J. E., 1978. Vegetated coastal dunes: growth detected from aerial infrared photography. *Remote Sensing of Environ.*, **7**, 73–76.

Streim, H. L., 1954. The seifs on the Israel-Sinai border and the correlation of their alignment. *Israel Res. Coun. Bull.*, **4G**, 195–198.

Swan, B., 1979. Sand dunes in the humid tropics: Sri Lanka. *Z. Geomorph. N.F.*, **23**, 152–171.

Tanaka, H., 1967. Ancient sand dunes on the coast of Japan. *Geol. Soc. Japan*, **73**, (2), 121.

Tanner, W. F., 1964. Eolian ripple marks in sandstone. *J. Sed. Petrol.*, **34**, 432–433.

Tanner, W. F., 1967. Ripple mark indices and their uses. *Sedimentology*, **9**, 89–104.

Tsoar, H., 1974. Desert dunes morphology and dynamics. El Arish (Northern Sinai): *J. Geomorph. N.F., Suppl. Bd. 20*, Berlin-Stuttgart, pp. 41–61.

Tsoar, H., 1983a. Internal structure and surface geometry of longitudinal (seif) dunes. *J. Sed. Petrol.*, **52**, 823–831.

Tsoar, H., 1983b. Wind tunnel modeling of echo and climbing dunes. *In*: Brookfield, M. E., and Ahlbrandt, T. S. (eds.), *Eolian Sediments and Processes*. Elsevier, Amsterdam, pp. 247–259.

Tsuriell, D. E., 1974. Sand dune stabilization in Israel. *Intl. J. Biometeor.* **18**, 89–93.

U.S. Army Corps of Engineers, 1977. *Shore Protection Manual*. Coastal Engineering Research Center, 3 Vol.

Vacher, L., 1973. Coastal dunes of younger Bermuda. *In*: Coates, D. R. (ed.), *Coastal Geomorphology*. Publ. in Geomorphology, Binghamton, NY, pp. 355–391.

Van Straaten, L. M. J. U., 1961. Directional effects of winds, waves and currents along the Dutch North Sea Coast. *Geol. en Mijnbouw*, **40**, 333–346, 363–391.

Van Straaten, L. M. J. U., 1965. Coastal barrier deposits in south and north Holland in particular in the areas around Scheveningen and Ljmuiden. *Mededelinger van de Geologische Stichting Neve Serie*, **17**, 41–75.

Verstappen, H. T., 1957. Short note on the dunes near Parang tritis (Java). *Tijd. Kon. Nederl. Aard, Gen.* **74**, 1–6.

Walsh, B. N., 1968. *Some Notes of the Incidence and Control of Drift Sands along the Caledon, Bredasdorp and Riversdale Coastline of South Africa*. S.A. Dept. Forestry Bull. 44, Pretoria, S.A., 79 pp.

Westgate, J. M., 1904. *Reclamation of Cape Cod Sand Dunes*. U.S. Dept. Agriculture, Bureau of Plant Industry Bulletin 65, 36 pp.

Williams, G., 1964. Some aspects of the eolian saltation load. *Sedimentology*, **3**, 257–287.

Woodhouse, Jr., W. W., and Hanes, R. E., 1967. *Dune Stabilization with Vegetation on the Outer Banks of North Carolina*. U.S. Army Cosatal Engr. Research Center, Tech. Memo. No. 22, 45 pp.

Woodhouse, W. W., Seneca, E. D., and Cooper, A. W., 1968. Use of sea oats for dune stabilization in the southeast. *Shore and Beach*, **36**, 15–21.

Yaalon, D. H., 1967. Factors affecting the lithification of eolianite and interpretation of its environmental significance in the Coastal Plain of Israel. *J. Sed. Petrol.*, **37**, 1189–1199.

Yaalon, D. H., 1975. Discussion of "Internal Geometry of Vegetated Coastal Sand Dunes." *J. Sed. Petrol.* **45**, 359.

Yaalon, D. H., and Laronne, J., 1971. Internal structures in eolianites and paleowinds, Mediterranean Coast, Israel. *J. Sed. Petrol.*, **41**, 1059–1064.

Zak, J. M., and Bredakis, E., 1963. Dune stabilization at Provincetown, Mass. *Shore and Beach*, **31** (2), 19–24.

Zeigler, J. M., Tuttle, S. D., Tasha, H. J., and Ciese, G. S., 1964. Residence time of sand composing the beaches and bars of Outer Cape Cod. *Proc. Ninth Coastal Engr. Conf.*, pp. 403–416.

Zeller, R. P., 1962. *A General Reconnaissance of Coastal Dunes of Calif.* U.S. Army Corps of Engineers Misc. Paper No. 1-62, 38 pp.

Zenkovich, Z. P., 1967. Aeolian processes on sea coasts. *In*: Steers, J. A. (ed.), *Processes of Coastal Development*. Interscience, New York, pp. 586–617.

Zingg, A. W., 1953. Wind tunnel studies of the movement of sedimentary material. Proc. Fifth Hydraulics Conf., Iowa Univ., *Stud. Engr. Bull.*, **34**, 111–135.

6

Beach and Nearshore Zone

Richard A. Davis, Jr.

Introduction

Although coastal environments are collectively characterized by change, the beach and nearshore zone is one of the most dynamic of these environments. The changes to which beaches are subjected may be seasonal or longer in duration; they may be as short as a single tidal cycle or even occur from one crashing wave to the next. Because of the beauty and romance of the beach environment, as well as its scientific interest, literally thousands of researchers have investigated the nature of beaches over the past two centuries. This environment has been studied more than any other of those discussed in this book. As a result, it is impossible to do more than cover the main points in this chapter. The reader who wishes greater depth on the subject is referred to numerous books on the subject of beaches (Hails and Carr, 1975; Komar 1976; Davis and Ethington, 1976; Leatherman, 1979; Davies, 1980; Greenwood and Davis, 1984).

Before proceeding to a discussion of the beach and nearshore zone, it is necessary to define these terms. A beach is here defined as the zone of unconsolidated sediment that extends from the uppermost limit of wave action to the low-tide mark. Commonly there is an abrupt change in slope and/or composition at the landward limit. This may be in the form of dunes, bluffs of glacial drift, bedrock, or in some cases a man-made structure. Beyond the beach in a seaward direction and extending across the bar and trough topography is the nearshore zone. In areas where the nearshore zone does not contain sand bars the seaward limit is placed at wave base. Although the beach and nearshore are intimately related, they are separated in this discussion because of the markedly different processes to which each is subjected.

Distribution

There are no real geographically imposed limits on beach development. A beach will form virtually any place where the land and sea meet, where sediment is available, and where a site is available for sediment accumulation. Beaches are by far the most widely distributed of any of the coastal sedimentary environments. According to one study (Dolan et al., 1972), 33% of the North American shoreline is beach; of this, 23% is beach-barrier islands, 2% is associated with rock headlands, and 8% is of the pocket beach type. Beaches can be found on small lakes covering less than a square kilometer as well as on the marine coast. With the exception of tides and fetch there is little significant difference between beach sedimentation on lakes and that along the ocean.

Extensive beach development which, as indicated above, is typically associated with barrier island coasts is also related to plate tectonics. Inman and Nordstrom (1971) have classified coasts on the basis of tectonics into three major categories: (1) collision coasts, where plates converge (west coast of North and South America), (2) trailing edge coasts where a coast faces a spreading zone (east cosat of North and South America), and (3) marginal sea coasts where a coast faces an island arc (east coast of Asia). Barrier islands, which contain extensive beaches, are distributed such that half are associated with trailing edge coasts and the remaining half are subequally divided between the other two categories (Glaeser, 1978).

Although beaches may occur in all climatic zones of the world, there are some obvious effects of certain severe climatic conditions. For example, in the very high latitudes there are only a few weeks when the water is ice-free and beaches are subjected to wave activity. As a result, the morphology and texture of such beaches are typically somewhat different from beaches in low latitudes. Another aspect of climate that may affect beach development is rainfall. In an area of bedrock coast and at least moderate precipitation, weathering and runoff would be expected to provide sufficient material for a well-developed beach. A similar area with an arid climate would depend almost solely on wave action to provide sediment from the bedrock coast in order to form a beach.

The best development of beaches is associated with low-lying coasts where great quantities of sediment are available. Both glacial drift and coastal plain areas provide ideal sources and locations for beach development. In addition, high relief and humid coastal areas, such as the west coast of the United States, also yield well-developed beaches. Even in areas of steep, rocky coasts, such as along Maine or Oregon, there are small pocket beaches in protected areas between bedrock prominences. In short, beaches occur on nearly all coasts of the world.

Geometry

Aerial Considerations

It is not possible to provide practical size limits for beaches. A beach is by its very nature a long and narrow feature in its usual development, with its length up to several hundred times its width. Along the Atlantic and Gulf coasts, the barrier island beaches are essentially continuous for hundreds of kilometers, the only interruptions being natural or man-made inlets. Mainland beaches along large stretches of the west coast of the United States and the Great Lakes are also essentially continuous. In both cases there are areas of bedrock headlands where the beach is interrupted; however, although deltas, bays, and other large-scale features are common, such coastal features also may have beaches developed on them.

Some beaches are so narrow that a person can step from the high-water to the low-water mark in one stride. These narrow beaches predominate in sediment-starved areas where the tidal range is small or absent. By contrast, the beaches associated with the flat intertidal zones of the Bay of Fundy and the North Sea coast of Holland and Germany may be kilometers wide. It is not unusual for open-coast beaches of the coastal plains or barrier islands to approach 100 m in width (Figure 6-1). For the most part, beach development and width are controlled by the slope of the inner shelf and coastal area, abundance of sediment, and tidal range. For example, the barrier island beaches of Texas are 100 m or so wide in many places. These areas are characterized by a gradual slope of the inner shelf and much available sediment, but the tidal range is only about 0.5 m. Areas of high relief adjacent to more steeply inclined shelves may have narrower beaches (Figure 6-2), even though the tidal range may be 3 m or more. The New England coast is an example. Any combination of these factors may exist from place to place.

Although straight and smoothly curved beaches with at least nearly parallel margins are the most widespread type of beach, a great variety of aerial shapes occurs. Deviations from the above generality are primarily caused by overall coastal configuration and bedrock influences. One of the most widespread of the nonlinear beach types is the pocket beach (Figure 6-3). Pocket beaches represent small and local interruptions in a normally continuous rocky headland area. They may be only a few meters in length and width or they may extend for a few hundred meters along a bedrock coast. In areas where there are numerous and rather long (~1 km) beaches developed along a rugged bedrock coast, the term pocket beaches is not appropriate.

Richard A. Davis, Jr.

Figure 6-1. Wide and gentle sloping beach on Mustang Island, a part of the Texas barrier island complex. The oblique aerial photograph was taken at low tide, and the beach width is nearly 100 m.

Figure 6-2. Rather narrow and steep beach comprised of cobble-sized gravel and located in Acadia National Park, Maine, where the tidal range is about 4 m.

Figure 6-3. Pocket boulder beach in small embayment of Precambrian bedrock at Cape Ann, Massachusetts.

The actual shape and orientation of the beach itself is dependent on a number of variables including the direction of wave-energy approach, littoral drift, the material comprising the beach, and the overall shape and composition of the coast. Contrasting opinions exist about the determining factor in beach orientation. There are some obvious coastal features that dictate the orientation including protruding headlands, river deltas, and man-made structures such as jetties, groins, and breakwaters. These are typically local in nature. Spits and their associated beaches may be aligned with littoral drift direction or with the crests of constructing waves and their generated swash (Davies, 1980).

A world-wide plot of the orientation of beaches and other constructional coastal features show some distinct trends (Figure 6-4). Although there is great local variability there are regionally significant patterns. For example, regardless of coastal trend, the beaches of the west coast of the United States and adjacent Baja California face the northwest but from there to central Chile they face to the southwest. Other similar patterns exist and major trends can conveniently be separated (Figure 6-4). It might appear that these trends are simply a reflection of the coasts of land masses and the limits of fetch imposed by their location. There are too many exceptions to this generalization, and detailed examination shows that beaches are not necessarily aligned to the orientation of the bedrock coast (Davies, 1980). Such factors as storm waves (Lewis, 1938), direction of maximum fetch and the resultant wind direction (Schou, 1945), and the dominance of long swells (Davies, 1959) have been suggested to explain beach orientation. Davies (1980) concludes that although the shape of continents and oceans is a major factor in determining wave approach and therefore beach orientation, other variables must also be considered. The global wave generation pattern and the direction of significant wave approach wave approach appears to be the most important (Davies, 1980), thereby supporting the work of Lewis (1938).

Beach and Nearshore Profiles

The topographic configuration displayed by most beach and adjacent nearshore profiles is rather easily generalized. There are three zones within the profile that are subjected to rather different conditions (Figure 6-5). The backshore extends from the normal high-tide level to the landward margin of the beach. Under extreme conditions, it may include the lower dune or cliff environment. The foreshore is equated with the intertidal zone. The nearshore represents the rather wide zone between the low-tide level and the seaward limit of the bar and trough topography.

Backshore

Practically all of the backshore zone of a sand beach is a rather flat and nearly horizontal to gently landward sloping area called the berm. The seaward limit of the berm is marked by an abrupt slope change at the berm crest. Beyond that line is the beachface, which is inclined rather steeply toward the sea (Figure 6-5). A beach that is undergoing erosion displays a much different backshore profile in that the berm is not developed. Under these conditions the backshore and upper foreshore zone are continuous in slope with a slightly concave upward profile. Beaches of gravel and cobbles may show a backbeach profile that differs markedly from that of a sandy

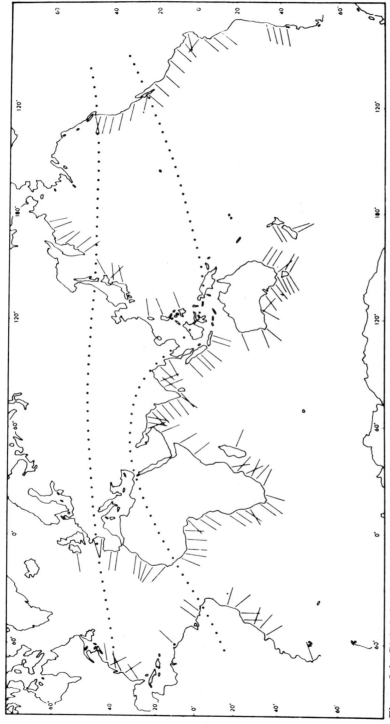

Figure 6-4. Global plot showing the orientation of beaches and other constructional coastal features. Dotted lines mark changes in major trends (from Davies, 1980, p. 147).

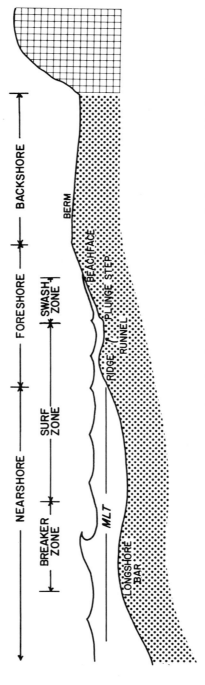

Figure 6-5. General profile diagram of beach and nearshore zones using terminology discussed in this chapter.

Figure 6-6. Storm ridge on gravel beach, Rye, New Hampshire.

beach. A high storm ridge usually characterizes such beaches (Figure 6-6), and some display multiple ridges. These ridges may rise several meters above the surrounding landscape.

Foreshore

The foreshore zone may display a variety of configurations. The simplest of these is a relatively steep beach face. The slope is dependent on both the composition of the foreshore and the processes acting upon it. It may be inclined only 1 to 3° or be as much as 30°. A small and commonly subtle step may be developed within the foreshore. This step is caused by the final plunge of waves as they break for the last time before surging up the beach face. The terms foreshore step (King, 1972), beach scarp (Shepard, 1973), and plunge step (Davis and Fox, 1972a) have been applied to this feature. The latter is preferred because of its combination of genetic and morphologic terms. On very gently inclined foreshore slopes the plunge step is quite subtle and its location may best be recognized by changes in texture and surface sedimentary structures only. It is typically marked by a concentration of shell debris and/or coarse sediment. The development of this feature is apparently directly related to both the tidal range and the foreshore slope. Maximum development occurs in areas of steep foreshore slope where tidal ranges are low.

Figure 6-7. Oblique aerial photograph showing well-developed shoreline sinuosity (rhythmic topography).

A sometimes overlooked but important aspect of foreshore beach topography is the regular sinuosity of the shoreline itself (Figure 6-7). This sinuosity may have a wavelength ranging from a few tens of meters to several kilometers (Dolan, *et al.*, 1974; Dolan, *et al.*, 1977), with the amplitude ranging from less than one to several tens of meters. Although the various wavelengths of the sinuosity of "rhythmic topography" (Hom-ma and Sonu, 1963) appear superficially alike, there are really various types with some overlap in the wavelength.

The smallest scale variety is beach cusps, which are generally spaced from 10 to 30 m. The horns or apexes are commonly characterized by coarse accumulations of sediment; however, there are also varieties that have a similar morphology but with no significant textural variation throughout their extent. Giant cusps (Shepard, 1952) typically have spacing or wave lengths in the 100–200 m range. These features are morphologically similar to beach cusps but without the striking contrast in grain size between the horn and adjacent areas.

Both the beach cusps and giant cusps display a similar morphology and are essentially subaerial beach features with no subtidal component (Komar, 1976). This distinguishes these features from rhythmic topography or rhythmic shorelines (Hom-ma and Sonu, 1963). Although similar to cusps in their general configuration and their uniform shape and spacing, the rhythmic shoreline is characterized by subtidal components in the form of sandbars. Smallest scale shoreline rhythms are comparable in wave length to giant cusps. The shoreline itself displays a smooth "sine curve" shape (Dolan,

1971; Davis and Fox, 1972b) whereas giant cusps have rather peaked apexes (Figure 6-8). The term *beach protuberances* has been applied to these convex-seaward rhythms.

These shoreline rhythms cover a fairly broad range in wavelength from about 100 m to many kilometers. Dolan *et al.* (1977) have discussed these features along the south Atlantic coast of the United States. They suggested that the cape features extending southward from the Outer Banks represent very large-scale shoreline rhythms. The origin of rhythmic topography is considered later in this chapter.

During certain coastal conditions there may be a small ephemeral sand bar developed within the intertidal foreshore zone. This bar has been termed a ridge (King and Williams, 1949; Hayes, 1969; Davis *et al.*, 1972, Owens and Frobel, 1977), swash bar (Russell, 1968), and repair bar (Kraft, 1971). Actually the term ridge and runnel, which applies to both the sand bar and the trough-like area landward of it, is the most widely used and will be applied in this discussion. The ridge forms during waning storm conditions and it migrates landward while low-energy conditions prevail. The ridge commonly welds to the beach, thereby restoring at least a portion of the sediment lost during the storm. A thorough discussion of these features is included in a subsequent portion of this chapter (see *Ridge and Runnel Structures*).

Nearshore

The zone between the low-water line and the outer limit of the nearshore zone is commonly at least a few hundred meters wide. There may be a broad range in width and configuration. At most locations this zone is characterized

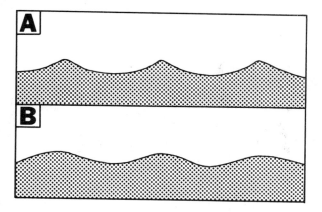

Figure 6-8. Generalized diagram showing the difference in plan view expression between (A) giant cusps and (B) beach protuberances.

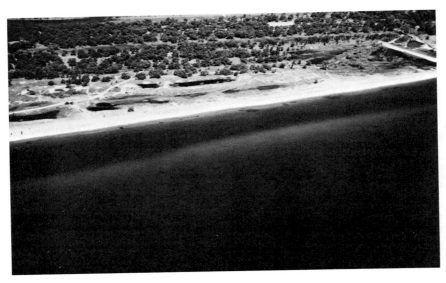

(A)

(B)

Figure 6-9. The position and general aerial configuration of longshore bars. In many instances the inner bar is discontinuous and ephemeral (A). Under relatively high-energy conditions the location and definition of longshore bars can be made by examination of the breaking wave patterns (B).

by subtidal sand bars (Figure 6-9) that are nearly parallel to the beach and are rather continuous. Only where there is a steep nearshore gradient and/or a dearth of sediment are the bars absent. The number of bars differs from one area to another but is related to the slope of the nearshore zone.

These sand bars were mentioned by de Beaumont (1845). They have been described and pondered by many subsequent authors, including Otto (1911–1912), Evans (1940), King and Williams (1949), Shepard (1950), Davis and Fox (1972b), Short (1975), Goldsmith *et al.* (1982), as well as others. Although a variety of terms has been applied to the same features, it seems appropriate to term these longshore bars (Shepard, 1950) because of their location. The term will be restricted to only those bars that are essentially parallel to the beach (Figure 6-9) and are submerged even under low tide. Along smoothly curved or straight coasts, longshore bars may be continuous for many kilometers. The inner or shallow bars commonly display lows or saddles (Figure 6-10a) that are rip current channels.

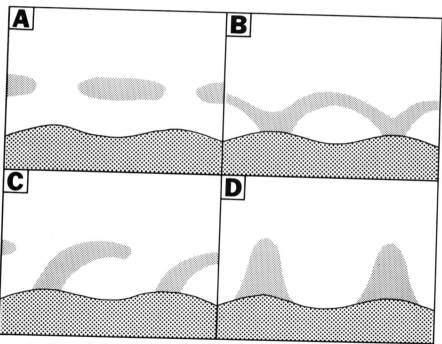

Figure 6-10. Generalized diagram showing the plan view configuration of four common types of nearshore bars and their relationships to the shoreline. Using the same sinuous shoreline as a reference surface the four types are: (A) the typical discontinuous and nearly parallel bars with rip channels opposite the protuberances, (B) crescentic bars that are convex seaward and commonly attached at protuberances, (C) lunate bars that are attached at protuberances and oriented with the convex side up the dominant longshore current direction, and (D) traverse bars that are essentially normal to the shore and attached at protuberances.

Along coasts where rocky headlands dominate with only local beach development, the longshore bars are absent or only one bar is present. Steep nearshore slopes adjacent to well-developed beaches may lack any longshore bar development, such as at Long Branch, New Jersey; Fort Ord, California (Bascom, 1964); Cedar Island, Virginia (Davis and Fox, 1978); and southern Oregon (Clifton et al., 1971). Most nearshore areas along barrier islands or other well-developed beaches have two or three prominent longshore bars.

Although there is a wide range in size and shape, the bars are gently sloped, commonly rising from less than a meter to 3 or 4 m above the adjacent troughs. At most locations the landward slope is slightly steeper than is seaward counterpart.

In addition to the common longshore bars, there are other types of submarine sand bars that are not continuous or parallel to the coast. One such type is the crescentic bars, which are convex seaward and exhibit a somewhat scalloped pattern along a given stretch of coast (Figure 6-10b). These bars form in conjunction with a sinuous shoreline characterized by cuspate topography. The horns or apexes of the cusps are the site of the tips of the crescentic bars. Studies of these bars indicate that the nearshore profile displays considerable range from one place to another within a single bar (Sonu and Russell, 1965; Short, 1979; Goldsmith, et al., 1982). At the beach protuberance the bar is rather flat and broad, whereas opposite the embayment the bar is narrow and well developed. Although there is a range in the size of crescentic bars, each one is commonly 100 m or more in length, and this type may extend for many kilometers along the coast.

Another type of discontinuous sand bar developed in combination with shoreline rhythms is the lunate bar (Figure 6-10c). These bars are connected to the beach protuberance by a shoaling zone in much the same fashion as crescentic bars; however, the bar is connected to the beach only at one end (Evans, 1939; Sonu et al., 1965; Sonu, 1969). The shoaling zone extends seaward at nearly a right angle and the well-developed portion of the bar bends to a near-parallel orientation with the shore. The bar terminates before reaching the next cusp apex, with rip channels commonly developed between the top of the bat and the next shoal. The convex seaward portion of the lunate bar is oriented nearly normal to the direction of wave approach.

Transverse bars (Tanner, 1960; Niedoroda and Tanner, 1970) may also develop in the inner nearshore zone. Like the crescentic and lunate bars, these also extend from the beach protuberances (Figure 6-10d). Although commonly essentially normal to the shoreline, transverse bars may be inclined in the direction of longshore transport (Sonu, 1969, 1973). Because of the topography associated with these features, wave are refracted such that the crests cross over the crest of the bar. Transverse bars are developed in areas of little wave energy and small tidal range (Niedoroda and Tanner, 1970).

Beach Materials

The material that comprises the beach and nearshore zone has an almost unlimited range in size, shape, and composition. In general, however, each beach area is characterized by a particular texture and composition with the great variation existing from place to place. That is not to say that temporal changes do not also take place at a given location; however, these variations are generally less spectacular than geographic changes.

Sources of Beach Sediment

A great many factors are involved in providing sediment for accumulation in the form of beaches. At most places this sediment is locally derived; however, it may travel great distances. First of all consider the possible sources. Erosion of the subaerial terrestrial environment may provide great quantities of sediment, much of which is carried seaward by rivers and eventually reaches the sea for its dispersal to the beach and beyond. In order for rivers to be direct suppliers of beach sand it is necessary for the gradient to be high enough that sand-sized material can be transported. An excellent example of such a situation is provided by most of the Pacific coast of North and South America. Problems have arisen in certain parts of California, however, because many rivers have been dammed, thus preventing the sediment from reaching the coast. Along much of the Atlantic and Gulf coast of the United States, the rivers have low gradients, and even if sand-sized material reaches the river mouth, it is retained by the numerous estuaries that characterize both coasts. These serve as very efficient sediment traps and are discussed in Chapter 2. The most direct source of beach materials is erosion of the coast itself. Areas of bedrock headlands and glacial drift provide much or all of the beach sediment in many areas. Pocket beaches located between rocky headlands are generally comprised of material that is essentially all derived from the adjacent bedrock exposures. The red beach and intertidal sands of the Bay of Fundy are derived from the adjacent Triassic red beds that border the bay. Glacial drift also is an important sediment contributor in the mid-to-high north latitudes where Pleistocene glaciation was prevalent. These deposits may be the prime contributor, as on the Long Island, New York, coast, or they may suppliment beach sediments, such as the coastal drumlins along the northern Massachusetts coast.

An important source of beach and nearshore material is the reworking and shoreward movement of sediment from the inner shelf. This is of particular significance as sea level rises such as it has during the Holocene transgression. The depth of effective wave activity is a controlling factor in the landward movement of sediment. This is generally restricted to < 10 m except during intense storms. Under nonstorm conditions landward movement appears to be the rule between this depth and the surf zone.

Perhaps the most volumetrically significant immediate source of beach sediment is by transport alongshore in the surf and intertidal zone. Although the sediment must ultimately be provided by one of the above-discussed sources, longshore transport is the final distributor of beach materials throughout most of the coastal beaches of the world. Only in small pocket beaches is there an absence of longshore sediment transport. The distance that beach sediment may travel along the coast ranges up to hundreds of kilometers. For example, sediment from the Mississippi Delta is carried westward along the Louisiana and Texas coast for great distances. Bedrock headlands interrupt the continuity of many California beaches so that the longshore transport is compartmentalized into zones of a few to many kilometers. In this situation the beach and nearshore sediment is often shunted offshore via submarine canyons (Shepard, 1973).

Detailed study of the composition of beach materials frequently makes it possible to determine the source of the sediment. Heavy mineral analysis is a frequently used technique in such studies, the principle being that each source is characterized by a rather unique suite of accessory minerals. By knowing the composition of likely sources it is then possible to determine which one or ones contributed the beach sediment along a given stretch of beach. Many such studies have been conducted, including those by Cherry (1965) and Judge (1970) along the California coast and by van Andel and Poole (1960) and Hsu (1960) along the Texas coast. In the latter study it was possible to distinguish between sediment supplied by the Mississippi and Rio Grande rivers at great distances from their respective deltas.

All of the above sediment sources are terrigenous; that is, they are ultimately derived from preexisting rocks. Biogenic or skeletal material may also be an important beach sediment constituent, and in some places it accounts for 100%. The nature of this skeletal sediment ranges widely from fine sand-sized Foraminifera to large clam and oyster shells or corals. For the most part, the distribution of skeletal beach sediment is restricted to low and middle-latitudes because of the general abundance of marine invertebrates in the waters of these areas. Notable exceptions do exist, however, with a striking example being a totally biogenic beach within the rugged basement complex headlands of Mount Desert Island in Maine (Figure 6-11). This pocket beach is only a few hundred meters long but it contains a high berm and is composed of up to 85% biogenic shell debris (Leonard and Cameron, 1981). The barnacle, *Balanus*, and the blue mussel, *Mytilus*, represent over 60% of the total carbonate. The lack of dilution by terrigenous sediments, extensive biologically productive intertidal zone, and the wave and current patterns provide this unusual situation (Leonard and Cameron, 1981). Hoskin and Nelson (1971) found beaches composed of 66% biogenic carbonate debris in Sitka Sound, Alaska. For the most part, beaches on middle-latitude land masses contain less than 25% biogenic debris, although

Figure 6-11. A biogenic pocket beach at Acadia National Park, Maine. The beach is composed almost entirely of sand-sized fragments of the mussel, *Mytilus*, which thrives in the shallow shelf and whose shells are broken and transported to this site by waves and currents.

locally it may be much higher. Storms frequently tend to concentrate large shell material by winnowing away the beach sand.

In tropical areas where reefs prevail, the beaches are entirely biogenic; constituents may include a wide variety of taxa or one species may dominate or be 100% of the sediment. Some coral and algae form the entire beach of atolls and other reef beaches.

The above examples consider skeletal debris derived from extant organisms, whereas at some locations beaches contain shell material that is being reworked from older sediments. Estuarine deposits containing brackish-water oysters and clams are being eroded along parts of the Texas coast, as are marsh deposits on the eastern shore of Virginia and Delaware, where the shoreline is transgressing. As a result the present-day beach contains a significant percentage of these large shells, with the estuarine muds being washed away.

Composition of Beach Sediment

The range in composition of beach materials is infinitely broad. However, most beaches are composed of quartz sand, with feldspar a distant second in abundance. Because of the wide range in compositions it is perhaps misleading to generalize about this aspect of beaches. Regardless of the materials, the composition tends to reflect that of the adjacent or nearby seas for reasons explained previously. Beaches are composed of whatever sediment is available.

Because of their chemical and physical durability with respect to other minerals, quartz and feldspar are dominant. Quartz may go through several cycles of deposition, erosion, and accumulation. Even in areas where these two minerals are dominant there are generally a few percent of accessory or heavy minerals. In some places the feldspar itself may show a wide variety of types and colors because of multiple sources.

Beach sands may be composed entirely of volcanic debris, such as in areas of Hawaii and western Central America, where obsidian, olivine, basalt, or other volcanic rock fragments may accumulate separately or in various combinations depending on the source. Other generally locally derived beach materials include the various biogenic carbonate grains that reflect the composition of the living biota adjacent to the beach, such as in areas of reef development. Beach sands may also be composed entirely of large percentages of ooliths, which are generated in the shallow shoaling environments of the Bahama Platform and other areas of carbonate deposition.

Texture of Beach and Nearshore Sediments

There are certain textural characteristics of beach sediment that can be generalized and others that show considerable range and variation. For example, the terrigenous component of beaches tends to be comprised of well-sorted and generally rounded grains. Sand beaches are by far the most common on a world-wide scale, with gravel or shingle beaches a distant second. Beach texture tends to reflect both the type of material comprising the beach and the intensity of the processes to which it is subjected.

Sand Beaches

A number of complex interactions between the aqueous and aeolian environments characterizes the backbeach. In general the sands here are fine and well sorted in comparison to the seaward zones of the beach and nearshore environment. Across the foreshore zone there is a gradual increase in mean grain size with differences of up to 0.3ϕ between the mid-foreshore and backbeach areas. Sorting follows the same pattern as grain size, with beach sands being characteristically well sorted to very well sorted ($< 0.35\phi$). Backbeach sands may be 0.10ϕ to 0.15ϕ better sorted than those in the foreshore (Mason and Folk, 1958; Hulsey, 1962; Ramsey and Galvin, 1971).

It has been shown that beach sediments tend to be negatively skewed, both on terrigenous beaches (Mason and Folk, 1958; Friedman, 1961; Chappell,

1967) and carbonate beaches (Folk and Robles, 1964). This may be the result of the high energy on a beach removing the fines or the addition of coarse particles, either in the form of terrigenous pebbles or shell material. In the case of the latter, abundant coarse material may cause the sediment to show a bimodal distribution that will yield a poor sorting value if both populations (modes) are combined in the textural analysis.

Grain size distributions have been divided into three components by Visher (1969) based on the transport mode. These are suspension, saltation, and rolling or surface creep. Only 1% or so of the total sample is necessary to define such a subpopulation on a probability plot. When a significant amount of shell or terrigenous gravel is present, the distribution shows a high percentage in the rolling subpopulation. Visher (1969) has also shown that a distinct break occurs in the saltation subpopulation in the swash or foreshore (Figure 6-12). This is explained as resulting from the uprush and backwash processes on this portion of the beach.

Absolute values for grain size on the foreshore zone seem to show a distinct relationship to the slope of the beachface—the coarser the sand, the steeper the beach (Bascom, 1951; Shepard, 1973) (Figure 6-13). For example, the coarse sand beaches of New England are steep, whereas the fine and very fine Gulf coast beaches are broad and flat. The above phenomenon is also controlled by variations in wave activity. The slope of an eroding beach tends to flatten, whereas the accreting beach slope steepens. The relationship between grain size and slope holds true under the latter circumstance.

A nonterrigenous sand beach that occurs on the Bahama Platform and other areas of carbonate precipitation is the ooid beach. The ooid shoals along the western margin of Great Bahama Bank near Cat Cay and to the east of Andros Island near Joulters Cay have pure ooid beaches adjacent to the islands (Ball, 1967). Beach sediments there are, of course, locally derived and are simply special types of shoals that occupy a beach environment of the island margins.

Nonsand Beaches

Mud beaches occur but are not common because the normal relatively high energy associated with beaches disperses fine sediment seaward. As a result, mud does not normally accumulate in the beach or surf zone. There are rather extensive mud plains and marshes along the southwest margin of Mississippi Delta. These were deposited while the major delta building took place in the area (Coleman, 1966). Similar conditions exist along the Surinam coast of South America where abundant mud and low wave energy permit mud "beaches" to prograde (Wells and Coleman, 1981).

Somewhat similar "beaches" are present along the currently eroding deltaic plains of the Brazos and Colorado rivers of the Texas coast and the

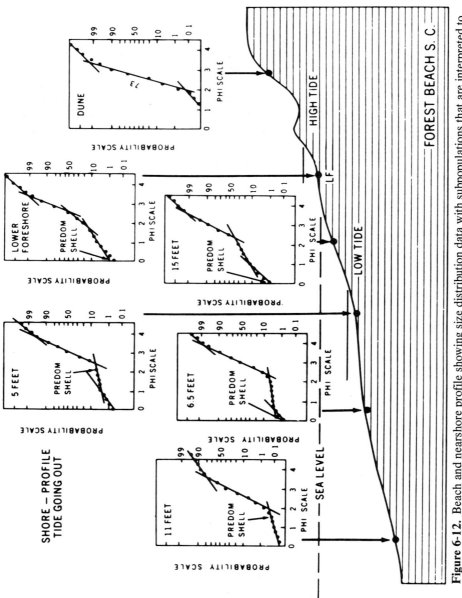

Figure 6-12. Beach and nearshore profile showing size distribution data with subpopulations that are interpreted to be due to different transport styles (from Visher, 1969, p. 1081).

Figure 6-13. Steep and coarse sand beach near Dakar, Senegal, in Western Africa (photo courtesy of L. H. Somers).

eastern shore of Virginia. At these locations, coastal erosion has exposed the delta or bay muds that crop out along the coast (Figure 6-14). The origin of the muds and general lack of typical beach geometry dictates that these features are beaches only if the broadest interpretation of the term.

Ephemeral mud "beaches" are present surrounding the mudlump islands (Morgan *et al.*, 1963) of the Mississippi Delta. These mudlump bodies are derived from sediments underlying the present delta and are squeezed up like "toothpaste" to form small islands. Because of their overall lack of resistance and fine sediment composition, the islands are severely modified or destroyed in just a few years by wave activity.

By far the most common nonsand beaches are those comprised of gravel and coarser sediment. These include gravel, shingle, cobble, or boulder beaches, all of which are generally not very laterally continuous. Gravel beaches are similar to sand beaches in that they are well sorted; however, the shape of the pebbles and cobbles shows a wide range within a single location. Composition may also be quite varied or all gravel may be of the same rock type depending on the source(s).

Gravel beach storm ridges typically have a zonation of particle sizes across the beachface, with the mid-beachface the finest and with clast size increasing in both directions (Timson, 1969). Apparently, however, the

Figure 6-14. Beach erosion exposing deltaic muds along the Texas coast near the Brazos River.

distribution of particle shape characteristics in the gravel beaches is more significant than grain-size parameters. Disks, spheres, and rod-shaped particles separate according to position on the beach and also show an imbricate fabric. Detailed studies by Bluck (1967) have shown distinct zones based on particle shape and arrangement. The base of the beachface is termed the outer frame zone and contains the best sorted gravels on the beach. Landward of this is an infill zone, where sand and fine gravel are mixed in the framework of spherical and rod-shaped pebbles (Bluck, 1967). An imbricate zone of discoidal pebbles and a wide zone of large disks complete the beach face zonation (Figure 6-15).

In order to access the processes of pebble shaping on gravel beaches, Dobkins and Folk (1970) restricted their studies to beaches composed entirely of basalt. This is in marked contrast to most gravel beaches that typically have a wide range of particle compositions (Bluck, 1967; Timson, 1969). It was found that the disk shape so characteristic of gravel beaches is developed in the beach and surf environment by abrasion as particles slide over sand and fine gravel.

Biogenic beach gravels are also rather common. They may comprise the entire beach sediment, as along coral reef beaches (Folk and Robles, 1964), or may be mixed with sand, such as on Padre Island, Texas (Watson, 1971), where pebble-sized shells comprise as much as 70% of the beach sediment.

The abundance of such biogenic debris on beaches reflects the source(s) of material and the processes acting on the beach. In the case of pure *Acropora* coral-stick beaches, it is a matter of that being the only sediment available. The shell concentrations of Padre Island appear to result from littoral processes that concentrate the shells (Watson, 1971).

Processes

The previous sections on beach and nearshore geometry and on materials have considered those aspects of the beach that are more or less descriptive. In order to understand how the materials are distributed, modified, and transported, it is necessary to consider the various processes acting along the beach and surf zone. Modifications of the beach and nearshore topography are responses to processes operating in these zones. Tides, waves, and currents are the primary factors in modifying the beach and nearshore environment, although wind also plays an important role. The interaction of these processes with one another and with the beach materials produces the various changes that make this environment so dynamic.

Tides

Changes in water level caused by lunar tides represent short-term cyclic changes in beach processes. The period is essentially 12 or 24 hr, depending on whether there is a semidiurnal or a diurnal tide; in some regions, such as

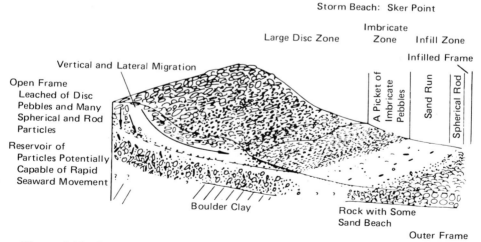

Figure 6-15. Composite diagram of Sker Point, storm beach showing zonation of large grain sizes and orientations (after Bluck, 1967).

the Gulf of Mexico, there is a mixed tide with the period varying within the lunar cycle.

Because of the wide range in tides, both temporally and geographically, there is a similar wide range in the effects on beach and nearshore areas. Open marine beaches typically have tidal ranges of 1 to 4 m, whereas estuaries and other coastal embayments may have tides that are several times as great, such as in the Bay of Fundy in Canada and the Bay of St. Malo in France.

Coasts have been categorized according to tidal range by Davies (1964, 1980). He designated three primary categories; microtidal (< 2 m), mesotidal (2–4 m), and macrotidal (> 4 m) (see Figure 7-2). These categories have served as basis for classifying coastal morphology (Hayes, 1975). Subsequent work by Hayes (1979) has further subdivided coastal types with some modifications of Davies' original categories (Table 6-1). These categories of tidal range appear to be better related to coastal morphologic types than those of Davies.

Actually tides play a rather passive or indirect role in sediment transport and changes in morphology of beaches. Environments of significant direct modification by tidal action include inlets, estuaries, tidal flats, and marshes (see Chapters 2, 3, 4, and 7). The primary role played by tides is to alternately expose and cover a large portion of the beach and inner surf zone. Depending on the tidal range, the change in the area subjected to various aqueous processes may be quite large. For example, some of the broad and gently sloping fine-sand beaches of the southern coast of 100 m or so in width is alternately subjected to waves and currents or to subaerial conditions. The large ridge and runnel features of the Massachusetts, Oregon (Figure 6-16), and Alaska coasts also are exposed during low tide. The net result of this movement of sea level is to retard the rate at which sediment transport and changes in morphology take place compared with areas where tides are small or essentially absent (Davis *et al.,* 1972). Tides therefore tend to slow rates of some processes by exposing the intertidal zone and its contained sediments and bedforms, whereas in nontidal environments similar features may be continually submerged and subjected to surf zone processes.

Table 6-1. Coastal types—medium wave energy ($H = 60$–150 cm) (after Hayes, 1979, p. 16).

Class	Tidal Range	Example
Microtidal	0–1 m	Gulf of St. Lawrence
Low mesotidal	1–2 m	New Jersey
High mesotidal	2–3.5 m	Plum Island, MA
Low macrotidal	3.5–5 m	German Bight
Macrotidal	> 5 m	Bristol Bay, AL

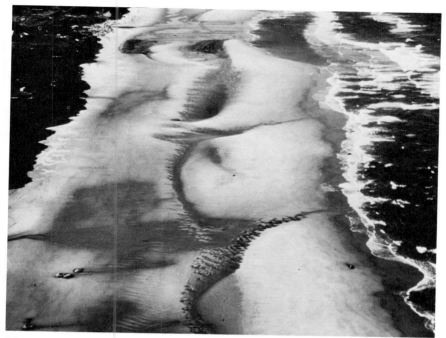

Figure 6-16. Broad low tide beach near Newport, Oregon, showing wide ridges and narrow runnel. Beach width is in excess of 300 m (photograph by W. T. Fox).

Effects on the Foreshore

The relationships that exist between groundwater level, tidal fluctuations, and foreshore processes show some direct influence of tides on beach sedimentation. A variety of investigations has considered this general problem, and results from most of the studies are in agreement as to the cause and effect relationships.

Position of the water table is a significant aspect of foreshore sedimentation because during high water table conditions there is a tendency toward erosion and during low water tables, accretion is more likely (Grant, 1948; Harrison et al., 1971; Waddell, 1976). As the relative positions of the water table and sea level change, therefore, there are changes in the foreshore processes. Duncan (1964) found that during low tide with the water table higher than the swash–backwash zone there was accumulation in the lower backwash. As the tide rose the accumulation zone shifted to the middle of the swash–backwash zone, and during high tides it shifted to the upper swash area. There was also a zone of erosion just seaward of the accumulation zone and it migrated up the foreshore (Figure 6-17). During falling tides the reverse took place (Duncan, 1964, p. 195). Somewhat similar observations

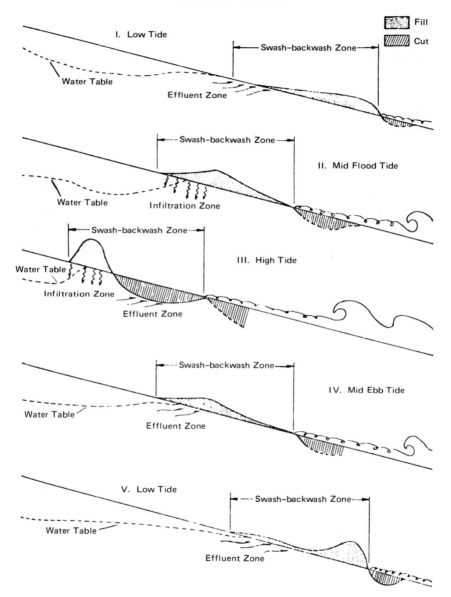

Figure 6-17. Interrelationships between water table and tidal stage with erosion and accumulation in the foreshore (after Duncan, 1964).

were noted by Harrison (1969) who developed empirical equations for foreshore changes. Midswash erosion and a steplike accumulation were observed by Strahler (1966) and Schwartz (1967) along the New Jersey and Nova Scotia coasts, respectively. The thickness of sediment eroded or deposited at any point during a single tidal cycle may exceed 30 cm (Duncan, 1964).

It should be remembered that the variation in waves and current activity to which the beach is subjected is great and tends to obscure those changes that are directly related to tidal fluctuations.

Effects of Ridge Migration

The development of ridge and runnel topography (King and Williams, 1949) in the inner surf zone is a rather common occurrence along most coasts. The position of the ridge at the time of formation varies in both space and time with respect to mean low water. In areas where tides are lacking or the range is small, it is common for the ridge crest to lie initially below mean low water, whereas the ridge is often developed above mean low tide in mesotidal areas (Figure 6-18) (Hayes, 1969; Hayes, 1972; Davis *et al.*, 1972; Owens and Froebel, 1977). The major effect these situations have on subsequent changes is with respect to the rate of migration of the ridge.

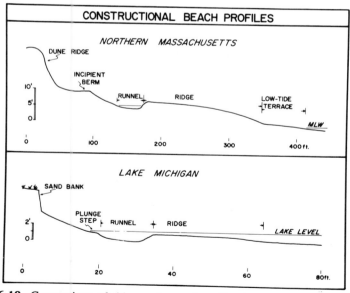

Figure 6-18. Comparison of ridge and runnel profiles between Massachusetts and Lake Michigan. Note that scale is the only significant difference (after Davis, *et al.*, 1972).

Richard A. Davis, Jr.

Initially ridges are convex upward with a nearby symmetrical profile. This profile soon changes to one exhibiting a steep shoreward face and a broad, gently sloping, but still convex upward seaward slope (Figure 6-19). Fluctuations in water level associated with tides control these morphologic modifications and the shoreward migration simply by permitting or preventing wave action acting on the ridge. Absence of tides permits continual modification and migration of such ridges, whereas in tide-dominated areas the ridge may be subaerially exposed for much of the tidal cycle and, as a result, its migration is correspondingly retarded (Davis *et al.*, 1972).

Role of Tides in Longshore Currents and Breaker Height

In areas of well-developed longshore bars, the tides cause a significant change in longshore current velocity by changing the depth of water over the bar crest. Under a given set of conditions, such a change in depth will cause slight responses in the wave height but more importantly will permit more of the wave's energy to pass over the bar and therefore increase the longshore current velocity. As the tide falls, the interference of the bar crest with wave motion will decrease the amount of wave energy reaching the zone shoreward of the bar. Observations of this effect are virtually absent from the literature because of a lack of closely spaced time-series data in the surf zone. This phenomenon has been observed along the barrier island coast of Texas (Davis and Fox, 1975), the eastern shore of Virginia (Davis and Fox, 1978), and on Plum Island, Massachusetts (Abele, 1973).

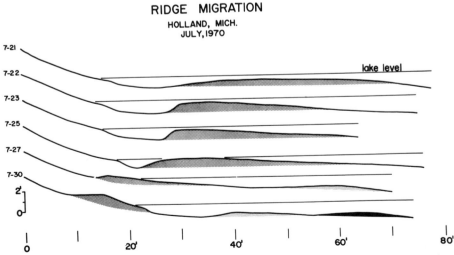

Figure 6-19. Sequential changes in shape and position of storm-generated ridge as it migrated onto the storm beach in Lake Michigan (after Davis, *et al.*, 1972).

Tides also have a marked influence on longshore current adjacent to inlets. In the inlet itself tides dominate and longshore currents are absent though they may be present seaward of the terminal lobe (see Chapter 7). On the open beach away from the inlet, tides have only indirect and minimal influence on longshore currents as stated above. In the transition between the zone of longshore current domination and tidal current domination, the interaction of processes is complex. A recent detailed study of this zone (Davis and Fox, 1981) shows that tidal currents are undetectable only a few hundred meters from the inlet. Longshore currents adjacent to inlets show decrease and increase in speed as tidal currents ebb and flood in the inlet. There is also a marked difference in longshore current speed on the upcurrent and downcurrent side of the inlet as tides flood and ebb (Davis and Fox, 1981).

Wind

Both direct and indirect effects of wind play a significant role in beach and nearshore sedimentation (Svasek and Terwindt, 1974). The subaerial beach (above mean high tide) is continually under direct influence of winds with few exceptions. Indirectly winds are perhaps more important in that they generate waves, which cause considerable modification themselves, and also produce longshore currents. Prevailing winds, winds associated with the passage of low-pressure systems, and, under extreme conditions, hurricanes, all have their respective effect on the beach environment.

Nonstorm Conditions

Temperature differences between the land mass and the adjacent water cause land breezes and sea breezes that show strong diurnal variations (Blair and Fite, 1965). These gentle breezes generally have little or no effect on beach sedimentation because of their low velocities and lack of competence. Such winds may have velocities of up to 12 knots and produce finer and better sorted backbeach sands in comparison to their foreshore counterparts. Prevailing winds also modify the backbeach texture and surface. During low tide or neap tide conditions, the upper foreshore zone is commonly dried by the winds, which also facilitates landward transportation of the sand and accumulation in the backbeach or foredune area (King, 1972).

Although onshore winds are prevalent along coastal beaches and the amount of sand transported landward is significant, the opposite effect may also take place. Prevailing offshore winds cause seaward transport of sediment from the foredunes and backbeach. In fact, one study (Rosen, 1979) shows that as much sediment is transported seaward by eolian processes as is transported landward by overwash.

Storm Conditions

Passage of low-pressure systems is accompanied by intensified wind velocities, which may cause considerable modification of the backbeach zone depending on the severity of the storm. The direction of the wind associated with such storms is dependent on the orientation of the coast with respect to the movement of the low-pressure system. When the shoreline is oriented normal to an onshore-moving low-pressure system, the cyclonic winds initially approach with a component out of the left as the observer faces the water. As the storm approaches, the wind velocity intensifies. When the weather system passes over the shoreline there is an abrupt shift in wind direction (Figure 6-20) so that it has a right-hand component as one faces the water (Fox and Davis, 1973). The most intense winds occur shortly after this shift in direction, and it is during these winds that much aeolian transport of sand occurs.

The most noticeable effects of this wind are the surface structures it produces. These include ripples and wind shadow features, both of which are discussed in detail in the section on sedimentary structures. In addition, landward transport of beach sediment in large quantities gives rise to coastal dunes, which are commonly associated with well-developed beaches (see Chapter 5).

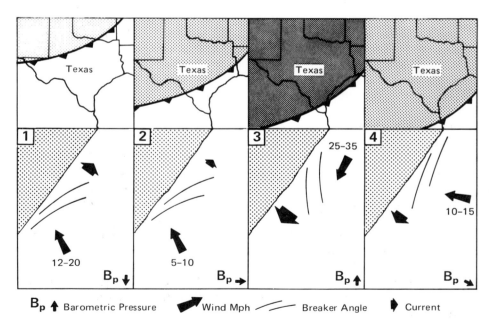

B_p ↑ Barometric Pressure   Wind Mph   Breaker Angle   Current

Figure 6-20. Weather system approaching (1, 2) and passing over (3, 4) the Texas coast showing the typical changes in coastal and related processes for each of the four stages in the cycle (after Davis and Fox, 1975).

Wind Tides and Storm Surge

Continual intense winds may cause significant water-level change along the coast. When associated with normal onshore winds, this phenomenon is called a wind tide; however, the term *storm tide* or *storm surge* is also applied when the effect is associated with hurricanes or other storm activity. In areas of a gently sloping inner shelf, such as the Texas Gulf Coast, normal conditions include positive wind tides of at least several centimeters. Such wind tides can be isolated by extracting all water level variations with a period 48 hr or longer (Davis and Fox, 1975). During intense offshore winds, the expected opposite phenomenon occurs, with sea level being lowered along the coast.

Storm tides associated with intense storms are often extremely destructive and in most cases are the cause of the majority of the property damage and loss during a storm. During Hurricane Carla in 1961, storm tides near 5 m were recorded in the Galveston, Texas, area (Fisher *et al.*, 1972) and 7 m at Port Lavaca (Hayes, 1967). Considerable erosion to the beach and foredune areas and breaching of barrier islands are associated with such storm tides. Depositional sites include the wind tidal flats and washover fans. After passage of the low-pressure system, the increase in atmospheric pressure and subsidence of the wind velocity permit coastal bays to be flushed of the normally large quantity of water pushed in by the storm tides.

Waves

Most of the dynamic nature of the beach and nearshore zone is the direct or indirect result of wave action. Waves move much sediment and consequently modify the bottom configuration as well as the distribution of sediment. They also generate currents that do likewise. Background information on wave mechanics may be obtained from Weigel (1964) and the U.S. Army Corps of Engineers (1973).

Nature of Surf Zone Waves

As a wave moves into shallow water where $d/L = < \frac{1}{2}$, where d is the depth of water and L is the wavelength, the wave begins to feel bottom. This causes friction between the bottom sediment and the orbital motion of the water particles. The surface effect of this phenomenon is to slow the wave velocity and cause a decrease in wavelength accompanied by an increase in wave steepness (H/L) with H being the wave height. Eventually the steepness is such that the orbital velocity exceeds the wave velocity, causing the wave to break. The great turbulence associated with breaking waves is of importance in coastal sedimentation because it causes considerable sediment to be temporarily placed in suspension and therefore available for transport by currents.

The nature of the wave in deep water and the slope of the nearshore bottom has much influence on the type of breaking wave that occurs. Swell characteristically has a sinusoidal profile with a small wave height as compared to the wavelength. After such waves begin to shoal and steepen they break, commonly as plunging breakers. Waves that are in the category called sea show a rather peaked crest with a broad trough. In shallow water these waves may break as plunging breakers or, more commonly, as spilling breakers. Surging breakers are commonly restricted to the beach face, where the slope is relatively steep. Breakers of all types form as the water depth approaches that of the wave height ($d = 1.0$ to $1.5H$) (Weigel, 1964).

Effects of Nonstorm Waves

During the vast majority of the time, the beach and nearshore zone is subjected to only moderate or low physical energy conditions. This situation is commonly associated with small locally derived waves or with breaking swell derived from some distance away. Under this regime there is an overall constructional character to this environment. The waves have a low steepness value and wave height, which produces a net transport of sediment in a landward direction. The amount and rate of transport is much less than during high-energy conditions. Consequently, although this phase tends to repair storm erosion, it is a slow process and if storms occur at rather frequent intervals, the net result is erosion.

During the waning period of energy after a storm, there is a build-up and shoreward migration of longshore bars. This period lasts at least several days (Davis and Fox, 1975; Owens and Frobel, 1977). Associated with these conditions may be ridge and runnel development in the intertidal or very shallow subtidal zone. The shoreward migration of the ridge and its eventual welding to the beach are caused by quite low-energy wave conditions (Hayes, 1972; Davis et al., 1972; Owens, 1977). In the event that a storm occurs during this accretion or healing period and prevents its completion, there is a sudden change to a storm profile with accompanying erosion.

Because most areas of the world have certain seasons or periods of time during which storms pass frequently, there are corresponding periods of beach erosion. When these storms pass repeatedly, the cost does not have time to recover between storms. If there is a rather severe storm during the usually low-energy season, there is probably sufficient time between storms to allow complete recovery so that little or no net loss is experienced by the beach.

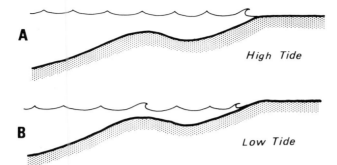

Figure 6-21. Generalized profile diagram showing the effect of tidal stage on nearshore wave conditions.

Tides may play a significant role in the wave climate of the surf zone. As the water level in the nearshore zone varies within the tidal cycle there may be a change in the size and shape of the breakers. More importantly, the position of the breakers can change greatly. As a result, the amount and location of wave-induced suspended sediment can change considerably within a tidal cycle. In barred beaches there may be a row of breakers over the longshore bar during low tide; however, at high tide the depth can be sufficient to allow the waves to pass over without breaking (Figure 6-21). Because of the short tidal cycle it might be expected that the above phenomenon would occur repeatedly under a particular wave climate.

Effects of Storm Waves

High-energy conditions caused by storm activity create considerable change in the beach and surf zone in a short period of time. Storm waves are characteristically steep and may also have rather long periods. This combination produces considerable erosion of the beach area and also within the surf zone, where longshore bars are displaced seaward. During such periods of high physical energy there is considerable sediment in suspension, with the greatest concentrations just landward of the breakers where the waves collapse (Brenninkmeyer, 1976). A very strong seaward surge occurs during these conditions, especially in areas where the nearshore slope is quite gentle. Detailed surveys of storms associated with the passage of winter weather fronts have shown that the inner sand bar may be planed off and displaced seaward more than 15 m in a single day (Davis and Fox, 1975).

Probably the most significant role played by storm waves is to provide great quantities of suspended sediment for transport by surf zone currents. This aspect is considered in a later section of this chapter.

A rather low-intensity hurricane (Fern) passed over the Texas coast during September of 1971 and caused a modest amount of erosion to the beach, yielding a typical storm profile (Davis, 1972). Because this was followed rather closely by the passage of the usual fall and winter weather fronts (northers), the beach did not recover until the following spring and summer (Davis, 1978) when typical low-energy conditions prevailed essentially without interruption.

Swash Processes

The foreshore zone or beach face is the location of the final energy dispersal of the waves; except for energy that is reflected from steep foreshores. It may be the site of considerable energy or very little (Waddell, 1973, 1976). The slope and configuration determines the amount of energy imparted on the beach and this in turn exerts an influence on the processes and responses in the swash zone. Areas where the nearshore slope is gentle or where longshore sand bars are present typically are characterized by little wave energy reaching the foreshore under most conditions. Large waves break at each of the sand bars or in the deeper water of the nearshore zone. Even though waves reform after each breaking in the surf zone, there is a considerable energy loss and as a result little energy reaches the foreshore. Obviously the effect and net result are quite different for each of these instances.

During high-energy conditions when waves are large and steep there is erosion in the foreshore accompanied by a decrease in slope (Weigel, 1964). There is also a planing off of the plunge step; however, there commonly remains a distinct textural change at this location. The opposite situation prevails during low-energy conditions.

After the final breaking of a wave in its progression across the nearshore zone, there is an uprush of water across the foreshore or beachface. Gravity and friction along with percolation into the beachface cause the water to slow and eventually stop. At this point a thin and narrow line of sediment (swash mark) is deposited. Backwash caused by gravity results in slow initial velocity; however, acceleration may be rapid with upper flow regime conditions achieved. Velocity of the backwash depends on foreshore slope, volume lost by percolation, volume of water in the backwash, and addition of water from the effluent zone (Schiffman, 1965; Waddell, 1976). Grain size of the beach face also is a factor in that it controls the amount of percolation. For example, fine sand will allow only slow percolation, thereby resulting in a rather large volume of water in the backwash and a high velocity. In contrast, there is nearly total percolation on a gravel beach, resulting in almost no backwash. This facilitates the development of steep gradients on coarse beaches and gentle gradients on fine sand beaches.

Currents

The foreshore and nearshore zones are almost continuously subjected to a complex variety of currents, which owe their origin dominantly to waves and wind. The constantly changing nature of these driving forces dictates that the currents vary in speed and direction in both space and time. There is, however, a definite relationship between their intensity, and therefore the amount and rate of sediment transport, and the wind conditions.

There are three basic types of currents operating in the nearshore zone, and all or any combination may be operating at a given time. Longshore currents move essentially parallel to shore and cause the greatest amount of sediment transport. Rip currents also carry a significant amount of sediment and move in a generally offshore direction. The third type of current is related to the type of wave activity and moves in an onshore or offshore direction depending on wave climate and location.

Currents Normal to Shore

A variety of laboratory studies has shown that currents oblique or normal to the shore have a large dependence on wave steepness (King, 1972). In general the steeper waves ($H/L = > 0.03$) generate seaward currents and rather flat waves cause shoreward currents (Johnson, 1949; Rector, 1954; Weigel, 1964). King (1972) cites a H/L value of 0.012 as critical. Field studies using tracer sediments have tended to substantiate this generality (Ingle, 1966; McArthur, 1969). It should be pointed out, however, that the grain size of the sediment and the nearshore gradient also play a role. In the breaker zone itself there are local currents that cause sediments on either side of the breaker zone to move toward the breaking wave (Figure 6-22), thus reflecting water particle motion in the wave (Ingle, 1966). As the wave collapses there is a landward surge, creating a bore as water moves across the surf zone. This causes a landward movement of sediment but is influenced by the offshore movement created by reflection of the preceding waves. Under certain conditions of wave steepness and nearshore gradient, the seaward surge may dominate and cause sediment transport in that direction. This has been observed by Ingle (1966) and is dependent on the dominance of oblique or normal currents over longshore currents (Figure 6-23).

Longshore Currents

The largest percentage of sediment movement is parallel to the shore and caused by longshore currents. As waves enter shallow water at an angle to the shoreline, they are refracted, providing a vector of energy parallel to the

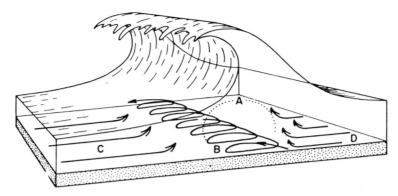

Figure 6-22. Diagram showing grain motion associated with a breaking wave. Arrows trace the general paths of the largest grains (B) and the grains shoreward (C) and seaward (D) of the wave, whereas (A) indicates the suspended particles (after Ingle, 1966, pp. 53).

shore. The current generated by this wave energy is commonly called a longshore current and is related to wave height and the angle that the approaching wave makes with the shoreline (Galvin, 1967). Because of their dependency on breaking waves, longshore currents are not significant beyond the breaker zone. The term *river of sand* is commonly applied to the zone bounded by the berm crest and the breaker zone. It is within this area that there is much longshore transport of sediment, and the river analogy is an appropriate one.

Generation of longshore currents is dependent on various wave parameters, including wave height (Komar, 1976), wave direction or angle with the shoreline, wave period, and the slope of the nearshore bottom. Reviews of formulas for longshore current velocities show lack of specific agreement on variables and constants used (Sonu *et al.*, 1965; Galvin, 1967; Longuet-Higgins, 1972; U.S. Army Corps of Engineers, 1973). Commonly an investigator finds it necessary to derive his or her own relationships in order to account best for variations in empirical data. This generally results from the location and instrumentation used in the initial measurements. It is therefore often difficult to compare longshore current data between various studies.

A time-series study by Fox and Davis (1973) has demonstrated a simple and generally usable relationship for predicting longshore current velocity and direction. Using closely spaced empirical data they found that about 70% of the variation in longshore current velocity can be accounted for by taking the first derivative of the 15-term Fourier Curve for barometric pressure. With modern computer techniques this is a simple process and requires data on only a single variable—barometric pressure.

Although it has long been recognized that longshore currents are the dominant factor in transporting great quantities of sediment parallel to the coast, it has been only in recent years that quantitative relationships have been established between wave parameters, longshore current velocity, and the rate of sediment transport. One of the first suggestions of the importance of these relationships was by Grant (1948). Since that time numerous field and laboratory investigations have considered this problem (Watts, 1953; Caldwell, 1956; Ingle, 1966; Komar and Inman, 1970; Komar, 1971a, 1975, 1976). Although all of these studies produced empirically derived relationships, there was lack of specific agreement. A workable model produced by Komar and Inman (1970) assumes that the wave power causes the sediment to be in suspension, with the longshore current producing the transport parallel to the coast.

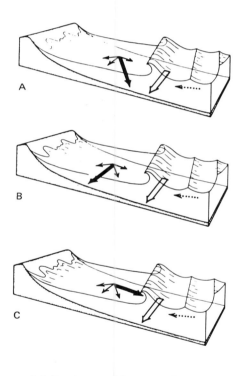

Figure 6-23. Diagram showing the predominant paths or vectors of tracer-grain movement when (A) longshore current velocity and wave motion exert equal influence, (B) when longshore current dominates, and (C) when wave motion dominates (after Ingle, 1966, p. 59).

↖↑← Dominant and Secondary Paths of Tracer Grains on the Foreshore Slope

⇐ Path of Tracer Grains within and Immediately Shoreward of the Breaker (Plunge) Zone

◁······· Path of Tracer Grains Seaward of the Breaker Zone

Extensive sediment transport data were collected from the breaker and surf zones by Ingle (1966). He found that the zones of greatest sediment transport did not coincide with the position of greatest longshore current velocity. This observation emphasizes the importance of breaking waves in making suspended sediment available for transport and also indicates that currents alone are somewhat ineffective in initiating sediment movement (Ingle, 1966, p. 183).

Rip Currents

The first comprehensive description of the phenomenon now referred to as rip currents was by Shepard *et al.* (1941). Prior to that study, rapid offshore-moving currents were called undertow. Careful observations, however, have shown that there are two distinct types of offshore mass movement of water: one that is rather slow, widespread, and confined to the near-bottom waters, and the other, which is more of a near-surface phenomenon and which is narrow and rapidly moving. The latter describes rip currents, whereas the term undertow is often used for the former.

In a more detailed investigation, Shepard and Inman (1950) determined that shallow-water topography and shoreline configuration play a role in rip current location and development. Rips may form as the result of water piling up between the shallow sand bar and the strand line as waves move shoreward. This water moves generally parallel to the shore and converges at a topographic low where it flows seaward through a saddle in the shallow sand bar (Figure 6-24). The converging currents are called feeder currents and supply the neck of the rip current system. Beyond the rip channel, through the breaker zone, the rip current disperses into a rip head (Figure 6-25).

Shepard and Inman (1950) also found that the shoreline configuration influenced rip current circulation in that protuberances or the apexes of giant cusps were commonly the foci of feeder currents, with the rip current moving seaward opposite the protuberance. Similar observations have been made by subsequent researchers (Komar, 1971a; Davis and Fox, 1972b) (Figure 6-26) although the opposite has also been observed (Sonu *et al.*, 1965; Komar, 1971b, 1976).

Rip currents may be recognized as turbid plumes or bubble trains extending across the surf and breaker zones. Although they are commonly a near-surface phenomenon, rips also have been shown to entrain and transport significant quantities of sediment (Cook, 1970). Visual evidence is provided by the sediment in the turbid plumes in rip heads. A striking example of the role of rip currents in sediment transport is provided by the detailed nearshore mapping presented by Davis and Fox (1972a). In this study rip currents developed during high wave conditions and moved over the

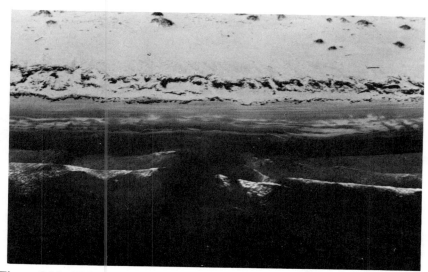

Figure 6-24. Oblique aerial photograph showing prominent rip channel formed by a saddle in a shallow longshore bar on Padre Island, Texas.

Figure 6-25. Nearly symmetrical plume of suspended sediment forming the head of a rip current system on eastern Lake Michigan.

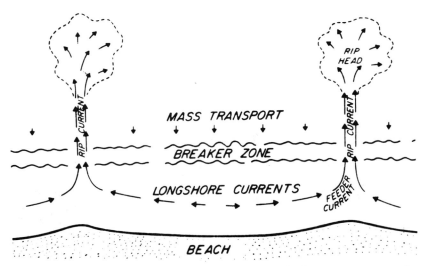

Figure 6-26. Diagram of rip current system showing feeder currents converging to form a rip current adjacent to a beach proturberance (after Komar, 1971, p. 2644).

crest of the existing nearshore sand bar. In doing so the rip excavated a channel with 0.5 m of sediment removed in a single day's time (Davis and Fox, 1972a, p. 406).

The overall configuration assumed by the rip current circulation cell may show some range. Ideally there is a rather symmetrical character to this as shown in the generalized diagram (Figure 6-26). The symmetric nature is largely controlled by the wave incidence and is generally restricted to waves that are nearly parallel to the shore. When waves approach at an angle, feeder currents are upcurrent in direction (Figure 6-27).

Shoreline configuration may also be an important factor in the location of rip currents. Sinuous shorelines cause a "meandering" nature to longshore currents. The deflection of swift longshore currents by the protuberances of cusp apexes results in rip currents (Sonu and Russell, 1965; Davis and Fox, 1972b). In some high-energy situations this type of coastal morphology causes rips to form in places where they would not form under low-energy conditions, namely across the sand bar crests instead of in the low saddles on the bars (Davis and Fox, 1972b). This indicates that under such conditions, the shoreline is more influential in rip currents than is the subaqueous topography.

The best development and definition of rip current systems appears to occur during moderate wave height conditions. Under storm conditions, rip currents commonly lack definition and are masked by strong wave surge and rapid longshore currents. Rip channels are also obscure under these conditions.

Rip current spacing may be related to wave conditions, with high physical energy producing fewer and more widely spaced rips than low-energy conditions (McKenzie, 1958; Cook, 1970). Although rip currents are commonly quite ephemeral they may be persistent (Davis and Fox, 1978) and may be related to the shoreline and submarine morphology (Shepard and Inman, 1950).

Edge waves have been shown to play an important role in coastal circulation, particularly in rip current development and location. These are standing waves that are oriented with crests normal to the shoreline. The height decreases seaward so that it is negligible only one wavelength offshore (U.S. Army Corps of Engineers, 1973). It has been shown that these standing edge waves have the same period as incoming gravity waves (Bowen and Inman, 1969). The combination of the incoming waves and the edge waves produces a regular alternation of higher and lower breakers along the shore, thus causing a regular and cellular pattern of circulation in the form of rip currents (Komar, 1976).

Because the wave length of the incoming waves and the standing waves have the same period there are nodal and antinodal lines along the surf zone where these waves intersect. At the antinodal line there is an increase or decrease in the wave height depending on whether the two wave types are in phase or 180° out of phase. At the nodal lines the edge wave does not contribute to the system; therefore, the true height of the incoming wave may

Figure 6-27. Photograph of large rip channels on the Oregon coast that are oblique in their orientation to the coast (photograph by W. T. Fox).

Richard A. Davis, Jr.

be observed (Komar, 1976). The result is a regularly spaced circulation pattern of rip currents with the rips located at antinodal positions where waves are smallest (Figure 6-28). Such circulation cells are typically developed along the west coast of North America where waves are large and tend to be refracted so as to approach parallel to the shoreline at and slightly seaward of the breaker zone.

Sedimentary Structures

Interaction of the water and wind with sediments of the beach and nearshore environment produces a wide variety of bedding and bedding plane sedimentary structures. The variety of physical processes that exists across the beach and adjacent nearshore area causes the formation of numerous different sedimentary structures, most of which are associated with a rather restricted zone within this environment. Some of these structures may

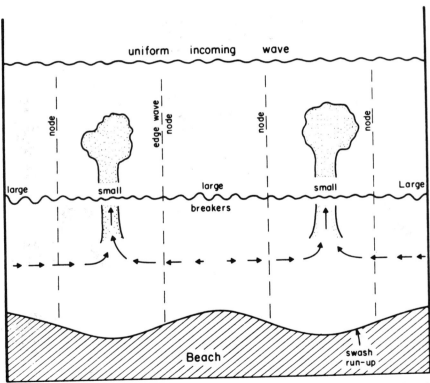

Figure 6-28. Schematic map view of relationship between rhythmic topography, rip currents, and edge wave nodes (from Komar, 1976, p. 179).

eventually be incorporated into the geologic record. Because of the diagnostic nature of some sedimentary structures, they are of significance in paleo-environmental reconstructions.

Backbeach Area

Surficial structures of the backbeach are formed primarily by the wind, particularly when a well-developed berm is present. Wind-generated ripples with high ripple indices (Tanner, 1967) prevail on the dry sand of the backbeach. Each time the wind shifts there is a corresponding reorientation of the ripples. Sand shadows and current crescents (Figure 6-29) may also be present in this zone. They are associated with pebbles, shells, or other coarse particles that may accumulate in the backbeach.

Stratification in the backbeach is thin bedded and generally parallel to the beach surface. Sediment is carried to this area by onshore winds or washover at the berm crest during high spring tide or storms. Burrows made by ghost crabs and sand hoppers abound in this zone (Hill and Hunter, 1973, 1976).

Beach Face Structures

From the berm crest area to the plunge step there is a wide variety of sedimentary structures, some of which are located at a distinct position within that narrow zone. The berm crest is commonly dotted with air holes or beach pits. Air holes (Emery, 1945) are small (< 1 cm) circular holes in the saturated upper foreshore. They are formed as water percolating from the upper swash zone forces air in the porespaces up to the surface of the beach. Beach pits (Davis, 1965a) are formed in completely dry sediment of the upper foreshore during low wave-energy conditions. Under high-energy conditions and wide swash, upper foreshore sediment is saturated and loosely packed. After it dries, this loosely packed sediment is easily disturbed, causing rearrangement of the grains in the form of small collapse features. These pits are irregularly circular and up to a few centimeters in diameter.

Most prominent of the surficial features of the foreshore zone are swash marks. These are small arcuate ridges of sediment marking the landward extent of each wave (Figure 6-30). They are concave seaward and the curvature is commonly related to swash energy, with gentle swash producing nearly straight swash marks. Rill marks may develop on the beach face and on the uniformly sloping storm beach. These small distributarylike drainage patterns are particularly common when the water table is intersected because of beach erosion. Current crescents may also be developed on the beachface.

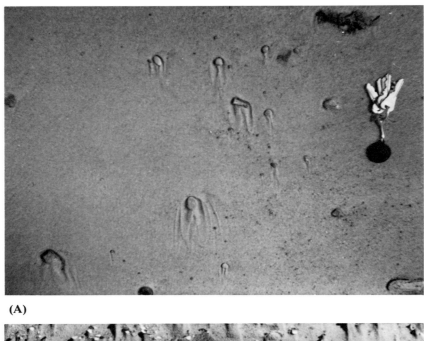

(A)

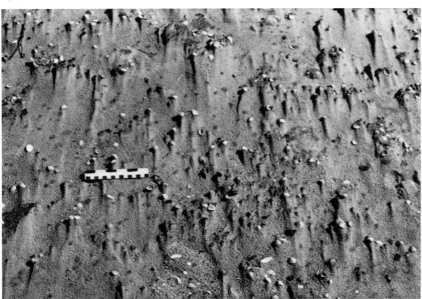

(B)

Figure 6-29. Beach structures caused by fluid motion; (A) current crescents and (B) sand shadows.

Figure 6-30. Swash marks, which are tiny arcuate ridges of sand that delineate in furthest landward advance of the wave by which they are formed.

Rather gently sloping beachface zones may be covered with low-amplitude antidunes generated by rapidly moving backwash over rather fine and generally impermeable sands (Figure 6-31). This contrasts with plane bed deposition, which dominates on coarser and more steeply inclined beaches. Detailed inspection of this beachface stratification shows that there is a distinct grading in each sedimentation unit from fine and abundant heavy minerals at the base to coarser with few heavies at the top (Clifton, 1969). This unique feature results from "bed flow" in wave backwash and is not restricted by beach sand texture as is the slope of the foreshore.

Foreshore beach stratification may also include layers of heavy minerals that developed as lag concentrates during beach erosion (Figure 6-32). These materials have a high specific gravity relative to quartz, feldspar, and carbonate. As a result, the grains of heavy minerals that may account for only a percent or so of the beach as a whole will be concentrated as the lighter particles are removed during high-energy periods. This placer-type accumulation may reach 25 cm.

Ridge and Runnel Structures

Surficial structures of ridge and runnel features consist of a complex of ripples. Ridges may be covered with landward-oriented ripples formed by lower flow regime conditions during high-tide conditions. As the tide rises

Figure 6-31. Widespread antidunes on the wide and gently sloping foreshore. These are common features on such beaches at or near low tide conditions.

Figure 6-32. Trench in back portion of storm beach showing veneer of heavy minerals resulting from erosion.

and falls, however, the thin layer of water moving over the ridge causes upper flow regime conditions and plane bed deposition. The runnel is covered with ladderback ripples (Figure 6-33). As water in the runnel moves parallel to shore toward rip channels, it generates small current ripples in the runnel. These are superimposed on the current ripples that were formed by waves and shoreward currents. Megaripples may also develop in the runnel with crests normal to the shoreline.

Mud or pelleted mud may accumulate in troughs of ripples in the runnel. As the ridge migrates landward over the runnel, the mud may be preserved in the ripple troughs forming flaser bedding.

Figure 6-33. Ladderbacked ripples in the runnel. These are formed by the near right-angle intersection of wave-generated ripples (parallel to shore) and ripples generated by currents flowing parallel to shore as the runnel is drained (photograph by M. O. Hayes).

Internal structures of the ridge show plane beds in the upper portion of the ridge that are nearly horizontal or dip slightly seaward parallel to the ridge surface. These overlie rather steeply dipping (nearly 30°), landward-oriented cross-laminations (Figure 6-34). The steeply dipping laminations represent landward migration of the slipface of the ridge.

As the ridge begins to weld to the beachface and build into a berm configuration there is a reworking of the ridge sediment by swash action. This causes accumulation of seaward-dipping stratification at inclinations of 10 to 15° (Davis *et al.*, 1972). A reconstructed section of such a welded ridge contains the complete sequence of internal features associated with the erosional or storm beach followed by the runnel, ridge, and welded beachface (Figure 6-35).

Nearshore Zone

The surface of the nearshore area is generally corrugated with ripples. These features respond rather rapidly to changes in physical energy and to wave direction. Because of their almost ubiquitous occurrence in modern near-

Figure 6-34. Internal stratification of a landward migrating ridge showing near horizontal plane bed deposition and relatively steeply landward-dipping slipface deposition. Pencil for scale (after Hayes, 1969, p. 310).

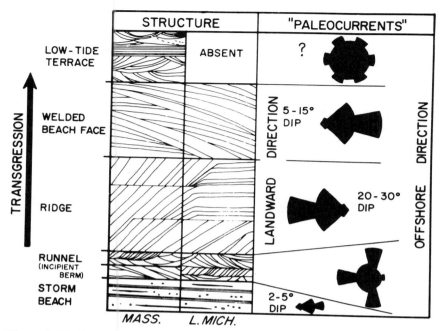

Figure 6-35. Sequence of cross-bedding types, orientations, and dips that form as a ridge migrates landward and welds onto the beach (after Davis *et al.*, 1972, p. 420).

shore areas and their rather common presence in rocks, number of detailed studies have been undertaken to understand better the nature and generation of ripples.

The orientation of the crests of wave-generated or symmetrical ripples has long been known to be related to the direction of wave approach. Although numerous geologists have used this to approximate the orientation of ancient shorelines, studies of modern ripples indicate that they may occur at large angles (up to 65°) with the shoreline in the inner nearshore zone. In addition, these ripples are quite ephemeral and may drastically change their orientation in a day's time to respond to shifts in wind direction (Vause, 1959; Davis, 1965b).

There are essentially three types of ripples formed in shallow water: current ripples, wave ripples, and combined flow ripples (Harms, 1969). The type that forms is related to wave and current parameters as well as to water depth and bottom slope. Although the nearshore area is dominated by wave ripples, both other types may also be present.

Areas without longshore sand bars yield a separate but distinct pattern of bedforms from those areas where bars are present. The surficial and internal character of nearshore bedforms has been investigated in great detail by

Clifton *et al.* (1971) In their study on the unbarred Oregon coast they found the nearshore zone to be characterized by four distinct facies, each having its own bedforms and internal structures. The inner planar facies corresponds to the swash zone and consists of the typical laminated plane beds of the beach face. It is interesting to note that Clifton *et al.* (1971) were able to find the same features as described above in Tertiary strata of California. The inner rough facies consisting of ripples and irregular beds with considerable variations shown as the slope differs (Clifton *et al.*, 1971). Internal bedding is a complex of differently oriented sets of cross-laminations that appears similar to the low-tide terrace deposits discussed in Hayes (1969). Under the breaker and outer surf zone is a planar facies that is dominated by sheet flow and near-horizontal lamination. This corresponds to the conditions over longshore bars on Mustang Island, Texas (Davis and Fox, 1975). In the outermost nearshore zone, where wave buildup begins, the surface is characterized by lunate megaripples (Clifton *et al.*, 1971). These features are landward oriented and have crests that are concave in that direction. Internally they contain steeply dipping landward cross-stratification.

This and related studies in other nonbarred coasts have resulted in the development of a predictable sequence of structures across this zone (Figure 6-36). Clifton (1976) has shown that as the bottom becomes active because of wave motion there is a regular variety of wave-formed bedforms that eventually culminates in upper plane beds (flat beds) in the breaker zone. Ripples go through numerous forms and eventually give way to megaripples (Figure 6-35). The wave climate and nearshore slope controls the relative widths of the zone in which each type will occur and also controls the extent to which the sequence is completed. For example, in Willapa Bay, a low wave-energy environment, the sequence proceeds only to irregular ripples (Figure 6-36).

Systematic studies of the barred nearshore zone demonstrate that there are also regular patterns of bedforms that correspond to subtle topographic variations, wave characteristics, and current parameters (Davidson-Arnott and Greenwood, 1974, 1976; Hunter *et al.*, 1979). The two studies used here as examples represent a microtidal, low- to moderate-energy coast on the Gulf of St. Lawrence (Davidson-Arnott and Greenwood, 1974, 1976) and a mesotidal, high wave-energy coast in Oregon (Hunter *et al.*, 1979).

Both nearshore areas are characterized by longshore bars and troughs with well-developed rip channels through the bars. In the low-energy area (Gulf of St. Lawrence), the bars are essentially shore-parallel with rip channels essentially normal to the shoreline. The high-energy coast (Oregon) has longshore bars that are oblique to the shore with similarly oriented rip channels (see Figure 6-10c). Surface bedforms and near-surface stratification show a variety of structures that can be categorized into distinct nearshore facies. Distinctions can be made between bars, trough, and rip channel environments on the basis of the size, nature, and orientation of the stratification.

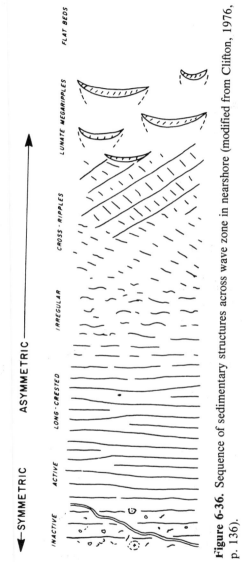

Figure 6-36. Sequence of sedimentary structures across wave zone in nearshore (modified from Clifton, 1976, p. 136).

The longshore bar actually contains three discernable facies; seaward slope, bar crest, and landward slope (Davidson-Arnott and Greenwood, 1976). The seaward slope facies contains wave and combined flow ripples and upper plane beds (Figure 6-37). The latter dip slightly seaward and form as waves shoal. On the bar crest, plane beds and lunate megaripples are prevalent although some ripples may be present. The landward slope of the bar is characterized by ripples and plane beds that are generated by avalanche accumulation if the slope is relatively steep (Davidson-Arnott and Greenwood, 1976).

The trough tends to be dominated by longshore currents and the resulting bedforms are dominantly current ripples oriented normal to the shoreline. Under high-energy conditions longshore currents may generate megaripples or even upper plane beds.

The rip channels are characterized by seaward-oriented bedforms—typically ripples but occasionally megaripples or upper plane beds. Cross-stratification in cores from these channels show seaward dips (Davidson-Arnott and Greenwood, 1976; Hunter et al., 1979).

Bedform orientation is the result of the orientation of waves and currents. These are in turn at least partly related to subtle variations in topography across this very complex and dynamic zone. As a consequence, plots of bedform orientation or cross-strata direction will be equally complex; however, a thorough knowledge on the dynamics of this zone and careful analysis of the data provide proper interpretation of the facies that occur.

In addition to complex physical structures, benthic organisms may generate bioturbation structures in the nearshore zone. Such structures tend to be most common in low latitudes where macrobenthos are abundant. The dominant biogenic structure is the burrows of the ghost shrimp, Callianasa, which may occur throughout the bar and trough system (Hill and Hunter, 1976). The sand dollar, Mellita, produces near-surface trails and may destroy physical structures in the upper few centimeters of sediment.

Process-Response Systems in the Beach and Nearshore Environment

Interaction between numerous environmental variables and the coastal and submarine morphology is rather complex in the beach and nearshore zone. These process-response relationships are the causes of the very dynamic character of this environment. In order to clearly understand and appreciate these complex interactions and to tie together the preceeding material in this chapter it is best to consider the system as a whole.

Variables that contribute to the dynamic nature of this environment include meteorologic, wave, current, and water-level phenomena. Barometric pressure, wind speed, and wind direction are the most important weather

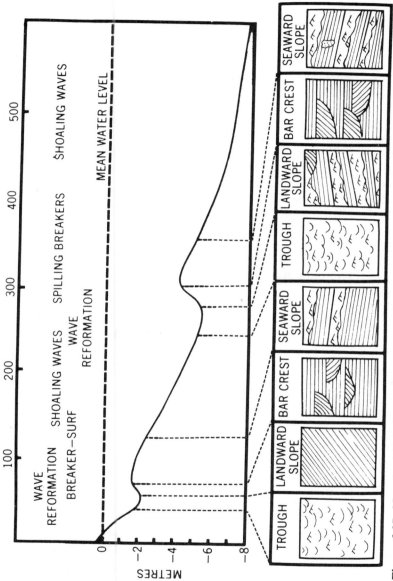

Figure 6-37. Nearshore barred topography showing characteristic sedimentary structures and wave transformation zones (from Davidson-Arnott and Greenwood, 1976, p. 154).

factors. The most critical wave parameters are wave height and angle of incidence in the breaker and surf zone. Although longshore currents are most important, both rip currents and wave-generated currents moving normal to the shore also play a significant role in sediment transport. Tidal fluctuations and the position of the groundwater table are also factors.

In order to determine which of the above are the most critical in the beach and nearshore environment, it is necessary to measure these variables systematically through time and try to relate them to responses exhibited by morphologic changes. The best approach to this type of investigation is by the time-series method. Studies of this nature have been completed on the Lake Michigan, central Texas, Virginia, and Oregon coasts (Davis and Fox, 1972b, 1975, 1978; Fox and Davis, 1978, 1979), in Massachusetts (Abele, 1973), and in the Magdalen Islands (Owens, 1977).

After the data are smoothed by Fourier analysis (Fox and Davis, 1971, 1973), it is apparent that distinct cyclic patterns exist among the more important environmental variables: barometric pressure, wind velocity, breaker height, and longshore current velocity and direction (Davis and Fox, 1971, 1972b). These relationships can be conveniently simplified into a conceptual process model (Figure 6-38). In this model the key variable is barometric pressure. Passage of a low-pressure system causes increase in wind velocity, wave height, and longshore current velocity as well as an abrupt change in wind direction that is reflected in the change of longshore current direction.

In order to relate this process model to morphologic changes it is necessary to have detailed data on beach and nearshore topography. Daily maps provide such data and show short-term temporal responses to changing conditions. These maps exhibit even subtle variations in such features as shoreline configuration and position of rip channels and sand bars (Davis and Fox, 1972b, 1975).

From comparisons of the process and response data it is obvious that the changes in morphology respond to the cyclic nature of the environmental variables. Storm conditions caused by the passage of low-pressure systems cause generation of large waves and swift longshore currents. These currents are deflected by the sinuous shoreline and form rip currents that move across sand bar crests and excavate a channel (Figure 6-39). The area of relatively low energy between rip currents is the site of sediment accumulation in the form of a sand bar. Under high barometric pressure, low-pressure, low-energy conditions dominate and sand bars move shoreward, with their breakwater effect causing modifications of the shoreline (Figure 6-39). The next low-pressure system brings about a similar set of conditions, and the cycle repeats itself (Davis and Fox, 1972b; Owens, 1977).

The above relationships and models were developed from studies on coastal Lake Michigan, where low-pressure systems move in an onshore direction. Similar time-series studies on the Texas and Virginia coasts have shown comparable relationships although the weather systems move offshore

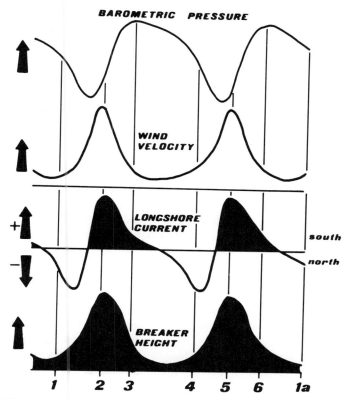

Figure 6-38. Conceptual model for locally generated coastal processes, including wind velocity, breaker height, and longshore current velocity and direction (after Davis and Fox, 1972b).

in these areas (Davis and Fox, 1975, 1978). In addition, large fetch in the Gulf of Mexico and the Atlantic Ocean provide considerable swell-wave influence that further complicates the system.

Owens (1977) conducted similar studies on the Magdalen Islands in the Gulf of St. Lawrence. These islands have the general configuration of barriers but are surrounded by open water. As the result, he was able to monitor the process-response system on both sides of the islands as storms approached and passed over. There is a significant difference between the two. Beach cycles on the east-facing protected side respond to storm and post-storm recovery periods, whereas the west-facing side experiences seasonal fluctuations resulting from changes in storm intensity and the relatively long fetch on that side (Owens, 1977).

Similar cyclic systems have been reported from the coast of Israel in the eastern Mediterranean (Goldsmith *et al.*, 1982). Time-series studies show sequential development of various bar morphologies as wave climate changes

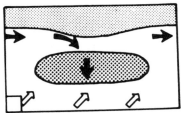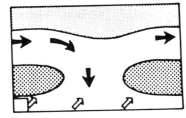

Figure 6-39. Diagram deflection of longshore current by protuberance causing rip current with cuts channel in sand bar (after Davis and Fox, 1972b).

during storm and post-storm recovery cycles. During a cycle, essentially all nearshore bar types appear in the sequence. Goldsmith *et al.* (1982) describe two different cycles, but there are similarities between them. During storm conditions the surf masks bar definition and large rips are obvious. As conditions wane there is a meandering inner bar and outer bar. Through time both become crescentic, although some intermediate steps may occur. The crescentic bars are rather stable during calm periods, generally the summer season.

Perhaps the most comprehensive treatment of the process-response mechanisms in the beach and nearshore zone has been conducted by L. D. Wright and his colleagues at Sydney University in Australia (Wright *et al.*, 1979; Wright, 1982; Short and Hess, 1982). Working in an area of great variation in coastal morphology along with a variable wind-wave climate and persistent high-energy swell, they have monitored and characterized beach morphology as dissipative and reflective with several morphologic varieties in a sequence from one to the other (Figure 6-40).

Dissipative beach and nearshore profiles have low overall gradients with waves breaking up to a few hundred meters from shore. Topography tends to be complex with longshore, crescentic, or oblique bars and a generally concave upward profile (Wright *et al.*, 1979). Rips are common, providing a three-dimensional topography (Figure 6-40, 1–5). The most dissipative morphology occurs immediately after storms (Figure 6-40, 1). Slow shoreward migration of the bar during quiescent periods causes the bar to steepen and the trough to become narrower and deeper. This permits reformation of waves landward of the bar developing a resonant zone where edge waves form beach cusps (Figure 6-40, 2–4). Dissipative conditions still persist seaward of the bar. Reflective beach and nearshore systems have steep, rather smooth beach faces with surging breakers. Bars and rip currents are absent and the topography is two-dimensional (Figure 6-40, 6). Much wave energy is reflected from the beach, and edge waves cause cusps to be widespread (Wright *et al.*, 1982).

Sediment supply, grain size, and wave climate all contribute to beach morphology. Abundant, fine sediment tends to produce dissipative conditions whereas coarse sediment causes steep reflective beaches.

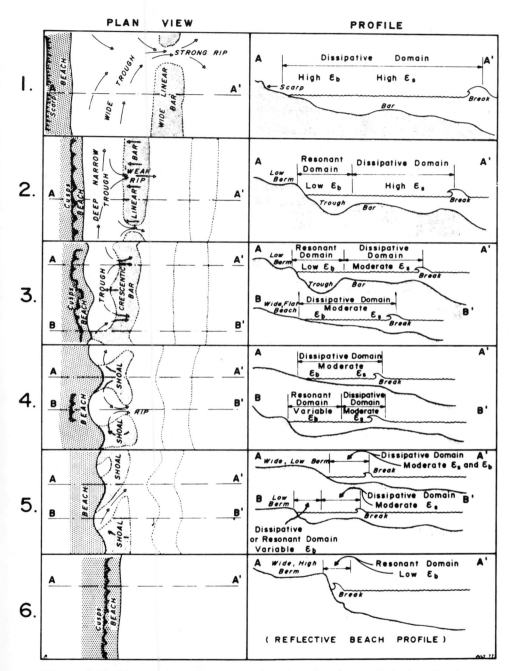

Figure 6-40. Map and profile views of various beach morphologies showing dissipative and reflective (resonant) zones (from Wright *et al.*, 1979, p. 122).

Stratigraphic Sequences

Preservation of beach and surf zone sediments in the stratigraphic record is typically confined to progradational conditions. Transgressive is not conducive to preservation in this zone; however, transgressive sequences do contain beach-related facies such as washovers, marshes, and flood tidal deltas (Kraft, 1971).

Progradational sequences develop according to Walther's Law and can be constructed using stratification and textural characteristics of the various subenvironments across the beach and nearshore zone such as are shown in Figure 6-41. Hunter *et al.* (1979) have presented examples of progradational stratigraphic sequences that were developed from their work along the Oregon coast. These serve as good examples for both barred and nonbarred coasts (Figures 6-41, 6-42).

In the nonbarred sequence that moves across the same sequence of environments and structures as shown in Figure 6-42, the base of the nearshore is characterized by a mixture of small-scale cross-strata and bioturbation. The overlying surf zone shows complex cross-strata that are formed by ripples, megaripples, and upper plane beds. The transition

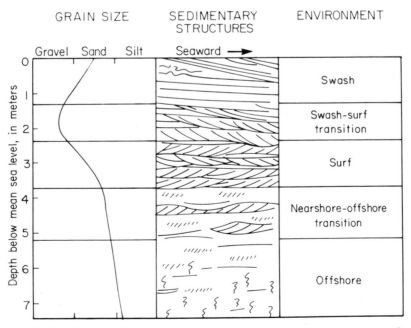

Figure 6-41. Theoretical stratigraphic sequence of sedimentary structures for a nonbarred coast (from Hunter *et al.*, 1979, p. 722).

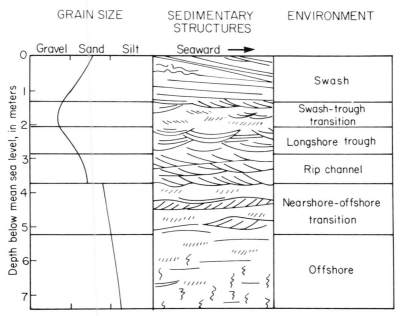

Figure 6-42. Theoretical stratigraphic sequence of sedimentary structures for a barred coast (from Hunter *et al.*, 1979, p. 721).

between surf and swash shows dominantly seaward-oriented cross-strata resulting from lunate megaripples. The entire sequence displays a fining-upward trend to this horizon but decreases up into the seaward-dipping swash or foreshore deposits (Figure 6-42a).

The presence of longshore bars complicates the sequence because of the three-dimensional topography. The basal portion of the section is similar to the nonbarred situation (cf. Figures 6-41 and 6-42). Rip channel facies with small to medium seaward-oriented cross-strata overlie the transition strata. The absence of the longshore bar facies is due to bar erosion or seaward migration as progradation takes place (Hunter *et al.*, 1979). Note also that there is a rather abrupt dislocation in the coarsening-upward trend at this horizon. Rip channels tend to contain coarse lag accumulations, thus showing a marked contrast with underlying transition beds. Longshore trough sediments show varied sizes and orientations of cross-strata because of the important component of shore-normal bedforms generated by longshore currents. The remaining overlying deposits are comparable to those in the nonbarred sequence (Figure 6-42).

These stratigraphic models represent rather simplified versions of beach–nearshore progradational sequences. Numerous bars, variations in wave approach and bar migration, sediment availability, and other variables can further complicate such sequences. The relative thickness of the various units may vary as well as the overall thickness. Rates of sea level changes as related to sediment supply can cause such variation.

References

Abele, R. W., 1973. *Short-Term Changes in Beach Morphology and Concurrent Dynamic Processes, Summer and Winter Periods, 1971–1972.* Plum Island, Massachusetts. Unpubl. M.A. Thesis, Univ. of Massachusetts, 166 pp.

Ball, M. M., 1967. Carbonate sand bodies of Florida and the Bahamas. *J. Sed. Petrol.*, **37**, 556–591.

Bascom, W., 1951. The relationships between sand size and beachface slope. *Amer. Geophys. Union Trans.*, **32**, 866–874.

Bascom, W., 1964. *Waves and Beaches.* Doubleday & Company Inc., New York, 267 pp.

Blar, T. A., and Fite, R. C., 1965. *Weather Elements*, 5th ed. Prentice-Hall, Inc., Englewood Cliffs, NJ, 364 pp.

Bluck, B. J., 1967. Sedimentation of beach gravels; examples from South Wales. *J. Sed. Petrol.*, **37**, 128–156.

Bowen, A. J., and Inman, D. L., 1969. Rip currents, 2: laboratory and field observations. *J. Geophys. Res.*, **74**, 5479–5490.

Brenninkmeyer, B., 1976. Sand fountains in the surf zone. *In*: Davis, R. A., and Ethington, R. L. (eds.), *Beach and Nearshore Sedimentation.* Soc. Econ. Paleont. Mineral. Spec. Publ. 24, pp. 69–91.

Caldwell, J. M., 1956. *Wave Action and Sand Movement Near Anaheim Bay, California.* Beach Erosion Board, U.S. Army Corps Engr. Tech. Memo. 68, 21 pp.

Chappell, J., 1967. Recognizing fossil strand lines from grain size analysis. *J. Sed. Petrol.*, **37**, 157–165.

Cherry, J., 1965. *Sand Movement along a Portion of the Northern California Coast.* Tech. Memo. No. 14, U.S. Army Corps of Engineers, Coastal Engineering Research Center, Washington, DC, 125 pp.

Clifton, H. E., 1969. Beach lamination—nature and origin. *Mar. Geol.*, **7**, 552–559.

Clifton, H. E., 1976. Wave-formed sedimentary structures: a conceptual model. *In*: Davis, Jr., R. A., and Ethington, R. L. (eds.). *Beach and Nearshore Sedimentation.* Soc. Econ. Paleont. Mineral. Spec. Publ. 24, Tulsa, OK, pp. 126–148.

Clifton, H. E., Hunter, R. E., and Phillips, R. L., 1971. Depositional structures and processes in the non-barred high-energy nearshore. *J. Sed. Petrol.*, **41**, 651–670.

Coleman, J. J., 1966. *Recent Coastal Sedimentation: Central Louisiana Coast.* Louisiana State Univ., Coastal Studies Institute, Tech. Rept. 29.

Cook, D. O., 1970. The occurrence and geologic work of rip currents off southern California. *Mar. Geol.*, **9**, 173–186.

Davidson-Arnott, R. G. D., and Greenwood, B., 1974. Bedforms and structures associated with bar and trough topography in the shallow-water wave environment, Kouchibouguac Bay, New Brunswick, Canada. *J. Sed. Petrol.*, **44**, 698–704.

Davidson-Arnott, R. G. D., and Greenwood, B., 1976. Facies relationships on a barred coast, Kouchibouguac Bay, New Brunswick, Canada. *In*: Davis, R. A. (ed.), *Beach and Nearshore Sedimentation*. Soc. Econ. Paleont. Mineral. Spec. Publ. 24, pp. 149–168.

Davies, J. L., 1959. Wave refraction and the evolution of shoreline curves. *Geogr. Studies*, **5**, 1–14.

Davies, J. L., 1964. A morphogenic approach to world shorelines. *Z. Geomorph.*, **8**, 27–42.

Davies, J. L., 1980. *Geographical Variation in Coastal Development*, 2nd ed. Longman, New York, 212 pp.

Davis, R. A., 1965a. Beach pitting; an unusual beach sand structure. *J. Sed. Petrol.*, **35**, 495–496.

Davis, R. A., 1965b. Underwater study of ripples, southeastern Lake Michigan. *J. Sed. Petrol.*, **35**, 857–866.

Davis, R. A., 1972. Beach changes on the central Texas coast associated with Hurricane Fern, September, 1971. *Contr. Mar. Sci.*, **16**, 89–98.

Davis, R. A., 1978. Sedimentology of Mustang and Padre Islands, Texas; a time-series approach. *J. Geol.*, **86**, 35–46.

Davis, R. A., and Ethington, R. L., 1976. *Beach and Nearshore Sedimentation*. Soc. Econ. Paleont. Mineral. Spec. Publ. 24, 187 pp.

Davis, R. A., and Fox, W. T., 1971. Beach and nearshore dynamics in eastern Lake Michigan. Kalamazoo: Tech. Rept. No. 4, ONR Contract 388-092, Western Michigan University, 145 pp.

Davis, R. A., and Fox, W. T., 1972a. Four-dimensional model for beaches and nearshore sedimentation. *J. Geol.*, **80**, 484–493.

Davis, R. A., and Fox, W. T., 1972b. Coastal processes and nearshore sand bars. *J. Sed. Petrol.*, **42**, 401–412.

Davis, R. A., and Fox, W. T., 1975. Process-response mechanisms in beach and nearshore sedimentation, I. Mustang Island, Texas. *J. Sed. Petrol.*, **45**, 852–865.

Davis, R. A., and Fox, W. T., 1978. Process-response mechanisms in beach and nearshore sedimentation, at Cedar Island, Virginia. *Southeastern Geol.*, **19**, 267–282.

Davis, R. A., and Fox, W. T., 1981. Interaction of beach and tide generated processes in a microtidal inlet. *Mar. Geol.*, **40**, 49–68.

Davis, R. A., Fox, W. T., Hayes, M. O., and Boothroyd, J. C., 1972. Comparison of ridge and runnel systems in tidal and non-tidal environments. *J. Sed. Petrol.*, **42**, 413–421.

de Beaumont, E. L., 1845. *Leçons de Geologie Pratique*. 7 me Leçon-Levees de Sables et Galets, P. Bertrand, Paris, 221–252.

Dobkins, J. E., and Folk, R. L., 1970. Shape development of Tahiti-Nui: *J. Sed. Petrol.*, **40**, 1167–1203.

Dolan, R., 1971. Coastal landforms: Crescentic and rhythmic. *Geol. Soc. Amer. Bull.*, **82**, 177–180.

Dolan, R., Hayden, B., Hornberger, G., Zieman, J., and Vincent, M., 1972. *Classification of the Coastal Environments of the World*, Part I, *The Americas*. Tech. Rept. No. 1, ONR Contract 389-159, Univ. of Virginia, Charlottesville, 163 pp.

Dolan, R., Vincent, L., and Hayden, B., 1974. Crescentic coastal landforms. *Z. Geomorph.*, **18**, 1–12.

Dolan, R., Hayden, B. O., and Heywood, J. E., 1977. Shoreline configuration and dynamics; a mesoscale analysis. *In*: Walker, H. J. (ed.), *Geoscience and Man*, Vol. 18, pp. 53–59.

Duncan, J. R., 1964. The effects of water table and tide cycle on swash-backwash sediment distribution and beach profile development. *Mar. Geol.*, **2**, 186–197.

Emery, K. O., 1945. Entrapment of air in beach sand. *J. Sed. Petrol.*, **15**, 39–49.

Evans, O. F., 1939. Mass transport of sediments on subaqueous terraces. *J. Geol.*, **47**, 324–334.

Evans, O. F., 1940. The low and ball of the east shore of Lake Michigan. *J. Geol.*, **48**, 476–511.

Fisher, W. L., McGowen, J. H., Brown, L. F., and Groat, C. G., 1972. *Environmental Geological Atlas of the Texas Coastal Zone—Galveston-Houston Area*. Univ. of Texas, Bur. Econ. Geol., 91 pp. + maps.

Folk, R. L., and Robles, 1964. Carbonate sands of Isla Rerez; Alacran reef complex, Yucatan. *J. Geol.*, **72**, 255–292.

Fox, W. T., and Davis, R. A., 1973. Simulation model for storm cycles and beach erosion on Lake Michigan. *Geol. Soc. Amer. Bull.*, **84**, 1769–1790.

Fox, W. T., and Davis, R. A., 1978. Seasonal variation in beach erosion and sedimentation on the Oregon coast. *Geol. Soc. Amer. Bull.*, **89**, 1541–1549.

Fox, W. T., and Davis, R. A., 1979. Surf zone dynamics during upwelling on the Oregon coast. *Est. Coastal Mar. Sci.*, **9**, 683–697.

Friedman, G. M., 1961. Distinction between dune, beach and river sands from their textural characteristics. *J. Sed. Petrol.*, **31**, 514–529.

Galvin, C. J., 1967. Longshore current velocity: A reviw of theory and data. *Rev. Geophys.*, **5**, 287–304.

Glaeser, J. D., 1978. Global distribution of barrier islands in terms of tectonic setting. *J. Geol.*, **86**, 283–297.

Goldsmith, V., Bowman, D., and Kiley, K., 1982. Sequential stage development of crescentic bars: Hahoterim Beach, southeastern Mediterranean. *J. Sed. Petrol.*, **52**, 233–249.

Grant, U. S., 1948. Influence of the water table on beach aggradation and deggradation. *J. Mar. Res.*, **7**, 655–660.

Greenwood, B. and Davis, R. A. (eds.), 1984. *Hydrodynamics and Sedimentation in Wave-Dominated Coastal Environments*. Developments in Sedimentology. v. 39, Elsevier, New York, 473 pp.

Hails, J. R., and Carr, A., 1975. *Nearshore Sediment Dynamics*. John Wiley and Sons, London, 316 pp.

Harms, J. C., 1969. Hydraulic significance of some sand ripples. *Geol. Soc. Amer. Bull.*, **80**, 363–396.

Harrison, W., 1969. Empirical equations for foreshore changes over a tidal cycle. *Mar. Geol.*, **7**, 529–551.

Harrison, W., Boon, J. D., Fang, C. S., Fausak, L. E., and Wang, S. N., 1971. *Investigation of the Water Table in a Tidal Beach*. Spec. Scientific Rept. No. 60, Virginia Institute of Marine Science, Gloucester Point, VA, 163 pp.

Hayes, M. O., 1967. *Hurricanes as Geological Agents: Case Studies of Hurricanes Carla, 1961, and Cindy, 1963*. Univ. of Texas, Bur. Econ. Geol., Rept. Inv. 61, 56 pp.

Hayes, M. O., 1972. *Forms of Sediment Accumulation in the Beach Zone*. Tech. Rept. No. 2–CRC, Coastal Research Center, Univ. of Massachusetts, Amherst, 58 pp.

Hayes, M. O., 1975. Morphology and sand accumulations in estuaries, *In*: Cronin, L. E. (ed.), *Estuarine Research*, Vol. 2, *Geology and Engineering*. Academic Press, New York, pp. 3–22.

Hayes, M. O., 1979. Barrier island morphology as a function of tidal and wave regime. *In*: Leatherman, S. P. (ed.), *Barrier Islands*, Academic Press, New York, pp. 1–29.

Hayes, M. O., 1969. Coastal Environments, Northeast Massachusetts and New Hampshire. Tech. Rept. No. 1–CRC, Coastal Research Center, Univ. of Massachusetts, Amherst, 469 pp.

Hill, G. W., and Hunter, R. E., 1973. Burrows of the ghost crab, *Ocypode quadrata* (Fabricius) on the barrier islands, south-central Texas. *J. Sed. Petrol.*, **43**, 24–30.

Hill, G. W., and Hunter, R. E., 1976. Interaction of biological and geological processes in the beach and nearshore, northern Padre Island, Texas. *In*: Davis, R. A. and Ethington, R. L. (eds.), *Beach and Nearshore Sedimentation*. Soc. Econ. Paleont. Mineral. Spec. Publ. 24, Tulsa, OK, pp. 169–187.

Hom-ma, M., and Sonu, C. J., 1963. *Rhythmic Pattern of Longshore Bars Related to Sediment Characteristics*. Proc. 8th Conf. on Coastal Engr., pp. 248–278.

Hoskin, C. M., and Nelson, R. V., 1971. Size modes in biogenic carbonate sediment, southeastern Alaska. *J. Sed. Petrol.*, **41**, 1026–1037.

Hsu, K. J., 1960. Texture and mineralogy of the recent sands of the Gulf Coast. *J. Sed. Petrol.*, **30**, 380–403.

Hulsey, J. D., 1962. *Beach Sediments of Eastern Lake Michigan*. Unpubl. Ph.D. Dissert., Univ. of Illinois, Urbana, 155 pp.

Hunter, R. E., Clifton, H. E., and Phillips, R. L., 1979. Depositional processes, sedimentary structures and predicted vertical sequences in barred nearshore systems, southern Oregon coast. *J. Sed. Petrol.*, **49**, 711–726.

Ingle, J. C., 1966. *The Movement of Beach Sand*. Elsevier, New York, 221 pp.

Inman, D. L., and Nordstrom, C. E., 1971. On the tectonic and morphologic classification of coasts. *J. Geol.*, **79**, 1–21.

Johnson, J. E., 1949. *Sand Transport by Littoral Currents*. Proc. 5th Hydraulics Conf. Bull. 34, State Univ. of Iowa Studies in Engineering, pp. 89–109.

Judge, C. W., 1970. *Heavy Minerals in Beach and Stream Sediments as Indicators of Shore Processes between Monterey and Los Angeles, California*. Tech. Memo. No. 33, U.S. Army Corps of Engineers, Coastal Engineering Research Center, Washington, DC, 44 pp.

King, C. A. M., 1972. *Beaches and Coasts*. Edward Arnold, London, 573 pp.

King, C. A. M., and Williams, W. W., 1949. The formation and movement of sand bars by wave action. *Geogr. J.*, **113**, 70–85.

Komar, P. D., 1971a. Nearshore cell circulation and the formation of giant cusps. *Geol. Soc. Amer. Bull.*, **82**, 2643–2650.

Komar, P. D., 1971b. The mechanics of sand transport on beaches. *J Geophys. Res.*, **76**, 713–721.

Komar, P. D., 1975. Nearshore currents: generation by obliquely incident waves and longshore variation in breaker height. *In*: Hails, J. R., and Carr, A. (eds.), *Nearshore Sediment Dynamics*. John Wiley and Sons, London, pp. 17–45.

Komar, P. D., 1976. *Beach Processes and Sedimentation*. Prentice-Hall, Englewood Cliffs, NJ, 429 pp.

Komar, P. D., and Inman, D. L., 1970. Longshore sand transport on beaches. *J. Geophys. Res.*, **75**, 5914–5927.

Kraft, J. C., 1971. Sedimentary facies patterns and geologic history of a Holocene marine transgression. *Geol. Soc. Amer. Bull.*, **82**, 2131–2158.

Leatherman, S. P., 1979. *Barrier Islands, from the Gulf of St. Lawrence to the Gulf of Mexico*. Academic Press, New York, 325 pp.

Leonard, J. E., and Cameron, B. W., 1981. Origin of high-latitude carbonate beach: Mt. Desert Island, Maine. *Northeastern Geol.*, **3**, 178–183.

Lewis, W. V., 1938. The evolution of shoreline curves. *Proc. Geol. Assoc.*, **49**, 107–127.

Longuet-Higgins, M. S., 1972. Recent progress in the study of longshore currents. *In*: Meyer, R. E. (ed.), *Waves on Beaches and Resulting Sediment Transport*. Academic Press, New York, pp. 203–208.

Mason, C. C., and Folk, R. L., 1958. Differentiation of beach dune and eolian flat environments by size analysis, Mustang Island, Texas. *J. Sed. Petrol.*, **28**, 211–226.

McArthur, D. S., 1969. *Sand Movement in Relation to Beach Topography*. Tech. Rept. No. 67, Coastal Studies Inst., Louisiana State Univ., Baton Rouge, 26 pp.

McKenzie, P., 1958. Rip current systems. *J. Geol.*, **66**, 103–113.

Morgan, J. P., Coleman, J. M., and Gagliano, S. W., 1963. *Mudlumps at the Mouth of South Pass, Mississippi River; Sedimentology, Paleontology, Structures, Origin and Relation to Deltaic Processes*. Baton Rouge: Louisiana State University Studies, Coastal Studies Series, No. 10.

Niedoroda, A. W., and Tanner, W. F., 1970. Preliminary study of transverse bars. *Mar. Geol.*, **9**, 41–62.

Otto, T., 1911–1912. Der darss und zingst. *Jahrbuch, Geogr. Gesell. (Greifswald)*, **13**, 235–485.

Owens, E. H., 1977. Temporal variations in beach and nearshore dynamics. *J. Sed. Petrol.*, **47**, 168–190.

Owens, E. H., and Frobel, D. H., 1977. Ridge and runnel systems in the Magdalen Islands, Quebec. *J. Sed. Petrol.*, **47**, 191–198.

Ramsey, M. D., and Galvin, C. J., 1971. *Size Analysis of Sand Samples from Three Southern New Jersey Beaches*. Unpubl. Rept., Coastal Engineering Research Center, Washington, DC, 50 pp.

Rector, R. L., 1954. *Laboratory Study of Equilibrium Profiles of Beaches*. Tech. Memo. No. 41, U.S. Army Corps of Engineers, Beach Erosion Board, 38 pp.

Rosen, P., 1979. Aeolian dynamics of a barrier island system. *In*: Leatherman, S. P. (ed.), *Barrier Islands*, Academic Press, New York, pp. 81–98.

Russell, R. J., 1968. *Glossary of Terms Used in Fluvial, Deltaic, and Coastal Morphology and Processes*. Baton Rouge: Coastal Studies Institute, Louisiana State Univ., Tech. Rept. No. 63, Contract 388-002, 97 pp.

Schiffman, A., 1965. Energy measurement in the swash-surf zone. *Limnol. Oceanogr.*, **10**, 255–260.

Schou, A., 1945. Det Marine foreland. *Folia Geogr. Danica*, **4**, 1–236.

Schwatz, M. L., 1967. Littoral zone tidal-cycle sedimentation. *J. Sed. Petrol.*, **37**, 677–683.

Shepard, F. P., 1950. *Longshore Bars and Longshore Troughs*. Tech. Memo. 41, Beach Eros. Board, U.S. Army Corps of Engineers.

Shepard, F. P., 1952. Revised nomenclature for depositional coastal features. *Amer. Assoc. Petrol. Geol. Bull.*, **36**, 1902–1912.

Shepard, F. P., 1973. *Submarine Geology*, 3rd ed. Harper & Row, New York.

Shepard, F. P., and Inman, D. L., 1950. Nearshore water circulation related to bottom topography and wave refraction. *Amer. Geophys. Union Trans.*, **31**, 196–212.

Shepard, F. P., Emery, K. O., and LaFond, E. C., 1941. Rip currents: a process of geological importance. *J. Geol.*, **49**, 337–369.

Short, A. D., 1975. Offshore bars along the Alaskan Arctic coast. *J. Geol.*, **83**, 209–221.

Short, A. D., 1979. Three dimensional beach stage model. *J. Geol.*, **87**, 553–571.

Short, A. D., and Hess, P. A., 1982. Wave, beach and dune interactions in southeastern Australia. *Mar. Geol.*, **48**, 259–284.

Sonu, C. J., 1969. *Collective Movement of Sediment in Littoral Environment*. Proc. 11th Conf. on Coastal Engr., London, pp. 373–400.

Sonu, C. J., 1973. Three-dimensional beach changes. *J. Geol.*, **81**, 42–64.

Sonu, C. J., and Russell, R. J., 1965. *Topographic Changes in the Surf Zone Profile*. Proc. 10th Conf. on Coastal Engr., Tokyo, pp. 502–524.

Sonu, C. J., McCloy, J. M., and McArthur, D. S., 1965. *Longshore Currents and Nearshore Topographies*. Proc. 10th Conf. on Coastal Engr., Tokyo, pp. 502–524.

Strahler, A. N., 1966. *Tidal Cycle of Changes in an Equilibrium Beach, Sandy Hook, New Jersey. J. Geol.*, **74**, 247–268.

Svasek, J. N., and Terwindt, H. J. T., 1974. Measurements of sand transport by wind on a natural beach. *Sedimentology*, **21**, 311–322.

Tanner, W. F., 1960. Expanding shoals in areas of wave refraction. *Science*, **132**(3433), 1012–1013.

Tanner, W. F., 1967. Ripple indices and their uses. *Sedimentology*, **9**, 89–104.

Timson, B. S., 1969. Post-storm profiles and particle-size changes of a New Hampshire gravel beach: a preliminary report. *In*: Hayes, M. O. (ed.), *Coastal Environments, Northeastern Massachusetts and New Hampshire*. Coastal Research Group, Univ. of Massachusetts, Amherst, Field Trip Guidebook, pp. 391–402.

U.S. Army Corps of Engineers (CERC), 1973. *Shore Protection Manual*. Coastal Engineering Research Center, Ft. Belvoir, VA, 3 Vols.

van Andel, T. H., and Poole, D. H., 1960. Sources of Holocene sediments in the northern Gulf of Mexico. *J. Sed. Petrol.*, **30**, 91–122.

Vause, J. E., 1959. Underwater geology and analysis of Recent sediments in the northern Gulf of Mexico. *J. Sed. Petrol.*, **30**, 91–122.

Visher, G. S., 1969. Grain size distributions and depositional processes. *J. Sed. Petrol.*, **39**, 1074–1106.

Waddell, E., 1973. *Dynamics of Swash and Implication to Beach Responses*. Coastal Studies Institute, Louisiana State University, Baton Rouge, Tech. Rept. No. 139, 49 pp.

Waddell, E., 1976. Swash–groundwater–beach profile interactions. *In*: Davis, R. A. and Ethington, R. L. (eds.), *Beach and Nearshore Sedimentation*. Soc. Econ. Paleont. Mineral. Spec. Publ. 24, Tulsa, OK, pp. 115–125.

Watson, R. L., 1971. Origin of shell beaches, Padre Island, Texas. *J. Sed. Petrol.*, **41**, 1105–1111.

Watts, G. M., 1953. *A Study of Sand Movement at South Lake Worth Inlet, Florida*. Tech. Memo. No. 42, U.S. Army Beach Erosion Board, Washington, DC, 62 pp.

Weigel, R. L., 1964. *Oceanographical Engineering*. Prentice-Hall, Englewood Cliffs, NJ, 532 pp.

Wells, J. T., and Coleman, J. M., 1981. Physical processes and fine-grained sediment dynamics, coast of Surinam, South America. *J. Sed. Petrol.*, **51**, 1053–1068.

Wright, L. D., 1982. Field observations of long-period, surf-zone standing waves in relation to contrasting beach morphologies. *Australian J. Mar. Freshwater Res.*, **33**, 181–201.

Wright, L. D., Chappell, J., Thom, B. G., Bradshaw, M. P., and Cowell, P., 1979. Morphodynamics of reflective and dissipative beach and inshore systems; southeastern Australia. *Mar. Geol.*, **32**, 105–140.

Wright, L. D., Guza, R. T., and Short, A. D., 1982. Dynamics of a high-energy dissipative surf zone. *Mar. Zeol.*, **45**, 41–62.

7

Tidal Inlets and Tidal Deltas

Jon C. Boothroyd

Introduction

Physical Processes and Parameters

This introductory section will serve to set the stage for later, more detailed investigations into the bedform patterns, sedimentary structures, detailed morphology, and stratigraphic sequences of tidal inlets and tidal deltas. Davies (1964) classifies shorelines on the basis of tidal range as follows: (1) microtidal, tidal range < 2 m; (2) mesotidal, 2–4 m; and (3) macrotidal, tidal range > 4 m. Hayes (1975), in a study of coastal charts of the world, found variations in morphology of depositional shorelines that could be related to variations in tidal range. His results are shown in Figure 7-1, a plot of tidal range versus the relative abundance of depositional coastal features.

Tides may be diurnal (one high and one low per day), semidiurnal (two highs and two lows per day), or mixed (a combination of diurnal and semidiurnal). Semidiurnal or mixed predominate in most areas (Davies, 1980). For instance, the east coast of the United States is characterized by semidiurnal tides, with a slight trend toward diurnal inequality (successive highs and lows of unequal magnitude). Of great importance to the study of tidal inlets is the amplitude difference between spring (full and new moon) and neap (first- and third-quarter moon) tides. Increased amplitude during times of spring tide leads to an increased tidal prism (the difference between estuarine volume at high tide and that at low tide), and this in turn leads to increased tidal current velocities because a greater volume of water must move in and out of the inlet within the same time span.

Jon C. Boothroyd

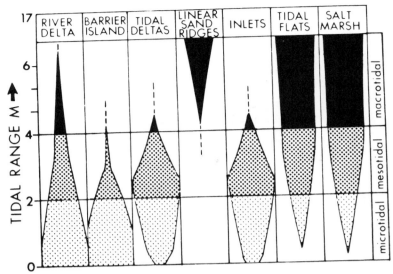

Figure 7-1. Variation in morphology of depositional shorelines with respect to variations in tidal range (from Hayes, 1975).

The other element of reservoir energy, waves, is generally considered to be the dominant process on microtidal coasts, whereas tidal-current energy dominates on macrotidal coasts. However, a combination of wave and tidal-current energy results in mixed-energy shorelines that are not strictly a product of either process. Hayes (1979) compared mean wave height versus mean tidal range of selected coastal-plain shorelines to develop the classification scheme shown in Figure 7-2. He further subdivided Davies' tidal classification as follows: microtidal, 0–1 m; low mesotidal, 1–2 m; high mesotidal, 2–3.5 m; low macrotidal, 3.5–5 m; macrotidal, > 5 m.

This discussion is limited to the inlets themselves plus the related and adjacent sediment bodies (Lauff, 1967; Schubel, 1971). The terminology followed is that developed by Hayes (1969, 1975, 1979), Hayes and Kana (1976), and a multitude of co-workers in the Coastal Research Group at the University of Massachusetts and the Coastal Research Division at the University of South Carolina. Other terms (e.g., Oertel, 1972, 1973, 1975; Oertel and Howard, 1972) have also been used to describe the same features.

Geographic Distribution

Tidal inlets and tidal deltas have the potential to develop on any depositional shoreline where sediment supply is adequate and where the antecedent topography and sea level fluctuations (or lack thereof) allow the correct geometry to develop. Figure 7-3 illustrates the distribution of tidal range world-wide, according to Davies (1980). The most studied examples of mesotidal inlets are those of (1) the Dutch Wadden Sea, actually a large lagoon behind the North Sea barrier islands; (2) the east coast of the United States including Maine, Massachusetts, southern New Jersey, and South Carolina; and (3) to a lesser extent, inlets on the northeast Gulf of Alaska and the northwestern contiguous United States. Locations of studied microtidal inlets on the east and Gulf coasts of the United States are: Rhode Island; Long Island, New York; northern New Jersey; North Carolina; Texas; and Florida. The reader is invited to inspect the large-scale coastal charts of the eastern United States for detailed examples. The National Ocean Survey (formerly U.S. Coast and Geodetic Survey) "brown and green" charts of 1:20,000–1:80,000 scale should be used for best results. Some good examples are charts numbered 13282, 13279 (northern Massachusetts), 13248 (Cape Cod, Massachusetts) 12318 (southern New Jersey), 11522 (southern South Carolina), and 11511 (northern Georgia).

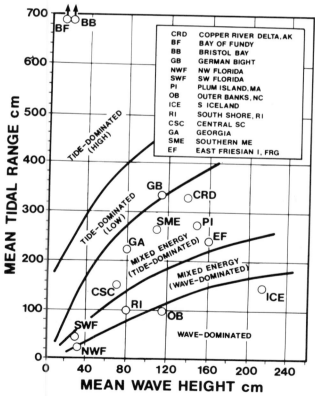

Figure 7-2. Shoreline classification of Hayes (1979) and Nummedal and Fischer (1978) based on wave and tidal-current energy. Additional data points from FitzGerald (personal communication).

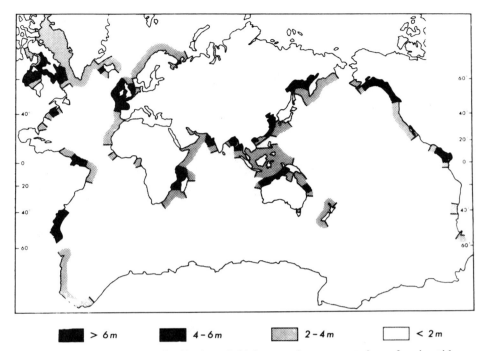

Figure 7-3. World-wide distribution of tidal range, in meters at time of spring tide (from Davies, 1980).

General Morphology

Tidal inlets are characterized by large sand bodies deposited and molded by tidal currents and by waves. The principal sand deposits are tidal deltas that form adjacent to the tidal inlets. The ebb-tidal delta is a sand accumulation seaward of the inlet throat, formed primarily by ebb-tidal currents but modified by wave action. The flood-tidal delta is an accumulation of sand landward of the inlet throat, shaped chiefly by flood-tidal currents. Further away from the inlet throat, meandering tidal channels with sandy point bars predominate. Small tributary channels and tidal flats distal to the inlet are sites of mud deposition and the whole sequence is capped by marsh deposits. A mesotidal estuary model is shown in Figure 7-4 and the estuary from which it is derived, Essex Bay, Massachusetts, is illustrated in Figure 7-5. The microtidal inlet model as developed by Hayes (1979) is generally more wave dominated than mesotidal inlets and commonly possesses a proportionally smaller ebb-tidal delta.

Of particular importance in both models are the deposits around tidal inlets. The tidal inlet portion of the models may be applied to inlets along the east and Gulf coasts of the United States, northeast Gulf of Alaska coast (in part), and the Wadden Sea, Netherlands, among others.

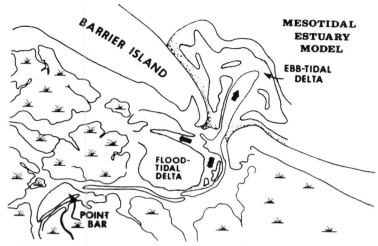

Figure 7-4. Model of a mesotidal lagoonal estuary. The flood- and ebb-tidal delta sandy deposits are particularly important (from Hayes, 1975).

Figure 7-5. Two mesotidal inlets and adjacent barrier islands on Ipswich Bay in northeastern Massachusetts. Essex Bay is in the foreground and the Parker River in the background. All photographs were taken at low tide. The mesotidal model (Figure 7-4) is based on Essex Bay.

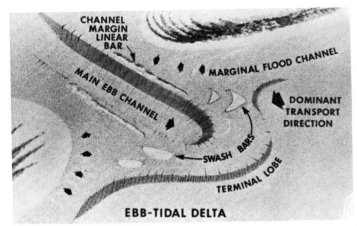

Figure 7-6. Morphologic model of ebb-tidal deltas. The arrows indicate dominant direction of the tidal-current flow (from Hayes, 1980).

Ebb-Tidal Deltas

A standard model of ebb-tidal delta morphology is shown in Figures 7-6 and 7-7. The main components of the model are (1) a main ebb channel, dominated by ebb-tidal currents; (2) channel-margin linear bars, leveelike features that flank the main ebb chanel and are built by the interaction of the ebb-tidal current and waves; (3) the terminal lobe, located at the distal end of the main ebb channel where the ebb jet diminishes to allow substantial deposition; (4) swash platforms, sand sheets deposited mainly by wave action between the main ebb channel and adjacent barrier islands; (5) swash bars that form and migrate across the swash platform by the action of wave-generated currents from breaking waves (King, 1972); and (6) marginal flood channels, which may occur between the swash platforms and both the updrift and downdrift barriers and are dominated by flood-tidal currents.

Flood-Tidal Deltas

A standard model for flood-tidal deltas formed in mesotidal inlets is illustrated in Figure 7-8 (Hayes, 1980) and by a photograph of Essex Bay (Figure 7-9). The major components are (1) flood ramps, the seaward-dripping surface dominated by flood-tidal currents with major sediment movement accomplished by the migration of sand waves landward, up the ramp; (2) flood channels, subtidal continuations of the flood ramp that separate the main intertidal flats of the delta and are dominated by flood-tidal currents; (3) an ebb shield, the topographically high, landward margin of the tidal delta that protects portions from modification by ebb-tidal currents; (4)

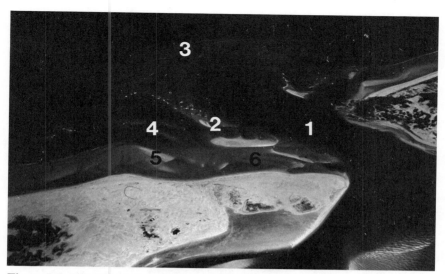

Figure 7-7. Ebb-tidal delta of Essex Bay in northeastern Massachusetts. The view is seaward with Castle Neck barrier island on the left. The numbered components are: (1) main ebb channel; (2) channel-margin linear bar; (3) terminal lobe; (4) swash platform; (5) swash bar; and (6) marginal flood channel.

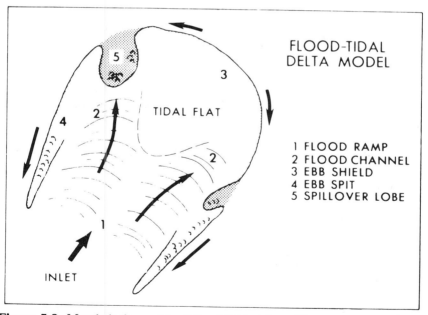

Figure 7-8. Morphologic model of flood-tidal deltas. The arrows indicate the dominant direction of tidal-current flow (from Hayes, 1980).

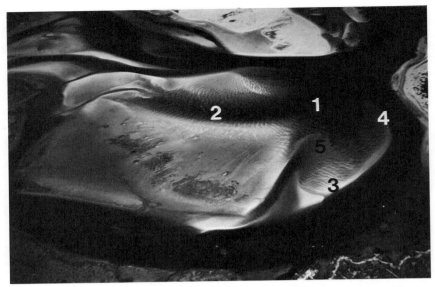

Figure 7-9. Flood-tidal delta in Essex Bay, a lagoonal estuary in northeastern Massachusetts. The view is seaward with Castle Neck barrier island at the top of the photograph. The numbered components are: (1) flood ramp; (2) flood channel; (3) ebb shield; (4) ebb spit; and (5) spillover lobe.

ebb spits, formed chiefly by ebb-tidal currents with some interaction by flood currents; and (5) spillover lobes, linguoid barlike features formed mainly as a result of ebb-tidal current flow over low areas of the ebb shield. Although these tidal–delta, tidal–inlet models were derived from mesotidal mixed energy conditions, later studies by Davis and Fox (1981), Gallivan and Davis (1981), Nummedal *et al.* (1977), and Hubbard *et al.* (1979) indicate that the morphologic models may be applied as well to more wave-dominated microtidal inlets.

This brief description of tidal delta morphology serves to set the stage for a discussion of tidal current flow and of bedform and sandbody genesis in inlets and adjacent estuaries. Tidal delta sediment patterns and variations are discussed in more detail in a later section. First, however, a brief discussion of sediments is followed by a discussion of bedforms in general. This will serve as an introduction to the special case of estuarine bedforms.

Sediment Distribution

The most comprehensive studies of sediment types and sediment-size distribution in small estuaries of the United States have been carried out in northern New England (DaBoll, 1969; Farrell, 1970; Hartwell, 1970; Hubbard, 1973) and Georgia (Greer, 1975; Oertel, 1972, 1975). These studies follow on the pioneering work of Van Straaten (1950, 1951, 1952, 1954, 1956, 1961), Van Straaten and Kuenen (1957), and Postma (1954, 1961, 1967) in the Dutch Wadden Sea which concentrated mainly on fine-grained tidal flats, with less work done on tidal channels and sand bodies.

Lagoonal estuaries can be divided into a number of subenvironments. Proceeding landward from the ebb-tidal delta they are: (1) major subtidal channels; (2) major intertidal flats, including tidal deltas; (3) large tidal creeks (for major secondary tidal channels); and (4) small tidal creeks (or minor secondary tidal channels). Generally, sediment size decreases up the estuary proceeding from the major channels to large creeks to small tidal creeks; sediment size also becomes finer with increasing elevation on the intertidal flats.

Informaton on mean size, sorting, and skewness of sediments from the various subenvironments of the Parker River estuary are shown in Figure 7-10 (DaBoll, 1969). These data show that the main channels (flood and ebb

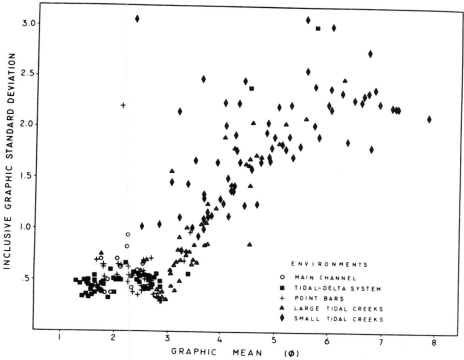

Figure 7-10. Scatter plot of Folk graphic mean versus sorting (inclusive graphic standard deviation) of 225 surface sediment samples from the Parker River estuary in northeastern Massachusetts. Tidal-delta sediments are coarsest and best sorted, whereas small tidal creek sediments are the finest and poorly sorted (from DaBoll, 1969).

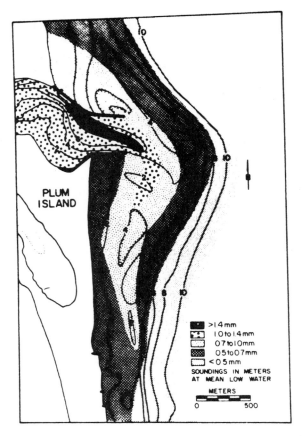

Figure 7-11. Mean grain-size distribution of surface sediments on, and adjacent to, the Merrimack ebb-tidal delta in northeastern Massachusetts. The coarsest material is in the main ebb channel and on the crest of the first swash bar immediately downdrift of the channel (from Hubbard, 1973).

dominated) and tidal deltas are primarily composed of medium to fine sand; the large tidal creeks consist of fine sand to coarse silt and the small tidal creeks are floored with fine silt and clay. Sediments of the tidal creeks become increasingly finer skewed as increasing amounts of silt and clay are deposited. The coarse-to-fine transition of mean size of sediments with increasing tidal flat elevation is generally adhered to but is subject to two important variations: (1) on flood-tidal deltas, the higher sandy portions (ebb shields) are usually coarser than the lower flood ramp areas (DaBoll, 1969); and (2) intertidal point bars in the tidal creeks are usually coarser (fine sand) than the adjacent muddy channels (DaBoll, 1969; Hartwell, 1970).

Very few sediment distribution maps of ebb-tidal deltas have been published. One is shown as Figure 7-11 (Hubbard, 1973); another (Oertel, 1972) is discussed in the section on tidal delta variation. Hubbard's map of the Merrimack River ebb-tidal delta illustrates that all of the delta is composed of coarse sand and that the coarsest sediments are found in the main ebb channel (deep subtidal) and on the surfaces of the swash bars on swash platforms.

Bedforms

Classifications

Even a casual observation of sandy tidal flats reveals that many surfaces are covered by wavy undulations or bedforms, usually called ripples. Ripples occur in a wide assortment of sizes and shapes and so it is convenient to devise some sort of classification scheme to characterize them. Such schemes are certainly not new; some earlier examples are those of Gilbert (1914), Simons and Richardson (1963), Allen (1968a, 1982d), Klein (1970), Boothroyd and Hubbard (1975), and Harms et al. (1975, 1982). The scheme presented below is descriptive but can be partially explained by variations in flow strength.

For the purpose of this discussion, bedforms will be classified on the basis of wavelength or spacing and not on the basis of amplitude or height. Figure 7-12 illustrates three basic scales of bedforms: ripples, megaripples, and sand waves. The smallest, ripples, range up to 60 cm spacing; megaripples, from 60 cm to 10 m; and sand waves, greater than 6 m spacing. This scheme generally follows that proposed by Hayes (1969) and is followed by Boothroyd and Hubbard (1975), Harms et al. (1975), and Dalrymple et al. (1978) (Table 7-1). Many geologists and engineers refer to megaripple-sized bedforms as dunes. This term is not used here because of the possible confusion with large-scale eolian dune forms (see Chapter 5). Others make no distinction between megaripples and sand waves, grouping both as large-scale ripples or dunes (e.g., Allen, 1968a, 1982d, Harms et al., 1982). A distinction is made here not only because there seems to be a distinct separation between the two of mean spacing or spacing "center of mass"; but also because different hydraulic conditions control the genesis of the two forms.

Plots of height versus spacing of individual bedforms are shown in Figure 7-13. These measurements were obtained on intertidal flats of flood- and ebb-tidal deltas and subtidal channels of two New England inlets. Ripples (not shown) range in spacing from 10 to about 60 cm, with a concentration between 20 and 40 cm. Heights vary between 1.5 and 8 cm, with a

(A)

(B)

Figure 7-12. (A) Megaripples on a channel-margin linear bar of the ebb-tidal delta in the Parker River estuary. Ripples are superimposed on the megaripple forms. Bedform spacing is about 2 m. (B) Flood-oriented low-energy sand waves on the flood-tidal delta of the Parker River estuary. Bedform spacing is about 14 m.

Table 7-1. Characteristics of Lower Flow Regime Bedforms (modified after Harms, *et al.*, 1975).

	Ripples	Megaripples	Low-energy sand waves	High-energy sand waves
Spacing	<60 cm	60 cm–10 m	> 6 m	> 10 m
Height/spacing ratio	variable	relatively large	relatively small	very small
Geometry	highly variable	sinuous to highly three-dimensional, prominent scour pits in troughs	straight to sinuous, uniform scour in troughs	straight to sinuous
Characteristic flow velocity	low (> 25–30 cm/sec < 40–50 cm/sec)	high (> 70–80 cm/sec, < 100–150 cm/sec)	moderate (> 30–40 cm/sec, < 70–80 cm/sec)	high (> 70–80 cm/sec, may be 150 cm/sec)
Velocity asymmetry	negligible to substantial	negligible to substantial	usually substantial	small to substantial

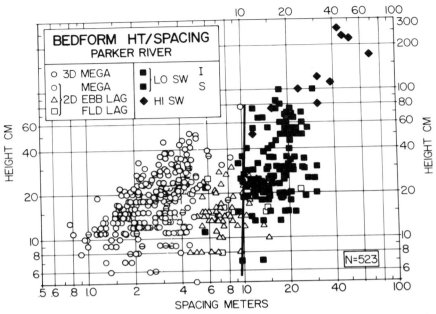

Figure 7-13. Height-spacing diagrams of megaripples and sand waves from the Parker River and Essex Bay estuaries in Massachusetts. The spacing "gap" present in the Parker River plot is not obvious in the plot of Essex Bay bedforms (from Boothroyd, unpublished).

concentration between 2 and 6 cm. Megaripples occur between spacing of 50 cm to 5 m, but the concentration is between 1 and 4 m, with considerable scatter. Heights range from 3 m to 50 cm, with most occurring between 8 and 30 cm. Sand waves, although defined as any bedform greater than 5 m, seem to fall in four categories that are discussed later. It suffices for now to say that the spacing range is from 5 m to 70 m with the concentration between 10 and 30 m. Heights vary from 6 cm to 2.5 m but are concentrated between 10 cm and 1 m. There is a spacing overlap between two-dimensional megaripples and low-energy sand waves, but the greatest concentration of megaripple spacings is of the three-dimensional forms of 1.5–4 m. Hydrodynamic distinctions are discussed in a later section. There is no distinct height difference between the two forms, particularly megaripples and intertidal sand waves. Subtidal sand waves are generally greater in height and spacing than are the intertidal variety, but the fields show considerable overlap (Figure 7-13).

It is therefore convenient to use a bedform classification system based on spacing, a parameter readily identifiable and measurable in the field. Although initially presented as a descriptive scheme, it can be demonstrated that bedform scale is also hydrodynamically controlled.

Flow Regimes

Observers of flow in natural channels and flumes from Gilbert (1914), through Simons and Richardson (1961), to the present have noted that sand- and silt-sized bed material is formed into a series of bedforms almost as soon as sediment begins to move along the bed. Flow strength may be defined as any variable that expresses the energy of the fluid that may be imparted to the bed. Two common expressions are flow velocity (u) and boundary shear stress (τ).

It is obvious that the ripples, megaripples, and sand waves discussed in the preceeding section are lower flow regime bedforms (Simons and Richardson, 1961). However, some important aspects of the flow must be kept in mind. Simons and Richardson's studies, in fact most flume experiments, have been carried out under conditions of steady and uniform unidirectional flow. That is, flow strength does not vary over time (steady); nor does it vary along the channel section from which measurements are obtained (uniform). In addition, the bed is allowed to attain equilibrium with the flow (no change in morphology), a process that ranges in duration from a few minutes to several days.

In the estuary, tidal current fluctuations lead to flow that is unsteady and nonuniform. Not only is flow nonuniform along the channel in the direction of sediment transport, but it also varies across the channel; succeeding sections of this chapter deal with this in detail. Also, of course, the tidal current reverses on either a semidiurnal or diurnal basis. There is some question, therefore, as to whether some bedforms are in equilibrium with the flow, particularly the large, subtidal sand waves.

Bedform Geometry and Flow Mechanisms

Ripples, megaripples, and sand waves all have a similar along-channel geometry or longitudinal section, differing only in overall scale and height-spacing ratios (ripple index). It is also useful to describe bedforms in map or plan view (Figure 7-14). The descriptive terms of Allen (1968a) are generally followed; the most important distinction is the grouping of lunate and linguoid forms under the general term of cuspate. The forms shown in the illustration are idealized; bedforms often exist in nature as complicated combinations of several forms.

The mechanics of water flow over ripples, megaripples, and sand waves shed much light on the migration of the bedforms and also on external morphology and internal structure. The external morphology can be a direct indicator of internal cross-stratification. External morphology can also be used as a predictor of the flow strength that has produced a given bedform.

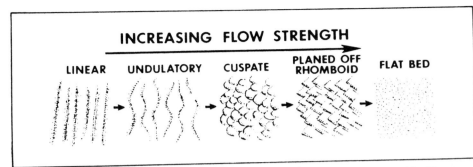

Figure 7-14. Change in shape from two-dimensional to three-dimensional bedforms with increasing flow strength (from Hayes and Kana, 1976).

A number of workers, among them Raudkivi (1963) and Southard and Dingler (1971), has shown that ripples propagate rapidly downstream from any irregularity of the bed because of flow separation over the original mound. Erosion or scour where the flow rejoins the bed generates a hole and another mound (ripple) further downstream. Jopling (1961, 1963, 1964, 1965a,b, 1966a,b, 1967) and Allen (1963, 1965, 1968a,b, 1982a,b) have extensively studied the nature of flow over megaripple-sized bedforms.

The flow proceeds across the stoss side of the bedform to the crest where it detaches or separates from the bed (Figure 7-15). In this complex flow separation zone there is an upper zone of no diffusion, or jet, where the fluid velocity is not affected by the turbulence below; a central, turbulent zone of mixing of water from the jet above with almost motionless water in the lee of the bedform below; and a lower zone of no flow or backflow, induced by separation turbulence above. The flow in the zone of mixing reattaches to the bed at a downstream distance of about six to eight times the height of the bedform. This reattachment zone may be subjected to intense turbulence and scour depending on the magnitude of the original flow velocity at the crest of the bedform.

When the flow separates from the crest of the bedform, the finer grains are carried downstream in the jet as suspended load, while the coarser grains avalanche down the lee side, or slipface, of the bedform. The larger grains in the suspended load settle through the zone of mixing to become deposited either in front of, or on, the slipface; the smaller grains are carried to the reattachment zone and thus to the next bedform further downstream. The collective avalanching and settling process results in downstream progradation of the slipface, while at the same time erosion at the point of flow reattachment removes material from the stoss side of the bedform. This combination of these two mechanisms results in the migration of a train of bedforms along a sandy bed.

Variation in the flow velocity will change the external form of the slipface as well as the shape of internal laminae (Figure 7-15). At low velocity, little avalanching occurs and no grains separate from the bed as suspended load. The slipface is a planar surface with an angular basal contact. As flow velocity increases, avalanching of bed material also increases; more importantly, however, grains swept into suspension settle on the lower portion and the toe of the slipface, resulting in a curved, concave-up form. Thus, internal lamination likewise varies in character with velocity increase.

In plan view the external form also changes with an increase in flow velocity. At low velocity the bedform crest of megaripples is straight to sinuous and scour pits formed by reattachment of flow are small and ill defined. As flow velocity increases, random variations in turbulence along the crest of the bedform cause short-lived variations in flow velocity near the

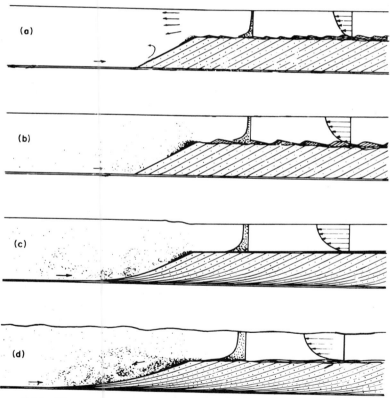

Figure 7-15. Change in bedform morphology and cross-stratification type in a laboratory megaripple-sized bedform with increasing flow velocity (from Jopling, 1965b).

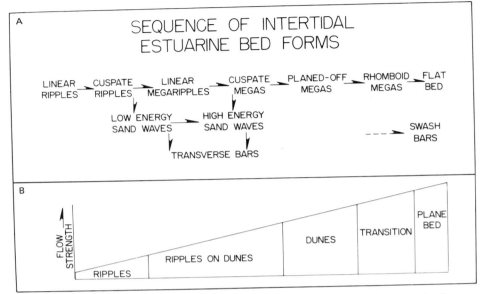

Figure 7-16. (A) Sequence of estuarine bedforms based on an increase in flow strength from left to right. Note that some bedforms are "stacked" with the one immediately above being the highest order form that may be superimposed on the lower. (B) Simon's and Richardson's bedform sequence arranged so that bedforms, stable at a given flow condition, are directly beneath those of part (a) that exist at similar conditions.

bed, which in turn results in a cuspate crestline. Intense erosion occurs at the zone of flow reattachment, resulting in deep scour pits in the megaripple troughs. Additional turbulence at the flow separation point (wake turbulence) is created by the unevenness of the bedform; the process is thus self-enhancing. Bedforms with a relatively straight crestline are sometimes termed *two dimensional*, whereas those with cuspate crestlines and scourpits that match a corresponding cusp are called *three dimensional*. These differences in morphology and internal structure are important because they allow deduction of the relative magnitude of paleoflow velocities.

An expanded classification scheme (Figure 7-16) depicts a bedform sequence that is based on increasing flow velocity, and Figure 7-17 illustrates individual bedforms within the sequence.

Lowest order forms, linear ripples, develop at a velocity just above the initiation of sediment motion and change rather rapidly to cuspate ripples. These ripples are termed low-energy and high-energy ripples, respectively, by Harms (1969). Under increasing flow velocity, rapid ripple migration leads to an amalgamation of ripple crests, an increase in slipface height, and the development of the next size class of bedforms, megaripples. The descriptive terms—linear, cuspate, planed-off, and rhomboid—characterize

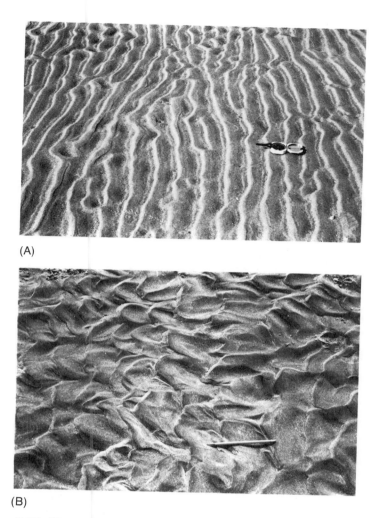

(A)

(B)

Figure 7-17. Illustrations of bedforms in the sequence of Figure 17. (A) Linear ripples formed by flood-tidal currents on the intertidal portion of a flood-tidal delta. Flow was from right to left; Brunton compass gives the scale. (B) Cuspate ripples formed by late-stage drainage of an intertidal flat. Flow was from right to left; ripple spacing averages about 10 cm. (C) Linear megaripples (ebb-oriented) on the Parker River flood-tidal delta. Flow was from left to right. (D) Cuspate megaripples (ebb-oriented) on the ebb shield of the Parker River flood-tidal delta. These bedfords are located about 20 m up-estuary from the linear megaripples of (C). Flow was from right to left; scale is 30 cm long. (E) Planed-off megaripples (ebb-oriented) on an ebb shield of the Parker River flood-tidal delta. Flow was from upper right to lower left; scale is 150 cm long. (F) Rhomboid megaripples on an intertidal flat. Flow was from left to right; scale is 30 cm long. (G) Plane bed formed by swash action. Note the grain lineation on the left. (H) Low-energy sand waves (flood-oriented) on an intertidal flood ramp of the Essex Bay flood-tidal delta. Flow was from upper right to lower left; spacing is 10–12 m. (I) Transverse bar on the intertidal flood ramp of the Parker River flood-tidal delta. Slipface is formed by the amalgamation of a number of sand waves. Flow is from right to left.

(C)

(D)

(E)   **Figure 7-17.** *(Continued.)*

(F)

(G)

(H)   **Figure 7-17.** *(Continued.)*

(l)   **Figure 7-17.** *(Continued.)*

morphologic changes in cross-section and plan view as flow velocity increases. Linear megaripples (2D) have a sinuous crestline, a planar slipface, and small scourpits in the trough of the bedforms. Cuspate megaripples (3D) exibit a highly irregular crestline that cannot be traced transverse to flow; a curved, spoon-shaped slipface, and well-developed scourpits. Slipface height is best developed in the cuspate forms. Planed-off megaripples usually have a diminished slipface height and a stoss side that is flat or plane. Rhomboid megaripples show almost no slipface development and are essentially in the plane bed flow regime. All of these aforementioned forms are well displayed in large flumes in the sequence shown in the top line of Figure 7-16 (Guy *et al.*, 1966, Bohacs, 1981; Costello and Southard, 1981). However, the last order of bedforms, sand waves, is too large to be adequately developed in the laboratory.

Sand waves are set apart (Figure 7-16) because of size considerations and also because they partially overlap the ripple and megaripple flow velocity fields. That is, both ripples and sand waves, or megaripples and sand waves, can exist together, with the smaller forms superimposed on the larger. Low-energy sand waves have superimposed ripples. Figure 7-18 illustrates some intertidal and shallow subtidal sand waves and some intertidal megaripples. There are two distinct spacing groups of sand waves illustrated, as well as the smaller megaripples. The reason for the spacing variation of the sand waves has to do with the asymmetry of flow velocity in the estuary. Low-energy sand waves with larger spacings are subjected to highly asymmetrical flow, whereas the sand waves with spacings between 6 and 10 m result from near symmetrical maximum flood and ebb velocities.

Figure 7-19 shows selected bottom profiles of various scales of bedforms (sand waves and megaripples), both intertidal and subtidal. The larger

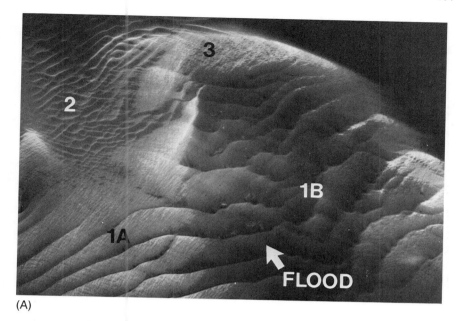

(A)

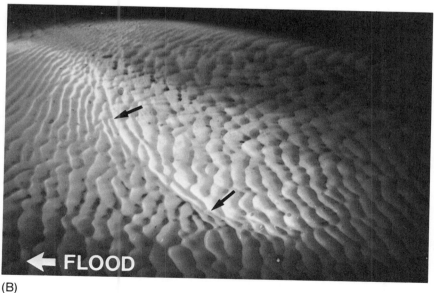

(B)

Figure 7-18. Aerial views of portions of the Parker River flood-tidal delta illustrating various scales of bedforms. (A) Bedforms are: (1A) flood-oriented sand waves, intertidal and subtidal, spacings are 12–18 m; (2) ebb-oriented sand waves; spacings are 6–10 m; (3) megaripples (ebb-oriented) spacings are 1.5–3 m. (B) Transverse bar (flood-oriented) on the Essex Bay flood-tidal delta. Arrows point to slipface. This is a nonrepetitive form.

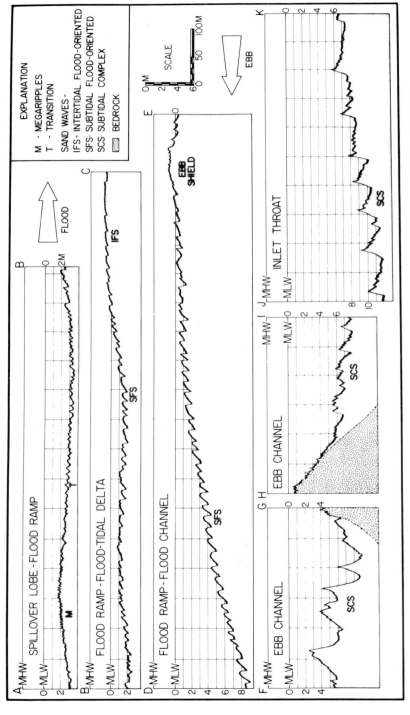

Figure 7-19. Bottom profiles of bedforms. Refer to the text for details (from Boothroyd and Hubbard, 1975).

subtidal sand waves (transect JK, Figure 7-19) are similar to high-energy intertidal sand waves on the Bay of Fundy tidal sand bars described by Klein (1970) and Dalrymple *et al.* (1975, 1978).

Transverse bars (Figure 7-18b) are large-scale forms, largely nonrepetitive, and are mantled by sand waves and/or megaripples. Most intertidal transverse bars migrate in a flood direction, although more slowly than sand waves. They possess an active slipface and perhaps they are formed from an amalgamation of sand waves. There is some justification for reclassifying some of the large subtidal sand waves as transverse bars, particularly line FG, Figure 7-19. The rationale is that the migration of these large forms is independent of the flow separation, reattachment process of the bedform immediately upcurrent.

Depth–Velocity Diagrams

Southard (1971; in Harms *et al.*, 1975, 1982; Costello and Southard, 1981) devised a method of illustrating empirical measurements of flow depth, flow velocity, and grain size that controls the development of bedforms in the flume. Boothroyd and Hubbard (1974, 1975) applied this technique to measurements of flow velocity and depth taken over various bedforms in the Parker River estuary, Massachusetts (Figure 7-20). Flow depths in the inlets are greater than those studied in flume runs, but the boundaries separating bedform types occur at similar flow velocities and the bedforms show greater sensitivity to changes in velocity than to changes in depth. Note that whereas linear megaripples and sand waves occupy the same basic stability field, it is true that as velocity increases, bedforms occur in the same general order as shown in the bedform sequence in Figure 7-16. This is not to say that there is a change of one bedform to another, but that each one exists for a given depth and velocity field. Mean grain size ranged from 0.38 to 0.44 mm for the depth–velocity curve shown in Figure 7-20.

Because bedforms are sensitive to flow velocity and somewhat independent of depth, bedform type is a powerful tool in estimating flow velocities in estuaries without actually making detailed measurements.

Internal Sedimentary Structures

Physical sedimentary structures are produced by traction transport of material along the bed and settling of material from suspension, and by mass movement and water-escape mechanisms during and after deposition. Biogenic sedimentary structures are produced by the movement and feeding of organisms. This discussion is mainly concerned with sedimentary structures resulting from the traction transport of sand-sized material and, to a much lesser extent, the deposition from suspension of silt.

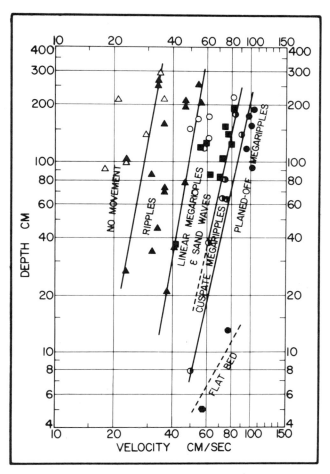

Figure 7-20. Depth-velocity diagram for tidal-current flow in a mesotidal estuary. General area of the current stations are shown in Figure 17-8(A) (from Boothroyd and Hubbard, 1975).

Many schemes have been developed to classify internal structures (McKee and Weir, 1953; Allen, 1963; Blatt *et al.*, 1980; Pettijohn *et al.*, 1973; Reineck and Singh, 1980; and Harms *et al.*, 1975, 1982). The general scheme of Harms *et al.* (1982) is followed, with some definitions from Blatt *et al.* (1980) and some other slight terminology changes. Stratification is used as a general term to describe layers of sediment that are termed beds if greater than 1 cm in thickness and laminae if less than 1 cm thick. Beds with internal structures (beds or laminae) inclined at an angle to bed boundaries (bedding planes) are said to exhibit cross-stratification or to be cross-stratified. A single layer of cross-strata is called a set.

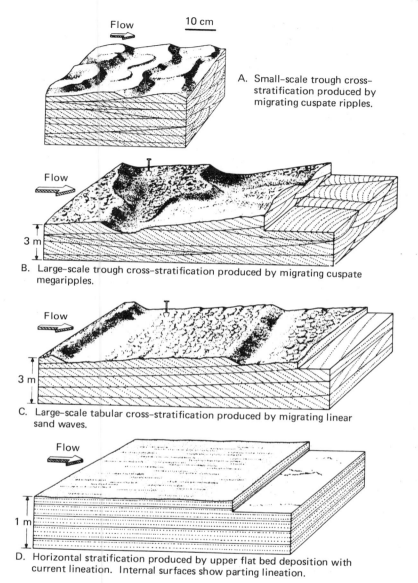

A. Small-scale trough cross-stratification produced by migrating cuspate ripples.

B. Large-scale trough cross-stratification produced by migrating cuspate megaripples.

C. Large-scale tabular cross-stratification produced by migrating linear sand waves.

D. Horizontal stratification produced by upper flat bed deposition with current lineation. Internal surfaces show parting lineation.

Figure 7-21. Internal sedimentary structures produced by the migration of various bedforms (from Harms *et al.*, 1975, as modified by Hayes and Kana, 1976).

Cross-stratification can also be classified by (1) the scale of sets; (2) the shape of individual cross-beds or cross-laminae; (3) the nature of the intersection of cross-stata with upper and lower bounding surfaces; and (4) the shape of the bounding surfaces themselves. It also makes a difference whether the internal structures are viewed in longitudinal section (parallel to flow direction), transverse section (perpendicular to flow direction), or horizontal section. Sets less than 5 cm thick are termed small scale; sets 5–50 cm are mesoscale; sets greater than 50 cm in thickness are large scale.

Tabular cross-stratification consists of planar, parallel laminae intersecting upper and lower bounding surfaces at high angles in longitudinal section, parallel to or intersecting bounding surfaces at low angles in transverse section, and straight or sinuous in horizontal section. Trough cross-stratification consists of curving, often nonparallel laminae intersecting the upper bounding surface at a high angle but becoming tangential to the lower bounding surface in longitudinal section. In transverse and horizontal section, the laminae are tangential and sometimes parallel to the set boundaries, which themselves are curved and nonparallel. Trough cross-stratification, to be correctly identified, must be viewed in all three sections. When seen in longitudinal section only, it may be confused with tabular cross-stratification. Often the terms planar-tangential or tangential are applied to curving cross-stratification seen in longitudinal section.

Mechanisms of Formation

The nature of stratification and cross-stratification is important because it can be directly linked to the bed geometry and the bedform can be linked to a certain flow regime or paleoflow intensity (flow strength). Thus by examining cross-stratification we can say something about the energy of the depositional environment that has formed the stratification.

Small-scale trough cross-stratification is produced by migrating cuspate ripples, whereas meso-scale trough cross-stratification is formed by the migration of cuspate (3D) megaripples (Figures 7-21 and 7-22). Small-scale planar cross-stratification results from the migration of linear ripples, whereas meso to large-scale tabular cross-stratification is produced both by linear (2D) megaripples and by sand waves. Plane lamination (called beds by Simons et al., 1965) is produced by upper flow regime plane bed deformation.

A change in stratification type from meso-scale tabular to trough with increase in flow strength is well illustrated by trenches cut in linear and cuspate megaripples (Figure 7-22 C–F). Also well shown by longitudinal cuts through megaripples is "herringbone" cross-stratification, an excellent indicator of reversing or bidirectional flow in estuaries (Figure 7-22G).

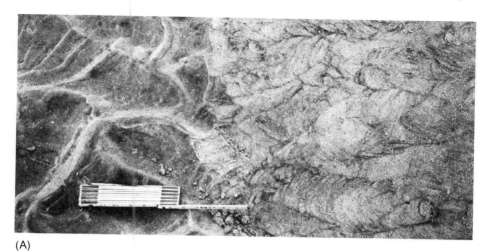

(A)

(B)

Figure 7-22. Cross-stratification produced by migration of estuarine bedforms. (A) Small-scale trough cross-stratification (horizontal surface) resulting from cuspate ripple migration. Flow was from left to right; scale is 30 cm long. (B) Meso-scale tabular cross-stratification produced by sand-wave migration. Flow was from left to right; scale is 30 cm long. (C) Tabular cross-stratification formed by linear (2D) megaripple migration. Longitudinal cut with 30 cm scale. (D) Horizontal surface cut across a linear (2D) megaripple exposing tabular cross-stratification. Scale is 30 cm long. (E) Tangential cross-stratification (longitudinal section) formed by cuspate megaripple migration; scale is 30 cm long. (F) Meso-scale trough cross-stratification exposed in longitudinal, transverse, and horizontal faces. Cuspate megaripple migration was from upper left to lower right; scale is 40 cm long. (G) "Herringbone" cross-stratification produced by alternate migration of flood- and ebb-oriented cuspate megaripples. Height of trench is 30 cm.

(C)

(D)

(E) **Figure 7-22.** *(Continued)*

(F)

(G)   **Figure 7-22.** *(Continued)*

If bedform type is combined with information obtained from depth–velocity diagrams and then this information is correlated with stratification type, it is not only possible to say something about the kind of bedform that has existed but it is also possible to estimate paleoflow velocities.

Genesis of Tidal Bedforms

Basic Flow Patterns

Patterns of flow in inlets and over tidal deltas and for that matter in most tidal-current situations follow two general principles: (1) horizontal segregation of flow into distinct flood and ebb channels, and (2) time-velocity asymmetry of tidal currents. Segregation of flow, as discussed by Robinson (1960), Price (1963), and many other workers, means that some tidal channels are dominated by flood-tidal current flow and others by ebb-tidal currents. It does not mean that any given channel carries only flood or ebb flow, but that one stage is dominant over the other, because it has either a higher maximum velocity or a longer duration of flow higher than a given velocity.

The concept of flood and ebb channels naturally leads to the principle of time-velocity asymmetry (Postma, 1961, 1967; Hayes, 1969). Essentially there are two components, time asymmetry and velocity asymmetry, and either one, neither, or both may occur at a given location in the estuary (Figure 7-23). Time asymmetry means that the maximum flow velocities occur not at midtide but at some other stage of the tidal cycle. Also, the duration or time span of the flood and ebb parts of the tidal cycle may be of unequal lengths. For instance, the maximum flood velocity may occur after the time of midflood in the marginal flood channel, whereas the maximum ebb velocity occurs near the time of low tide in the main ebb channel (Figure 7-23). The reasons for time asymmetry are not well understood but have to do with the geometry and topographic elevation of channels and flats within the estuary (Boon and Byrne, 1981; FitzGerald and Nummedal, 1983). Velocity asymmetry of tidal currents means that the maximum flow velocity or duration above a given velocity is greater for either the flood or the ebb state. The velocity curve for the marginal flood channel may be flood dominant and the curve for the main ebb channel, ebb dominant; hence the titles flood and ebb channels. Flood velocity asymmetry is caused by topographic shielding of flood channels to divert ebb flow. Flood flow tends to move in a straight line up shallow channels and over flats. Ebb velocity asymmetry may be caused either by topographic shielding or, more importantly, by the fact that drainage off tidal flats concentrates flow in the

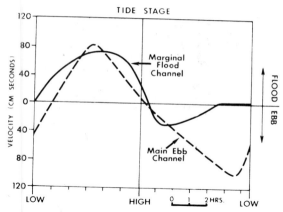

Figure 7-23. Tidal current velocity curves for an ebb-tidal delta that illustrate time-velocity asymmetry of flow in mesotidal inlets. Flow in the marginal flood-channel is flood dominant, whereas the main ebb channel is ebb dominant with maximum flow velocity occurring late in the tidal cycle (from Hayes and Kana, 1976).

deeper channels later in the ebb cycle. Thus, shallow channels and elevated flats are flood dominant. The process is self-enhancing as flood currents transport sand into the estuary and the ebb currents attempt to flush out the estuary.

Intertidal Bedforms

Time-Velocity Asymmetry

Both sand waves and megaripples exist on the intertidal portions of flood-tidal deltas of lagoonal estuaries (Figure 7-24), whereas megaripples are the type of bedform that predominates on ebb-tidal deltas. Linear and cuspate ripples are found in both environments. Numerous studies have been conducted in New England, Long Island, South Carolina, Georgia, Florida, and the West German North Sea coast (DaBoll, 1969; Greer, 1969; Hayes, 1969, 1980; Farrell, 1970; Hartwell, 1970; Kaczorowski, 1972; Oertel, 1972, 1973, 1975; Boothroyd and Hubbard, 1974, 1975; Hine, 1975; Hubbard, 1975; Hubbard and Barwis, 1976; Finley, 1976, 1978; Fitz-Gerald, 1976; Hubbard et al., 1977, 1979; Nummedal et al., 1977; Oertel, 1977; Davis and Fox, 1981; Gallivan and Davis, 1981; Nummedal and Penland, 1981; FitzGerald and Nummedal, 1983). Inspection of sand body morphology and bedform patterns of inlets in New Jersey and North Carolina show basic similarities in bedform distribution and tidal delta

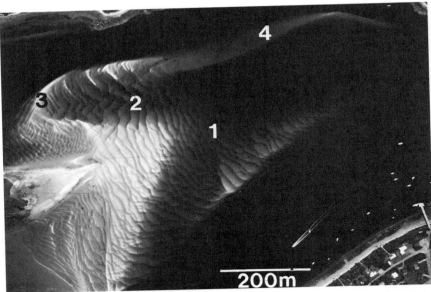

Figure 7-24. Aerial view of a portion of Middle Ground, the flood-tidal delta of the Parker River estuary. Numbering system of the components is the same as for Figure 7-8. Some areas are shown in more detail in Figure 7-19.

morphology. Details can differ widely, but the following specific examples should serve as a general model for lagoonal estuaries. Intertidal sand waves are found mainly on flood ramps and are flood oriented; that is, they are migrating in the direction of flood-tidal current flow and are of the low-energy variety (Figure 7-16 and 7-17I). A representative time-velocity curve exhibits a very large flood asymmetry, with a maximum flood velocity of 80 cm/sec but a maximum ebb velocity of only 30 cm/sec (Figure 7-25). The flood dominance is achieved by topographic shielding provided by the well-developed ebb shield (Figure 7-24).

Intertidal megaripples change orientation with the tidal stage; they are flood oriented during flood tide and ebb oriented during ebb tide. This change is effected within the first hour after flow has begun for cuspate (3D) megaripples, and somewhat later for linear (2D) megaripples. The time-velocity curve (Figure 7-25) shows a maximum velocity of about 100 cm/sec during both flood and ebb, but very little velocity asymmetry (refer also to Figure 7-12A and 7-18A). Of great importance is the maximum velocity differential between cuspate megaripples and low-energy sand waves. This is believed to account for variation in spacing (Figure 7-13). The linear megaripple–low-energy sand wave separation is not so distinct. Many linear or two-dimensional megaripples have spacings larger than 4 m and overlap the low-energy sand-wave field (Figure 7-13).

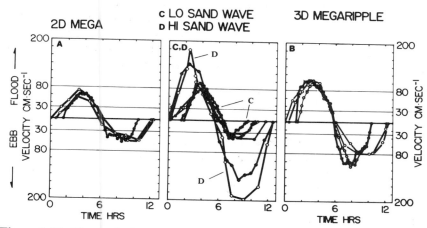

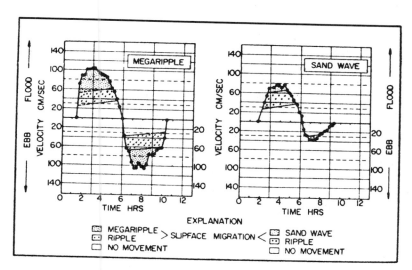

Figure 7-25. Time-velocity curves for various bedform types in the Parker River, Massachusetts (adapted from Boothroyd and Hubbard, 1975, plus additional data).

Figure 7-26. Migration habit for low-energy sand waves and cuspate (3D) megaripples in lagoonal estuaries. Megaripples are high-velocity, velocity-symmetric forms. Low-energy sand waves are lower velocity and velocity asymmetric (flood dominant). Note the time and duration of slipface migration of each bedform. Consult the text for details (from Boothroyd and Hubbard, 1975).

An inspection of the time-velocity curves (Figure 7-25) reveals that these bedforms are subject to a maximum velocity of about 60 cm/sec for both the flood and the ebb stages of the tidal cycle. Maximum flow velocity is similar to that of low-energy sand waves but there is little or no velocity asymmetry. This current pattern allows substantial modification of the form inherited from the preceding tidal stage to take place but may not permit total reorientation as is always the case for 3D megaripples.

Migration Habit

In order to assess the migration of bedforms in an estuary it is necessary to remember that any given location is subject to an increase, and then a decrease, in flow strength for each half-tidal cycle. The bedform we actually see exposed at low water on the intertidal flats is therefore the net result of a combination of modifying influences. It is difficult to study bedform migration over a single tidal cycle in the field because of (1) lack of clarity of the water and (2) inability of divers to maintain station during high current velocities. One of the few studies of this type was by Boothroyd and Hubbard (1974); some of their results are given here.

Inspection of the time-velocity curves for sand waves (Figure 7-26) indicates that there is a 1–2 hour time span at the beginning of the tidal cycle when the rising tide has not yet reached the bedforms in question and therefore no current flow is recorded. As the sand waves are covered, ripple migration does not begin until a current velocity of 25–30 cm/sec is reached. Above this velocity, flood-oriented ripples migrate across the stoss side of the bedform up to the crest and dump sand over the slipface of the larger form. Ripple migration increases in intensity until, at about 60 cm/sec, continuous migration of the sand wave slipface begins. This migration continues through the maximum velocity phase of the flood cycle and terminates when the velocity drops below 60 cm/sec. Ripple migration continues until the flood velocity drops below 30 cm/sec; then all bed movement ceases. Bed material transport does not begin again until the ebb-tidal current reaches a velocity of 25–30 cm/sec. At this point the flood-oriented ripples on the stoss side of the sand waves are destroyed and reconstituted as ebb-oriented cuspate ripples. However, because of the extreme velocity asymmetry, the maximum ebb velocity reached is 40 cm/sec, still well within the ripple stability field. No reorientation of the sand wave takes place, therefore, and as the sand wave emerges at the end of the tidal cycle it has a flood-oriented slipface and a stoss side mantled with ebb-oriented cuspate ripples.

For cuspate megaripples, the migration habit is quite different. Most megaripples in lagoonal estuaries change orientation with change in flow direction (flood to ebb) during a tidal cycle and are so ebb oriented when the tidal flats are uncovered at low water. As the flats become flooded by the rising tide, the ebb-oriented bedforms are destroyed and new flood-oriented megaripples are constituted from the eroded remnants of the last half tidal cycle. This process requires approximately one-half to 1 hr. Ripple migration begins at the familiar flow velocity of 15–30 cm/sec, and megaripple slipface migration commences at about 60 cm/sec (Figure 7-26). The sequence of events continues much like that of sand waves: Slipface migration of the large bedforms ceases; then ripple migration ceases until the corresponding minimum velocity for bottom transport is reached during ebb, when ebb ripple migration begins. The difference is that as the ebb-tidal current velocity passes upward through 60 cm/sec, the bedforms reorient once again to yield the ebb-oriented field visible at low water.

It is clear from the foregoing discussion that the velocity field of cuspate megaripples allow the complete reorientation of these bedforms, whereas the low energy sand wave velocity field dictates that the flood-oriented form is the stable bed feature. It is also clear that despite the extremely unsteady flow within half a tidal cycle phase, the bedform in equilibrium with the maximum flow velocity is the one that is preserved. The bedform is said by Allen (1973, 1974, 1982d,e) to lag behind the changes in flow regime; that is, relaxation time (Allen, 1974; Allen and Friend, 1976) is greater than the one-half tidal cycle.

The morphology and spacing of the larger 2D megaripples is explained in the foregoing context as being in the low-energy sand-wave velocity field for both flood and ebb flow. The maximum flood or ebb velocity, or velocity duration about 60 cm/sec, leads to a relaxation time of about one-half tidal cycle.

Migration rates for various kinds of bedforms are usually monitored by placing stakes at the crests of the bedforms and measuring how far the crest migrates away from the stake during successive time periods. Inspection of Figure 7-28 reveals that: (1) there is a very large difference in the rates of migration of cuspate megaripples and low-energy sand waves, and (2) there also is a large variation in rates of migration from spring tide to neap tide. Within the cuspate megaripple velocity field, slipface migration rates range from 1 to 8 cm/min (Figure 7-27B), whereas within the low-energy sand-wave velocity field the rate of slipface migration averages less than 0.5 cm/min, or 25 cm/hr, or 40 cm per tidal cycle (Figure 7-27B).

Longer term monitoring (3 months) of low-energy sand-wave migration reveals the profound influence the spring–neap tidal range variation has on tidal bedforms (Figure 7-27A). Tidal range variation means that the tidal prism of a lagoonal estuary varies from a larger (spring) to a smaller (neap) volume. Because the time of flushing is constant (semidiurnal tidal period for New England), flow velocity must increase during times of spring tide in order to exchange the enlarged tidal prism. Sand-wave migration rates are 20–30 cm per tidal cycle during full-moon spring tide but drop to 0.5 cm per tidal cycle during neap tide and new moon spring tide for the estuary studied (Parker River, Massachusetts). This resulted in an overall migration of the wave forms of 15–18 m over the 3-month period. It was obvious that during

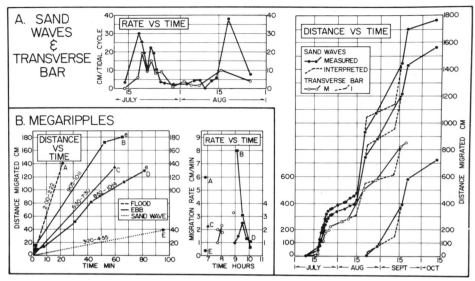

Figure 7-27. Migration rates for various intertidal bedforms; (A) sand waves and a transverse bar, (B) megaripples (from Boothroyd and Hubbard, 1975).

some neap tidal periods, maximum tidal current velocity during the flood portion of the tidal cycle is depressed below the velocity needed to effect slipface migration (low-energy and-wave velocity field of Figure 7-26).

What can be summarized about time-velocity asymmetry and bedform migration rates? First, bedform type and orientation at a given site are powerful predictors of the nature of the tidal-current curve for that site. Aerial photographs of bedforms can therefore be used to estimate tidal current curves. Second, the migration rate reveals details on the variation in the rate of bed-load transport of sand. Bedform migration does not give a total bed-load transport rate but is a good qualitative measure. Third, to get a true picture of estuarine bed-load processes, measurements should be taken during times of spring tide, not during mean conditions as is often done.

Subtidal Bedforms

Low-Energy Sandwaves

In subtidal channels and on the subtidal portions of flood- and ebb-tidal deltas, bedform type and orientation are also controlled by complex intertidal and subtidal topography (Figure 7-19). Figure 7-13 illustrates the height and spacing of some of the bedforms depicted in the longitudinal profiles (Figure 7-19). In addition to megaripples, there appear to be two types of subtidal

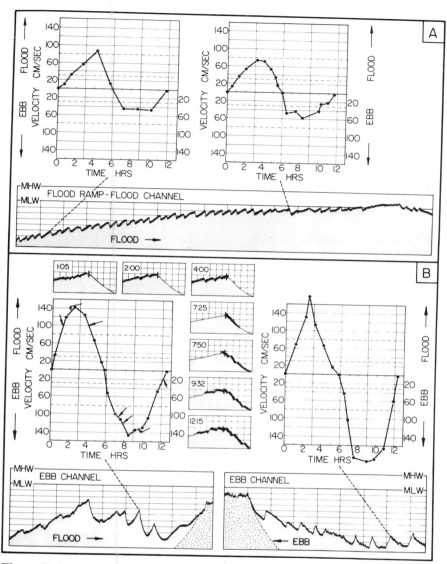

Figure 7-28. Time-velocity curves for selected subtidal bedforms. (A) Low-energy sand waves on a flood ramp and in a flood channel. (B) High-energy sand waves with superimposed megaripples in an ebb channel (from Boothroyd and Hubbard, 1975).

sand waves: low-energy sand waves on flood ramps and high-energy sand waves in deeper ebb channels.

Figure 7-28A shows a longitudinal profile of flood-oriented sand waves of a flood ramp–flood channel complex and two time-velocity curves for selected subtidal sand waves. The velocity asymmetry and maximum flood velocity are quite similar to the intertidal low-energy sand wave (Figure 7-26). Therefore, we can say that flood-oriented sand waves, shielded from ebb flow, are similar in morphology and migration habit and differ only in scale, whether in 7 m of water (at mean low water), less than 2 m of water, or intertidal.

High-Energy Sand Waves

In deep ebb channels (carrying a large volume of ebb-tidal flow), flow velocities are very high, up to 200 cm/sec, and the dominant bedforms are sand waves (Boothroyd and Hubbard, 1975). Where velocity asymmetry is low, longitudinal bottom profiles show the sand waves to be nearly symmetrical (Figure 7-28B, left); where velocity asymmetry is high, however, the sand waves are oriented (e.g., ebb oriented in the profile on the right in Figure 7-28B). Observations by divers confirm that when the mean flow velocity (measured either at the surface or at 1 m above the sand wave crest) exceeds about 80 cm/sec, megaripples are superimposed on the sand-wave form; hence the term high-energy sand waves. Major bed-load movement of sand in deep ebb channels is by migration of megaripples, even though the principal forms are the large sand waves. Sketches made by divers (Figure 7-28B) illustrate bedform movement over a large symmetrical sand wave. During the flood-tidal cycle, megaripples migrate up to the sand-wave crest and deposit sand down a flood-oriented, avalanche slipface. During the ebb-tidal cycle, the flood-oriented slipface of the sand wave is modified by ebb-megaripple migration up the slipface and over the sand wave crest. This mechanism of sand movement and the appearance of the large bedforms in deep, subtidal channels of lagoonal estuaries is similar to that of large sand waves on macrotidal sand flats (Dalrymple et al., 1975, 1978, 1982).

Dalrymple's studies of the intertidal sand flats in the Bay of Fundy indicate that his high-energy sand waves and 3D megaripples overlap the Parker River subtidal high-energy sand waves as shown on Figure 7-29, a depth–velocity diagram modified from Costello and Southard (1981). Likewise, his low-energy sand waves plot as a deepwater extension of the 2D megaripple/low-energy sand-wave field of the mesotdial environment. The similarity of hydraulic conditions and morphology of the Fundy forms and those of subtidal lagoonal estuaries is important because it allows the connection to be made between internal sedimentary structures of high-energy sand waves of the less easily studied subtidal channels of lagoonal estuaries.

Allen (1980a,b, 1982a,b,c,d) has recently presented an exhaustive study of sand waves of all types from many different environments. His speculative model for the internal structure of sand waves as governed by the velocity asymmetry of tidal currents is shown in Figure 7-30.

The internal structure of the more velocity symmetric sand waves is similar to that of many Bay of Fundy high-energy sand waves. The complex sets of planar-tangential cross-stratification are formed by the migration of 3D megaripples during both flood and ebb parts of the tidal cycle. It is important to note that the internal structure of high-energy sand waves often does not mimic the external morphology. The deep subtidal channels would have the highest preservation potential of any subenvironment of lagoonal estuaries.

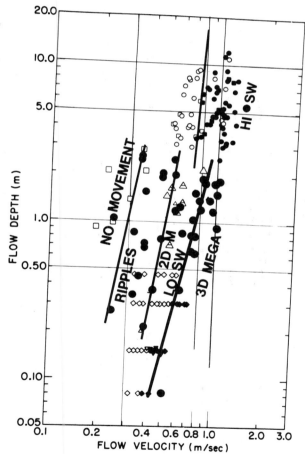

Figure 7-29. Depth-velocity diagram illustrating flume and field data to flow depths greater than 10 m (modified from Costello and Southard, 1981).

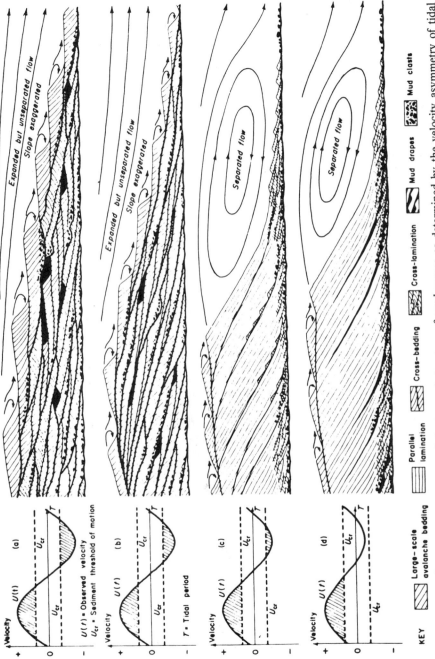

Figure 7-30. Hypothetical examples of the internal structure of sand waves as determined by the velocity asymmetry of tidal currents (from Allen, 1982a).

Large Accumulation Forms

Topographic forms of a larger scale than sand waves, but smaller in scale than the tidal deltas themselves, exist in lagoonal estuaries. Two forms of interest are transverse bars and spillover lobes. (Figures 7-8, 7-9, and 7-24).

Spillover lobes may be either flood oriented or ebb oriented, depending on the geometry and channeling of flow in the estuary, but they usually occur in flow-velocity fields where megaripples are the stable bedform. These features are similar to the spillover lobes described by Ball (1967) and Hine (1977) on Bahamian oolite shoals. They are also similar to chute bars in fluvial coarse-grained meander belts (McGowen and Garner, 1970) and to linguoid bars in sandy braided streams (Collinson, 1970). They form where a constricted jet undergoes flow expansion into a less restricted area, such as from the breaching of an ebb shield and consequent flow expansion into a flood channel (Figure 7-31).

Transverse bars in intertidal and shallow subtidal environments occur on flood ramps, where they appear as large sediment forms mantled with flood-oriented sand waves (Figures 7-17I, 7-18B). They migrate in a flood direction but at a lower rate than individual sand waves. Rate and distance of slipface migration at the leading edge of a transverse bar is shown in Figure 7-27. The migration rate, 20 cm per tidal cycle during spring tide, is

Figure 7-31. An ebb-oriented spillover lobe in the Parker River estuary.

about half that of the sand waves on the stoss side of the bar; recorded migration has been 8 m for a 2-month period or about 12 cm/day. These forms may be an important agent for sand transport landward on flood-tidal deltas but need to be studied over years and not just months. A photographic series qualitatively illustrates the migration of a transverse bar in the Parker River, Massachusetts (Figure 7-32).

There is justification for classifying the large, solitary, subtidal, high-energy sand waves (Figures 7-19, 7-26) as transverse bars because they appear to migrate independently of one another. Allen's work, cited earlier, included these bedforms under the general category of sand waves. However, more work remains to be done on defining the shapes and charting the migration of these large forms.

Tidal Deltas

Introduction

So far the discussion has concerned the development and internal sedimentary structures of bedforms, including some specific localities where the various types exist on flood-tidal deltas. The general morphology of flood- and ebb-tidal deltas and the terminology for the various segments of those deltas have also been introduced. The morphologic characteristics are reviewed in Figures 7-6 to 7-9.

(A)

Figure 7-32. A sequence illustrating the migration of a flood-oriented transverse bar up the flood ramp of the Parker River flood-tidal delta. (A) July 27, 1967. (B) October 10, 1969. (C) May 28, 1970. Two bar slipfaces are visible. (D) January 21, 1972. Slipface has been degraded by ebb flow over the ebb shield.

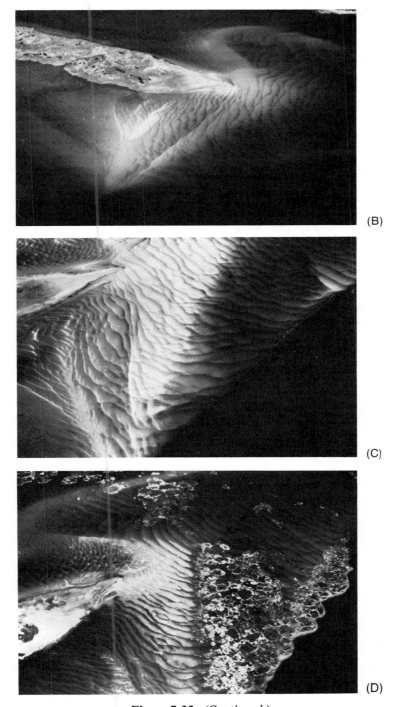

(B)

(C)

(D)

Figure 7-32. *(Continued.)*

As inspection of charts of the eastern United States and Gulf of Mexico coasts (National Ocean Survey, 1:20,000–1:80,000 scale) reveals variations in tidal delta morphology that qualitatively can be related to differences in tidal range, and thus the tidal prism, and to differences in wave-energy flux impinging on the site of the inlet. Three such tidal delta variations for the shorelines of New England, South Carolina, and Texas, as envisioned by Hayes (1980), are illustrated in Figure 7-33. Microtidal areas, such as the Texas coast, are thought to have poorly developed ebb-tidal deltas and relatively larger flood-tidal deltas because of the dominance of wave-energy flux over the small tidal prism. Ebb-tidal deltas of the mesotidal South Carolina coast are more elongate and extend further seaward than those of the mesotidal New England coast because of decreased wave energy in South Carolina. An increased tidal prism caused by freshwater discharge also may contribute to the increased size of the South Carolina ebb-tidal deltas. The diminished size of the flood-tidal deltas in this area, as compared to New England, may be caused by the ratio of open water to marsh in the estuaries and lagoons that creates an ebb-dominant inlet flow and does not allow large flood-tidal deltas to develop.

The foregoing considerations should serve to set the stage for discussion of tidal deltas in greater detail, including morphology, bedform distribution, tidal current flow patterns, and sedimentary structures.

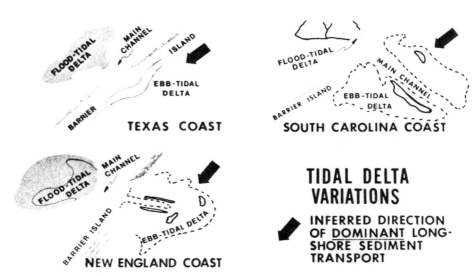

Figure 7-33. Tidal-delta variation for the coasts of Texas (microtidal), New England (mesotidal, higher wave energy), and South Carolina (mesotidal, low-wave energy) (from Hayes, 1980).

Flood-Tidal Deltas

Mesotidal Example

To properly assess the development and characteristics of flood–tidal deltas, it also is necessary to consider the channels flanking the deltas (Figure 7-34). Hubbard and Barwis (1976), primarily using data presented earlier (Figure 7-26) and from DaBoll (1969), show the importance of time-velocity asymmetry toward creating and maintaining delta morphology. Flanking channels are likely to show an even greater ebb dominance than the tidal current curve of Figure 7-34B-3. The hypothetical topography of a general model (Figure 7-33) is based on actual measurements, including the Parker River (DaBoll, 1969).

The distribution of bedforms of the intertidal area of the Parker River flood-tidal delta is illustrated in Figure 7-35. This tidal delta complex

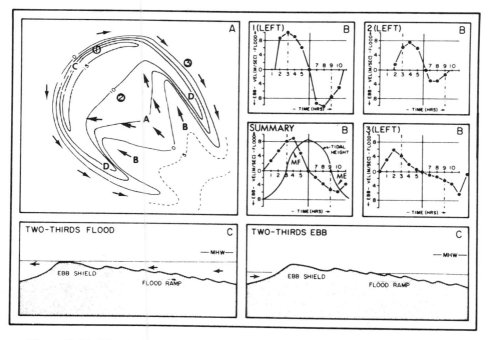

Figure 7-34. Morphology and flow pattern of a typical mesotidal flood-tidal delta. (A) Components are: (A) flood ramp; (B) flood channels (C) ebb shield; (D) ebb spits. Dominant bedform orientations are shown by arrows. (B) Time-velocity curve for various locations shown in (A). (C) An illustration of the concept of topographic shielding allowing the preservation of flood-oriented sand waves at low tide (from Hubbard and Barwis, 1976).

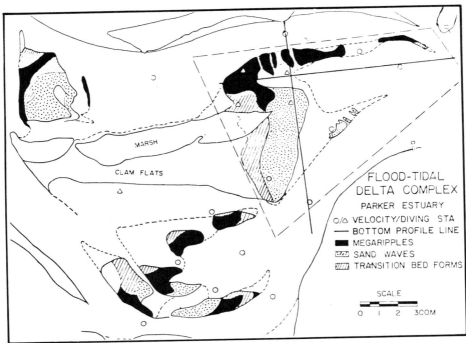

Figure 7-35. Intertidal bedform map of the Parker River estuary flood-tidal delta complex. The area outlined by the dashed line is shown in Figure 7-25 (from Boothroyd and Hubbard, 1974).

actually is divided into three parabolic segments, separated by shallow subtidal channels (Figure 7-36). Megaripples occupy the ebb shields, ebb spits, and ebb-oriented spillover lobes; flood-oriented sand waves are situated on the multiple flood ramps; and ebb-oriented, short-spaced sand waves are found on the up-estuary side of incompletely formed ebb shields and as transition zones between flood-oriented sand-wave and megaripple areas. As the photograph clearly illustrates, the subtidal flood ramps and channels in any bedform distribution pattern must be considered, even though bottom features are not visible if suspended sediment is present in the water column.

The long intertidal finger (Figure 7-36) comprising the central segment of the tidal delta extends up estuary as the topographically highest and most mature area of the complex. The marsh cap extends up to the elevation of mean high water. The marsh is expanding in areal extent by actively growing out over the clam flat surface. The clam flat, in turn, occupies the highest surface of an older, intertidal flood ramp. This succession should be kept in mind, for it is important to the development of stratigraphic models (Chapter 9). The highest marsh-covered part of a flood-tidal delta, or the complete delta, often is identified on coastal charts as a "middle ground," as is the case in the Parker River.

Microtidal Example

Many flood-tidal deltas that develop in microtidal lagoons of large tidal prism possess morphologic features similar to those just described for the Parker River. However, flood-tidal deltas that form in lagoons with small tidal prisms appear much more flood-dominant than the large prism microtidal examples. The Ninigret Pond, Rhode Island, flood-tidal delta will serve as a small tidal prism, microtidal example (Figure 7-37).

The terminology for flood-tidal delta subenvironments generally follows that of Hayes (1980) and Hayes and Kana (1976), with additional terms added to describe features not covered in the Hayes classification. The main channel appears to have formed as a result of barrier breaching during storms and did not migrate, but rather closed and reopened within a few hundred meters of the older inlet. This led to a large sand accumulation of overlapping delta lobes (Figures 7-37 and 7-38). The terminal lobes and channel-margin flats range from shallow subtidal (< 1 m deep) to intertidal (5–10 cm above mean low water). The lagoonward margins of inactive lobes are covered with eelgrass (*Zostera marina*).

Accretion of the sandy tidal-delta lobes upward into the intertidal zone has allowed the establishment of salt marsh on the inactive lobes. *Spartina alterniflora* is succeeded by *Spartina patens* and *Distichlis spicata* in the

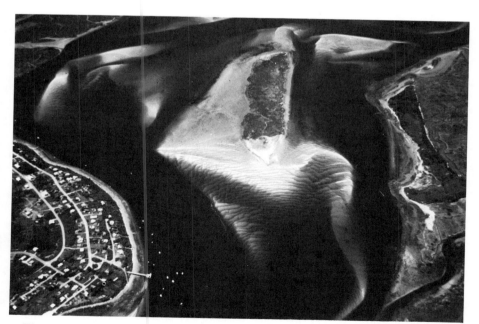

Figure 7-36. An aerial view of the Parker River flood-tidal delta illustrating the three parabolic segments separated by shallow subtidal channels. Compare this with Figure 7-35.

Figure 7-37. The Ninigret Pond flood-tidal delta and adjacent barrier spits; East Beach to the left, Charlestown Beach to the right (Charlestown, Rhode Island). Inlet was stabilized in 1952; an abandoned inlet is to the left. Note the lack of an ebb-tidal delta.

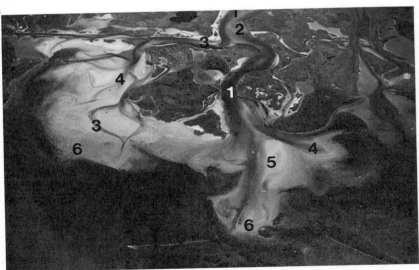

Figure 7-38. Lagoonward portion of the Ninigret flood-tidal delta. Numbers refer to the following features: (1) main tidal channel; (2) point bars; (3) tidal channel branches; (4) flood ramps; (5) channel-margin flats; and (6) terminal lobes.

process of marsh maturation (Frey and Basan, Chapter 4). This growth is similar to that reported by Godfrey and Godfrey (1974) for microtidal flood-tidal deltas in North Carolina. The marsh cover, although locally extensive, is only 1–2 m thick. The marsh growth, together with outcrops of glaciofluvial material that exist within the delta complex, govern tidal channel location based on interpretation of sequential aerial photographs and old charts. The principal difference between the microtidal and mesotidal examples is the abundance of flood terminal lobes and the lack of ebb spits in the microtidal case (Figure 7-38).

Tidal Delta Variation

Discussion of flood-tidal delta morphology, bedform distribution, and tidal current flow patterns has used examples from northern New England mesotidal estuaries, or lagoons, where ebb-tidal current return flow is directed around the flanks of the flood delta, creating the prominent ebb spit–ebb shield topography. These estuaries have no wide expanses of open water at high tide, which means there is a limited wave fetch; therefore waves are small and wave-induced modification of the tidal-delta morphology and internal structures is relatively unimportant. Also, the inlet throats are narrow and the flood-tidal deltas are not immediately adjacent to the inlet throat; therefore, open-ocean wave action is almost nil in the standard model. If the flood-tidal delta is built into a wide bay or is close to the inlet throat ($<$ 1 km), or the inlet throat is wide enough ($>$ 1 km) to admit open-ocean waves inside the inlet, the tidal delta characteristics vary from the standard model.

Figures 7-39 through 7-41 illustrate the flood-tidal deltas of Chatham Harbor, Massachusetts, North Inlet, South Carolina, and Little River, South Carolina, respectively. These photographs, together with Figures 7-8, 7-35, and 7-36, depict the variation of morphology and bedform distribution, and hence internal structure, present in flood-tidal deltas of mesotidal and, to a certain extent, microtidal inlets. The Essex and Chatham examples can be considered as simple end members of deltas relatively unaffected by wave action, and with an abundant sediment supply. The Essex delta has substantial tidal current return flow, whereas the Chatham delta has very little return flow. The tidal current curves of Figure 7-42 indicate the extreme flood dominance of the Chatham delta (Hine, 1975). These should be compared with the suite of curves from the Parker River flood-tidal delta (Figures 7-26 and 7-28). Wave action does form swash bars on the leading edge of the Chatham delta, but it is not an important agent in the modification of the overall surface. Flood-oriented sand waves are the most prevalent large-scale bedform, but flood-oriented megaripples (nonreversing) do exist (Figure 7-43). Transverse bars also are very important (Figure 7-39).

Figure 7-39. Flood-tidal delta of Chatham Harbor, Massachusetts, an extremely flood-current dominated example with little ebb-return flow. Note the well-developed transverse bars.

Figure 7-40. Flood-tidal delta of North Inlet, South Carolina. The delta exists as a ramp built against a large marsh and is subject to extensive ebb-current modification due to a poorly developed ebb shield (photo courtesy of Robert J. Finley).

Figure 7-41. Flood-tidal delta of Little River, South Carolina. This example is subject to both wave modification and ebb-tidal current modification (photo courtesy of Dennis K. Hubbard).

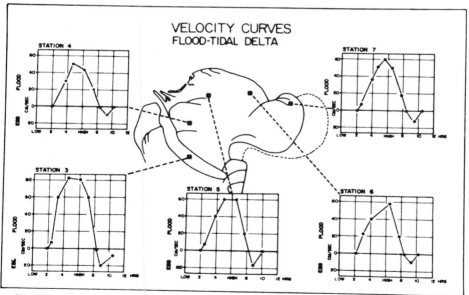

Figure 7-42. Tidal-current time-velocity curves for the Chatham Harbor flood-tidal delta illustrating the pronounced flood asymmetry (from Hine, 1975).

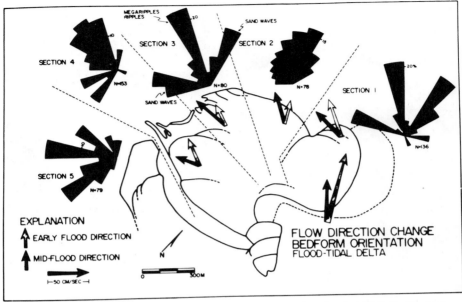

Figure 7-43. Bedform orientation on the Chatham Harbor flood-tidal delta illustrating the dominance of flood-oriented features. The measurements were obtained at low water (from Hine, 1975).

The small flood-tidal deltas of North Inlet and Little River, South Carolina, serve as representatives of an end member having a sparse sediment supply and subject to extensive wave modification. These two deltas vary in that one (Little River) is situated in mid-lagoon, whereas the other (North Inlet) exists as a ramp built against an extensive marsh. These small deltas do not have extensive fields of flood-oriented sand waves but are subject to extensive ebb-current modification because of poorly developed ebb shields. Figure 7-44 illustrates the ebb orientation of bedforms on what is, morphologically, a flood ramp (Finley, 1975). The leading edge of the North Inlet tidal delta is essentially a series of large, lobate, swash bars being driven over the marsh sediments by waves entering the inlet at high tide (Finley, 1975). These swash bars have a 180° reverse orientation compared with those formed on the leading edge of the Chatham flood delta. Wave modification also results in the formation of plane lamination or flat beds, which is otherwise rare as a flood-tidal delta internal sedimentary structure but is volumetrically important in ebb-tidal deltas (see the section on ebb-tidal deltas).

This discussion has encompassed merely a small sample of representative flood-tidal delta types; all manner of variation in lagoon size, return flow, wave modification, and sediment supply is possible. Table 7-2 qualitatively

Table 7-2. Characteristics and Major Features of Selected Flood-Tidal Deltas.

| | | | | | | | Major features | | | | | | |
Flood-ramp	Ebb spits	Ebb shield	T bar	F-0 sand waves	Ebb spillover lobe	Flood spillover lobe	Lagoon size	Throat width	Dist. to throat	Wave action, lagoon	Wave action, ocean	Sediment supply	Example
Well developed	A[a]	A	C-R	A	C	R	Small	Narrow	Far	Nil	Nil	Abundant	Essex, Parker
Well developed	R	C	A	A	C-R	C	Large	N-mod	Far	Mod	Nil	Abundant	Chatham
Mod-poorly developed	A	C-R	AB	C-R	C	R	Small	M	Close	Nil	Important	Mod-poor	North Inlet, Little River
Mod-well developed	C-A	C-R	C-A	C	C-R	C	Large	N	Mod	Mod	Nil	Abundant	Moriches
Mod-well developed	R	AB	AB	R	R	A	Small	N	Close	Nil	Nil	Mod-poor	Ninigret
Poorly developed	C	C-R	AB	R	C	A	Large	N	Close	Nil	Nil	Poor	Redfish Pass

[a] A, abundant; C, common; R, rare; AB, absent

summarizes some of these variations and compares them with the relative abundance of major elements of some actual examples of flood-tidal deltas.

Ebb-Tidal Deltas

At first glance, flood- and ebb-tidal deltas of mesotidal inlets appear quite similar to one another in external morphology, especially, when the two deltas are of similar size and inlet and lagoon geometry is simple and uncomplicated. However, on closer inspection, it is apparent that there are major differences in external morphology, tidal current flow paths, bedform distribution, and, most important, the influence of wave action. Figures 7-4, 7-5, and 7-6 review ebb-tidal delta terminology.

Open-ocean wave action, along with tidal currents, is a dominant factor affecting ebb-tidal deltas. Longshore transport of material by wave-generated currents in the surf and swash zones is the major sediment provider to ebb deltas. The alongshore transport of sand is interrupted by the ebb jet issuing

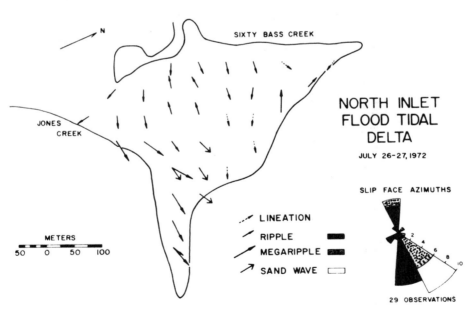

Figure 7-44. Bedform orientations on the North Inlet, South Carolina, flood-tidal delta. Ebb-oriented forms are common because of the poorly developed ebb shield (from Finley, 1975).

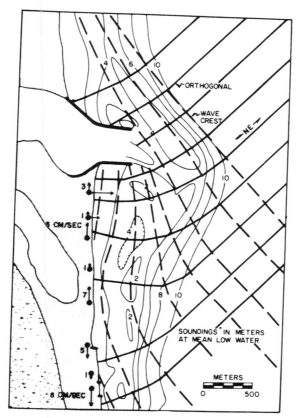

Figure 7-45. Wave refraction pattern around the Merrimack River ebb-tidal delta with a drift reversal on the downdrift (south) side of the inlet. The wave period equals 10 seconds and current velocities and directions are from field measurements (from Hubbard, 1975).

from a tidal inlet, and consequently the sand is deposited around the inlet throat. Incoming waves are refracted both by the ebb jet and by the tidal delta itself, resulting in a shift in wave approach direction around the margin of the delta and a reversal in the direction of sediment transport (Figure 7-45). The dominant wave approach direction is from the northeast for the Merrimack River, Massachusetts, ebb-tidal delta, but waves are refracted so that surf zone transport of sediment on the south side of the inlet is to the north (Hubbard, 1975). The updrift side of the inlet is to the north and the downdrift side is to the south.

Tidal current flow patterns and the importance of time-velocity asymmetry for ebb-tidal deltas are shown in a general model for mesotidal ebb deltas (Figure 7-46) discussed by Hubbard and Barwis (1976). The main ebb

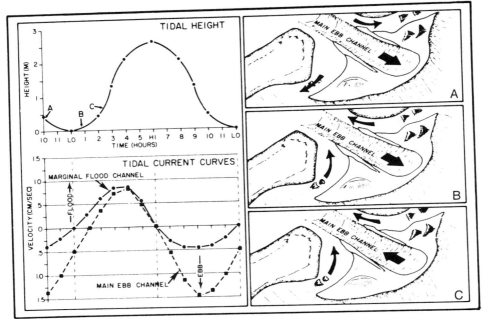

Figure 7-46. A general model for segregation of flow on mesotidal ebb-tidal deltas and the feedback effect on delta morphology. Representative tide stage and tidal-current velocity curves are on the left; flow patterns and delta morphology are on the right (from Hubbard and Barwis, 1976).

channel has a strongly ebb-dominant tidal current velocity curve, whereas the marginal flood channels are strongly flood dominant. Time asymmetry is also important in that the main ebb channel carries ebb-tidal current flow past the time of low water (Figure 7-46) and into the flooding portion of the tidal cycle. The flooding open-ocean water skirts the edges of the ebb jet and enters the inlet throat via the marginal flood channels. It should be pointed out that wave-generated currents have an additive effect on flood-tidal currents in the marginal flood channels and contribute greatly to the flood velocity asymmetry.

Breaking waves, and the resulting wave-generated current (swash), are also the most important agent in the generation and migration of swash bars on the swash platform of the ebb-tidal delta. Swash bar morphology, processes, and internal sedimentary structures are shown in Figure 7-47A–C. Swash bars are synonymous with ridges, the beach accretional feature referred to by Hayes and Boothroyd (1969) and Davis *et al*. (1972), who cited the original reference of King and Williams (1949). Swash bars are morphologically similar to the transverse bars of flood-tidal deltas in plan and in scale and even migrate in the same general direction (i.e., landward). Internal structure is similar, consisting mainly of tabular cross-stratification,

(A)

(B)

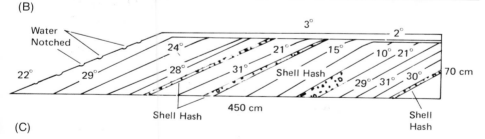

(C)

Figure 7-47. Swash-bar morphology, processes, and internal sedimentary structures. (A) Flat-bed surface of a swash bar on the Parker River ebb-tidal delta shown at low tide. Antidune bedforms exist as thin dark stripes (center). (B) Plane bed and antidune bed configurations resulting from swash flow during a flooding tide stage. Acceleration of flow from the high point of the bar (right) allows antidune water waves to form (center). Flow separation at the bar crest (left) allows sediment to avalanche down the steep slipface. (C) Longitudinal cut in a swash bar on the Price Inlet, South Carolina, ebb-tidal delta. Dips are given in degrees (from FitzGerald, 1976, Figure 8).

Figure 7-48. Aerial view of the recurved spit, welding swash bars, and marginal flood-channel system of the Chatham Harbor ebb-tidal delta and adjacent barrier (Nauset spit).

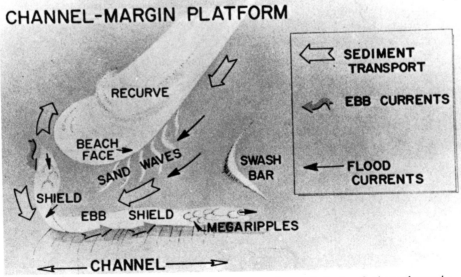

Figure 7-49. Model of sediment transport and morphology of channel-margin platforms of mesotidal ebb–tidal deltas (from Hayes, 1980).

but with one major difference: The tabular cross-stratification is capped by plane lamination, a structure rare to nonexistent inside the estuary on flood-tidal deltas.

Hayes (1980) has further refined the model of sediment transport in, and in the vicinity of, marginal flood channels, using in part the delta of Hine (1975) from Chatham Harbor, Massachusetts. An aerial photograph of the Chatham Harbor–Nauset spit area shows the recurved barrier spit (Figure 7-48), a large continuous swash bar with smaller subsidiary swash bars welding to the seaward margin, and a well-developed marginal flood channel containing sand waves migrating toward the inlet throat. Slipfaces of major and minor bedforms, together with internal sedimentary structures, show a predominant landward orientation. This information is compiled into a model for channel-margin platforms (Figure 7-49), which concludes that the dominant sediment transport direction is landward and that marginal flood channels are a major, perhaps the major, pathway by which sand enters a mesotidal estuary.

The bedforms on the total intertidal portion of a mesotidal ebb-tidal delta (Price Inlet, South Carolina) are primarily megaripples and flat beds (Figure 7-50). The dominant slipface orientation of the bedforms, and the

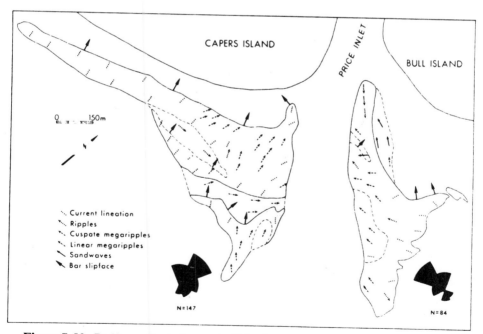

Figure 7-50. Bedform orientation on the Price Inlet, South Carolina, ebb-tidal delta. The measurements taken at low tide indicate that the dominant slipface orientation is toward the inlet (from FitzGerald, 1976).

orientation of grain lineation on the flat beds, is landward (grain lineation is, of course, bidirectional) (FitzGerald, 1976) and is attributed to the combined result of flood-tidal current flow and wave swash. This observation of ebb-oriented bedforms in the main ebb channel (Hine, 1972), combined with the aforementioned tidal current curves showing marked ebb velocity asymmetry in the main ebb channel, suggests that net sediment transport is seaward in the deeper, main ebb channel and landward over the shallower swash platforms. This pattern is summarized in Figure 7-51, a model developed for the Chatham ebb-tidal delta (Hine, 1975).

Tidal Delta Variation

There is a greater variation of morphology, bedform distribution, and internal sedimentary structures in ebb-tidal deltas than in flood-tidal deltas, because wave action is such an important modifying process and also because the ebb-tidal delta tends to react with, and to, changes in the adjacent barrier islands or spits. The inlet itself can be classified according to the geographic relationship of the barriers on either side. If the barrier on either side is situated further seaward than the one on the other side, the inlet is said to be

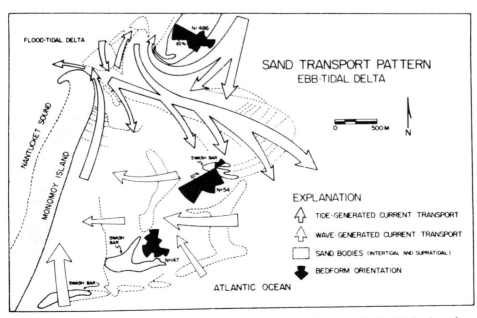

Figure 7-51. Sand transport pattern for the Chatham Harbor, ebb-tidal delta based on slipface orientations (from Hine, 1975).

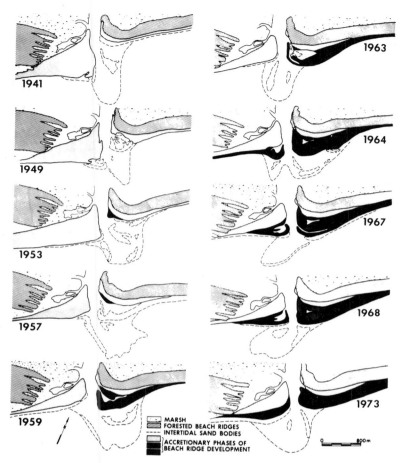

Figure 7-52. Variation through time of the Price Inlet, South Carolina, ebb-tidal delta and adjacent barrier islands based on sequential vertical aerial photograph (from FitzGerald, 1976).

offset. If one barrier extends in front of the other, the inlet is said to be overlapped, or they may have no asymmetry at all.

Earlier literature considered many tidal inlets to be updrift offset and that this offset indicated the alongshore sediment transport direction. That is, if an inlet is offset, the offset is on the updrift side or in the direction from which the sediment is derived. Hayes *et al.* (1970) have pointed out that many mesotidal inlets are downdrift offset according to studies made of the dominant wave approach direction. They attributed the downdrift offset to the wave-refraction drift-reversal process. This appears to be true for some New England inlets and northeast Gulf of Alaska inlets (Hayes *et al.*, 1970) and some New Jersey inlets (Lynch-Blosse and Kumar, 1976). Other reasons for asymmetric inlets are: main ebb channel eroded in cohesive mud (Wachepreague, Virginia, downdrift offset; DeAlteris and Byrne, 1975); tidal flushing paths and velocity asymmetry (Fire Island, New York, updrift overlap; Kaczorowski, 1972); and large sediment supply (Chatham, Massachusetts, updrift overlap; Hine, 1972). The offset is usually a supratidal part of the barrier itself but may be intertidal or subtidal as part of the swash platform.

A time-series study of Price Inlet, South Carolina, by FitzGerald (1976), utilizing sequential, vertical aerial photographs, documented a shape change from downdrift offset, to updrift offset, to nearly symmetric over a 32-yr period (Figure 7-52). The accretional changes were in response to spit growth caused by landward migration of swash bars across the swash platforms and subsequent welding to the beach face, whereas the erosional changes were caused by a shift in the main ebb channel, which allowed a shift in the accompanying swash platform that in turn left the barrier spit exposed to storm wave erosion (FitzGerald, 1976).

A combination of the factors just discussed, location of the main ebb channel, shift of that channel, and landward migration of swash bars leads to sediment by-passing the inlet instead of being permanently deposited on the tidal-inlet sand body. That is, sand from the updrift barrier is transferred over the ebb-tidal delta to the downdrift barrier. Figure 7-53, Norderney Seegat in

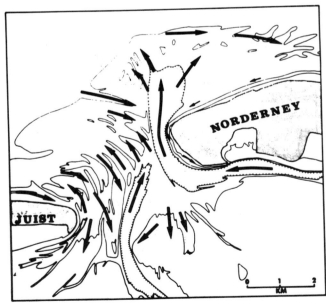

Figure 7-53. Sediment transport pattern at Norderney Seegat, East Friesian Islands, FRG. Sand by-passes the inlet from west to east (from Nummedal and Penland, 1981).

the East Friesian Islands, FRG, illustrates the complexity of the dispersal pattern leading to the by-passing (Nummedal and Penland, 1981; Nummedal and FitzGerald, in press). Compare the Nordenney Seegat pattern with that of Chatham Harbor Inlet (Figure 7-51).

Recently, comprehensive studies by FitzGerald, Nummedal, Hubbard, and co-workers (FitzGerald and Nummedal, 1983; Nummedal and Penland, 1981; Hubbard et al., 1979; Nummedal et al., 1977, FitzGerald et al., 1976) have quantified the relationship of tidal inlet processes to the development of ebb-tidal delta morphology and sediment bodies in the Georgia embayment. One aspect of the studies documented the conditions governing the ebb dominance of inlets such as Price Inlet (Figures 7-50 and 7-52); other work investigated the role of waves versus tidal currents governing tidal inlet type.

FitzGerald and Nummedal (1983) determined that the ebb dominance at Price Inlet was due to the relatively less efficient flooding and drainage of marsh areas at elevations near mean high water compared with the greater efficiency of flow in subtidal channels at times near low water. This relationship is similar to that investigated by Boon and Byrne (1981) for Virginia tidal inlets. The result is that the flood cycle is longer than the ebb cycle by 30 minutes; maximum ebb-tidal current velocity is higher and occurs relatively late in the tidal cycle (the well-known time-velocity asymmetry). The difference between maximum flood and ebb velocities is shown in Figure 7-54a.

The importance of ebb dominance, or flood dominance, to the understanding of tidal delta size and internal structure is knowing where the sediment will finally be deposited. FitzGerald and Nummedal have used the Maddock (1969) bed-load equation to demonstrate the potential dominance of bed-load sediment transport in an ebb direction (Figure 7-54b). Maddock (1969) states that total bed-load transport is proportional to the cube of the current velocity as follows:

$$U = 4.03 \sqrt[3]{L/\rho} \qquad (7\text{-}1)$$

where U is velocity (m/sec); ρ is fluid density (kg/m^3); and L is sediment load (kg/m/sec). The relationship is considered valid for a velocity range of 0.1–2 m/sec, and for mean grain sizes of 0.3–0.7 mm. Potential ebb transport of sediment is greater than potential flood transport except for low neap tidal ranges, and, what is even more important, ebb transport exhibits greater dominance during the spring phase of the tidal cycle. This helps explain the robust ebb-tidal delta and the lack of a flood-tidal delta at Price Inlet.

Hubbard et al. (1979) investigated the relative importance of waves versus tidal currents on the morphology and growth of tidal delta sediment bodies in the Georgia Embayment. Following up the work of Nummedal et al. (1977), they found that, north to south, tidal range increased in southern South Carolina and Georgia, then decreased in northern Florida (Figure 7-55a). Deepwater wave energy as determined from SSMO data decreases from a

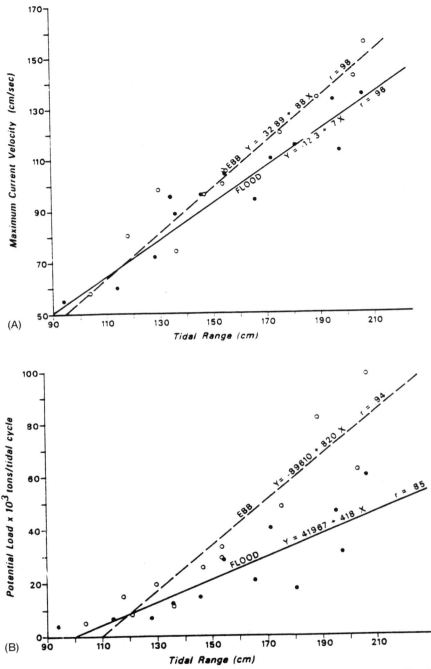

Figure 7-54. Tidal-current velocity and sediment transport, Price Inlet, South Carolina. (A) Maximum current velocity versus tidal range (from FitzGerald, 1977). (B) Potential sediment transport calculated using Maddock (1969) relationship (from FitzGerald and Nummedal, 1983).

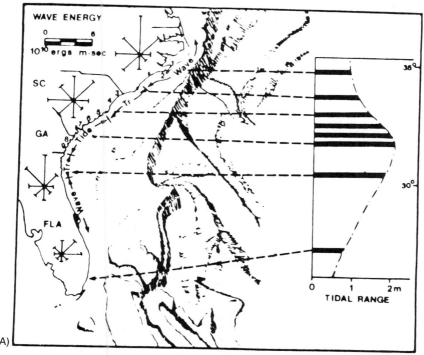

Figure 7-55. Regional trends in tidal inlet parameters in the Georgia Embayment. (A) Deepwater wave energy and tidal range. Inlet classification also shown (from Hubbard *et al.*, 1979). (B) From Nummedal *et al.*, (1977).

high in North Carolina to a low in Florida (Figure 7-55B), but nearshore wave height, from visual observations and wave gauges operated by the U.S. Army Corps of Engineers, shows an increase in Florida from the low in Georgia. Hubbard and his co-workers consider these latter wave data a more important factor in controlling tidal inlet processes.

The overall morphology of the ebb-tidal delta is due to the response to open-ocean waves and to tidal currents. Tidal currents respond primarily to open-ocean tidal range but also to open-water area in the lagoon and to tidal prism. The results give characteristic shapes to ebb-tidal deltas (Figure 7-56). Oertel (1975) has presented a model for the shapes of ebb-tidal deltas of nonstratified Georgia estuaries that can be used for tidal inlets in general (Figure 7-56). Essentially, if wave energy is high the tidal delta is blunted, with the terminal lobe close (1–2 km) to the barrier spit. If wave energy is low, the terminal lobe is a few kilometers out on the shelf, with consequent greater development of channel-margin linear bars. The main ebb channel may also be deflected to the left or right in response to longshore currents. It must be remembered, however, that a main-ebb channel offset

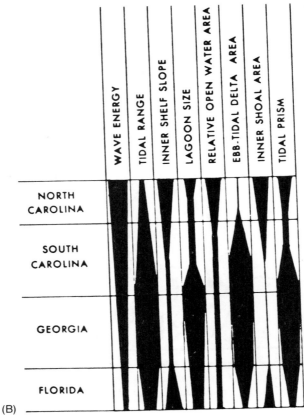

Figure 7-55. *(Continued.)*

can be a temporary orientation, as illustrated by FitzGerald (Figure 7-52), and does not necessarily represent the long-term, regional direction of dominant drift.

The variation between wave-dominated and tide-dominated ebb-tidal deltas is shown in Figure 7-57. Matanzas Inlet (Figure 7-57A) is wave dominated, with a small ebb-tidal delta consisting mainly of a swash platform in the inlet throat that is cut through by many ebb channels. Fripp Inlet (Figure 7-57B), in contrast, is further north in the Georgia Embayment where wave energy is diminished because of a lower gradient continental shelf. Tidal current energy dominates and the tidal delta is elongated seaward with well-developed channel-margin linear bars that are modified by flood- and ebb-oriented spillover lobes. The main ebb channel splits at its seaward terminus to form a many-lobed terminal lobe (Figure 7-57B). New England mesotidal ebb-tidal deltas, the basis for the general model, appear to be midway between the two Georgia Embayment end members of Hubbard *et*

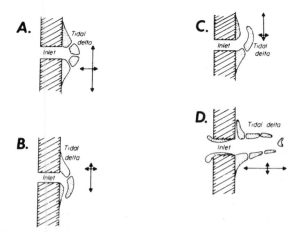

Figure 7-56. Ebb-tidal delta variation for Georgia mesotidal estuaries. (A) Wave dominated. (B,C) Longshore-current dominated. (D) Tide dominated (from Oertel, 1975).

al. (1979). It should be noted that all of the Georgia Embayment tide-dominated examples, and the New England mesotidal group, fall within the mixed-energy (tide-dominated) classification of Hayes (1979) (Figure 7-2).

A final variation in ebb-tidal deltas is the variation in grain size or, to be more specific, the amount of mud versus the amount of sand present. The New England ebb-tidal deltas are constructed wholly of sand-sized material. By contrast, the southern South Carolina and Georgia estuaries are characterized by ebb-tidal deltas in which mud is an important constituent. The textural and sedimentary structure distribution of a model Georgia estuary is illustrated in Figure 7-58 (Oertel, 1975). The swash bars are mostly sand, but mud lenses occur in lower energy zones on the swash platform. Interbedded mud is deposited in lower energy areas of the ebb channel, adjacent to channel-margin linear bars, and in topographically shielded areas along the swash platform and in the marginal flood channels.

Other Sand Bodies

This discussion has considered flood- and ebb-tidal deltas separately and has discussed the variation in each. Also discussed, but in less detail, was the bedform distribution and tidal current flow patterns in channels adjacent to the tidal deltas. In some cases, for example Matanzas Inlet (Figure 7-57A), there is little distance separating the flood and ebb deltas, but in some New

(A)

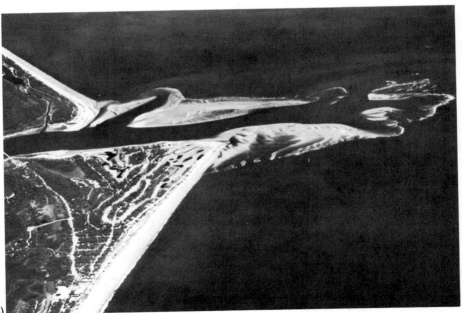

(B)

Figure 7-57. Ebb-tidal delta variation in the Georgia Embayment. (A) Aerial view of Matanzas Inlet, Florida, a wave-dominated delta (photo courtesy of R. A. Davis, Jr.). (B) Aerial view of Fripp Inlet, South Carolina, a tidal-current dominated delta. Compare both with Figure 7-53 (photo courtesy of Dennis K. Hubbard).

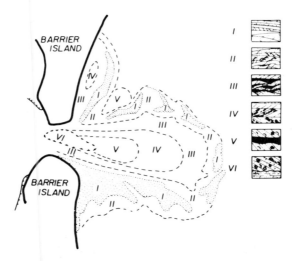

Figure 7-58. Textures and sedimentary structures of Georgia ebb-tidal deltas. Note that there is a mud facies that is absent in New England ebb-tidal deltas. Zones are: I. clean sand; II. sand with mud lenses; III. interbedded sand and mud; IV. mud pebbles in sand; V. interbedded mud and poorly sorted sand; and VI. pebbly sand over Miocene sandstone outcrop (after Oertel, 1975).

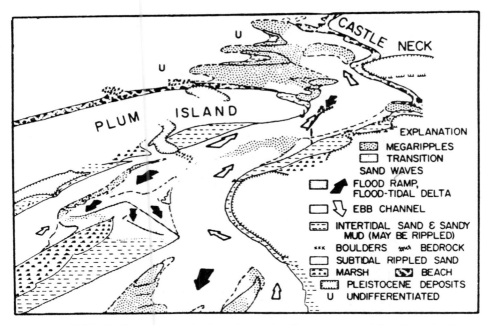

Figure 7-59. Bedform type and orientation and sediment transport directions for the lower Parker River estuary, including both the flood- and ebb-tidal deltas (from Boothroyd and Hubbard, 1975).

England inlets the distance is greater. Figure 7-59 (Boothroyd and Hubbard, 1975) is a summary diagram that shows bedform distribution, sand-wave orientation, and inferred net sediment transport for a continuum from ebb-tidal delta to flood-tidal delta in the Parker River estuary, Massachusetts. It indicates that the deep channel connecting the flood and ebb deltas is an ebb-dominated channel floored mainly by high-energy sand waves.

Sand bodies also occur landward of the principal flood-tidal delta, mainly as point bars within the main or subsidiary channel systems. These point bars have all the attributes of flood-tidal deltas, such as flood ramps, ebb spits, and flood- and ebb-oriented spillover lobes (Figure 7-60). A discussion of point bars is beyond the scope of this chapter, but for an excellent, in-depth discussion of tidal-creek point bars, see Barwis (1978).

Figure 7-60. Point bars (bottom) in the Parker River estuary landward of the flood-tidal delta (top) (photo courtesy of Dennis K. Hubbard).

Stratigraphic Sequences and Depositional Models

It is necessary to combine the areal bedform and resulting sedimentary structure distributions into vertical successions in order to generate stratigraphic sequences of tidal inlet deposits. The purpose is twofold: (1) to properly assess the vertical sequences of Recent deposits for such varied uses as sand resource estimation, channel dredging operations, foundation stability, and water supply pipeline paths for nuclear power plants; and (2) to construct depositional models to interpret better sedimentary rock sequences that can aid in such tasks as the search for petroleum reservoirs and the distribution of coal deposits. It is not the intent of this section to produce an exhaustive or definitive set of models of stratigraphic preservation of inlet deposits, but to comment on some examples and offer ideas for future thought.

Stratigraphic sequences are presented as either regressive or transgressive successions. In a regressive sequence, sedimentary structures and textures are presented at much the same elevation (or depth), and in the same order upward, in which they are observed within the active inlet today. That is, in an ebb-tidal delta example, features of the main ebb channel are at the bottom of a sequence that passes stratigraphically upward to features of intertidal swash bars at the top of the sequence. In a transgressive sequence, the supposed regressive stratigraphic succession is often presented inverted, as if a rising sea level has allowed the deposition and preservation of each bathymetrically deeper feature over the top of the shallower. The most difficult aspect to deal with is preservation potential, the chance that any given feature or interval is preserved and recognizable in the stratigraphic record.

Some Existing Examples

Figure 7-61 (Hayes, 1980) presents a stratigraphic sequence for an inlet-affiliated recurved barrier spit that records the preservation of the inner ebb-tidal delta system. It is essentially a regressive sequence and includes barrier spit dunes and beachface depositional subenvironments that have not been discussed in this chapter. This model is similar to that of Kumar and Sanders (1974, 1975) for Fire Island inlet, New York (Figure 7-62). It is especially applicable to updrift-overlap inlets, or in any location where rapid spit migration and consequent shifting of inlet location can be documented, as in the case of Chatham inlet, Massachusetts (Figures 7-48, 7-49, and 7-51).

A regressive sequence proposed by Greer (1975) for Ossabaw Sound, Georgia, is shown in Figure 7-63. This inlet sequence illustrates a stratigraphic succession of ebb-tidal delta deposits, overlain by tidal flat and

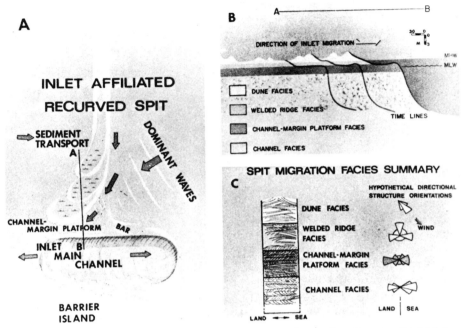

Figure 7-61. Hayes' proposed stratigraphic columns for a migrating mesotidal inlet. (A) Morphologic and sediment transport model. (B) Preserved facies during migration along line AB in (A). (C) Stratigraphic sequence (from Hayes and Kana, 1976).

marsh sediments deposited inside the tidal inlet. More mud and bioturbated units are present because of the decreased wave energy in the central part of the Georgia Embayment.

Hubbard and Barwis (1976) presented two hypothetical sequences based on South Carolina inlet systems (Figures 7-64 and 7-65). Figure 7-64 shows a regressive sequence that includes deposits of the ebb-tidal delta, the barrier spit island, and the flood-tidal delta open-bay system. Figure 7-65, a transgressive sequence, is actually two separate sequences: a regressive flood delta open-bay succession, overlain by a transgressive ebb-tidal delta, shoreface, open-shelf succession. Figures 7-44 and 7-57 present information germaine to these sequences.

Hubbard *et al.* (1979) presented hypothetical block diagrams of sand bodies produced by wave-dominated and tide-dominated mixed energy inlets of the Georgia Embayment (Figure 7-66). These diagrams are a logical follow-on of the Hubbard and Barwis (1976) sequences. However, note that the faces of the blocks indicate transgression with 100% preservation.

Boothroyd *et al.* (in press) illustrates with a series of vibracores a microtidal, wave-dominated flood-tidal delta (Figure 7-67). The stratigraphic sequences indicate a stack of delta lobes formed by transgressive on-

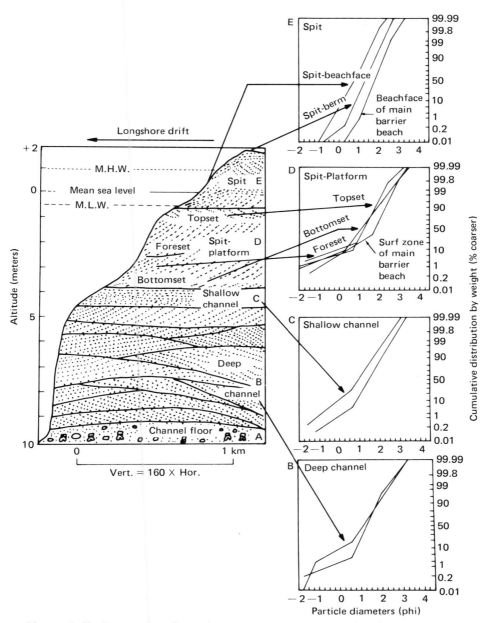

Figure 7-62. Kumar and Sander's stratigraphic sequence (based on box-core information) deposited by westward migration of Fire Island spit, New York, into Fire Island inlet (from Kumar and Sanders, 1974).

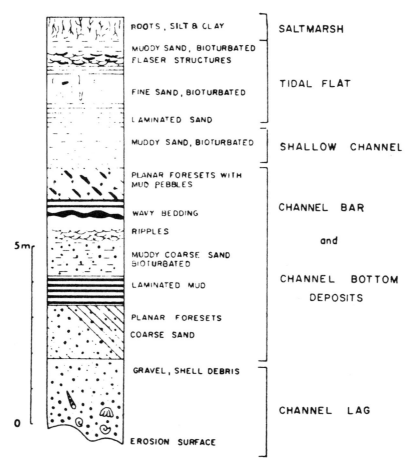

Figure 7-63. Greer's proposed regressive estuarine sequence for Ossabaw Sound, Georgia (based on box-core information) (from Greer, 1975).

lap over low-energy basin deposits. Refer to Figures 7-36 and 7-37 for further illustration.

Lastly, proposed stratigraphic sequences for a mesotidal flood-tidal delta and associated ebb channels, barrier island, and ebb-tidal delta are shown in Figure 7-68. These regressive sequences are based on a composite of information obtained from northern Massachusetts inlets, principally the Parker and the Essex (Figures 7-35, 7-36, and 7-59). The flood delta sequence is similar to that presented by Hayes (1980, Figure 12) except that the dominant internal sedimentary structure in ebb channels is trough cross-stratification generated by megaripples on sand waves as was suggested by Hubbard *et al.* (1979). The ebb-tidal delta sequence is similar to Hayes (1980, Figure 8) but differs in the inclusion of large-scale tabular cross-stratification on the swash platform.

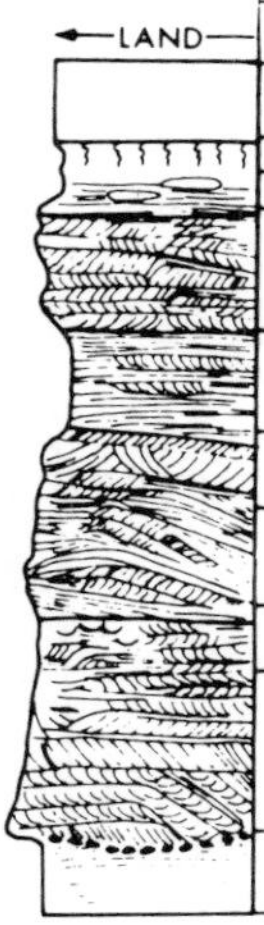

DEPOSITIONAL ENVIRONMENT	PALEOCURRENT DIRECTIONS	DOMINANT SEDIMENTARY STRUCTURES	FAUNAL DIVERSITY
←—LAND—	←— LAND —		
NONMARINE	?	?	?
MARSH		rooting, burrows	low
TIDAL FLAT		burrows, small scale cross beds	
FLOOD DELTA	(symbol)	planar cross beds / planar & trough cross beds	low
OPEN BAY	(symbol)	varies	low
DUNE & WASHOVER	(symbol)	large scale planar & festoon cross beds (dune) / landwarddipping plane beds & small planar crossbeds	low
BEACH*	(symbol)	shallow seaward dipping plane beds; landward dipping high angle planar cross beds	low
EBB DELTA	(symbol)	plane beds, multidirectional trough cross beds	low
SHALLOW ** CHANNEL / DEEP CHANNEL **	(symbol)	large scale planar crossbeds grading upward to bidirectional trough cross beds	low
CHANNEL ** BOTTOM	?	coarse sand & shell lag / varies	moderate

*After Hayes et al (1969)   **Kumar & Sanders (1974)

Figure 7-64. Hubbard and Barwis' hypothetical regressive sequence of mesotidal inlets based on South Carolina examples (from Hubbard and Barwis, 1976).

DEPOSITIONAL ENVIRONMENT	PALEO-CURRENT DIRECTIONS	DOMINANT SEDIMENTARY STRUCTURES	FAUNAL DIVERSITY
←— LAND—	←— LAND—		
OFFSHORE MARINE	?	varies	high
NEARSHORE MARINE; lower contact disconformable	varies*	variety of small scale cross beds	moderate
TIDAL INLET; lower contact disconformable	(symbol)	SHALLOW: small scale cross beds DEEP: large scale planar and trough cross beds BOTTOM: coarse channel lag	low
MARSH; point bars & tidal deltas in channels	(symbol)	MARSH: rooted bioturbated SAND: bimodal trough cross beds, sets to 1m	low
FLOOD-TIDAL DELTA; fanshaped sheet sand	(symbol)	trough and planar cross beds, sets to 1m	low
PROTECTED LAGOON; shallow open bay behind newly formed barrier	?	ripples, bioturbated	low

* Clifton et al (1971)

Figure 7-65. Hubbard and Barwis' hypothetical transgressive sequence of mesotidal inlets and bays based on South Carolina examples (from Hubbard and Barwis, 1976).

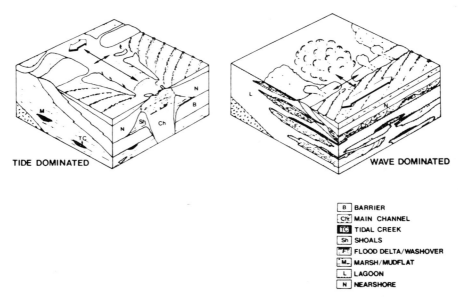

TIDE DOMINATED   WAVE DOMINATED

B	BARRIER
Ch	MAIN CHANNEL
TC	TIDAL CREEK
Sh	SHOALS
FF	FLOOD DELTA/WASHOVER
M	MARSH/MUDFLAT
L	LAGOON
N	NEARSHORE

Figure 7-66. Hypothetical block diagrams of inlet sand bodies in the Georgia Embayment. Sequences assume 100% preservation. Arrows indicate relative migration of inlet position (white) and sediment transport, waves versus tides (black). Scale is 100–200 m top to bottom, 4 km ocean to lagoon (from Hubbard *et al.*, 1979).

Concluding Remarks

It is evident from the preceding sequences that even though there are copious maps of sedimentary structures, depicting an instant in geologic time, there is limited information on stratigraphic sequences deposited over time, based on Recent depositional models. "Hypothetical" and "inferred" are the terms often used, even for sedimentary structure distribution. The studies by Greer (1975), Kumar and Sanders (1974, 1975), and Boothroyd *et al.* (in press) are quite useful because they are based on information from cores. Another problem is the matter of preservation potential; Hubbard and Barwis (1976) specifically address this problem in the variation between their two models (Figures 7-64 and 7-65). Perhaps the best way to approach the regressive versus transgressive and preservation potential problems is to consider geologic processes and resulting distribution of inlet features as they exist today and have existed during the late Holocene transgression.

Presently, sea level is rising at the relative rate of 30 cm per 100 yr for most of the east coast United States (Hicks and Crosby, 1974). In addition, most barrier islands and spits show long-term (40–150 yr) trends of 1–10 m/ yr of beach erosion caused by storm wave attack and lack of sediment supply

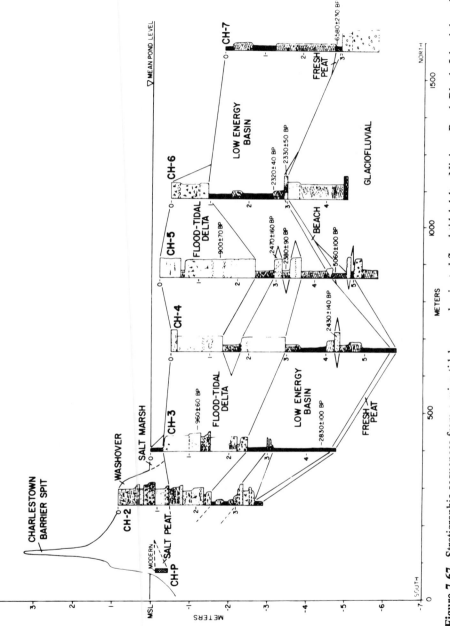

Figure 7-67. Stratigraphic sequences from a microtidal, wave-dominated flood-tidal delta, Ninigret Pond, Rhode Island, based on vibracores. Numbers (960±60BP) indicate ^{14}C radiometric dates (from Boothroyd et al., in press).

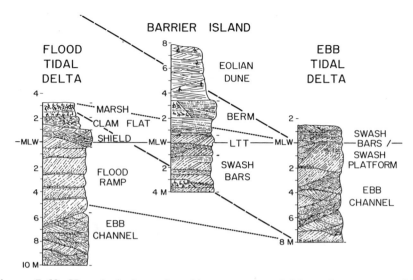

Figure 7-68. Hypothetical stratigraphic sequences of Massachusetts mesotidal mixed-energy inlets, partly based on data of Hayes (1980). Sets of dashed lines show migration paths of the ebb-tidal delta over the flood-tidal delta with transgression and sea level rise. Barrier preservation would be minimal.

(Stephen *et al.*, 1975; Fisher and Simpson, 1979). Last, there is some documented evidence to support the general observation that flood-tidal deltas show a long-term accretional trend (Hine, 1972; Kaczorowski, 1972; Godfrey and Godfrey, 1975).

Computations by Finley (1978) indicate the ebb-tidal delta of North Inlet, South Carolina, is enlarging at the expense of the eroding, nearby barrier spits. Unpublished data of Hine (personal communication, 1977) indicate the preservation of inlet sequences (ebb-tidal deltas) within the shoreface and inner continental shelf stratigraphy of southern North Carolina. Observation of the beaches of many east coast United States barrier islands shows salt marsh peat cropping out on the beachface near mean low water, particularly after storm events. This means that the barrier sediments are rolling back over the uppermost unit of the flood-tidal delta bay sequence. Rhodes (1973), in a seismic refraction and washbore-drilling study of Plum Island, Massachusetts, the area from which much of the tidal inlet information discussed above was drawn, indicated that there is a 10–15 m thick sequence of estuarine sand overlain by about 5 m of barrier sand. The thickness measurements are significant because they provide a scale dimension for at least the preserved flood-tidal delta sequences. Note the scale, or lack thereof, of the stratigraphic models discussed above.

What conclusions can be drawn from the foregoing facts and observations? They are as follows: (1) Ebb-tidal deltas may be preserved as regressive sequences within shoreface successions; (2) the process of barrier island retreat perhaps removes the topmost units of the flood-tidal sequence and most likely all of the upper barrier spit sequence; and (3) flood-tidal delta and bay-lagoon sediments are preserved as regressive sequences. Figure 7-66 (Hubbard *et al.*, 1979) is good because it shows the preservation of ebb delta regressive sequences, but not so good in that it assumes total preservation of the barrier island. Figure 7-68 allows generation of one's own stratigraphic sequence by following any given set of dashed lines that indicate transgression and sea level rise. Figure 7-64 (Hubbard and Barwis, 1976) is favored as probably the best overall predictive model of those that have been offered for inspection. However, there is a great need for additional stratigraphic studies.

References

Allen, J. R. L., 1963. Asymmetrical ripple marks and the origin of water-laid cosets of cross-strata. *Liverpool Manchester Geol. J.*, **3**, 187–236.

Allen, J. R. L., 1965. Sedimentation to the lee of small underwater sand waves: an experimental study. *J. Geol.*, **73**, 95–116.

Allen, J. R. L., 1968a. *Current Ripples: Their Relation to Patterns of Water and Sediment Motion.* North-Holland, Amsterdam, 433 pp.

Allen, J. R. L., 1968b. The diffusion of grains in the lee of ripples, dunes, and sand deltas. *J. Sed. Petrol.*, **38**, 621–633.

Allen, J. R. L., 1973. Phase differences between bed configuration and flow in natural environments, and their geological relevance. *Sedimentology*, **20**, 323–329.

Allen, J. R. L., 1974. Reaction, relaxation, and lag in natural sedimentary systems: general principles, examples, and lessons. *Earth Sci. Rev.*, **10**, 263–342.

Allen, J. R. L., 1980a. Large transverse bedforms and the character of boundary-layers in shallow water environments. *Sedimentology*, **27**, 317–323.

Allen, J. R. L., 1980b. Sandwaves: a model of origin and internal structure. *Sed. Geol.*, **26**, 281–328.

Allen, J. R. L., 1982a. Simple models for the shape and symmetry of tidal sand waves: (1) statically stable equilibrium forms. *Mar. Geol.*, **48**, 31–49.

Allen, J. R. L., 1982b. Simple models for the shape and symmetry of tidal sand waves: (2) dynamically stable symmetrical equilibrium forms. *Mar. Geol.*, **48**, 51–73.

Allen, J. R. L., 1982c. Simple models for the shape and symmetry of tidal sand waves: (3) dynamically stable asymmetrical equilibrium forms without flow separation. *Mar. Geol.*, **48**, 321–326.

Allen, J. R. L., 1982d. *Sedimentary Structures, Their Character and Physical Basis*, Vol. 1: *Developments in Sedimentology 30A.* Elsevier, New York, 593 pp.

Allen, J. R. L., 1982e. *Sedimentary Structures, Their Character and Physical Basis*, Vol. 2: *Developments in Sedimentology 30B*. Elsevier, New York, 663 pp.

Allen, J. R. L., and Friend, P. F., 1976. Relaxation time of dunes in decelerating aqueous flows. *Geol. Soc. London Quart. J.*, **132**, 17–26.

Ball, M. M., 1967. Carbonate sand bodies of Florida and the Bahamas. *J. Sed. Petrol.*, **37**, 556–591.

Barwis, J. H., 1978. Sedimentology of some South Carolina tidal creek point bars, and a comparison with their fluvial counterparts. *In*: Miall, A. D. (ed.), *Fluvial Sedimentology*. Canadian Society of Petroleum Geologists, Calgary, AB, pp. 129–160.

Blatt, H., Middleton, G., Murray, R., 1980. *Origin of Sedimentary Rocks*, 2nd ed. Prentice-Hall, Englewood Cliffs, 782 pp.

Bohacs, K. M., 1981. Flume studies on the kinematics and dynamics of large-scale bedforms, Massachusetts Institute of Technology, Cambridge, Sc.D. Dissertation, 178 pp.

Boon, III, J. D., and Byrne, R. J., 1981. On basin hypsometry and the morphodynamic response of coastal inlet systems. *Mar. Geol.*, **40**, 27–48.

Boothroyd, J. C., Friedrich, N. E., and McGinn, S. R., 1985. Geology of microtidal coastal lagoons: the Rhode Island example. *Mar. Geol.*, **54**.

Boothroyd, J. C., and Hubbard, D. K., 1974. *Bedform Development and Distribution Pattern, Parker and Essex Estuaries, Massachusetts*. Misc. Paper 1-74, Coastal Engineering Research Center, Ft. Belvoir, VA, 39 pp.

Boothroyd, J. C., and Hubbard, D. K., 1975. Genesis of bedforms in mesotidal estuaries. *In*: Cronin, J. E. (ed.), *Estuarine Research*, Vol. 2, *Geology and Engineering*. Academic Press, New York, pp. 217–234.

Collinson, J. D., 1970. Bedforms of the Tana River, Norway. *Geogr. Ann.*, **52A**, 31–56.

Costello, W. R., and Southard, J. B., 1981. Flume experiments on lower-flow-regime bedforms in coarse sand. *J. Sed. Petrol.*, **51**, 849–864.

DaBoll, J. M., 1969. *Holocene Sediments of the Parker River Estuary, Massachusetts*. Cont. No. 3-CRG, Dept. of Geology, Univ. of Massachusetts, 138 pp.

Dalrymple, R. W., Amos, C. L., and McCann, S. B., 1982. Excursion 6A: Beach and nearshore depositional environments of the Bay of Fundy and Southern Gulf of St. Lawrence. Int. Assoc. Sedimentologists, 11th Int. Congr. on Sedimentology, *Field Excursion Guidebook*, 116 pp.

Dalrymple, R. W., Knight, R. J., and Lambiase, J. J., 1978. Bedforms and their hydraulic stability relationships in a tidal environment, Bay of Fundy, Canada. *Nature*, **275**, 100–104.

Dalrymple, R. W., Knight, R. J., and Middleton, G. V., 1975. Intertidal sand bars in Cobequid Bay (Bay of Fundy). *In*: Cronin, L. E. (ed.), *Estuarine Research*, Vol. 2, *Geology and Engineering*. Academic Press, New York, pp. 293–308.

Davies, J. L., 1964. A morphogenic approach to world shorelines. *Z. Geomorph.*, **8**, 27–42.

Davies, J. L., 1980. *Geographical Variation in Coastal Development*, 2nd ed. Longman, New York, 212 pp.

Davis, Jr., R. A., and Fox, W. T., 1981. Interaction between wave- and tide-generated processes at the mouth of a microtidal estuary: Matanzas River, Florida (U.S.A.). *Mar. Geol.*, **40**, 49–68.

Davis, R. A., Fox, W. T., Hayes, M. O., and Boothroyd, J. C., 1972. Comparison of ridge-and-runnel systems in tidal and non-tidal environments. *J. Sed. Petrol.*, **32**, 413–421.

DeAlteris, J. T., and Byrne, R. J., 1975. The recent history of Wachapreague Inlet, Virginia. *In*: Cronin, L. E. (ed.), *Estuarine Research*, Vol. 2, *Geology and Engineering*. Academic Press, New York, pp. 167–182.

Farrell, S. C., 1970. *Sediment Distribution and Hydrodynamics, Saco River and Scarboro Estuaries, Maine*. Cont. No. 6-CRG, Dept. of Geology, Univ. of Massachusetts, 129 pp.

Finley, R. J., 1975. Morphologic development and dynamic processes at a barrier island inlet, North Inlet, S.C. *In*: Cronin, L. E. (ed.), *Estuarine Research*, Vol. 2, *Geology and Engineering*. Academic Press, New York, pp. 277–292.

Finley, R. J., 1976. *Hydraulics and Dynamics of North Inlet, S.C., 1974–75*. GITI Report 10, U.S. Army Coastal Engineering Research Center, 188 pp.

Finley, R. J., 1978. Ebb-tidal delta morphology and sediment supply in relation to seasonal wave energy flux, North Inlet, S.C.: *J. Sed. Petrol.*, **48**, 227–238.

Fisher, J. J., and Simpson, E. J., 1979. Washover and tidal sedimentation rates as environmental factors in development of a transgressive barrier shoreline. *In*: Leatherman, S. P. (ed.), *Barrier Islands from the Gulf of St. Lawrence to the Gulf of Mexico*: Academic Press, New York, pp. 127–148.

FitzGerald, D. M., 1976. Ebb-tidal delta of Price Inlet, S.C.: geomorphology, physical processes, and associated inlet shoreline changes. *In*: Hayes, M. O., and Kana, T. W. (eds.), *Terrigenous Clastic Depositional Environments*. Rept. No. 11-CRD, Coastal Res. Div., Dept. of Geol., Univ. of South Carolina, II-143-157.

FitzGerald, D. M., 1977. Hydraulics, Morphology and Sediment Transport at Price Inlet, South Carolina. Ph.D. Dissert., Univ. of South Carolina, 84 pp.

FitzGerald, D. M., and Nummedal, D., 1983. Response characteristics of an ebb-dominated tidal inlet channel. *J. Sed. Petrol.*, **53**, 833–845.

FitzGerald, D. M., Nummedal, D., and Kana, T., 1976. Sand circulation patterns at Price Inlet, South Carolina. *Proc. 15th Coastal Engineering Conference*, Amer. Soc. Civil Engr., pp. 1868–1880.

Gallivan, L. B., and Davis, Jr., R. A., 1981. Sediment transport in a microtidal estuary: Matanzas River, Florida (U.S.A.). *Mar. Geol.*, **40**, 69–83.

Gilbert, G. K., 1914. *The Transportation of Debris by Running Water*. U.S. Geol. Survey Prof. Paper 86, Washington, DC.

Godfrey, P. J., and Godfrey, M. M., 1974. The role of overwash and inlet dynamics in the formation of salt marshes in North Carolina barrier islands. *In*: *Ecology of Halophytes*. Academic Press, New York, pp. 407–427.

Godfrey, P. J., and Godfrey, M. M., 1975. Some estuarine consequences of barrier island stabilization. *In*: Cronin, L. E. (ed.), *Estuarine Research*, Vol. 2, *Geology and Engineering*. Academic Press, New York, pp. 485–516.

Greer, S. A., 1969. *Sedimentary Mineralogy of the Hampton Harbor Estuary, New Hampshire and Massachusetts*. Univ. of Massachusetts, Coastal Research Group, Cont. No. 1, pp. 403–414.

Greer, S. A., 1975. Sandbody geometry and sedimentary facies at the estuary-marine transition zone, Ossabaw Sound, GA: a stratigraphic model. Senckenberg. Marit., **7**, 105–136.

Guy, H. P., Simons, D. B., and Richardson, E. V., 1966. *Summary of Alluvial Channel Data from Flume Experiments, 1956–61*. U.S. Geol. Survey Prof. Paper, 462-I, 96 pp.

Harms, J. C., 1969. Hydraulic significance of some sand ripples. *Geol. Soc. Amer. Bull.*, **80**, 363–396.

Harms, J. C., Southard, J. B., Spearing, D. R., and Walker, R. G., 1975. Depositional environments as interpreted from primary sedimentary structures and stratification sequences. Lecture Notes, *Soc. Econ. Paleont. Mineral. Short Course 2*, Dallas, TX, 161 pp.

Harms, J. C., Southard, J. B., and Walker, R. G., 1982. Structures and sequences in clastic rocks. *Soc. Econ. Paleont. Mineral. Short Course 9*, 249 pp.

Hartwell, A., 1970. *Hydrography and Holocene Sedimentation of the Merrimack River Estuary, Massachusetts*. Cont. No. 5-CTG, Dept. of Geology, Univ. of Massachusetts, 166 pp.

Hayes, M. O., 1969. *Coastal Environments, NE Massachusetts and New Hampshire, Eastern Section*. Soc. Econ. Paleont. Mineral. Field Trip Guidebook, 462 pp.

Hayes, M. O., 1975. Morphology of sand accumulations in estuaries. *In*: Cronin, L. E. (ed.), *Estuarine Research*, Vol. 2, *Geology and Engineering*. Academic Press, New York, pp. 3–22.

Hayes, M. O., 1979. Barrier island morphology as a function of tidal and wave regime. *In*: Leatherman, S. P. (ed.), *Barrier Islands from the Gulf of Mexico*. Academic Press, New York, pp. 1–27.

Hayes, M. O., 1980. General morphology and sediment patterns in tidal inlets: Sed. Geol., **26**, 139–156.

Hayes, M. O., and Boothroyd, J. C., 1969. Storms as modifying agents in the coastal environments. *In*: *Coastal Environments, NE Massachusetts and New Hampshire*. Soc. Econ. Paleont. Mineral. Field Trip Guidebook, Cont. No. 1-CRG, Univ. of Massachusetts, pp. 245–265.

Hayes, M. O., Goldsmith, V., and Hobbs, C. H., 1970. Offset coastal inlets. *Proc. 12th Coastal Engr. Conf.*, Washington, DC, pp. 1187–1200.

Hayes, M. O., and Kana, T. W., 1976. *Terrigenous Clastic Depositional Environments*. Tech. Rept. No. 11-CRD, Coastal Research Division, Dept. of Geology, Univ. of South Carolina, 302 pp.

Hicks, S. D., and Crosby, J. E., 1974. *Trends and Variability of Yearly Mean Sea Level, 1893–1972*. NOAA Tech. Memo., NOS 13, Rickville, MD, 14 pp.

Hine, A. C., 1972. *Sand Deposition in the Chatham Harbor Estuary and on Neighboring Beaches, Cape Cod, Massachusetts*. Unpubl. M.S. Thesis, Univ. of Massachusetts, 154 pp.

Hine, A. C., 1975. Bedform distribution and migration patterns on tidal deltas in the Chatham Harbor Estuary, Cape Cod, Massachusetts. *In*: Cronin, L. E., (ed.), *Estuarine Research*, Vol. 2, *Geology and Engineering*. Academic Press, New York, pp. 235–252.

Hine, A. C., 1977. Lilly Bank, Bahamas: history of an active oolite sand shoal. *J. Sed. Petrol.*, **47**, 1554–1583.

Hubbard, D. K., 1973. *Morphology and Hydrodynamics of Merrimack Inlet, Massachusetts*. Final Report, Coastal Engineering Research Center, Cont. DACW72-72-CO 0032, 162 pp.

Hubbard, D. K., 1975. Morphology and hydrodynamics of the Merrimack River ebb-tidal delta. *In*: Cronin, L. E. (ed.), *Estuarine Research*, Vol. 2, *Geology and Engineering*. Academic Press, New York, pp. 253–266.

Hubbard, D. K., and Barwis, J. N., 1976. Discussion of tidal inlet sand deposits: example from the South Carolina coast. *In*: Hayes, M. O. and Kana, T. W. (eds.), *Terrigenous Clastic Depositional Environments*. Tech. Rept. No. 11-CRD, Univ. of South Carolina, Columbia, pp. II-158–II-171.

Hubbard, D. K., Barwis, J. H., and Nummedal, D., 1977. Sediment transport in four South Carolina inlets. *Proc. Coastal Sediments*, Amer. Soc. Civil Engr., pp. 582–601.

Hubbard, D. K., Oertel, G., and Nummedal, D., 1979. The role of waves and tidal currents in the development of tidal-inlet sedimentary structures and sand body geometry: examples from North Carolina, South Carolina and Georgia. *J. Sed. Petrol.*, **49**, 1073–1092.

Jopling, A. V., 1961. *Origin of Regressive Ripples Explained in Terms of Fluid-Mechanic Processes*. U.S. Geol. Survey Prof. Paper 424-D, pp. 15–17.

Jopling, A. V., 1963. Hydraulic studies on the origin of bedding. *Sedimentology*, **2**, 115–121.

Jopling, A. V., 1964. Laboratory study of sorting processes related to flow separation. *J. Geophys. Res.*, 69, 3403–3418.

Jopling, A. V., 1965a. Laboratory study of sorting processes in crossbedded deposits. *In*: Middleton, G. V. (ed.), *Primary Sedimentary Structures and Their Hydrodynamic Interpretation*. Soc. Econ. Paleont. and Mineral Spec. Publ. 12, Tulsa, OK, pp. 53–65.

Jopling, A. V., 1965b. Hydraulic factors controlling the slope of laminae in laboratory deltas. *J. Sed. Petrol.*, 35, 777–791.

Jopling, A. V., 1966a. Origin of cross-laminae in a laboratory experiment. *J. Geophys. Res.*, **71**, 1123–1133.

Jopling, A. V., 1966b. Some applications of theory and experiments to the study of bedding genesis. *Sedimentology*, **7**, 71–102.

Jopling, A. V., 1967. Origin of laminae deposited by the movement of ripples along a stream bed. *J. Geol.*, **75**, 287–305.

Kaczorowski, R. I., 1972. *Offset Tidal Inlets, Long Island, New York*. Unpubl. M.S. Thesis, Univ. of Masachusetts, 150 pp.

King, C. A. M., 1972. *Beaches and Coasts*, 2nd ed., St. Martins Press, New York, 570 pp.

King, C. A. M., and Williams, W. W., 1949. The formation and movement of sand bars by wave action. *Geogr. J.*, **113**, 70–85.

Klein, G. deV., 1970. Depositional and dispersal dynamics of intertidal sand bars. *J. Sed. Petrol.*, **40**, 1095–1127.

Kumar, N., and Sanders, J. E., 1974. Inlet sequence: a vertical succession of sedimentary structures and textures created by the lateral migration of tidal inlets. *Sedimentology*, **21**, 491–532.

Kumar, N., and Sanders, J. E., 1975. Inlet sequence formed by the migration of Fire Island Inlet, Long Island, New York. *In*: Ginsburg, R. N. (ed.), *Tidal Deposits*. Springer-Verlag, New York, pp. 75–83.

Lauff, G. H., 1967. *Estuaries*. Amer. Assoc. Adv. Sci. Publ. 83, Washington, DC.

Lynch-Blosse, M. A., and Kumar, N., 1976. Evolution of downdrift-offset tidal inlets: a model based on the brigantine inlet system of New Jersey. *J. Geol.*, **84**, 165–178.

Maddock, T., 1969. The behavior of straight open channels with moveable beds, U.S. Geol. Surv., Prof. Paper 622-A, 70 pp.

McGowen, J. H., and Garner, L. E., 1970. Physiographic features and stratification types of coarse-grained point bars; modern and ancient examples. *Sedimentology*, **14**, 77–111.

McKee, C. E., and Weir, G. W., 1953. Terminology of stratification and cross-stratification. *Geol. Soc. Amer. Bull.*, **64**, 381–390.

Nummedal, D., Oertel, G. F., Hubbard, D. K., and Hine, III, A. C., 1977. Tidal inlet variability—Cape Hatteras to Cape Canaveral. *Coastal Sediments*, Amer. Soc. Civil Engr., pp. 543–562.

Nummedal, D., and Fischer, I. A., 1978. Process-response models for depositional shorelines: the German and the Georgia Bights. *Proc. 16th Coastal Eng. Conf.*, *Amer. Soc. Civil Engrs.*, 1215–1231.

Nummedal, D., and Penland, S., 1981. Sediment dispersal in Norderneyer Seegat, West Germany. *In*: Nio, S. D., Schuettenhelm, R. T. E., and van Weering, T. C. E. (eds.), *Holocene Marine Sedimentation in the North Sea Basin*. Int. Assoc. Sedimentologists Spec. Publ. 5, 515 pp.

Oertel, G. F., 1972. Sediment transport of estuary entrance shoals and the formation of swash platforms. *J. Sed. Petrol.*, **42** (4), 858–863.

Oertel, G. F., 1973. Examination of textures and structures of mud in layered sediments at the entrance of a Georgia tidal inlet. *J. Sed. Petrol.*, **43**, 33–41.

Oertel, G. F., 1975. Ebb-tidal deltas of Georgia estuaries. *In*: Cronin, L. E. (ed.), *Estuarine Research*, Vol. 2, *Geology and Engineering*. Academic Press, New York, pp. 267–276.

Oertel, G. F., and Howard, J. D., 1972. Water circulation and sedimentation at estuary entrances on the Georgia coast. In Swift, D. J. P., Duane, D. B., and

Pilkey, O. H. (eds.), *Shelf Sediment Transport*. Dowden, Hutchinson and Ross, Stroudsburg, PA, pp. 411–428.

Pettijohn, F. J., Pottern, P. E., and Siever, R., 1973. *Sand and Sandstone*. Springer-Verlag, New York, 618 pp.

Postma, H., 1954. Hydrography of the Dutch Wadden Sea. *Arch. Neerl. Zool.*, **10**, 405–511.

Postma, H., 1961. Transport and accumulation of suspended matter in the Dutch Wadden Sea. *Neth. J. Sea Res.*, **1**, 148–190.

Postma, H., 1967. Sediment transport and sedimentation in the estuarine environment. *In*: Lauff, G. A. (ed.), *Estuaries*. Amer. Assoc. Adv. Sci., Washington, DC, pp. 158–179.

Price, W. A., 1963. Patterns of flow and channeling in tidal inlets. *J. Sed. Petrol.*, **33** (2), 279–290.

Raudkivi, A. J., 1963. Study of sediment ripple formation. *J. Hydr. Div. Amer. Soc. Civil Engr. Proc.*, **89**(HY6), 15–33.

Reineck, H. E., and Singh, I. B., 1980. *Depositional Sedimentary Environments*. 2nd ed., Springer-Verlag, New York, 549 pp.

Rhodes, E. G., 1973. *Pleistocene-Holocene Sediments Interpreted by Seismic Refraction and Wash-Bore Sampling, Plum Island-Castle Neck, Massachusetts*. Tech. Memo. No. 40, U.S. Army Coastal Engineering Research Center, 75 pp.

Robinson, A. H. W., 1960. Ebb-flood channel systems in sandy bays and estuaries. *Geography*, **45**, 183–199.

Schubel, J. R., 1971. *The Estuarine Environment—Estuaries and Estuarine Sedimentation*. Wye Institute, MD, Amer. Geol. Inst. Short Course Lecture Notes, 30–31 October, 1971.

Simons, D. B., and Richardson, E. V., 1961. Forms of bed roughness in alluvial channels. *Amer. Soc. Civil Engr. Proc.*, **87**(HY3), 87–105.

Simons, D. B., and Richardson, E. V., 1963. Forms of bed roughness in alluvial channels. *Amer. Soc. Civil Engr. Trans.*, **128**, 284–323.

Simons, D. B., Richardson, E. V., and Nordin, Jr., C. F., 1965. Sedimentary structures generated by flow in alluvial channels. *In*: Middleton, G. V. (ed.), *Primary Sedimentary Structures and Their Hydrodynamic Interpretation*. Soc. Econ. Paleont. and Mineral. Spec. Publ. 12, Tulsa, OK, pp. 34–52.

Southard, J. B., 1971. Representation of bed configurations in depth-velocity-size diagrams. *J. Sed. Petrol.*, **41**, 903–915.

Southard, J. B., and Dingler, J. R., 1971. Flume study of ripple propogation behind mounds on flat sand beds. *Sedimentology*, **16**, 251–263.

Stephen, M. F., Brown, P. J., Fitzgerald, D. M., Hubbard, D. K., and Hayes, M. O., 1975. *Beach Erosion Inventory of Charleston County, South Carolina*. South Carolina Sea Grant Tech. Rept. No. 4, 79 pp.

Van Straaten, L. M. J. U., 1950. Environment of formation and facies of the Wadden Sea sediments. *Koninkl. Ned. Aardrijkskde Genoot.*, **67**, 94–108.

Van Straaten, L. M. J. U., 1951. Longitudinal ripple marks in mud and sand. *J. Sed. Petrol.*, **21**, 47–54.

Van Straaten, L. M. J. U., 1952. Biogene textures and the formation of shell beds in the Dutch Wadden Sea. *Proc. Koninkl. Ned. Akad. Wetenschap. Amsterdam, Ser. B*, **55**, 500–516.

Van Straaten, L. M. J. U., 1954. Composition and structure of recent marine sediments in the Netherlands. *Leidse Geol. Mededelin.*, **19**, 1–110.

Van Straaten, L. M. J. U., 1956. Composition of shell beds formed in tidal flat environment in the Netherlands and in the Bay of Arcachon (France). *Geol. en Mijnbouw n.s.*, **18**, 209–226.

Van Straaten, L. M. J. U., 1961. Sedimentation in tidal flat areas. *Alberta Soc. Petrol. Geol. J.*, **9**, 203–226.

Van Straaten, L. M. J. U., and Keunen, P. H., 1957. Accumulation of fine grained sediments in the Dutch Wadden Sea. *Geol. en Mijnbouw n.s.*, **19**, 329, 354.

8

The Shoreface

*Alan W. Niedoroda, Donald J. P. Swift,
and Thomas S. Hopkins*

Introduction

The shoreface (Barrell, 1912) is a relatively simple and easily distinguished coastal environment. It appears as a concave upward surface with a slope on the order of 1:200 and is just seaward of the surf zone. This slope decreases seaward until the shoreface blends into the continental shelf with a slope on the order of 1:2000 (Figure 8-1).

Early students of coastal geomorphology recognized that the shoreface is a dynamic surface, a response to the fluid power expenditure of waves and currents (Fenneman, 1902). On unconsolidated coasts exposed to the full force of marine processes, shorefaces are surfaces curved about an axis parallel to the shoreline and exhibit little change in the alongshore direction (Figure 8-1). On rocky coasts, however, the time required for the profile to be incised into the substrate is long compared to the rate of *relative* sea level change; hence shorefaces are poorly developed on rocky coasts. Local sand accumulations on rocky coasts, such as spits, barriers, and tombolos, develop well-defined shorefaces, but these shoreface fragments are irregularly distributed in plan view. There are many variations in the width, depth, shape, and longshore extent of the shoreface. Although it is a common feature, it is not always present. In general, the shoreface environment makes up the seaward boundary of what is considered the coastal zone.

The boundaries between the shoreface and such related coastal environments such as the surf zone, inlets, and the inner shelf floor are difficult to define. For instance, the surf zone may be defined by the morphologic features that are formed and maintained by processes related to breaking waves. On the other hand, the dynamic definition of the surf zone is the zone of breaking waves. The offshore extent of the surf zone defined by short-term dynamics often does not correspond to the surf zone defined by morphologic

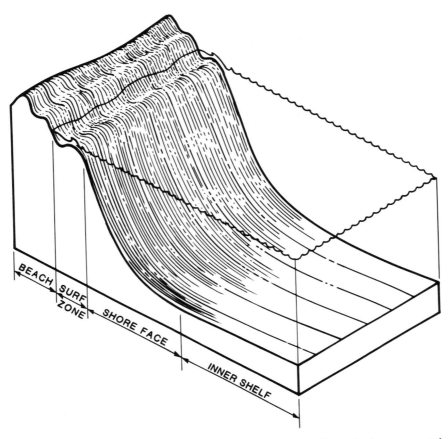

Figure 8-1. Schematic diagram of the morphologic zones of a typical open coastal ocean.

features. Is the surf zone morphology "tuned" to average wave conditions, wave conditions characterizing common storm conditions, or wave conditions characterizing extreme storm conditions? Tanner (1960) has related gross coastal features to the mean annual breaker height, but more detailed refinement of this concept has not been developed. Price (1955) demonstrated the relationship between beach and shoreline types with the general offshore bottom slope. The problem of matching the dynamic length and time scales to the morphological scales of a coastal environment exists in discussing the shoreface as well as the surf zone.

Fluid Dynamics of the Shoreface

Coastal processes occur on time scales that vary from nearly instantaneous to geologic intervals. A comprehensive discussion of the processes that dominate and control a coastal environment, such as the shoreface, needs to consider all of these scales and to distinguish the combination of processes that are most relevant to operation of the physical system at each time scale. Subgeologic time-scale processes are discussed in this and the two following chapter sections. Geologic time-scale processes are discussed in the last section.

Local sea level changes and variations in local sediment supply have relatively small effects on subgeologic time-scale shoreface processes. Observations and empirical data suggest that the shoreface of open-ocean sandy coasts is a dynamic feature whose shape is adjusted to bring the combination of forces (gravitational and fluid forces from waves and currents) that act on the shoreface sediments into a time-averaged equilibrium. If this is true, then the morphology of the shoreface should result from this time-averaged balance of forces. Differences in the width, depth, slope, and shape of the shorefaces off different coasts should be explainable in terms of differences in the oceanographic environments. The processes that dominate the sediment transport at different times and places on the shoreface should be controlled by the relative action of waves and currents. The significant features of waves and currents in shoreface regions are discussed in this section to provide a basis for understanding sediment transport and morphologic response in this environment.

Waves

The shoreface corresponds to an important transition region for ocean waves. Surface gravity waves are not significantly influenced by the bottom as they pass over the continental shelf. The effect of the bottom increases as waves propagate across the shoreface and is very important where they are breaking in the surf zone. The shoreface region is the transition zone where the decreasing depths cause the form of the waves to change to more peaked crests, to increase in height (the shoaling effect), and to alter their propagation toward a more shore-normal direction (wave refraction). Because these changes in gravity wave behavior influence the sediment transport processes acting on the shoreface, it is relevant to discuss them briefly.

Gravity waves in the ocean occur in a wide range of shapes that propagate as apparently regular undulations. As physical theory was developed to define and describe these wave motions, a series of approximations was made that permitted the mathematical solution of governing differential equations. Airy (1845) developed a widely used wave theory by simplifying the apparent behavior of gravity waves to a regular long-crested sinusoidal form whose amplitude is very small compared with the wave length. The solution of the governing equation requires linearization of one of the free surface boundary conditions and hence Airy wave theory is also referred to as linear wave theory.

Linear wave theory is relatively easy to apply and remarkably useful in representing many aspects of real gravity wave behavior. However, it fails to represent some features that are quite relevant to considerations of shoreface processes. For example, the surface profile and time-history of horizontal orbital velocities are regular sinusoidal functions in linear wave theory, whereas they are generally observed to have peaked crests and broader, flattened troughs in nature.

Stokes (1847, 1880) developed a more complete theory for waves of finite amplitude. The solution of the governing equation utilizes a series expansion. The solution based on truncating this series after the second term is referred to as Stokes' second-order theory. Subsequent researchers extended this theory to third-order (Borgman and Chappelear, 1958; Skjelbreia, 1959) and fifth-order (Skjelbreia and Hendrickson, 1961). Stokes' second- and higher order wave theories yield a more realistic surface profile and time history of orbital velocities than linear wave theory but are also more cumbersome to apply. Stokes' theory does not represent behavior of gravity waves under breaking and near-breaking conditions very well.

Other theories, such as the cnoidal wave theory (Korteweg and de Vries, 1895; Keulegan and Patterson, 1940; Keller, 1948; Littman, 1957), solitary wave theory (Boussinesq, 1872; Rayleigh, 1876; McCowan, 1891), and numerical stream function (Chappelear, 1961; Dean, 1965, 1970; Dean and Asce, 1965), have been developed. Each theory has advantages with respect to its ability to represent certain aspects of gravity wave behavior while presenting disadvantages usually caused by the relative complexity of the mathematical representations. Researchers applying wave theories to geophysical problems must select the wave theory that best represents the physical behavior that is important to the situation. In general, the simplest wave theory applicable to the problem is used in order to avoid undesired mathematical complexity.

Numerous researchers have compared observed wave behavior to that predicted by various wave theories to establish guidelines for their relative range of applicability (Miche, 1944; Keller, 1948; Ursell, 1953; Longuet-Higgins, 1956; Housley and Taylor, 1957; Littman, 1957). Komar (1976, pp. 60–62) provides an excellent summary of this work and a nomograph that distinguishes fields of applicability for Airy, Stokes, Cnoidal, and Solitary wave theories based on the ratios of wave height to depth and depth to wave length while maintaining preference to the application of the simplest theory to the widest range of conditions.

Figure 8-2 shows a replotting of these criteria for waves of 6-sec period (a common period for ocean waves) in depths representative of most shorefaces. From this figure it is seen that for depths greater than approximately 20 m the boundary between wave heights corresponding to the applicability of linear or Stokes theory is independent of depth. At shallower depths the waves become progressively more peaked in the crests and flattened in the troughs so that the range of applicability of linear wave theory is reduced.

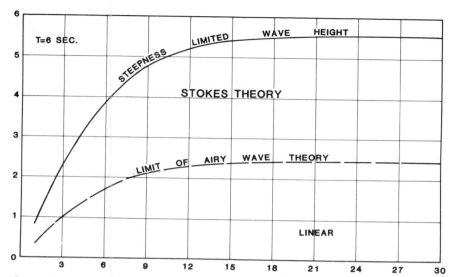

Figure 8-2. Example of the range of conditions where Stokes and linear wave theories are applicable. The abscissa is depth in meters; the ordinate is wave height in meters. The wave example period is 6 sec.

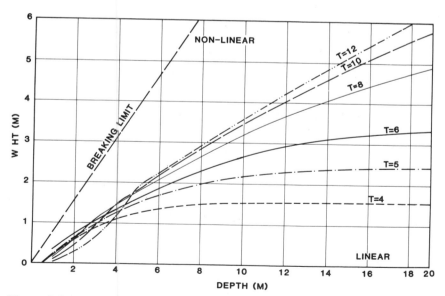

Figure 8-3. Ranges of conditions where nonlinear and linear (Stokes) wave theories apply for different wave periods.

Figure 8-3 shows a replotting of the field of applicability criteria from Komar's figure for a range of wave periods that is characteristic of open-ocean wave spectra. This figure illustrates that gravity wave behavior best represented by Stokes' theory is increasingly more important as the waves approach and pass over the shoreface.

Wave refraction also becomes significant in depths characteristic of the shoreface. This phenomena is covered in many basic textbooks (e.g., Kinsman, 1965; Ippen, 1966; Barber, 1969; and Komar, 1976). For the purposes of discussing shoreface sediment dynamics, it suffices to point out that wave refraction is generally not sufficient to bring the wave propagation direction parallel to the slope normal. Thus, ocean waves usually propagate across the shoreface at some angle.

Shoreface Currents

The shoreface zone corresponds to an important transition zone for ocean currents as well as for waves. Inshore of the shoreface, the currents within the surf zone are primarily driven by energy imparted by breaking waves. These wave-driven longshore currents overwhelm wind-driven and tidal currents in the surf zone. Seaward of the surf zone, the effects of wind stress, astronomical tides, the Coriolis effect, and internal pressure gradients resulting from small slopes of the mean sea surface and the internal density field become important. Currents outside of the surf zone can result from local forcing, such as the wind stress acting on a given area, or from external forcing that is generally characterized by slopes of the mean surface level or internal stratification that propagate to an area as long wavelike motions.

The present level of understanding of currents in the coastal ocean is modest at best. Unlike deep ocean areas where currents are relatively constant in magnitude and pattern over long periods of time, the coastal ocean currents are highly variable. Vertical profiles of currents in this zone frequently demonstrate reversed directions between surface and bottom flows. Similar reversals in the direction of alongshore currents can occur within a few kilometers of the shoreline. The reduced depths of the coastal ocean promote faster response of the currents to changes in the forcing. Where forcing changes rapidly, as often happens with changes in the local winds, the currents require specific response times to adjust. If the forcing is weak or rapidly changing, the resulting pattern of currents is generally incoherent. On the other hand, strongly forced flows that have existed for sufficient durations demonstrate regular and coherent current patterns.

This section will examine coastal ocean or shoreface current patterns in detail. This discussion will be entirely descriptive. The concept of dynamic components is used to separate different types of current patterns as a convenient method of explanation. These consist of tidal current components, wind-driven components, etc. Such an expedient separation of these phenomena assists in developing a description of the dynamics of this zone. This division is, however, somewhat hazardous because it ignores higher order interactions between the dynamic components.

In the following discussion of wind-driven currents in the coastal ocean, it is important to consider the time required to establish steady flow conditions. This time is usually taken to be that of a half pendulum day, which falls between the diurnal and semidiurnal periods for mid-latitude coasts. Essentially it is the cycling time of a perturbed parcel of water in a rotating system. Friction tends to dampen the cycling time. Over the shoreface friction is greater so that perturbations come to dynamic equilibrium quicker than offshore, i.e., on the order of several hours instead of a half pendulum day. Because of this greater frictional coupling, the response has less lag to the changes in the forcing than offshore.

In the following portions of this chapter section, tidal currents will be discussed separately from other coastal ocean currents because their dynamics are fundamentally different. Tidal currents are considered first.

Tidal currents in the coastal ocean

Tidal currents arise as a result of astronomical forcing that causes the periodic rise and fall of sea level, generally referred to as the tide level. The internal density structure of a stratified shelf water mass can co-oscillate and cause the baroclinic tide. Although the baroclinic tide may be locally important, it is not as ubiquitous as that resulting from the vertical excursions of sea level known as the barotropic tide. In this section only barotropic tidal effects are discussed.

Early work on tidal currents in shallow seas conducted by Sverdrup (1927), Fjeldstad (1929, 1936), Van Veen (1938), Fleming (1938), Rattray (1957), and Redfield (1958). More recent work by Clarke and Battisti (1981) and Battisti and Clarke (1982a, 1982b) has shown that the bottom slope, local coastal tide range, angular frequency of the tidal species under consideration, a measure of the longshore derivative of the local tide height, and empirical data on the phase of coastal tides. Their work shows that currents tend to vary over a tidal cycle in a well-known elliptical pattern. The ellipticity of this pattern as well as its phase and orientation relative to the coastline are highly variable. Tide heights and tidal currents tend to be amplified by wide continental shelves. Close to the shore, within a water depth of 20 m, the tidal currents tend to diminish rapidly as a result of relatively enhanced frictional effects. Figure 8-4 shows examples of M_2 barotropic tidal currents off several U.S. coastal locations.

Very little work has been devoted to studying the characteristic vertical profile of tidal currents. Sverdrup (1927) and Fjeldstad (1936) showed that

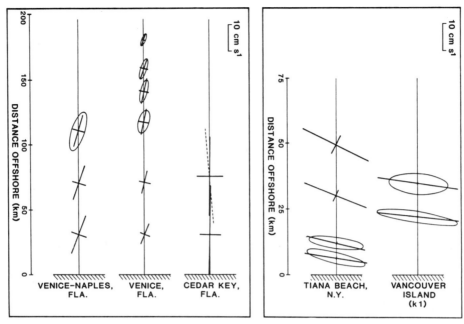

Figure 8-4. Calculated versus measures M_2 tidal current ellipses for several North American coastal locations (from Battisti and Clarke, 1982).

tidal currents over continental shelves are typically effected by friction only in the bottom layer. The depth of this frictional boundary layer is a function of the tidal species, local bottom conditions, and characteristic turbulence. Van Veen (1938) used data from the Dover Straits to show that the vertical velocity of the tidal currents varies as a power of the height over the bottom. Niedoroda *et al.* (1984) showed data that indicated that tidal currents may be fitted with a logarithmic velocity profile in the coastal boundary layer.

Tidal currents acting over the shoreface are often of limited importance in controlling net sediment transport. The fluid shear stress acting on the sea bottom as a result of tidal currents combines with that resulting from other currents and contributes to the entrainment and movement of the bed load. Frictionally induced turbulence in tidal currents is important in maintaining suspended sediment transport. On the other hand, tidal currents in many coastal areas are very nearly symmetrical so that on the average no net sediment transport results solely from the effect of this current component. Where local bathymetry produces asymmetry in the tidal currents they can play a significant role in producing net shoreface sediment transport.

Wind-driven shoreface currents

The wind acts on the coastal ocean in three ways. Energy is transferred directly from the wind to surface currents through the air–water interface.

These wind-driven surface currents can converge against the shoreline and produce a set-up of the mean water surface against the shore. Conversely, surface currents with an offshore component diverge at the shoreline and yield a set-down of the mean water surface. The set-up or set-down results in pressure gradient forces within the coastal water mass that drive alongshore currents. The wind-driven currents also deform the pattern of internal density stratification in the coastal ocean, resulting in internal pressure gradient forces that oppose those generated by the sea level distortions and reduce the alongshore flow.

Coastal ocean dynamics have been the subject of increasing interest in recent years. Distinct features of currents have been identified over a wide range of spatial scales and frequencies. Papers reporting observations of coastal ocean currents and density fields have been published by Blanton and Murthy (1974), Scott and Csanady (1976), Bennett and Magnell (1979), Wiseman and Rouse (1980), and Schwing et al. (1983). The dynamics of the coastal boundary layer have been discussed by Csanady (1972, 1977a,b, 1978a), and Allen (1980). More general information is given in textbooks such as Pedlowsky (1979), Gill (1982), and Csanady (1982). In spite of this attention the complexity of coastal ocean kinematics and dynamics has not been completely described. This is in part an observational problem, since accurate, comprehensive data are uncommon.

From a comprehensive data set Hopkins and Swoboda (in press) are able to demonstrate the distinctiveness of a nearshore regime determined by rapid changes in the dynamic balances as the water column depth decreases. On this basis, Niedoroda (1980), Niedoroda and Swift (1981), Niedoroda et al. (1984) and Swift et al. (1985) have shown that the processes of the nearshore dynamic zone can lead to significant patterns in shoreface sediment transport.

The coastal nearshore is exposed to great variations in forcing, i.e., in wind stress, sea level, and internal density structure. What we observe in nature is a composite superposition of the responses to these forces. To make matters worse, the response to each forcing varies with depth. Fortunately, the degrees of freedom are somewhat limited, e.g., alongshore flow is preferred and characteristic flow patterns do exist. It is convenient to use the artifice of steady flows to discuss these patterns. Ekman boundary layers and geostrophic currents will also be referenced. The reader not familiar with these subjects may wish to consult a physical oceanography textbook (such as Neumann and Pierson, 1966; von Arx, 1962; or Dietrich, 1963).

Areas of convergence (Figure 8-5) and divergence of currents are also an important aspect of the shoreface current patterns. A convergence is defined as a negative divergence. A convergence of horizontal current components can result in a rise in the local sea level and vice versa for a divergence. That is, sea level rises if water flows together or against a boundary.

In order to define characteristic wind-driven current patterns over the shoreface, it is convenient to divide the inner portion of the coastal ocean into three zones. These are a friction-dominated zone, a transition zone, and a

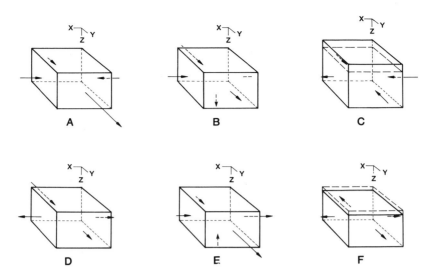

Figure 8-5. Convergence of horizontal current components in the upper three control volumes are related to: acceleration of a component (A), generation of a vertical component (B), and a rise of the height of the control volume (C). Divergence of horizontal current components in the bottom three control volumes (D,E,F) are related to converse effects.

geostrophic zone, illustrated schematically in Figure 8-6. The geostrophic zone is of sufficient depth so that the surface Ekman boundary layer and bottom Ekman boundary layer are separated by an interior region of unaffected geostropic flow. The boundaries and dimensions of this zone are not fixed. The thickness of the layers is controlled by the intensity of the wind forcing and bottom-induced turbulence. Under typical conditions the geostrophic zone seldom extends shoreward of depths ranging between 20 and 40 m. The outer limits of what we are calling the geostrophic zone (or more appropriately the geostrophic zone of the coastal boundary layer) blend into a zone of mid-shelf flow.

Figure 8-7 shows a schematic diagram of characteristic flow patterns in the geostrophic zone. This diagram illustrates that the depth mean current in the surface Ekman boundary layer is perpendicular and cum sole (to the right of, in the Northern Hemisphere) to the direction of the wind. Depending on the orientation of the wind and coast, the wind-driven transport can cause a convergence or divergence leading to a sea level set-up or set-down. The resulting pressure gradient gives rise to flow throughout the water column directed such that sea level rises (high pressure) to the right of the flow. Bottom friction reduces the velocity of the current and thereby the magnitude of the Coriolis force in what is called the bottom Ekman layer causing the bottom boundary flow to veer more in the direction of the horizontal pressure

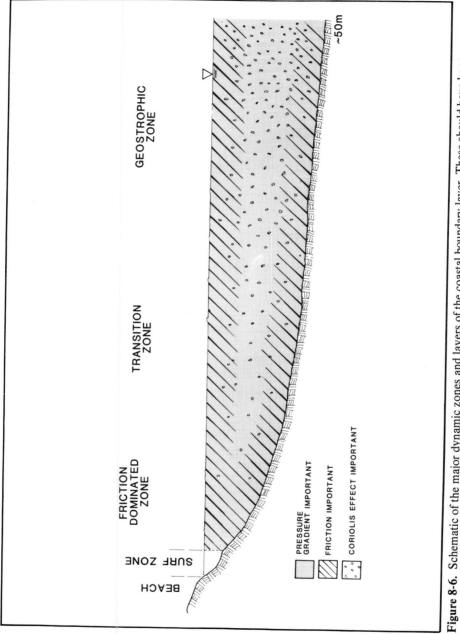

Figure 8-6. Schematic of the major dynamic zones and layers of the coastal boundary layer. These should have been seen as grading from one to the other depending on the relative magnitude of the different forcing.

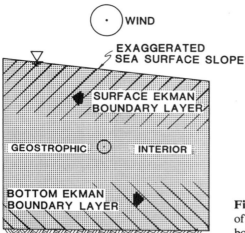

Figure 8-7. The major dynamic layers of the geostrophic zone of the coastal boundary layer.

gradient force. Thus, the layer-averaged flow in the bottom Ekman layer is 90° contra sole (to the left in the Northern Hemisphere) of the flow in the geostrophic layer.

Where the coastal ocean water mass is stratified, the patterns of flow schematically depicted in Figure 8-7 are approximately the same except that the geostrophic interior flow is sheared (reduced) as is, consequently, the magnitude of the bottom Ekman layer. The surface Ekman transport is compensated by both the (reduced) Ekman transport and an ageostrophic transport, which is generated by internal friction in the pycnocline.

To reiterate, the major points concerning wind-driven currents in the geostrophic zone are that coast-parallel winds cause shore-normal currents in the surface boundary layer. These result in developing a set-up or set-down of the mean water surface and internal horizontal pressure gradient forces directed in the shore-normal direction. These, in turn, result in shore-parallel flows in the geostrophic layer and shore-normal flows in the bottom boundary layer. On the other hand, shore-normal winds cause shore-parallel flow in the surface Ekman boundary layer. In the absence of other effects this does not result in flow in the geostrophic and bottom Ekman boundary layers.

In the inner portion of the shoreface zone, from just seaward of the surf zone to a depth on the order of 10 m, the scale of the friction forces exceeds the scale of the Coriolis effect. In this friction-dominated zone the boundary layer flows are not characterized by appreciable turning from the direction of the applied forces. Thus, surface currents tend to be aligned with the surface wind stress. The proximity of the shoreline causes surface currents with shore-normal components to rapidly develop shore-normal slopes of the mean surface level. This results in shore-normal pressure gradients that affect the flow at all levels and drive bottom currents. This situation is illustrated in Figure 8-8. Winds with onshore components result in onshore surface components and a net mean sea surface set-up against the shoreline. The offshore-directed horizontal pressure gradient force opposes the surface currents. Close to the bottom, this offshore-directed horizontal pressure gradient dominates over the effect of wind stress and propels bottom currents with offshore flow components. The circulation pattern is then onshore at the surface and offshore at the bottom, or downwelling, as shown in the top of Figure 8-8. Conversely, wind stresses acting in an offshore

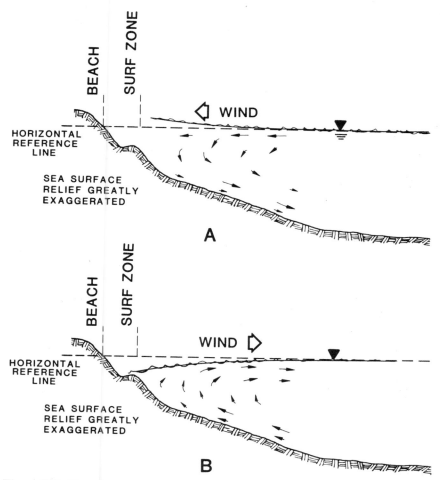

Figure 8-8. Shore-normal and vertical current components in the friction dominated zone. Sea level distortions have been greatly exaggerated.

direction cause offshore surface currents, a set-down of the coastal sea level, and onshore flow at the bottom, or an upwelling circulation.

It is evident from the above discussion that the same surface wind stress causes different responses in the geostrophic and friction-dominated zones. For example, an onshore wind results in a shore-parallel current in the surface Ekman boundary layer of the geostrophic zone, while it produces an onshore surface current and offshore bottom current in the friction-dominated zone. Continuity and dynamics require that these current patterns be matched. In order to illustrate this matching, an artifact is introduced that will be called the transition zone. Strictly speaking, this zone does not possess unique characteristics but merely represents a gradation between the offshore geostrophic zone, and the shoreward friction-dominated zone. The dimensions of the transition zone and the relative importance of its different processes depend upon the strength of the forcing

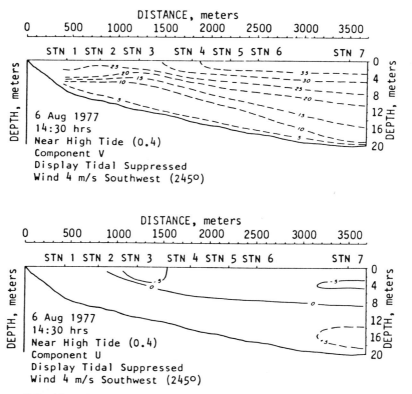

Figure 8-9. Alongshore (upper) and cross-shore (lower) components of shoreface currents during a mild upwelling situation measured on the Tiana Beach, N.Y., shoreface during the INSTEP project. Contours are in cm/sec and dashed contours represent either onshore or eastward flow (from Niedoroda, 1980).

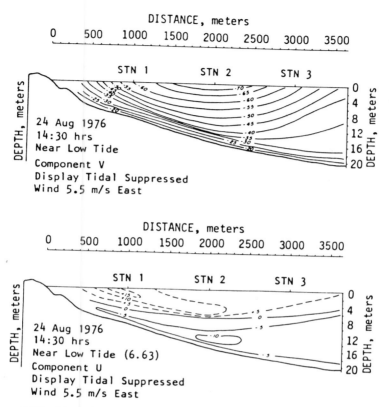

Figure 8-10. Alongshore (upper) and cross-shore (lower) components of shoreface currents during a mild downwelling coastal jet measured on the Tiana Beach shoreface during the INSTEP project. The contours represent the same values as in Figure 8-9 (from Niedoroda, 1980).

and the duration of the conditions. This is best illustrated through examples of shoreface current kinematics and dynamics under typical wind conditions.

Figures 8-9 and 8-10 show data on shoreface currents that resulted from measurements made during project INSTEP (Inner Shelf Sediment Transports Experiment). One of the aims of this project was to measure waves and currents in a typical open-ocean shoreface environment with sufficient resolution of permit computation of bed-load sediment transport using available equations. These measurements were conducted during the summers of 1976 and 1977. The INSTEP project was closely coordinated with project COBOLT (Coastal Boundary Layer Transect) resulting in measurements from fixed instruments and boat stations over a shore-normal transect extending across the entire coastal boundary layer (a total distance of 12 km offshore, Figure 8-11). The data presented in Figures 8-9 and 8-10

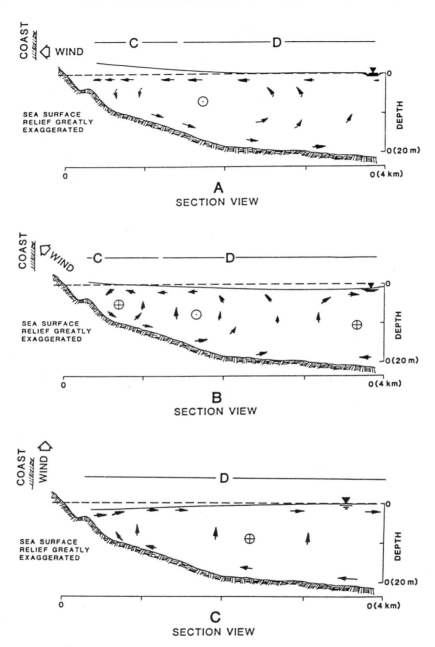

Figure 8-11. Schematics of shoreface transports under different orientations of the wind relative to the coast in the Northern Hemisphere. A small auxilliary figure above and to the left of each panel represents a map view of the wind direction relative to the coast. Sea level distortions have been greatly exaggerated.

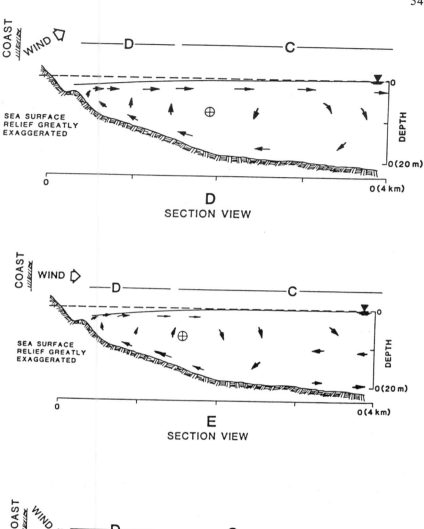

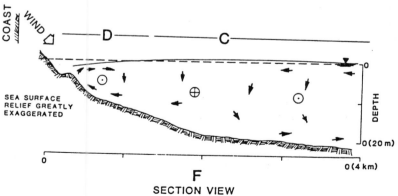

Figure 8-11. *(Continued.)*

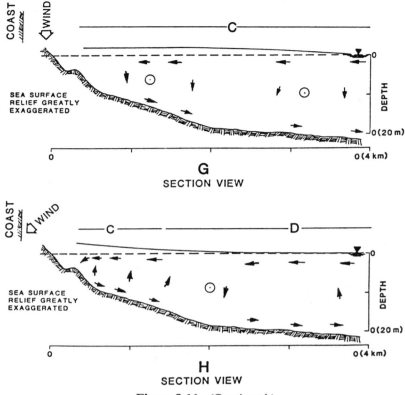

Figure 8-11. *(Continued.)*

result from boat stations that were located at 3-km intervals across the shoreface during 1976 and at approximately 500-m intervals during the summer of 1977. Currents were measured daily at these boat stations using a Bendix Q-15 side-case current meter. Time-averaged readings were taken at 2-m depth intervals. More details concerning these experimental programs and measurement procedures are in Niedoroda (1980).

Figure 8-9 shows a typical coastal upwelling event over the shoreface. The wind from the southwest was nearly parallel to the Long Island coast and resulted in a pattern of northeastward currents. Note that the difference in flow structure between stations 2 and 4 is an indication of the transition zone. The alongshore component is approximately an order of magnitude greater than the cross-shore component; as is typically observed from other locations, e.g., Blanton and Murthy (1974), Murray (1975), and Schwing *et al.* (1983). The primary divergence of the offshore surface current component occurs between the outer margin of the surf zone at approximately 1500 m. Horizontal current components were measured so that the locus of principal upwelling can only be inferred to lie this close to the shore.

Figure 8-10 shows the results of similar measurements on the Tiana Beach shoreface under an easterly wind. The cross-shore components indicate a classic downwelling situation while the alongshore components show the existence of a coastal jet-like flow. The speed contours for the alongshore component indicate that the cross-shore length scale of this jetlike flow is similar to that of the shoreface. Strong coast-parallel flows of similar cross-shore length scale are shown from data taken in Lake Ontario in Csanady (1972) and for the south shore from Long Island from Csanady (1977b).

The data shown in Figures 8-9 and 8-10 suggest that consistent and unique current patterns can be found in open ocean shoreface environments. These data bear some resemblance to the "classical" description of coastal up- and downwelling. However, most "classical" treatments have the maximum vertical flow at the coastal boundary. Here it occurs at the seaward portion of the transition zone. It is fortuitous because the transition zone somewhat uncouples the geostrophic zone from the friction-dominant zone making "classical" treatments credible if the coastal boundary is artificially placed near the 20-m depth contour.

Figure 8-11 A–H show schematic patterns of shoreface current components under varying orientations of the surface wind stress for an unstratified coastal ocean. A bathymetric cross-section similar to that which exists off Tiana Beach, Long Island, was used, so the vertical and horizontal scales shown on these figures are only rough estimates of what may be found at other locations. Field data from the INSTEP project, the COBOLT project, and results of numerical modeling are combined to produce these schematic diagrams.

The orientation of the wind stress relative to the coast that produced the flow shown in Figure 8-9 is identical to the schematic diagram shown in Figure 8-11C. The primary cross-shore divergence of the surface transport occurs over the upper shoreface. However, this divergence continues across the entire shoreface for this orientation of the surface wind stress. Thus, upwelling exists over the entire shoreface and an onshore flow component develops near the bottom. The effect of the divergence of wind driven surface transport in the cross-shore direction is to produce a set-down of the mean surface level over the entire shoreface. This slope with a mean surface level is primarily responsible for the onshore bottom flow in the friction-dominated flow and for a shore-parallel transport further offshore where geostrophic processes become more dominant. In this offshore zone Ekman turning in the bottom boundary layer contributes to the onshore bottom transport. Thus, the overall upwelling nature of the flow over the shoreface results from somewhat different dynamics in the friction-dominated zone over the upper shoreface in comparison to the geostrophic zone over the lower shoreface.

The cross-shore components of flow shown in measured data in Figure 8-10 are similar to the flow shown schematically on Figure 8-11H. The orientation of the surface wind stress relative to the coastline is also similar. Figure 8-11H shows that over the outer shoreface the surface transports diverge. From the middle of the shoreface to the upper shoreface the surface transports converge. This convergence is primarily the result of a reduction of the Ekman-like turning and secondarily the result of enhanced friction. The divergence of the cross-shore near-surface transport component over the outer shoreface results in a local set-down of the mean surface level. The convergence of the cross-shore surface transport component causes a set-up of the mean surface level over the upper shoreface. This results in an offshore-directed slope of the mean surface level that contributed to the offshore bottom flow over the upper shoreface and a shore-parallel flow component over the lower shoreface. Ekman turning in the bottom boundary layer of the offshore transport contributes to the offshore bottom flow. Continuity is satisfied by downwelling in the area of convergence of surface transports over the upper shoreface and upwelling over the lower shoreface. Thus, a weak nearshore vertical circulation cell results.

Figure 8-11A–H shows a complete set of schematic diagrams for shoreface flows under systematically varying orientations of the wind stress with respect to the coast. The patterns shown on Figures 8-11B and 8-11F are particularly interesting. Figure 8-11B corresponds to a southeast wind blowing against a north-south oriented coast with the land to the west. Wind-driven surface transports are divergent over the entire shoreface with the exception of a marked convergence at its very upper portion. The cross-shore circulation is characterized by two vertical circulation cells. Classical upwelling theory, which fails to account for the existence of an inner friction-dominated zone, also fails to predict this more complex pattern of cross-shore current components.

Figure 8-11F shows the cross-shore current components resulting from a northwest wind and the same coastal orientation as discussed for Figure 8-11B. In this case the cross-shore flow components are convergent over the entire shoreface with the exception of a strong divergence close to the surf zone. This pattern of convergence and divergence results in two circulation cells that are opposite to those shown in Figure 8-11B. Upwelling characterizes a narrow zone immediately seaward of the surf zone while downwelling develops over the middle region of the shoreface. Both surface and bottom transports are reversed over the upper and lower portions of the shoreface.

The various patterns of cross-shore current components shown in Figure 8-11B–H represent distinct circulation patterns that develop as a result of steady winds in a predictable manner. In some cases, the pattern of cross-shore transport components corresponds closely with that predicted by classic upwelling discussions while in other cases more complex circulation patterns are shown. In all cases, the primary zone for convergence or divergence of the surface currents is over the upper shoreface. Upwelling and bottom currents are similarly stronger in this location. Under certain combinations of wind and coast orientation, a secondary zone of convergence or divergence of wind-driven surface components develops over the middle portion of the shoreface. The development of this secondary zone of divergence is often related to vertical current components and the development of vertical circulation cells.

The position of maximum vertical current components is associated with the position of convergence or divergence of the horizontal components of near-surface and bottom currents. Where the cross-shore length of the divergence zones is large, the up- or downwelling covers most of the shoreface whereas narrow divergence zones concentrate the up- or downwelling. Thus, the on- or offshore gradients of bottom currents are closely related to the overall circulation pattern for each wind orientation.

The scale matching between the distinctive nearshore current patterns and the shape and dimensions of the shoreface is important. It has been shown in this section that the development of these organized flow patterns is in large part due to dynamic effects caused by the shoaling depth. A following section covers the effect of these distinct circulation patterns in controlling bed-load transport and shaping the shoreface.

The cross-shore component currents shown in the schematic diagrams represent flow that develops in unstratified coastal waters. Where a density stratification exists, as is common in most summertime conditions, the flow patterns are qualitatively similar, but become more complex.

Sediment Dynamics on the Shoreface

Marine Bedload Mechanics in the Zone of Shoaling Waves

Bedload transport on the shoreface results from a combination of wave and current processes. These processes often act together in nature although their relative magnitudes vary across the shoreface and with time. Several bedload transport processes of particular importance in the shoreface environment can be isolated. Particular attention is given to those that are significant in controlling cross-shore transport.

Wave theories describing different representations of gravity wave behavior were discussed earlier in this chapter. Consideration of shoreface bed-load transport will begin through examination of a very simplified case. Picture an area, perhaps on the lower shoreface, where low amplitude swells are acting on a small slope. This situation may be adequately represented by assuming the slope to be negligible and applying Airy wave theory. The orbital velocities acting on the bottom are solely horizontal with a symmetric on- and offshore oscillation. Provided that the maximum orbital velocity exceeds that necessary to entrain the bottom sediment and that the scale of the sediment inertia is small relative to the entrainment forces, the sand will be moved back and forth with no net transport. This simple set of conditions has been studied with laboratory apparatus by Bagnold (1947), Kalkanis (1964), and Abou-Seida (1965), among others, in order to evaluate sediment entrainment criteria and transport rates under oscillating flow. These studies showed surprisingly good agreement between predictions based on steady flow relationships for entrainment and transport when the wave orbital velocities were appropriately represented and bedforms were absent (plane-bed) or small. Manohar (1955) and Carstens et al. (1969) have used laboratory experiments to show that the presence of ripple marks complicate the bed-load transport process. The sediment is often advected upward through and above the wave boundary layer by vortices that form over the crest of the ripples (Inman and Bowen, 1963). This probably explains the relatively high concentration of suspended sand well above the bed that has been measured during periods of high waves in several field programs (Vincent et al., 1983; Vincent et al., 1981; Young et al., 1982; Brenninkmeyer, 1976). The symmetry of sand transport is altered in these conditions because the next half orbital oscillation advects the sediment entrained by the previous half oscillation.

Bailard and Inman (1981) and Bailard (1981) have examined the case of wave-induced bed-load transport in plane bed conditions on a simple slope. They show that the presence of the slope invariably yields downslope transport if only symmetrical wave orbital velocities are considered. Although their work was confined to surf zone environments, it illustrates an important point for all coastal marine bedload processes. Slopes in unconsolidated granular sediments that are subject to vigorous wave action require application of net or time-averaged upslope fluid forces to maintain equilibrium. A variety of processes can yield time-averaged upslope fluid forces on the sediment bed in the marine environment.

The role of asymmetric wave orbitals on bedload transport has been studied experimentally by Manohar (1955) and in shoreface environments by Niedoroda (1980) and Niedoroda and Swift (1981). As gravity waves propagate across the shoreface, they respond to the shoaling depths by developing higher and more peaked crests with broader, flatter troughs. Stokes wave theory is a better representation of these waves than Airy theory. The time history of the near-bottom orbital velocities is asymmetric with a brief, faster onshore velocity beneath the crest and a longer, slower offshore velocity beneath the trough. This suggests that on a flat bed more sediment is moved in the direction of wave propagation than in the opposite direction during each wave cycle.

Niedoroda (1980), Niedoroda and Swift (1981), and Swift et al. (1985) explored the effect of asymmetric wave orbital motion on shoreface bed-load transport using a simplified approach. This work shows that the asymmetry in gravity wave orbital velocities becomes important in depths that correspond to most open-ocean shorefaces. Stokes second-order wave theory was applied to represent the asymmetric orbital velocity. A simplified form of Bagnold's sediment transport theory was used (Bagnold, 1946, 1947,

1956, 1963, 1966). All the constants and coefficients not related to the orbital motion were lumped into a single dimensional constant (K). To avoid dealing with variations in bed-load transported as a result of different values of bottom slope, a horizontal bottom was assumed. This analysis explores the dependence of onshore bed-load transport owing to asymmetric orbital motions on depth and wave parameters but does not show the balance between the increase in bottom slope with decreasing depths that characterizes most shorefaces.

The Bagnold relationship is simplified to:

$$M_s(t) = K u_0(t)^3 \qquad u_0(t) \geq U_{crit} \qquad (8\text{-}1)$$

$$M_s(t) = 0 \qquad u_0(t) < U_{crit} \qquad (8\text{-}2)$$

where $u_0(t)$ is the instantaneous orbital velocity at the sea floor and $M_s(t)$ is a relative measure of the instantaneous mass flux of bed-load per unit with wave crest. This relationship can be integrated and normalized by the wave period to yield a measure of the time-average bed-load mass flux. If second-order Stokes theory is used to represent the orbital velocity and it is assumed that U_{crit} is negligible compared with characteristic maximum orbital velocities associated with vigorous bed-load transport, the equation can be integrated to yield:

$$\overline{M}_s = \frac{9}{8} K \frac{(\pi H)^4}{L T^3} \frac{1}{\sinh^6 kh} + \frac{189}{564} K \frac{(\pi H)^6}{L^3 T^3} \frac{1}{L^2 \sinh^{12}(kh)} \qquad (8\text{-}3)$$

where $\overline{M}_s$ is a measure of the time-averaged bed-load mass flux per unit width of wave crest; L is the wavelength; T is the wave period; H is the wave height; h is the depth; and k is the wave number ($2\pi/L$). Clearly the magnitude of bed-load transport is very sensitive to wave height and the ratio of local depth to the wave length.

If the value of U_{crit} is not taken to be non-negliglible, a numerical solution must be used. The results of such a numerical solution are shown in Figures 8-12 and 8-13. Figure 8-12 shows the strong dependence of the relative bed-load transport on depth, wave height, and wave period. Representative ranges of ocean-wave parameters have been used to generate Figure 8-13, which shows that the effect of asymmetric wave orbital velocities become important in depths ranging from 5 to 15 m. The asymmetry in bottom fluid shear stress associated with peaked waves over the shoreface is in the direction of wave propagation and is thus generally onshore.

Another feature of gravity waves that is predicted by Stokes wave theory is a net current caused by the waves. Eagleson *et al.* (1958) and Dean and Eagleson (1966) show that integration of the second-order Stokes theory equation for horizontal orbital velocity over the wave period results in a slow mean current in the direction of wave propagation in a deep unbounded ocean. This effect is commonly called Stokes' mass transport. Longuet-

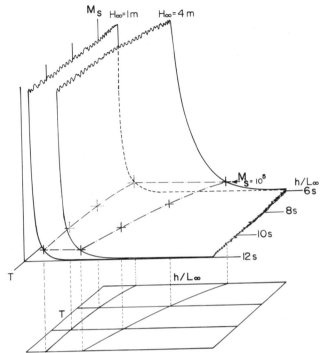

Figure 8-12. Change in the relative bed-load transport resulting from asymmetric wave orbital velocities (M_s) on a flat bottom as a function of wave period (T) and the ratio of depth to the deepwater wave length L_∞ (from Niedoroda, 1980).

Higgins (1953) studied this effect in a more realistic situation where the presence of a shoreline, finite depth, and viscosity were included. His solution is limited by assumptions concerning laminar flow. It predicts vertical velocity profiles with currents in the direction of propagation near the surface and bottom with an opposed flow at mid-depths. Longuet-Higgins' theory shows good agreement with wave tank measurements made by Bagnold (1947) and Russell and Orosio (1958). In an appendix to the latter paper, Longuet-Higgins shows that his theory remains valid for turbulent wave boundary layers if the turbulent eddy viscosity increases with height above the bed. His equation predicts a near-bottom current resulting from Stokes' mass transport that is in the order of centimeters per second to tens of centimeters per second for characteristic wave conditions over depths typical of open-ocean shorefaces. However, the existence of this slow wave-driven onshore bottom flow has not been demonstrated with field measurements. Such flow may not develop on open coasts where circulation in the horizontal plane often dominates.

Another form of wave-driven circulation that influences shoreface sediment dynamics has been proposed by Larsen (1982) and Shi and Larsen (in press). They point out that natural gravity waves tend to propagate in series that appear as regular to irregular groups of alternating higher and lower waves. They refer to this as amplitude modulation. Using the radiation-stress theory of Longuet-Higgins and Stewart (1964), they show that amplitude-modulated wave trains induce changes in the variations in the mean sea surface and a mean current. The mean current is directed opposite to the propagation direction during the portion of the amplitude-modulation cycle when the individual waves are high. They argue that the higher waves entrain more bottom sediment that can then be transported offshore by the associated mean current. They point out that this effect can be particularly

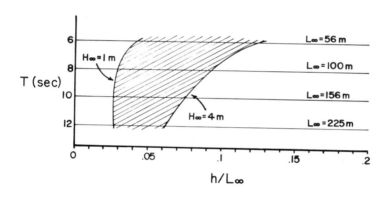

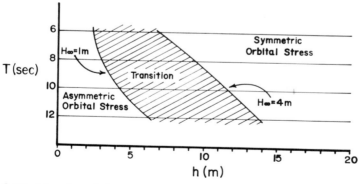

Figure 8-13. Mapping of the transition zone between significant and insignificant effects of asymmetrical wave orbital velocities on coastal ocean bed-load transport. These regions are shown as functions of wave period (T) and the ratio of depth to deep water wavelength (h/L_∞) in the upper panel and as functions of wave period and depth in the lower-panel (from Niedoroda, 1980).

effective in selectively transporting fine sand that tends to be suspended higher off the bottom.

When waves and currents act together they can produce subtle but significant bed-load transport effects. Vincent *et al.* (1983) have pointed out that slowly varying shoreface currents that generally result from the effects of wind and tide are dominantly shore-parallel while waves usually propagate across the shoreface in an onshore direction that does not coincide with slope normal. When the geometry of the vector combination of wave orbital and slowly varying bottom boundary layer current velocities is considered, there is a preference for stronger onshore combined flow over offshore combined flow (Figure 8-14). The critical or threshold velocity for bed-load entrainment is exceeded earlier in the onshore orbital half cycle and transport occurs for longer durations. Figure 8-15 demonstrates that this effect will add to a tendency for onshore bed-load movement provided that the wave propagation direction is not perpendicular to shore and that the slowly varying current does not have a large cross-shore component. The turning of the combined wave-current bottom fluid shear stress noted by Grant and Madsen (1979a,b, 1982) may also influence this behavior.

Rip currents are also a contributor to granular sediment transport in a shoreface environment. Early studies of these features include Shepard *et al.* (1941), Shepard (1948), Putnam *et al.* (1949), Shepard and Inman (1951), Russell and MacMillan (1952), Inman and Quinn (1952), and Arthur (1962). Bowen (1967, 1969) and Bowen and Inman (1969) developed a comprehensive physical analysis of these narrow, strong, shore-normal currents that penetrate through, and are spaced along the outer margins of, the

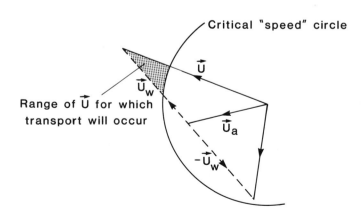

Figure 8-14. Vector addition of symmetric onshore $(\bar{U}_w)$ and offshore $(-\bar{U}_w)$ wave orbital velocities with a steady bottom current $(\bar{U}_a)$ generally produces resultants that favor onshore bed-load transport. The waves propagate nearly, but not exactly, toward the shore in most shoreface regions and the steady currents are nearly alongshore (from Vincent *et al.*, 1981).

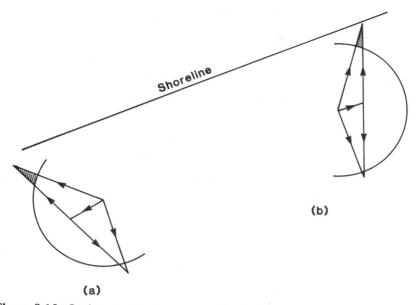

Figure 8-15. Onshore bed-load creep resulting from the combined effect of waves and currents for two wave propagation and current directions over a shoreface (from Vincent *et al.*, 1981).

surf zone. They show that the rip currents are driven by momentum flux exchange processes ultimately driven by breaking waves in the surf zone. Cook (1970), Langford-Smith and Thom (1969), and Cook and Gorsline (1972) have concluded that the band of relatively fine sand often found mantling the upper shoreface results from rip current fall out. Thus, some sand is moved offshore over the shoreface by rip currents.

The Shoreface Bedload Transport System

The general shape and dimensions of open-ocean shorefaces tend to remain nearly constant over long periods of time, even when their positions shift as a consequence of general shoreline erosion or submergence. This implies that the time average of shoreface sediment transport is adjusted to preserve an equilibrium morphology. It also suggests the existence of positive feedback between transport processes that keep the system stable. Thus, an understanding of the shoreface physical system depends on determining the relative importance of the processes and how they are balanced to create the dimensions, shapes, and histories that are observed in this coastal ocean environment.

This section will consider the patterns of shoreface bedload transport, distinguish the relative importance of the bedload transport processes on different portions of the shoreface, and describe how the magnitude of these sediment transporting processes are altered by changes in the strength of wind forcing. At this time there is no comprehensive theory available to predict shoreface currents, and only a small amount of data on shoreface sediment transport exists. Therefore, this discussion will center on limited field data and results of previously published analyses of these data.

Shoreface bedload transport on the long, straight, sandy southeast coast of Long Island has been studied in the INSTEP project. Results given in Niedoroda (1980), Niedoroda and Swift (1981), and Swift *et al*. (1985) show patterns of bedload movement that result from the combined effects of waves and currents (to avoid repetition these are designated the N-S results). The combined data on wind, waves, currents, and shoreface bathymetry were coupled with numerical modeling to analyze the general response of the shoreface sediments to changing wave and current conditions. Spatial resolution of the current measurements (2 m in the vertical and 500 to 1000 m in the horizontal) was sufficient to define detailed patterns of shoreface current patterns. The method for computing bedload sediment transport developed by Madsen and Grant (1976) was used as the basis of a numerical model. This method is not as sophisticated as those given in Grant and Madsen (1979a, b, 1982), but it was judged to be capable of yielding adequate results. Niedoroda *et al*. (1982) used a similar numerical model to compute bedload siltation rates in test pits at depths of 12 m and 16 m offshore of San Francisco. The siltation rates in these test pits were monitored through repetitive, detailed bathymetric surveys. In spite of relatively low-quality wave and current data the computed siltation rates generally agreed with the measured rates within the limit of accuracy of the bathymetric data over 2-week intervals. This confirmed the value of the 1976 Madsen and Grant technique for computing bed-load transport in a shoreface environment.

The pattern of computed shoreface bed-load transport for a westward upwelling flow in the presence of 1.5 m high, 13-second waves on the Tiana Beach shoreface profile from the N-S study is shown in Figure 8-16. Under these conditions bedload is entrained to a depth of approximately 16 m. The largest onshore transport occurs near the middle of the upper shoreface where the depth-dependent wave agitation of the sediment is combined with the strongest onshore bottom current. While wave orbital velocities increase on the upper shoreface, the onshore bottom current is reduced and onshore bedload transport decreased. Although the maximum rate of alongshore shoreface bedload transport was more than an order of magnitude greater than the maximum cross-shore rate, the rates of alongshore and cross-shore bedload transport are similar over the middle shoreface. If the waves and currents were uniform in the alongshore direction, then divergence and convergence of onshore bedload transport causes erosion on the middle

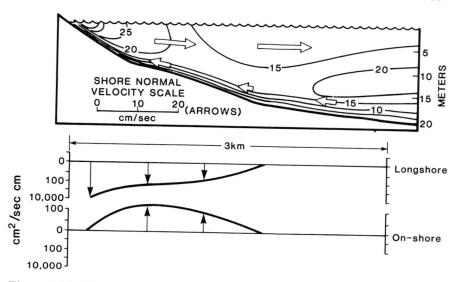

Figure 8-16. Alongshore (contoured in cm/sec) and cross-shore (arrows; short < 5 cm/sec, medium 5–10 cm/sec, long 10–15 cm/sec) shoreface current components and computed bed-load transport for 1.5 m significant wave heights on the Tiana Beach shoreface. The lower graphs show the longshore (westward) and cross-shore (onshore) bed-load transport for an upwelling flow (from Niedoroda and Swift, 1981).

shoreface and deposition on the upper shoreface. Niedoroda and Swift (1981) point out that alongshore bedload transport tends to be one or two orders of magnitude smaller on the shoreface than in the surf zone.

Figure 8-17 shows another set of N-S results for a downwelling coastal jetlike flow. The shoreface bedload transport was computed for two wave conditions. A 0.65 m high, 9-second wave train is insufficient to entrain bottom sediment over most of the shoreface in spite of the strong mean current. In contrast, a 1.5 m high, 13-second wave train results in bedload transport across the entire shoreface. Conditions illustrated in Figure 8-17 probably resemble those that exist during northeaster storms on the east coast. The downwelling current results in transfer of bed load from the surf zone across the shoreface.

Comparison of Figures 8-16 and 8-17 illustrates that although the magnitude of bottom wave orbital velocities is most important in causing entrainment of shoreface sediments, the fluid stress resulting from bottom currents contributes to sediment entrainment. The stronger bottom currents shown on Figure 8-16 help to entrain sediment further down the shoreface in the presence of identical wave conditions.

Storm-induced shoreface bedload transport has been measured, computed, and studied by Vincent *et al.* (in press), Swift *et al.*, (1985, in press,b), and

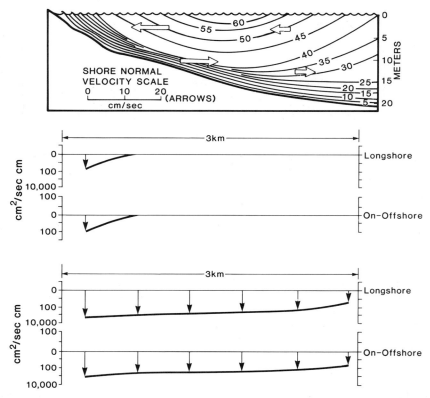

Figure 8-17. Shoreface current components and computed bed-load transport for a downwelling coastal jet. Current and bedload parameters are displayed as defined in Figure 8-16. The upper pair of bed-load transport graphs reflect computations based on a 0.7 m significant wave height and the lower pair represent the same based on a 1.4 m significant wave height (from Niedoroda and Swift, 1981).

Niedoroda *et al.* (1984). The former paper reports on data collected from an 18-day deployment of a CV-probe at a 10 m depth on the Tiana Beach shoreface. This instrument measures waves and tides with a pressure gauge, horizontal current components with an electromagnetic current meter located at a height of 1.1 m above the sea floor, and suspended sediment concentration with an acoustic profiling system (Huff and Friske, 1980; Young *et al.*, 1982). Shoreface sediment was entrained by two moderate storm events. The alongshore component of bed-load transport tended to be a factor of five greater than the cross-shore component during these events, and the typical combined effect of combining wave orbital velocities with the mean bottom currents was to favor onshore bed-load transport. Figure 8-18 shows progressive vector diagrams of shoreface bed-load transport during these storm events.

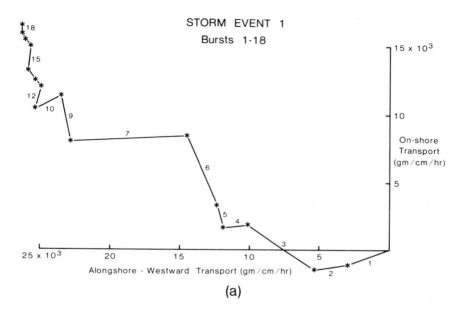

(a)

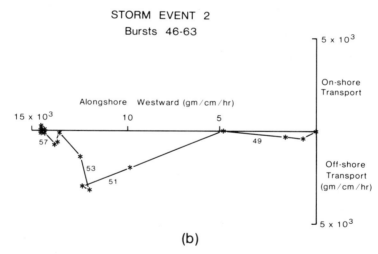

(b)

Figure 8-18. Progressive vector plots of shoreface bed-load transport during two storms measured with the Project INSTEP CV-probe (from Vincent *et al.*, 1983).

Niedoroda *et al.* (1982) combined a time series of CV-probe measurements at a 10 m depth on the Tiana Beach shoreface for a northeast storm during March 1978, with previous analyses of characteristic shoreface current patterns to produce an estimated, synthetic series of wave and current conditions across the shoreface during the storm. They used a numerical model of marine bedload transport (based on Madsen and Grant, 1976) to compute estimated patterns of sediment movements during 17 periods in the storm event (Figure 8-19). The results show a dominance of alongshore sediment transport with significant cross-shore components. During the early stages of the storm, the strong easterly wind caused downwelling westward currents and rapidly building waves. The resulting cross-shore bedload transport was offshore with a strong divergence across the shoreface. As the storm center passed the area, the wind switched to a general northwesterly direction. The strong offshore wind rapidly diminished the wave heights and caused the shoreface currents to change to an eastward-upwelling pattern. The association of lower waves with the upwelling reduced the onshore shoreface bedload transport during the late stages of the storm. Niedoroda *et al.* (in press) point out that the hysteresis in storm-induced shoreface bedload transport that results in a net offshore transport should be a general phenomenon along many portions of the U.S. coast, because downwelling currents and high waves tend to be associated during coastal storms.

These observations of the Long Island shoreface show that sand moves landward across the shoreface most of the time. This movement occurs landward of the 10-m isobath at rates up to 1×10^4 g/cm/sec in response to asymmetric wave orbital currents. On the lower shoreface and inner shelf floor, rates are an order of magnitude lower; computed values are on the order of 10^2 g/cm/sec. Observations further show that during storms, complex, wind-driven circulations develop. Jetlike alongshore flows may extend for several kilometers seaward of the beach, across the entire shoreface and adjacent inner shelf. When these flows are accompanied by downwelling and storm-intensified wave orbital currents, sediment is moved seaward across the shoreface in significant quantities.

Numerous studies have shown that most sand transferred by storms from the beach to the upper shoreface will eventually be returned to storage in the beach prism (summary in Komar, 1976). These studies suggest that the fair-weather landward asymmetry of wave orbital currents, as described in the proceeding portion of this chapter, may play a major role in this transfer. However, during strong storms such as the one analyzed, the upper shoreface is not a closed system. A significant portion of the sediment was transported across the shoreface and lost to deposition on the inner continental shelf. Thus the shoreface transport regime of the Long Island coast consists of long periods of time (months) during which sand moves slowly toward the beach, punctuated by short, intense periods (hours or days) during which the sand is transferred down the shoreface and some may be lost to the adjacent inner shelf. The long-term sense of movement of the shoreface, whether retreating

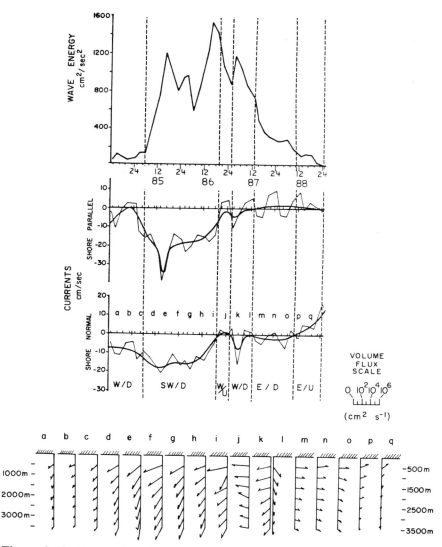

Figure 8-19. Temporal and spatial distribution of shoreface bed-load sediment transport during the March 1978 northeaster; Tiana Beach shoreface. Upper graph shows wave energy (cm²/sec²) during Julian Days 85–88. Middle graphs show alongshore (upper) and cross-shore bottom currents (cm/sec) at a 10 m depth for the same period. The lower figures show computed bed-load transport for 7 locations across the shoreface at 17 intervals (a–q) during the storm. The hashed line represents the beach and the vertical line represents distance across the shoreface in each of these 17 figures. The highest waves are associated with downwelling westward shoreface bottom currents, which results in a net offshore sediment transport indicating deposition of a tabular sand body across the shoreface that is thickest on the upper shoreface (from Niedoroda *et al.*, 1984).

or prograding, must depend on the loss or gain of sand by the shoreface as a whole, with respect to the inner shelf.

The shoreface bedload transport system has been identified as a distinct regime by Swift (1976) and in somewhat more detail by Niedoroda (1980) and Niedoroda and Swift (1981). The dynamics of both waves and currents that control sediment transport are different for the shoreface environment than for the adjoining surf zone and continental shelf environments. The morphology of the shoreface is adjusted to time-averaged bedload transport equilibria. Sediment transport depends on wave agitation and is more episodic than surf zone sand transport. The magnitude of shoreface bedload transport is one to two orders of magnitude smaller than the magnitude of longshore drift in the surf zone in the few examples where they have been compared. However, shoreface processes can provide an important control on coastal erosion and deposition.

Figure 8-20 summarizes shoreface bedload transport processes. These processes are largely forced by the wind as opposed to corresponding surf-zone processes that are largely controlled by breaking waves. Strong offshore convergences and divergences of cross-shore mean current components and asymmetric wave orbital bottom fluid stresses dominate the bedload transport balance of the more steeply sloped upper shoreface. This is also the locus of rip current fall out. The lower shoreface is less affected by asymmetric wave orbitals. Bedload transport results from entrainment largely because of symmetric wave orbital motions with the transport speed and direction being controlled by the mean or slowly varying bottom current patterns. Significant cross-shore current divergences and convergences occur over the lower shoreface under certain orientations of the wind relative to the coast.

Most of the data and results presented in this chapter section come from a series of measurements made on an open-ocean shoreface off the southeast shore of Long Island, New York. This is a relatively long, straight, sandy shore with uncomplicated local bathymetry. Many of the wave, current, morphology, and sediment transport features that have been discussed are sufficiently general to be expected in most shoreface environments. On the other hand, shoreface sediment dynamics research is a new field and much more is to be learned.

Morphodynamics of the Shoreface

Morphodynamics Time and Space Scales

The preceeding sections have looked at the fluid and sediment dynamics of the shoreface. While the relationships presented are basic to an understanding of shoreface processes, many of them were in fact first described by engineers concerned with problems of coastal management. The time scales

SHOREFACE SAND SEDIMENT SHIFTER

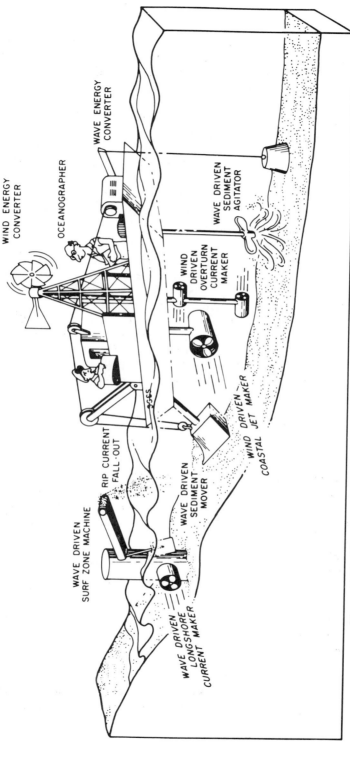

Figure 8-20. The shoreface sand sediment shifter. Cartoon model of shoreface sediment dynamics.

of interest have been minutes, hours, days, or months. In this section, consideration is given to large-scale, slow changes in the shoreface as a whole that are associated with coastal progradation and retreat. The time scales of interest here are months, years, centuries, and millenia.

Concern here is with a larger spatial scale as well. The coast retreats and advances as a whole, and in the context of morphodynamics, the beach and surf zone, shoreface, and inner shelf are considered as a whole.

As indicated in the preceding section, the intensities and durations of landward-directed wave-generated currents and seaward-directed currents induced by storm downwelling must control the sense of shoreface translation, determining whether the shoreface progrades seaward or undergoes erosional retreat. Curray *et al.* (1969) and Krumbein and Sloss (1963) have described the process variables governing coastal behavior as the rate of sediment input, the rate of relative sea level rise, and the "energy level." This last variable is more exactly the mean annual fluid power input in excess of the transport threshold (Swift, 1976). It is a function of both the wave climate and the frequency and intensity of storms. If sea level is falling, a coast must prograde seaward. If sea level is rising, as it is relative to most modern coasts, then the sense of shoreface translation, whether landward or seaward, must depend on the relationship between the rate at which sand is supplied to the shoreface by the wave-driven littoral current, and the fluid power applied to the shoreface to remove it seaward. The mechanisms of shoreface transport that were described previously constitute the short-term "kinetics" of the geological variable of fluid power expenditure.

The Shoreface in Profile: Coastal Advance and Retreat

During the modern period of rising sea level, most, but not all unconsolidated coasts are undergoing erosional shoreface retreat. The kinematics of this process were first described by Bruun (1962). Bruun argued that if the coastal profile was accepted as an equilibrium response of the sea floor to the coastal fluid power expenditure, then the effect of sea level rise could be deduced as a landward and upward translation of the profile. The geometric consequences of this conclusion are quite rigid (Figure 8-21). The shoreface must retreat by erosion, while there is a site of potential aggradation under the rising limb of the profile. Bruun suggested that if there were no along-coast sand discharge gradient (if there were as much coming out the downdrift face of the coastal box as going in the up-current face), then there should be an equal-volume transfer of sediment from the eroding shoreface to the adjacent shelf floor.

The erosional retreat process has been well documented on the middle Atlantic Coast where it is pushing the shoreline back at rates varying from 0.1 to 3.0 m/yr (Taney, 1961; Caldwell, 1969). Site evaluation studies at a number of points indicate that the shoreline retreat is part of a more general problem of erosional shoreface retreat as described by Bruun (1962). In a study of the Long Branch, New Jersey, dredge spoil dump site, Harris (1954) showed that during a 4-yr period the shoreface underwent between 5 and 26 cm of erosion, while in accord with the Bruun hypothesis, an irregular pattern of deposition prevailed on the sea floor (Figure 8-22).

In a somewhat longer time series, Kim and Gardner (1974) showed that two of the three profiles examined along proposed sewage outfall routes

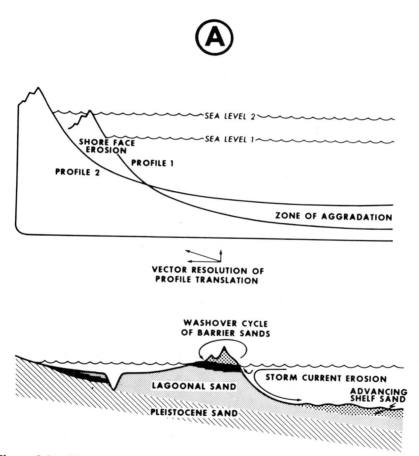

Figure 8-21. Models for the response of the shoreface to sea level changes. (A) Kinematic (above) and stratigraphic (below) model for erosional shoreface retreat in response to a rising sea level. (B) Kinematic (above) and stratigraphic (below) model for progradation of a coast when sediment input exceeds the rate of sea level rise (from Swift, 1976).

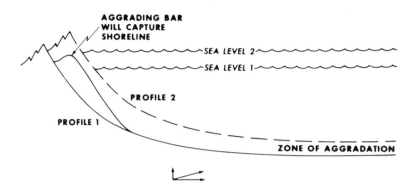

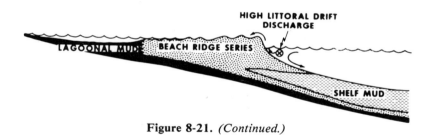

Figure 8-21. *(Continued.)*

across the New Jersey shoreface exhibited 1.5 to 2.0 m of erosion over 20 yr (Figure 8-23). The third profile is immediately south of a shoreface-connected sand ridge; here aggradation has occurred as a consequence of southward ridge migration. A 39-yr time series in the vicinity of the proposed Suffolk County, Long Island, N.Y., sewage outfall routes presents a somewhat similar picture. The western edge of the study area borders on the ebb tidal delta of Fire Island inlet and is undergoing deposition. The shoreface to the east, however, has undergone erosion, as has a series of sand ridges of the inner shelf floor.

The most detailed observations of shoreface retreats are those conducted by David Moody along the Delaware Coast (Moody, 1964, pp. 142–154). Moody notes that over a 33-yr period, the shoreface steepened toward the

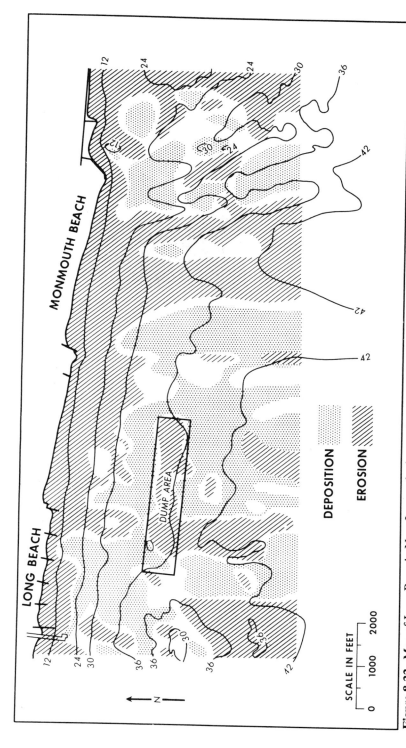

Figure 8-22. Map of Long Branch, New Jersey, dredge spoil dumpsite, showing 0.4 to 2.0 feet of deposition, 0.4 to 2.0 feet of erosion, or less than 0.4 feet of change. Contour interval is 6 feet (after Harris, 1954).

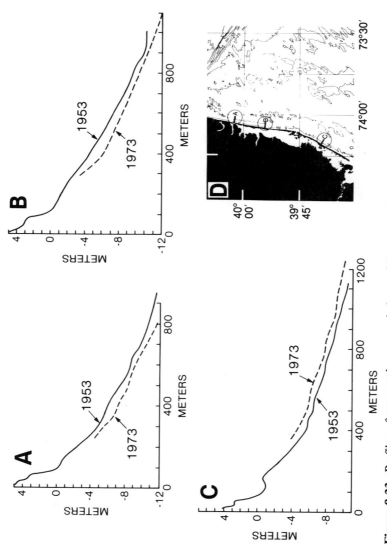

Figure 8-23. Profiles of proposed sewage sludge outfall on the New Jersey Coast. Sites A and B have undergone erosion over a 20-yr period. Site C is immediately downdrift from a shoreface-connected sand ridge and has upgraded (after Kim and Gardner, 1974).

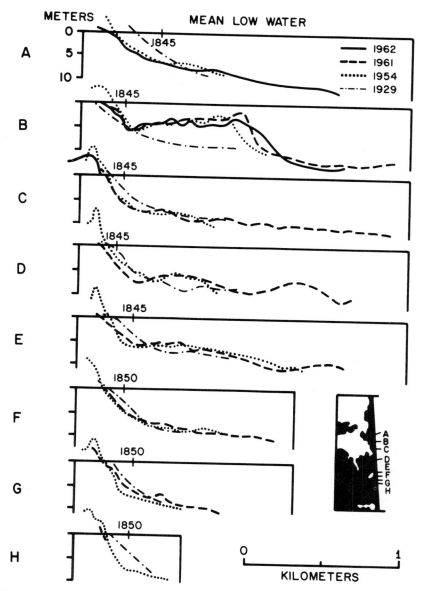

Figure 8-24. Erosional retreat of the Delaware Coast over a 34-yr period, based on U.S. Coast and Geodetic Survey records and a survey of Moody (after Moody, 1964).

ideal wave-graded profile, during which time the shoreline remained relatively stable (Figure 8-24). The steepening process was not continuous, but varied with the frequency of the storms and the duration of the intervening fair-weather periods. The slope of the upper shoreface steepened from 1:40 to 1:25 between 1929 and 1954, but erosion on the shoreface between 1954 and 1961 regraded the slope to 1:40.

The steepening process was terminated by a major storm, during which time the gradient was reduced and a significant landward translation of the shoreline occurred. Moody (1964, p. 199) describes the great Ash Wednesday Storm of 1962 as having stalled for 72 hr off the Central Atlantic Coast. Its storm surge raised the surf into the dunes for six successive high tides and the shoreline receded 18 to 75 m during the storm. While some of the sand was transported over the barrier to build washover fans over 1 m thick, much more was swept back onto the shoreface by large rip currents and by the storm-driven, seaward-trending bottom flow of the shoreface (Moody, 1964, p. 114). During the 33-yr period, shoreface erosion on the Delaware Coast was, in fact, nearly compensated by sea-floor deposition in accordance with the Bruun principle. The deficit, about 25% of the erosion total, is attributable to barrier washover and along-coast loss (Table 8-1).

In an erosional regime such as that of the Atlantic shelf of North America, sea level is rising sufficiently rapidly with respect to the rate of river sediment input that the river mouths have become estuaries. The estuaries trap not only all fluvial sand, but also littoral drift (Meade, 1969). Storm downdwelling currents must move sand seaward faster than asymmetrical wave orbital currents and combined wave-current effects return it, because the coast is retreating almost everywhere. Cores of most shorefaces reveal a thin veneer of modern sand (several decimeters thick) over older back-barrier strata (Figure 8-25). In this kind of coastal system, estuaries and the inner shelf floor become the ultimate sinks for sand. The primary source is the eroding shoreface. Storms may strip off the entire beach prism and back-barrier tree stumps, clays, and peats become briefly exposed along hundreds of kilometers of beach before fair-weather waves return the sand (Harrison and Morales-Alamo, 1964). This denudation extends down the entire shoreface during major events (Charlesworth, 1968), with ancient back-barrier deposits exposed at the sea floor to 10 or 15 m water depth. The Holocene sand sheet on the Atlantic shelf appears to have been generated in this fashion by shoreface erosion, as the shoreface retreated back across the shelf surface in response to post-glacial sea level rise (Swift, 1976; Swift et al., 1985).

In a second coastal system described by Curray et al. (1969), the rate of river sand input overwhelms the effects of sea level rise, and river mouths become deltas that inject sand directly into the surf zone (Figure 8-21). The beach, bar, and shoreface all receive sand more rapidly than they can exchange it with adjacent environments. During neap tides, bars are captured as new berms and the coast progrades seaward as a series beach ridges or strand plain (Figure 8-26). In more mud-rich prograding coasts, muddy tidal flats build

Table 8-1. Sediment budget from the Delaware coast (Moody, 1964).

Sediment source	Period	Average volumetric change[a] (m/yr)
Barrier (mean low water to tow or sand barrier)	1929–1961	−148,000
Sand dunes (mean low water to top of sand dunes)	1954–1961	−100,000 estimated
Offshore erosion (principally on northwest side of ridges)	1919–1961	−100,000
Erosion from bay inside Indian River Inlet	—	− 69,000
	Total erosion	−417,000
Site of deposition		
Tidal delta	1939–1961	+120,000
Barrier south of Indian River Inlet	1939–1961	+ 5,700
Offshore accretion	1919–1961	+256,000
	Total accretion	+381,700

Total erosion	−417,000
Total accretion	+381,700
Net erosion	−25,300 m³/yr

[a] + indicates accretion; − indicates erosion.

episodically seaward. During periods of low sediment supply, erosion and storm washover create sand or shell beaches (cheniers); then, when the rate of sediment supply is restored to its former value, progradation is reserved (Figure 8-27). Prograding and retrograding coastal regimes are contrasted in Figure 8-28.

The Shoreface in Plan View: Shoreface Retreat and Barrier Formation

Open ocean shorefaces display an innate tendency toward two-dimensionality. That is, shorefaces that are not initially straight tend to become so with time. Wave refraction is an important agent in the straightening process; wave rays are focused on headlands, accelerating headland retreat with

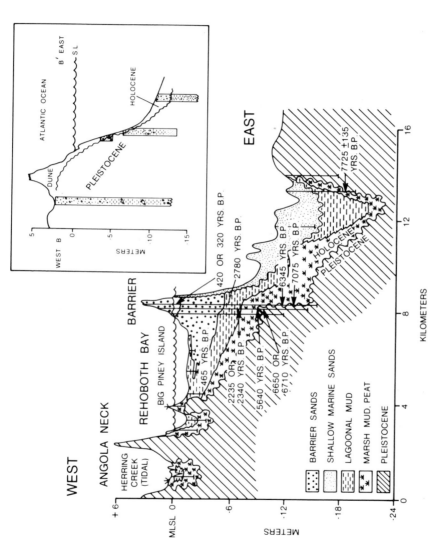

Figure 8-25. Cross section of the Delaware Coast. Based on seismic profiles and vibracores. Inset reveals the thin, veneer-like nature of upper shoreface sands (from J. Kraft, University Delaware, Unpublished).

respect to the adjacent bays (Figure 8-29). Johnson (1919) has described a cycle in which a coast, rejuvenated by a sudden submergence, develops cliffed headlands, then bay mouth bars, then, at last, a straight shoreline. Davies (1980) has summarized the work of many authors to show that beaches tend to rotate into one of two characteristic orientations; a swash orientation, in which the beach is normal to the direction of wave approach, and a swash alignment, in which the beach is oriented at 40 to 50° to the direction of wave approach. In this case, any increase in the angle causes a slacking of drift potential and therefore of deposition, so that the alignment is restored; any decrease in the angle causes deposition for the same reason, but in this case a change to a swash alignment occurs. Davies also describes the interaction of wave approach direction and littoral drift supply to create offset or zetaform coasts, in which each successive bay takes the form of a logarithmic spiral (Figure 8-30). These various mechanisms for coastal

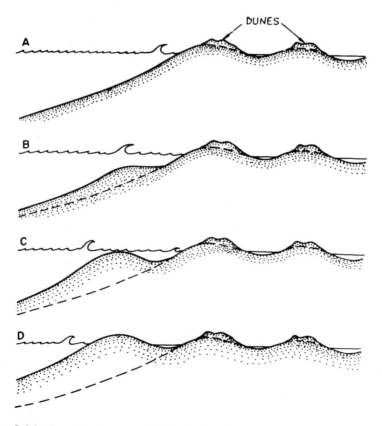

Figure 8-26. Growth of a strand plain. Point bar is captured by eolian processes during period of neap tides (from Curray *et al.*, 1969).

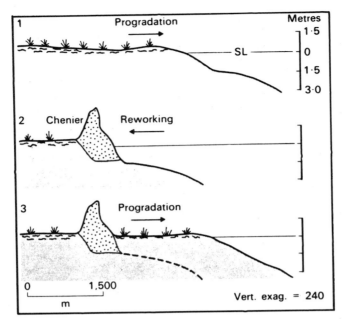

Figure 8-27. Episodic development of a chenier plain (from Elliott, 1978).

straightening are dominated by events in the surf zone, and here the mechanism is most clearly understood, but the effects are felt throughout the shoreface as well.

As has been discussed in the preceding sections, alongshore sediment transport on the shoreface, unlike that in the surf zone, is not directly linked to fluid motions associated with shoaling and breaking waves. Waves are important in producing the oscillating fluid shear that serves to entrain shoreface sediments, but the primary mechanism for alongshore transport comes from the wind-driven coastal ocean currents. In the surf zone, the interaction of wave refraction pattern with the shoreline configuration results in littoral drift reversals that will serve to smooth the configuration. Such reversals are uncommon in the alongshore component of coastal currents, which result from large-scale forcing. Coastal winds remain nearly uniform over scales of tens to hundreds of kilometers, and tidal forcing is constant over similar scales. The large-scale alongshore uniformity in forcing itself acts to cause the two-dimensionality of the shoreface. Irregularities in the shoreface will cause expansion and contraction of the cross-sectional area of the flow. As the flow accelerates over highs and decelerates over lows on the shoreface, it will erode the former and deposit on the latter. Wake formation downcurrent from headlands on highly embayed coasts and the development of underwater spits has been studied mainly in areas of strong tidal currents (Pingree and Maddock, 1979) but would be expected to occur in response to

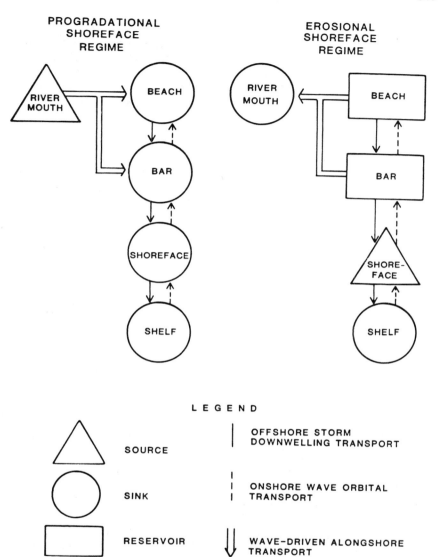

Figure 8-28. Schematic diagram illustration and budgets of retreating and advancing coasts (from Niedoroda *et al.*, 1983).

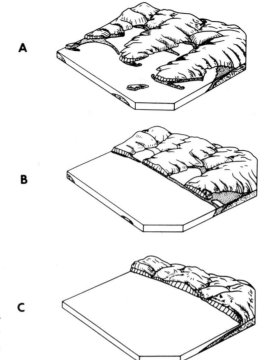

Figure 8-29. Evolution of a sub-merged coast through youth, maturity, and old age (from Johnson, 1919).

storm currents also; such deeper processes would work hand-in-hand with the surf zone processes that lead to the building of subaerial spits by littoral drift. However, the sediment mass balance may be the most important mechanism that links the surf zone and the shoreface in the coastal straightening process (Figure 8-28). The surf zone is the main conduit for alongshore sand transport, and when the surf zone suffers a sand surplus, so must the adjacent shoreface.

The innate tendency of an open-ocean shoreline toward two-dimensionality is one of the important processes responsible for forming barrier spits and islands during periods of coastal submergence. When a coastline of appreciable relief (bay-headland coast) begins to undergo transgression, shorefaces are first incised into the seaward faces of promontories and will propagate by constructional means in the downdrift direction as long as material is available with which to build and a foundation is available to build on (Figure 8-29). We give the lateral propogation of the shoreface into coastal voids the descriptive term *spit building by coastwise progradation* (Gilbert, 1885; Fisher, 1968).

However, the tendency of the shoreface to maintain lateral continuity also acts to prevent discontinuities as well as to seal them off after they have

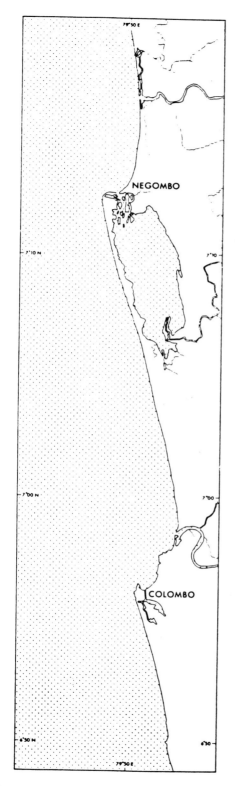

Figure 8-30. Zetaform coast in western ceylon. Wave approach from southwest (from Davies, 1980).

formed. In order to illustrate this, one may consider another set of initial conditions; a low coastal plain with wide, shallow valleys after a prolonged stillstand during which processes of coastal straightening by headland truncation and spit-building have gone to completion. Bay-head beaches have prograded to the position of adjacent headlands. Most of the estuaries have been sealed off by spits and filled in by marshes. As this coastline submerges, the water will invade valleys more rapidly than headlands can be cut back. The oceanic shoreline, however, cannot follow, for if it should start to bulge into the flooding stream valleys, the bulge would become a zone of longshore sediment transport convergence. The local rate of sedimentation would increase and this deposition would forestall any further shoreline retreat at this sector. Eventually, continuity along the coast would be restored. Instead, the oceanic shoreline detaches from the mainland. In this process, the beach and dune are nourished by littoral drift, grow upwards at the same rate as sea level rise, but the swale behind the dune cannot. Shallow water bodies would extend into these swales from the sides of estuaries. Thus a straight or nearly straight oceanic shoreline must detach from an irregular inner shoreline as a barrier island and be separated from it by a lagoon of varying width (Figure 8-31). The detachment of a beach is not an instantaneous process but instead a serial one; a lagoon creeps in from an estuary, and the beach is progressively detached, like the skin being peeled from a banana (Figure 8-32). The process of barrier island formation by mainland-beach detachment was first proposed by McGee (1980) and later described in detail by Hoyt (1967) and Hails and Hoyt (1968).

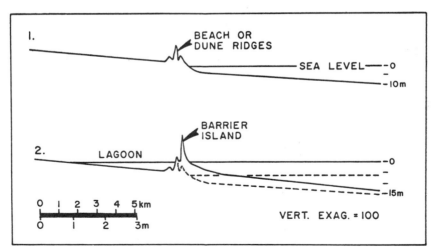

Figure 8-31. Two-dimensional model for mainland beach detachment (from Hoyt, 1967).

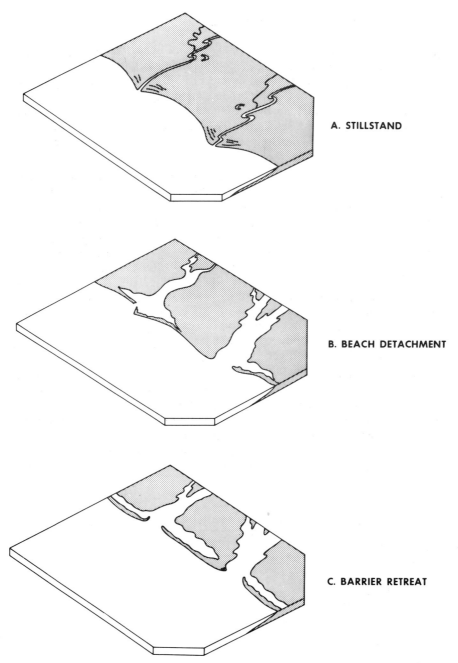

A. STILLSTAND

B. BEACH DETACHMENT

C. BARRIER RETREAT

Figure 8-32. Three-dimensional model for mainland beach detachment (from Swift, 1975).

Shoreface Retreat and Barrier Migration

During this modern era of eustatically rising sea level, barrier islands constitute 10 to 13% of the world's continental coastline (Cromwell, 1971, in Schwarz, 1973); primarily in areas undergoing coastal retreat. Barriers are linear coastal features of recent origin, composed of sand or gravel and separated from the mainland by a shallow water body known as a lagoon (definition modified from Schwarz, 1973). Barriers may be attached at one end to the mainland (barrier spits) or may be islands (barrier islands); they are typically a few kilometers to 500 m wide and lie a few kilometers offshore. Schwarz (1973) has noted the wide disparity in theories of barrier island origin. He has suggested that barriers are due to multiple causes and has proposed a genetic classification of barrier islands (Table 8-2). Primary barriers were first described by McGee (1890). Their mode of formation was analyzed by Hoyt (1967) in his presentation of the theory of mainland beach detachment, in which beaches and dunes tend to grow upward and landward during a transgression, while the swale behind them floods. Spit building and breaching is a second, less-frequent method of barrier island formation. Barrier island genesis by the upward growth of submarine bars and their consequent emergence has been advocated by Otvos (1970). This pheno-menon builds strand plains consisting of beach ridges 50 m wide (Figure 8-26), but its applicability to the much larger spatial scale of barriers (1–2) km wide) has not met with general acceptance.

The mainland beach detachment mode of barrier island formation is favored by low unconsolidated coasts in which the shorefaces can maintain themselves without interruption along many tens of kilometers of open coast and appears to be the predominant mode of barrier island formation. On the Atlantic Coast of North America, many of the coastal compartments between major estuaries contain barriers that are morphologically spits in the sense that they are attached at one end to the mainland, but they are not

Table 8-2. Genetic classification of barrier islands (Schwarz, 1973).

I. Primary
  1. Engulfed beach ridges

II. Secondary
  1. Breached spits
  2. Emergent offshore bars
    a. Sea level rise
    b. Sea level fall

III. Composite
  (Combinations of two or more of the above)

genetically spits, since they cannot have formed by coastwise progradation. The nodal point of littoral drift divergence tends to occur not on the mainland beach but on the spit near its base (Figure 8-33). These barriers must, therefore, derive their sediment from shoreface erosion rather than from coastwise littoral drift. They are ultimately of mainland beach detachment origin (Swift, 1975; Field and Duane, 1976; Leatherman, in press; Niedoroda et al., 1984), but their most immediate origin is landward migration from a slightly more seaward position, by erosional shoreface retreat, eolian and storm washover, and flood tidal delta formation (Godfrey and Godfrey, 1976; Leatherman et al., 1977).

The process is a complex one in which repeated breaching of the barrier builds a pavement of flood-tidal deltas on the seaward side of the lagoon. Storm and eolian washover drives the barrier crest onto the new platform. Shoreface erosion releases sand for delivery to flood-tidal deltas, to storm washovers, and also seaward, to the leading edge of the advancing shelf sand sheet (Figure 8-34). The advancing shelf sand sheet is deposited directly on back-barrier deposits; the barrier itself is represented by a disconformity. Every grain of sand in the shelf sand sheet has been in the barrier at one time or another. However, the primary structures are not beach or upper shoreface structures, but instead those of the shelf floor.

The stratigraphy shown in Figure 8-34 raises an embarrassing issue for coastal scientists, in that the barrier appears as a stratigraphy-generating machine that runs without fuel. If the migrating barrier lays a pavement of flood-tidal delta deposits before it and caps this pavement with a shelf sand sheet as it passes, where does all the sand come from? Some of it can be provided from eroding headlands, but as indicated in the preceding section, this is demonstrably not the case for important stretches of the North Atlantic coast of the United States. Another important source is inlet scour; inlet floors are generally two to three times as deep as the adjacent lagoons. If neither source is locally available, the barrier must draw on its own capital as it migrates landward and is in danger of being overstepped. On the Atlantic shelf, the main episode of mainland beach detachment would have occurred as the Pleistocene low stand ended and the Holocene transgression began (Swift, 1975). The character of the shelf floor suggests that overstep events were rare or lacking and that the barriers retreated more or less continuously across the shelf as sea level rose. However, recent studies of the present coastline show that pauses as long as several thousand years may occur. During these pauses, the barriers become narrower in response to both shoreface and lagoon-side erosion, until a critical width is reached and retreat resumes. The process does not result in overstep (Leatherman, in press, Leatherman and Joneja, in press).

In contrast, it has been recently suggested that the barrier systems of the Mississippi delta are commonly overstepped by the sea shortly after formation (Boyd and Penland, 1984). In this area, the rate of subsidence and sedimentation are high. Barriers such as the Chandeleur Islands have

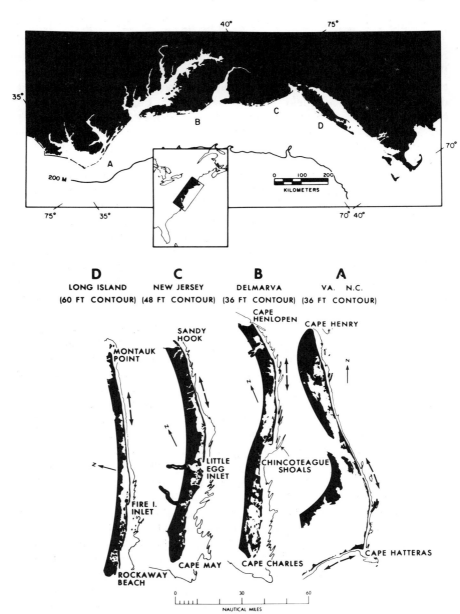

Figure 8-33. Coastal compartments of the Middle Atlantic Bight (from Duane *et al.*, 1972). Paired arrows indicate points of littoral drift divergence.

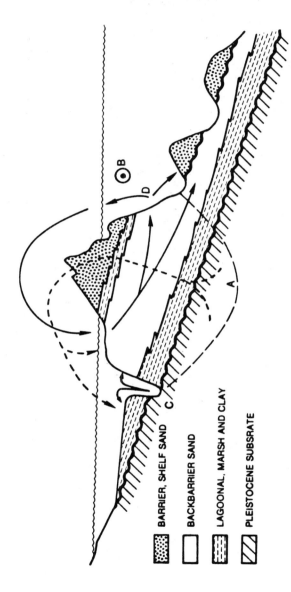

Figure 8-34. Schematic sediment budget for a landward migrating barrier on the Middle Atlantic shelf. (a) Sand eroded from an inlet throat (ultimate source) is deposited in flood-tidal delta; mud eroded from inlet is deposited on lagoon floor. (b) Sand eroded from Pleistocene headland (ultimate source) is supplied to system via littoral drift. (c) Sand and mud is also eroded by tidal channel in lagoon (ultimate source). (d) Sand eroded from shoreface (immediate source) moves onto beach or toward shelf sand sheet.

BARRIER, SHELF SAND

BACKBARRIER SAND

LAGOONAL, MARSH AND CLAY

PLEISTOCENE SUBSRATE

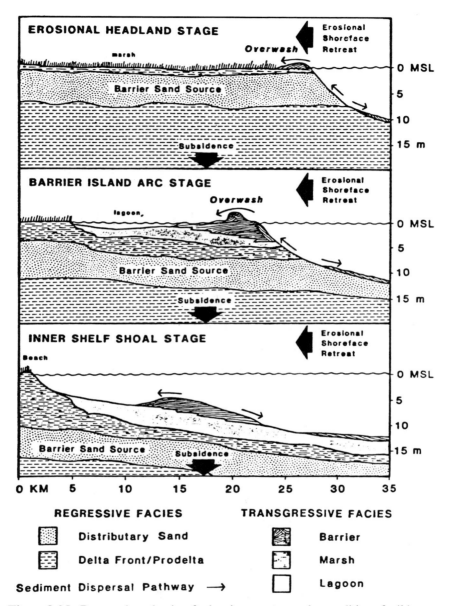

Figure 8-35. Proposed mechanism for barrier overstep under condition of mild wave climate and high sediment input (from Boyd and Penland, 1984).

retreated so slowly across the low-gradient delta plain that a broad lagoon has developed between the barrier and the mainland beach. Owing to a very mild wave climate, the barrier shorefaces are half the depth of Atlantic coast shorefaces. Should the lagoonal muds over which they are retreating become thick enough, they will become unable to erode the underlying delta top and be overstepped for lack of nourishment (Figure 8-35).

Shoreface Advance and Strand Plain Formation

Prograding shorefaces make up perhaps 7% of the world's coasts (Heward, 1981). They occur in areas where coastal retreat in response to the post-glacial sea level rise has been reversed by a high rate of sediment input and are generally therefore found in the vicinity of major rivers, such as the Mississippi, Orinoco, or Rhone. Prograding coasts comprise a spectrum from sand-rich strand plains (Curray et al., 1969) through sand and mud chenier plains (Otvos and Price, 1979) to coasts built by prograding tidal mud-flats (Rhodes, 1982; Wells and Coleman, 1981). Strand plains are made up of successive sand prisms, each capped by a beach ridge (Figures 8-36 and 8-37). In chenier plains, sand or shelly beach ridges rest on, and are separated by, flat-lying mud strata (Figures 8-38 and 8-39), while mud-flat coasts consist of mud strata with few or no beach ridges present. The coastal structure depends on the grain-size mix of the river supplying the sediment input, whether sand-rich or mud-rich, and on the wave climate, with a mild wave climate favoring mud-flat coasts. Because of their proximity to a fluvial source, prograding coasts tend to be incorporated into the complexes of depositional environments known as deltas. In the spectrum of delta types of Fisher et al. (1969), strand plains appear as the flanks of "high destructive deltas." However, in modern strand plains such as the Costa de Nayarit of Mexico (Curray et al., 1969), the deltaic facies per se appears to make up a relatively small part of the total sand mass.

Curray et al. (1969) have shown that strand plains grow by capture of breakpoint bars during periods of neap tides and by the successive accretion of these bars as beach ridges (Figure 8-36). Classical explanations for chenier plain growth have stressed cyclic variation in the rate of sediment supply. In this model mud flats prograde during periods of a high rate of sediment supply, while the coast is eroded and a beach ridge driven back over it during periods of reduced sediment supply (Gould and McFarlan, 1959). However, Woodruff et al. (1983) have shown that in some settings, erosion, deposition, and migration of a chenier can occur simultaneously, and long-term variations in sea level or sediment supply do not appear necessary to explain the episodic formation of chenier ridges.

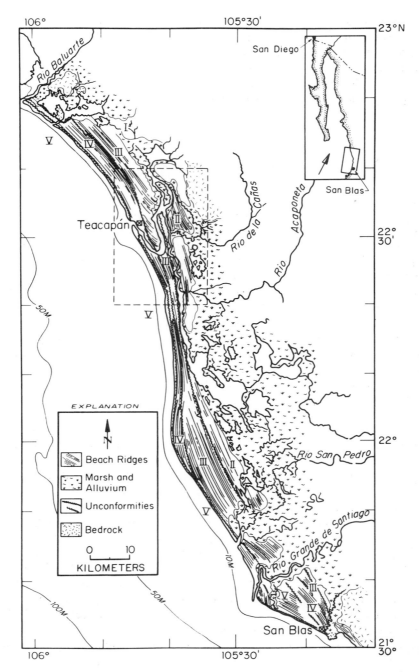

Figure 8-36. Costa de Nayarit strand plain (from Curray *et al.*, 1969).

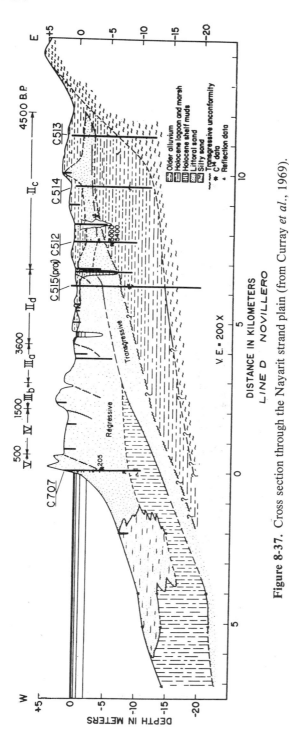

Figure 8-37. Cross section through the Nayarit strand plain (from Curray *et al.*, 1969).

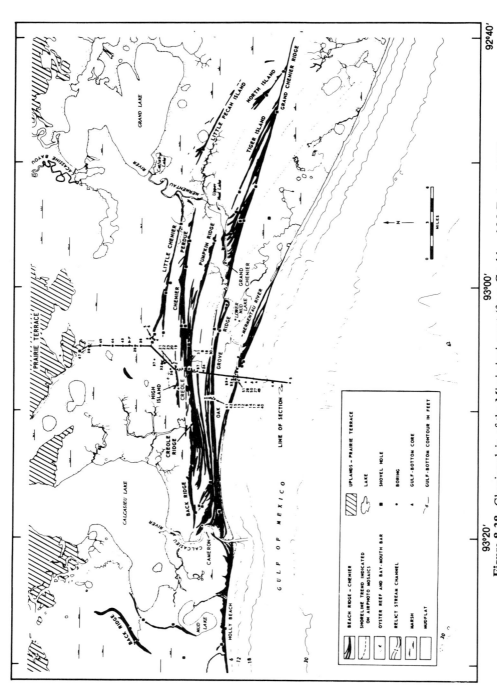

Figure 8-38. Chenier plain of the Mississippi coast (from Gould and McFarlan, 1959).

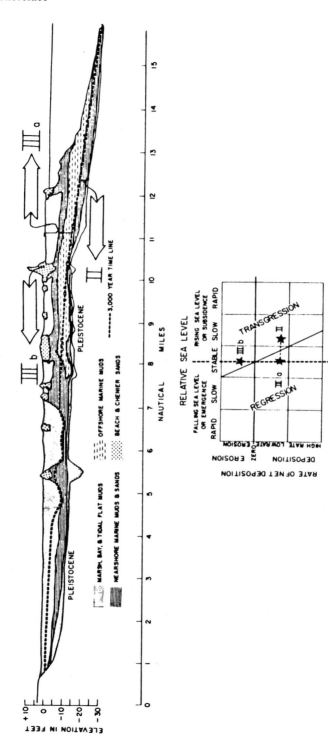

Figure 8-39. Cross section through the Louisiana coastal plan (from Curray, 1964).

Shoreface Facies

Sediment Distribution on the Shoreface

The character of shoreface sands varies from coast to coast as a function of: (1) the mean annual fluid power expenditure, (2) the mix of grain sizes contributed to the littoral drift system from the eroding substrate or from adjacent rivers, and (3) water temperature and salinity, which in turn affect faunal community structure.

On almost all coasts, the character of upper shoreface deposits differs from that of lower shoreface deposits. The character of the coast, whether retreating or prograding, is also an important variable. Shoreface deposits on retreating coasts are similar in facies types to deposits on prograding coasts, but occur in different sequence. The contrast between the two coastal types can be generalized as follows: Inner shelf deposits of retreating coasts consist of lag sand if the sedimentation rate is low and mud overlying lag sand if it is high. Inner shelf deposits on prograding coasts are muds, which become overlain by shoreface sand as the coast continues to advance.

Shoreface deposits on coasts undergoing erosional shoreface retreat have been more frequently studied than those of prograding shorefaces. These deposits are generally not preserved in the rock record, but their lithologies are probably representative of shoreface sediments in general, including those of prograding shorefaces.

Lithofacies on Retreating Coasts

Upper Shoreface Sands

On retreating coasts, modern sediment forms an ephemeral blanket overlying the eroding lagoonal and Pleistocene strata. Up to several meters of sand periodically accumulate in the beach prism and on the breakpoint bar. These are lag deposits and constitute a temporary storage of the coarser fraction of the sediment load that is undergoing intermittent alongshore transit in the littoral current. A seaward-fining and -thinning blanket of sand extends from the crest of the breakpoint bar to approximately the 10 m isobath. Here it thins to a feather edge over a thin (5 cm) veneer of coarse lag sand that mantles the eroding strata underneath (Figure 8-40). This pattern has been reported from the North Carolina coast (Swift *et al.*, 1971, Figure 8-40b) and from the coast of Holland (Van Straaten, 1965), where bimodal sands have been sampled at the contact between the seaward-fining fine sand and the coarse material that underlies it and extends seaward of it. It has also been reported from the mesotidal coast of Georgia (Howard and Reineck, 1972) and occurs intermittently on the macrotidal Yorkshire coast (Jago, 1981) in areas where wave processes locally dominate over tidal processes. Upper shoreface deposits on retreating coasts are periodically stripped off by major storms, and the underlying back-barrier or older coastal deposits are subject to erosion. As storm currents wane, and in the intervening fair-weather periods, the upper shoreface sands reaccumulate.

The origin of the seaward-fining grain-size gradient of the shoreface has been the subject of some debate. Cornaglia (1887, in Munch-Peterson, 1950) was the first to suggest that a seaward-fining grain-size gradient would result from the balance between the net landward-directed fluid force applied to a grain and the average downslope component of gravitational force. Toward the top of the shoreface, the wave orbital velocities become increasingly intense and more asymmetrical, and the grain size of sediment must become correspondingly coarser up the slope of the shoreface for the equilibrium to prevail. This idea, the null point theory, has been explored in some detail by Ippen and Eagleson (1955) and more recently by Jago and Barusseau (1981a) and Bowen (1980).

It seems probable, however, that other mechanisms are also operating. Shi and Larson (in press) have described a wave-induced sorting in which finer grains move seaward. Seaward transport of sand by rip currents may be a major or even dominant mechanism. Turbid rip current heads have been observed carrying sand seaward for many hundreds of meters (Cook and Gorsline, 1972; Reimnitz et al., 1976). The seaward-fining grain-size gradient may thus be analogous to the down-wind grain-size gradient observed in volcanic ash and loess deposits (Krumbein and Sloss, 1963). Of the several mechanisms described above, this is the only one that accounts adequately for the geometry of the deposit in question (a seaward-fining apron, thinning to a sharp boundary, overlying sand of dissimilar size).

As noted by Jago and Barusseau (1981b), the grain size at a given isobath on the shoreface generally does not reflect the maximum fluid power expenditure observed. This suggests that a supply term is significant in the equilibrium relationship, affecting it in perhaps two ways: (1) by providing an upper limit on the range of grain sizes being provided (if, for example, supplying rivers are trapping out everything coarser than fine sand, then only this grain size could be deposited), and (2) by "flooding" the system so that the shoreface cannot "winnow out" to the maximum stable grain sizes permissible under null point theory (because rip currents are constantly raining out fine sand as fast as wave agitation and alongshore flow can remove it).

Howard and Reineck (1981) have recently investigated the shoreface stratigraphy of the high wave-energy coast of southern California by means of vibracores, box cores, and can cores (Figure 8-41). They find an upper shoreface facies extending seaward to −9 m. Parallel laminated sand is the

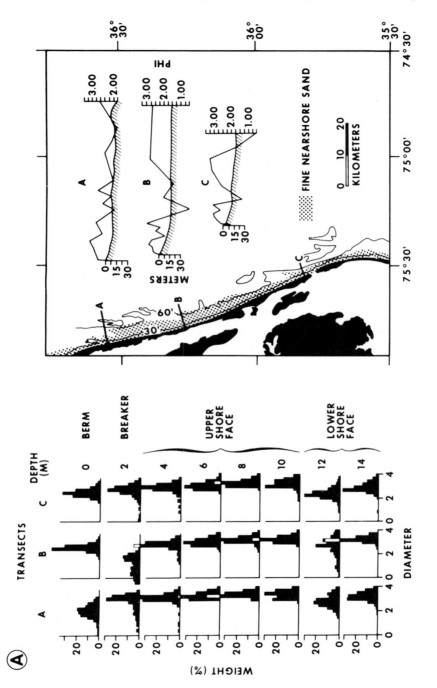

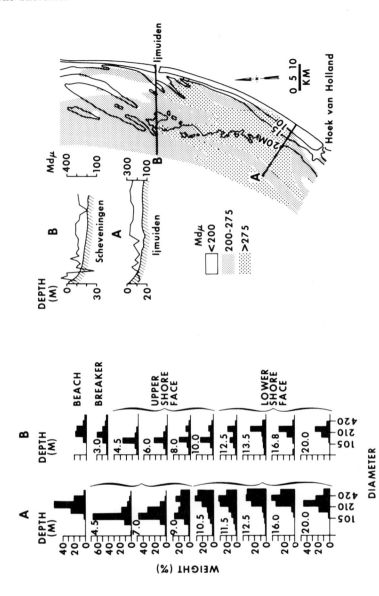

Figure 8-40. Shoreface grain-size gradients from the North Carolina and Dutch Coasts (from Swift *et al.*, 1971; Van Straaten, 1965).

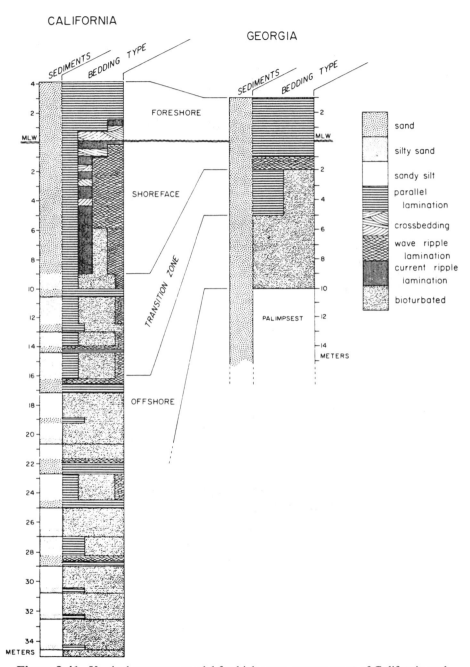

Figure 8-41. Vertical sequence model for high wave-energy coast of California and low wave-energy Georgia coast (from Howard and Reineck, 1981).

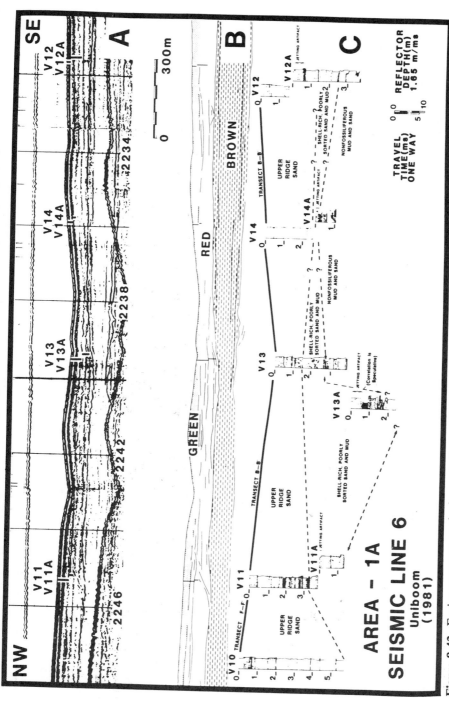

Figure 8-42. Facies sequence on the inner shelf of a coast undergoing erosional shoreface retreat (New Jersey coast). Inner shelf sand (green seismic unit) overlies back-barrier sand (red seismic unit) and lagoonal clay (brown seismic unit) (from Figueiredo, et al., 1983).

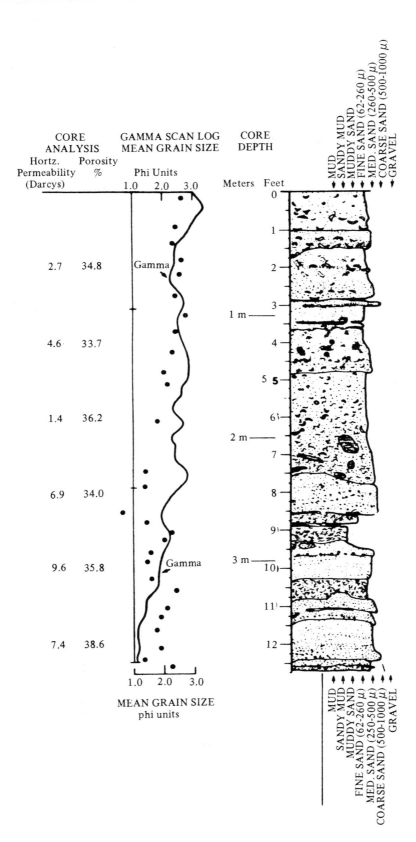

dominant sedimentary structure. Individual laminae commonly pinch out at erosional contacts, suggesting that they are organized into very gently dipping wedge-shaped sets. High-angle cross-strata sets are most abundant near the low-water line and become less frequent with increasing depth. Alongshore orientations prevail over onshore- and offshore-oriented sets (which are of about equal frequency). A small-scale oscillation ripple lamination is also present. A similar upper-shoreface facies has been described from the low wave-energy coasts of Georgia (Howard and Reineck, 1972; Figure 8-41). Here high-angle cross-stratification is absent, and the facies only extends to about 2-m water depth.

Lower Shoreface and Inner Shelf Deposits on Retreating Coasts

Beneath the seaward-fining sand, sediments on retreating shorefaces consist of a thin relatively coarse lag developed on top of back-barrier or older strata. This lag thickens in a seaward direction and becomes a clean, well-sorted sand that varies from 0 to 10 m thick on the adjacent inner shelf (Figures 8-42 and 8-43). Grain size is variable. If coarse clasts are available, a terrigenous gravel may occur sporadically at the base of the sequence (Belderson et al., 1966; Figueiredo et al., 1981a). Otherwise coquina, fish bones and teeth, or other intraclasts may accumulate. These constitute the basal gravel of the Holocene transgression. Upward-fining sequences also occur within the shelf sand sheet. They form during storm episodes. Accelerating currents during the beginning of the storm erode the sand sheet until it is armored with a coarse sand or pebble lag; then, as the storm flow wanes, a graded sand bed accumulates (Figueiredo et al., 1981a). Such storm beds are called tempestites (Seilacher, 1982).

Microtidal and mesotidal coasts such as the Atlantic coast of North America, the Argentine coast, and on the west Fresian coast of Holland, grain size varies systematically across the inner shelf sand sheet as a consequence of the development of sand ridges in response to storm current regime (Swift et al., 1979; Figure 8-44). On macrotidal coasts, the mean annual fluid power expenditure on the sea floor is yet greater. Tidal currents increase in intensity in a seaward direction and a seaward-coarsening grain-size gradient results (Jago, 1981; Figure 8-45).

Subsiding modern coasts with high rates of sediment input may be undergoing sufficiently slow transgression that the inner shelf facies can exhibit some characteristics of a regressive setting; that is, they can accommodate mud. The Valencia coast of Mediterranean Spain, for example, is undergoing erosional shoreface retreat (Maldonado et al., in

---

Figure 8-43. Detail from a core through the inner New Jersey shelf, showing the contact between the inner shelf sand sheet and the back-barrier deposits (Modified from Swift et al., in press from data obtained from Cities Service Atlantic shelf coring project, 1979; R. Tillman, principle investigator. Core analysis by J. Rine.)

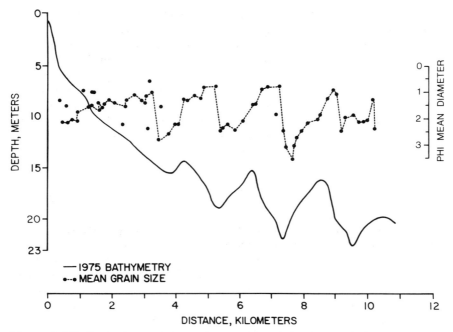

Figure 8-44. Grain-size variation across the Long Island inner shelf. Variation is controlled by ridge and swale topography, with coast sand and up-current side of ridges and finest sand on down-current side (from Swift *et al.*, in press).

press). A transgressive basal sand is being created by this process and is advancing landward as the shoreface retreats. However, a little further seaward, mud is accumulating over the sand, forming a mud blanket that becomes several inches thick further out on the shelf. Piston cores through the mud reveal graded sand beds alternating with mud near the basal contact (Figure 8-46). Maldonado *et al.* (in press) interpret these beds as deposits formed by downwelling coastal storm currents that periodically sweep shoreface sand out over the inner shelf sands, so that the basal sand has intertongued with the overlying mud as their respective environments have transgressed landward. Curray (1960) has described a similar relationship between the basal sand and overlying mud of the Gulf of Mexico shelf.

Bedforms of the lower shoreface and inner shelf have been studied in some detail on the Atlantic coast of North America (Swift and Freeland, 1978; Swift *et al.*, 1979; Swift *et al.*, in press, a; Amos and King, 1984). Flow-transverse bedforms are present at three spatial scales. Megaripples are flow-transverse bedforms with spacings on the order of 2 to 3 m (Figure 8-47). Sand waves are flow-transverse bedforms with spacings in the order of 100 m (Figure 8-48). Sand ridges are flow-oblique bedforms with spacings on the order of 2 to 4 km. (Figure 8-49).

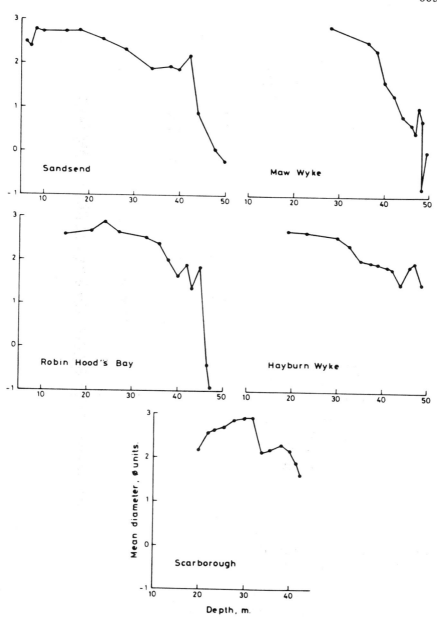

Figure 8-45. Seaward increase in grain size on the macrotidal Yorkshire coast (from Jago, 1981).

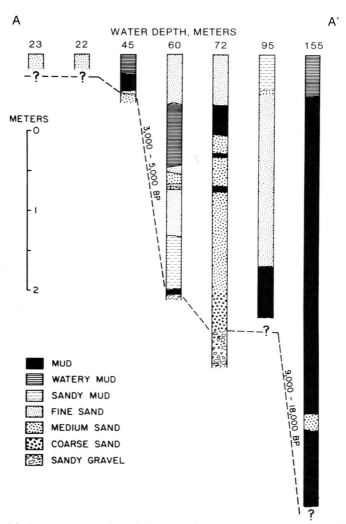

Figure 8-46. Intertonguing of basal transgressive sand and overlying mud facies on the Valencia Shelf of Mediterranean Spain (from Maldonado *et al.*, in press). Dashed line represents the top of the basal transgressive gravel.

The relationship between megaripples and sand waves is fairly well understood as a consequence of numerous studies in the intertidal zone (e.g., Dalrymple *et al.*, 1978) and in laboratory flumes (summary in Harms *et al.*, 1982). Megaripples on the lower shoreface and inner shelf may resemble the "dunes" of flume workers and the megaripples of the intertidal zone in being sharp-crested (having avalanche faces), short-crested (having poor crestal continuity), and sinuous. It has recently been recognized, however, that inner-shelf megaripples on the Atlantic coast are often "hummocky" in appearance; that is, they are oval to circular in plan view and dome-like in cross section (Figures 8-48 and 8-50) with side slopes of 15° or less (Swift *et*

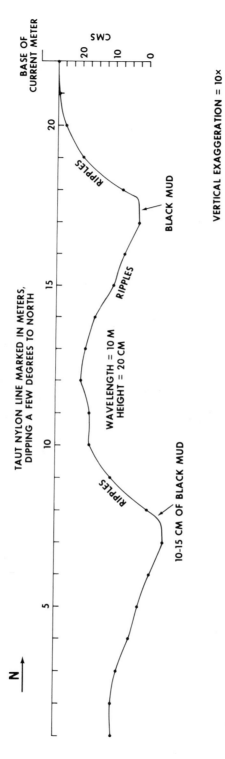

Figure 8-47. Profile across sharp-crested megaripples collected by scuba divers on the Virginia Shelf (from Swift *et al.*, 1979).

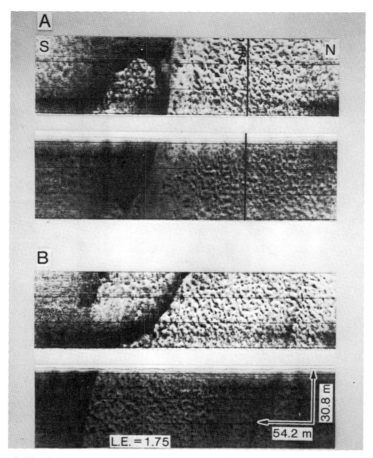

Figure 8-48. Side scan sonar record from the inner Long Island Shelf, showing hummocky megaripples on the backs of sand waves. Dark bands are the coarse sands of sand wave troughs (from Swift *et al.*, 1983).

al., 1983). Both types of megaripples appear to be responses to the combined flow regime characteristic of storms, in which a high-frequency wave orbital current component is superimposed on a mean flow. Hummocky megaripples are the preferred configuration in fine or very fine sand, when the flow components are subequal in value. Sharp-crested megaripples form in coarser sand and are favored when the wave orbital current component is weak relative to the mean flow component. The hummocky megaripples of the modern Atlantic coast appear to be the modern equivalent of the hummocky cross-strata sets of ancient inner-shelf and shoreface deposits (Dott and Bourgeois, 1982), while the sharp-crested megaripples appear to be the modern equivalent of ancient trough cross-strata sets (Allen, 1963).

Sand waves on modern retreating coasts with storm-dominated regimes tend to be low in amplitude (a meter or less) relative to their wavelength (100 m or more) except in areas where the shelf shoals or becomes narrow in the down-current direction, thus accelerating storm flows (e.g., Hunt *et al.*, 1977; Swift *et al.*, 1978). Avalanche faces are generally not present. Sand waves appear on side scan sonar records as dark bands. The bands are the troughs of the sand waves that tend to penetrate to the basal coarse sand, shell hash, and gravel of the Holocene transgressive sand sheet; such coarse

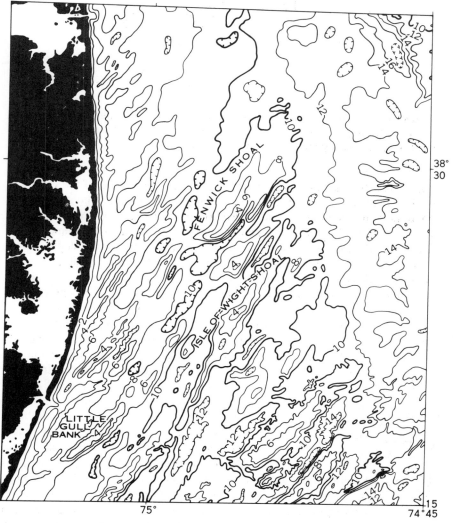

Figure 8-49. Coast of Delaware showing field of sand ridges (from Swift *et al.*, 1973).

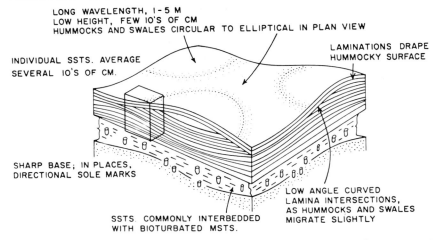

Figure 8-50. Schematic diagram showing characteristics of hummocky cross-strata sets (unpublished diagram of R. Walker). Rectangular solid illustrates difficulty of identifying hummocky cross-strata in box cores.

materials reflect sound more effectively than the finer sand of the crest and back of the sand wave. On tide-dominated coasts, however, the more nearly continuous application of fluid power to the sandy sea floor results in sand waves with avalanche faces and heights up to 7 m. Tidal sand waves commonly do not extend closer to shore than about 20 m water depth, where they tend to be suppressed by the more intense wave orbital currents characteristic of shallow water (McCave, 1971).

Sand ridges occur in the storm-dominated Atlantic coast of North America (Swift *et al.*, 1973), on the Buenos Aires coast of Argentine (Parker *et al.*, 1982), on the west Fresian coast of the North Sea (Swift *et al.*, 1978), on the tide-dominated coasts of Nantucket Island (Mann *et al.*, 1981), and the Anglian coast of Great Britain (Robinson, 1980). They rise above the sea floor as high as 10 m, but side slopes rarely exceed 1°. Their crestlines can often be traced for 10 km or more and may start in water as shoal as 2 m. On modern coasts they typically make angles of 10 to 45° with the shoreline, with the angle opening into the direction from which the storm (or dominant tidal) flows come. The ridges commonly have steeper down-current sides, although this asymmetry tends to reverse very near the shoreline, where the onshore portion of the trough between the ridge and the shoreline tends to become a sort of blowout and is steeper than the down-current flank. The up-current flank is always coarser grained than the down-current flank. Time series of coastal maps and radiocarbon dates indicate that the ridges migrate slowly offshore and down coast in the prevailing direction of storm flow while the eroding shoreface retreats out from under them. The fluid dynamics of

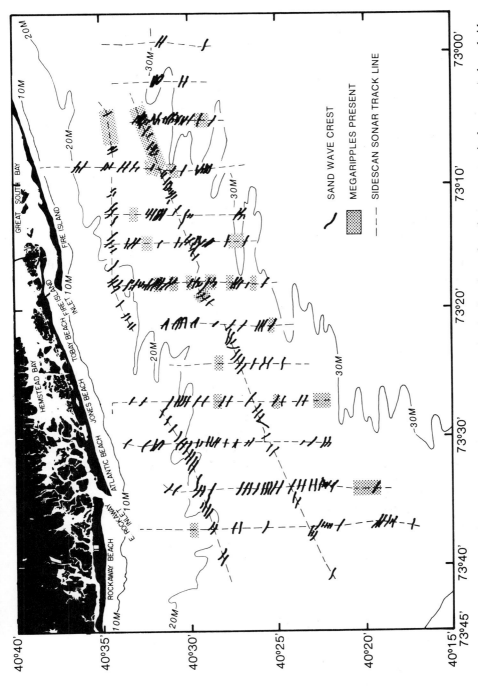

Figure 8-51. Side scans sonar map of the Long Island coast, showing locations of sand waves, megaripple zones, and sand ridges (indicated by sands in contours) (from Swift *et al.*, 1979).

tidal sand ridge formation is now reasonably well understood (Huthnance, 1982a,b); and the origin of sand ridges on storm-dominated coasts is believed to be similar (Huthnance, personal communication; Figueiredo et al., 1981b).

Megaripples, sand waves, and sand ridges occur together on the storm-dominated Atlantic coast of North America. Sand waves and megaripples trend at 80–85° to the shoreline with the acute angle opening to the northeast, reflecting the offshore component of southwest-tending storm bottom flows (Figure 8-51).

Lithofacies on Prograding Coasts

On prograding coasts, the seaward-thinning and -fining apron of shoreface sand commonly passes seaward into mud deposits. In cases of extremely high mud supply, no sand is present and chenier plains, whose beach ridges may consist of coquina rather than sand, pass seaward into mud flats (Greensmith and Tucker, 1976).

The studies of Bernard et al. (1962) of the shoreface of Galveston Island and of Reineck and Singh (1971) of Gulf of Gaeta on the Tyrrhenian coast of Italy seem to be the only published examinations describing the shoreface facies of prograding modern coasts. They suffice to show that the facies are the same as those retreating coasts.

Facies of prograding shorefaces are much better known from studies of the rock record, because of their greater preservation potential (e.g., Howard and Scott, 1983; Hunter, 1980; Campbell, 1979). The upper shoreface facies are much the same as those described from modern retreating coasts. However, the stratigraphic pattern of the lower shoreface and inner shelf of prograding shorefaces is more complex than that of retreating shorefaces in that inner shelf muds intertongue with the advancing shoreface sand above them. Prograding shoreface deposits appear in outcrop as distinctive sequences in which sandstone tempestites with hummocky cross-strata sets become thicker, coarser, and more closely spaced up-section and are capped in turn by: (1) a zone of amalgamated beds of hummocky cross-strata, (2) a zone of sandstone with trough cross-strata sets (surf facies), and (3) a zone of low-angle, large-scale wedge-shaped cross-strata sets (swash and beach facies, Figure 8-52).

Engineering Implications

Because research on shoreface sediment dynamics is relatively new, especially when compared with the great body of literature concerning surf zone sediment dynamics, there have been only a few applications of our knowledge of shoreface processes to engineering problems. However, shoreface sediment dynamics and morphodynamics have application to problems ranging from siting structures, such as municipal outfalls, cooling water intakes, pipelines, or open-coast piers, to mitigating beach erosion and forecasting the effects of major offshore structures (offshore "super-ports" or offshore power stations) on adjacent shorelines. These subjects are briefly discussed to provide final application for the results of research on shoreface processes.

Previous chapter sections have shown that there is significant but episodic entrainment and transport of sand across the entire shoreface. Vincent *et al.* (in press) and Young *et al.* (1982) have shown that some of this sand is suspended at least a meter above the sea floor. This movement of shoreface sand can result in its ingestion into cooling water intakes. Periods of local deposition and erosion on the shoreface can lead to clogging of sewer outfalls and exposure of buried pipelines. It has been pointed out that coupling between the beach and the shoreface is weaker than the coupling that exists between the surf zone and the beach. Nevertheless it is logical to expect a relationship between long-term transgression (regression) of the shoreface and erosion (deposition) of the beach. Beach erosion control structures that do not take into account shoreface processes may be defeated by long-term shoreface erosion. Mitigation of beach erosion by artificial sand fill is a common practice. In many areas there are shortages of local terrestrial sand deposits, and offshore dredging of sand to nourish the beach is becoming more prevalent. Clearly these operations should be sensitive to local shoreface processes because if the dredging is conducted at too shallow a depth it will oversteepen the equilibrium shape of the shoreface and accelerate the erosion of the beach fill.

The field measurement and sediment dynamics analysis routines that have been applied to research on shoreface sediment transport are available for use in engineering problems. It is necessary to understand that a significant element of physical coastal processes exists beyond the classically considered surf zone.

Summary

The coastal environment called the shoreface is morphologically and dynamically distinct from the surf zone and the continental shelf. It is a common morphologic feature of sandy open coasts. The characteristic offshore scale of shoreface morphology corresponds to a cross-shore length scale where particular features of waves and nearshore coastal boundary layer currents are enhanced. Cross-shore bedload transport processes, resulting from features of wave orbital behavior and cross-shore current pattern, are unique to the shoreface. They appear to balance the offshore

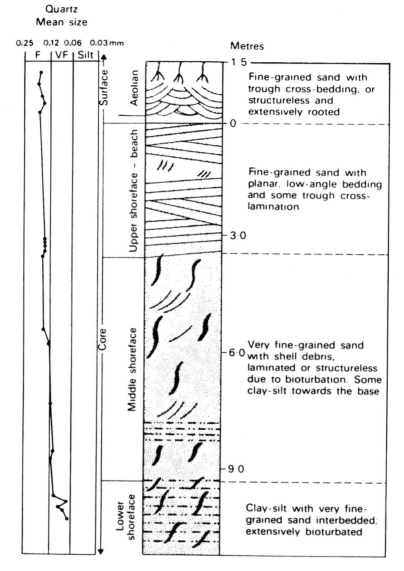

Figure 8-52. Vertical sequence and sand body produced as the Galveston barrier island underwent local transformation into a strand plain (from Elliott, 1978, after Bernard *et al.*, 1962).

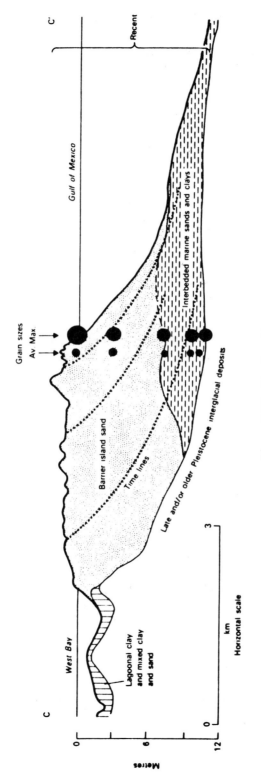

Figure 8-52. *(Continued.)*

slope to yield a time-averaged dynamic equilibrium of the shoreface shape on annual scales although no complete quantitative relationship has yet been developed. Wave orbital asymmetry, wave-current interactions, repetitive patterns of upwell and downwelling cross-shore current components and perhaps reverse transport in amplitude modulated waves are important processes in the overall annual-scale cross-shore sediment balances. Cross-shore sediment transport is dominantly controlled by the balance of wave-driven onshore processes reacting with the slope on the upper shoreface while cross-shore mean bottom current components are most important to the cross-shore sediment balance on the middle and lower shoreface.

Major shoreface sediment transport is even more episodic and storm related than surf zone sand transport. Storms tend to result in an offshore bedload transport hysteresis because onshore winds produce larger waves and downwelling conditions. Shoreface bedload transport can extend between the surf zone and the inner shelf during even moderate storms. Net offshore sand transport during storms is generally characterized by an offshore divergence resulting in deposition across the shoreface. This deposition is not uniform but tends to concentrate as a tabular sheet on the upper shoreface. The near equilibrium over annual time scales suggests that onshore bed-load transport is favored during nonstorm conditions.

The features and processes associated with alongshore shoreface sediment transport have not been well studied. The contrast between wave-driven currents in the surf zone and wind-driven and tidal currents on the shoreface suggests that sharp alongshore sediment transport gradients are less likely in typical open-ocean shoreface conditions. Thus, shoreface processes should tend to suppress irregularities in the coastline shape if the sea floor is erodable, a sufficient sand supply exists, and the processes have been acting for sufficient time.

The cross-shore bedload transport processes associated with the open-ocean shoreface environment provide mechanisms for offshore nourishment of barrier islands. Under appropriate conditions of marine transgression, the development of a relatively steep shoreface slope on a more gentle regional slope may initiate barrier island formation.

The dynamic equilibrium of the shoreface that characterizes the annual time scale is readily converted to a shifting equilibrium on longer time scales by sea level changes and relative sediment supply. Geologic deposits formed by prograding shoreface processes are commonly the most volumetrically significant coastal sand formations. Most modern deposits of this type are associated with an excess sand supply but can also be produced during a sea level fall. Rising sea level results in a transgressing shoreface that generally destroys beach and dune deposits. Under such conditions the history of shoreline migration is preserved in sand deposits formed on the lower shoreface. When coastal barriers (spits and barrier islands) are present, the transgressive sand sheet that forms over the lower shoreface often lies unconformably on lagoonal sedimentary deposits.

Much research lies ahead before a complete and adequate understanding of the shoreface environment is developed. The material presented in this chapter is meant to provide a comprehensive framework for conceptualizing and understanding this environment, but many important elements are unavailable. Although the processes that control the implied time-averaged balance of cross-shore sediment transport that must exist to preserve the demonstrated shoreface equilibrium shape have been enumerated and discussed, a quantitative relationship is lacking. Most research on shoreface bedload sediment dynamics was conducted on the open coast of Long Island, which is only assumed to be representative of other sandy coasts. The alongshore interaction of sediment transport on sandy shorefaces with rocky or muddy coasts has not been explored. Similar alongshore interactions with estuaries, major shoals, and tidal inlets are also unstudied.

References

Abou-Seida, M. M., 1965. *Bedload Function Due to Wave Action*. Berkley Hydraulics Engr. Lab., Rept. HEL-2-11, 78 pp.

Airy, G. B., 1845. Tides and waves. *Encycl. Metrop., Art.*, **192**, 234–396.

Allen, J. R. L., 1963. The classification of cross-stratified units with notes on their origin. *Sedimentology*, **2**, 93–114.

Allen, J. S., 1980. Models of wind-driven currents on the continental shelf. *Ann. Rev. Fluid Mech.*, **12**, 389–433.

Amos, C. L., and King, E. L., 1984. Bedforms of the Canadian eastern seaboard: A comparison with global occurrences. *Mar. Geol.*, **57**, 167–208.

Arthur, R. S., 1962. A note on the dynamics of rip currents. *J. Geophys. Res.*, **67**, 2777–2779.

Bagnold, R. A., 1946. Motions of waves in shallow water; interaction between waves and sand bottoms. *Roy. Soc. London Proc., Ser. A*, **187**, 1–15.

Bagnold, R. A., 1947. Sand movement by waves: some small scale experiments with sand of very low density. *J. Intl. Civil Engr.*, No. 4, 1946–47, pp. 447.

Bagnold, R. A., 1956. The flow of cohesionless grains in fluids. *Roy. Soc. Phil. Trans. London, Ser. A*, **249**, 235–297.

Bagnold, R. A., 1963. Beach and nearshore processes, part I: mechanics of marine sedimentation. *In*: Hill, M. N. (ed.), *The Sea*, Vol. 3. John Wiley and Sons, New York, pp. 507–528.

Bagnold, R. A., 1966. *An Approach to the Sediment Transport Problem from General Physics*, U. S. Geol. Survey Prof. Paper 422-I, U. S. Govt. Printing Off., Washington, DC, 37 pp.

Bailard, J. A., 1981. An energetics total load sediment transport model for a plane sloping beach. *J. Geophys. Res.*, **86**, 10938–10954.

Bailard, J. A., and Inman, D. L., 1981. An energetics bedload model for a plane sloping beach: local transport. *J. Geophys. Res.*, **86**, 2035–2043.

Barber, N. F., 1969. *Water Waves*. Wykeham, London, 142 pp.

Barrell, J., 1912. Criteria for the recognition of ancient delta deposits. *Geol. Soc. Amer. Bull.*, **23**, 377–446.

Battisti, D. S., 1982. Estimation of nearshore tidal currents on nonsmooth continental shelves. *J. Geophys. Res.*, **87**, 7873–7878.

Battisti, D. S., and Clarke, A. J., 1982. A simple method for estimating barotropic tidal currents on continental margins with specific application to the M_2 tide off the Atlantic and Pacific coasts of the United States. *J. Phys. Oceanogr.*, **12**, 8–16.

Belderson, R. H., Kenyon, N. H., and Stride, A. H., 1966. Tidal current fashioning of a Basal Bed. *Mar. Geol.*, **4**, 237–257.

Bennett, J. R., and Magnell, B. A., 1979. A dynamical analysis of currents near the New Jersey Coast. *J. Geophys. Res.*, **84**, 1165–1175.

Bernard, H. A., LeBlanc, R. J., and Major, Jr., C. F., 1962. *Recent and Pleistocene Geology of Southeast Texas*. Geology of the Gulf Coast and Central Texas and Guidebook of Excursion, Houston Geological Society, pp. 175–225.

Blanton, J. O., and Murthy, C. R., 1974. Observations of lateral shear in the nearshore zone of a Great Lake. *J. Phys. Oceanogr.*, **4**, 660–663.

Borgman, L. E., and Chappelear, J. E., 1958. The use of the Stokes-Struik approximation for waves of finite height. Proc. 6th Conf. Coast. Engr., pp. 252–280.

Boussinesq, J. 1872. Théorie des ondes et de remous qui se propagent le long d'un canal rectangulaire horizontal, en communiquant au liquide contenu dans ce canal des vitesses sensiblement parallèles de la surface au fond. *J. Math. Pures et Appliquées* (Lionvilles, France), **17**; 55–108.

Bowen, A. J., 1967. Rip Currents. Dissert., Univ. of California, San Diego.

Bowen, A. J., 1969. Rip currents, 1. Theoretical investigations. *J. Geophys. Res.*, **74**, 5467–5478.

Bowen, A. J., 1980. Simple models of nearshore sedimentation: beach profiles and alongshore bars. *In*: S. B. McCunn (ed.), *The Coastline of Canada*. Geol. Survey Canada Paper 80-10, pp. 1–11.

Bowen, A. J., and Inman, D. L., 1969. Rip currents, 2. Laboratory and field observations. *J. Geophys. Res.*, **74**, 5479–5490.

Boyd, R., and Penland, S., 1984. Shoreface translation and the Holocene stratigraphic record. Examples from Nova Scotia, the Mississippi Delta and Eastern Australia. *Mar. Geol.*, **60**, 391–412.

Brenninkmeyer, B., 1976. Sand fountains in the surf zone. *In*: Davis, R. A., and Ethington, R. L. (eds.), *Beach and Nearshore Sedimentation*. Soc. Econ. Paleont. Mineral. Spec. Publ. 24, Tulsa, pp. 69–91.

Bruun, P., 1962. Sea level rise as a cause of shore erosion. *J. Waterways and Harbors Div., Amer. Soc. Civil Engr. Proc.*, **88**, 117–130.

Caldwell, J. M., 1969. Coastal process and bead erosion. *J. Soc. Cir. Engl.*, **53**, 142–157.

Campbell, C. V., 1979. *Model for Beach Shorefire in Valley Sandstone (Upper Cretaceous of North Western New Mexico*. New Mexico Bureau of Mines and Mineral Resources Circ. 164, 29 pp.

Carstens, M. R., Neilson, F. M., and Altinbilek, H. D., 1969. *Bed Forms Generated in the Laboratory Under Oscillatory Flow: Analytical and Experimental Study.* U.S. Army Corps of Engr., CERC Tech. Memo. 28, 39 pp.

Chappelear, J. E., 1961. Direct numerical calculation of wave properties. *J. Geophys. Res.*, **62**, 501–508.

Charlesworth, L. J., 1968. *Bay, Inlet and Nearshore Marine Sedimentation: Beach Haven-Little Egg Inlet Region, New Jersey.* Ph.D. Dissert., Univ. of Michigan, Dept. of Geology, 614 pp.

Cook, D. O., 1970. *Occurrences of Bottom Sediment Movement Due to Wave Action.* U.S. Army Corps of Engr., Beach Erosion Board Tech. Memo. 75, 121 pp.

Cook, D. O., and Gorsline, D. S., 1972. Field observations of sand transport by shoaling waves. *Mar. Geol.*, **13**, 31–55.

Cornaglia, P., 1887. *Sol Regime Delta Spiagge e sulla regulazine dei porti.* Turin.

Csanady, G. T., 1972. The coastal boundary layer in Lake Ontario, part I, the spring regime. *J. Phys. Oceanogr.*, **2**, 41–53.

Csanady, G. T., 1977a. The coastal jet conceptual model in the dynamics of shallow seas. *In*: Goldberg, E. D., McCave, I. N., O'Brien, J. J., and Stede, J. H. (eds.), *The Sea*, Vol. 6. John Wiley and Sons, New York, pp. 117–114.

Csanady, G. T., 1978a. Wind effects on surface to bottom fronts. *J. Geophys. Res.*, **83**, 4633–4640.

Csanady, G. T., 1978b. The arrested topographic wave. *J. Phys. Oceanogr.*, **8**, 47–62.

Csanady, G. T., 1982. Circulation in the coastal ocean. Reidel Publ. Co., Boston, 280 pp.

Curray, J. R., 1960. Sediments and the history of the Holocene transgression, continental shelf, Northwest Gulf of Mexico. In: Shepard, F. P., Phleger, F. B., and Van Andel, T. H. (eds.), *Recent Sediments, Northwest Gulf of Mexico.* American Association of Petroleum Geologists, Tulsa, OK, pp. 221–226.

Curray, J. R., 1964. Transgressions and regressions. *In*: Miller, R. L. (ed.), *Papers in Mar. Geol.* Shepard Vol.; Macmillan, New York, 175–203.

Curray, J. R., Emmel, F. J., and Crampton, P. J. S., 1969. Lagunas Costeras, un simposio. *In*: *Mem. Simp. Int. Lagunas Costeras.* UNAM-UNESCO Nov. 28–30, 1967, Mexico, pp. 63–100.

Dalrymple, R. W., Knight, R. J., and Lambiase, J. J., 1978. Bedforms and their hydraulic stability relationships in a tidal environment, Bay of Fundy, Canada. *Nature*, **275** (5676), 100–104.

Davies, J. L., 1958. Wave refraction and the evolution of curved shorelines. *Geogr. Studies*, **5**, 1–14.

Davies, J. L., 1980. *Geographical Variation in Coastal Development*, 2nd ed. Hafner, New York, 212 pp.

Dean, R. G., 1965. Stream function representation of nonlinear ocean waves. *J. Geophys. Res.*, **70** (18), 4561–4572.

Dean, R. G., 1970. *Relative validities of water wave theories. Amer. Soc. Civil Engr. Proc.*, **96** (WW1), 105–119.

Dean, R. G., and Asce, A. M., 1965. *Stream Function Wave Theory: Validity and Application*. Coast. Engr., Santa Barbara Speciality Conf., Amer. Soc. Civil Engr., pp. 269–297.

Dean, R. G., and Eagleson, P. S., 1966. Finite amplitude waves. *In*: Ippen, A. T. (ed.), *Estuary and Coastline Hydrodynamics*. McGraw-Hill Book Co., Inc., New York, 744 pp.

Dietrich, G., 1963. *General Oceanography*. Wiley-Interscience, New York, 588 pp.

Dott, Jr., R. H., and Bourgeois, J., 1982. Hummocky stratification: significance of its variable bedding sequences. *Geol. Soc. Amer. Bull.*, **93**, 663–680.

Duane, D. P., Field, M. E., Miesberger, E. P., Swift, D. J. P., and Williams, S. J., 1972. Linear shoals on the Atlantic Continental Shelf, Florida to Long Island. *In*: Swift, D. J. P., Duane, D. B., and Pilkey, O. H. (eds.), *Shelf Sediment Transport, Process and Pattern*. Dowden, Hutchinson and Ross, Stroudsburg, PA, 676 pp.

Eagleson, P. S., Dean, R. G., and Peralta, L. A., 1958. *The Mechanics of the Motion of Discrete Spherical Bottom Sediment Particles Due to Shoaling Waves*. U. S. Army Beach Erosion Board, Tech. Memo. 104, 41 pp.

Elliott, T., 1978. Clastics shorelines. *In*: Reading, H. T. (ed.), *Sedimentary Environments and Facies*, Elsevier, New York, 143–177.

Fenneman, N. M., 1902. Development of the profile of equilibrium of the subaqueous shore terrace. *Geology*, **10**, 1–32.

Field, M. E., and Duane, D. B., 1976. Post-Pleistocene history of the United States inner continental shelf: significance to origin of barrier islands. *Geol. Soc. Amer. Bull.*, **87**, 692–702.

Figueiredo, A. G., Jr., 1983. Submarine sand ridges: geology and development, New Jersey, U.S.A. Ph.D. dissertation, U. Miami, Coral Gables, Florida, (not conseq. paged).

Figueiredo Jr., A. G., Sanders, J., and Swift, D. J. P., 1981a. Storm-graded layers on inner continental shelves: examples from Southern Brazil and the Atlantic Coast of the Central United States. *Sed. Geol.*, **31**, 171–190.

Figueiredo, Jr., A. G., Swift, D. J. P., Stubblefield, W. L., and Clarke, T., 1981b. Sand ridges on the inner Atlantic Shelf of North America Morphometric comparisons with Huthnance stability model. *Geo-Marine Letters*, **1**, 187–191.

Fisher, J. J., 1968. Barrier Island formation: discussion. *Geol. Soc. Amer. Bull.*, **79**, 1421–1426.

Fisher, W. L., Brown, L. F., Scott, A. J., and McGowen, J. H., 1969. Delta systems in the explorationship to occurrence of oil and gas. Bur. Econ. Geol. Univ. Texas, Austin, 78 pp.

Fjeldstad, J. E., 1929. Contribution to the dynamics of free progressive tidal waves, Norwegian North Polar Expedition with the *Maud*, 1918–1925. *Sci. Results*, **4** (3), 1–80.

Fjeldstad, J. E., 1936. Results of tidal observations, Norwegian North Polar Expedition with the *Maud*, 1918–1925. *Sci. Results*, **4** (4), 1–88.

Fleming, R. H., 1938. Tides and tidal currents in the Gulf of Panama, *J. Mar. Res.*, **1**, 192–206.

Gilbert, G. K., 1885. *Lake Bonniville*. U. S. Geol. Survey Monograph 1, 438 pp.

Gill, A. E., 1982. *Atmosphere-Ocean Dynamics*. Academic Press, New York, 662 pp.

Godfrey, P. J., and Godfrey, M. M., 1976. *Barrier Island Ecology of Cape Lookout National Seashore and Vicinity, North Carolina*. Natl. Park Service Scientific Monograph Series 9, 160 pp.

Gould, H. R., and McFarlan, Jr., E., 1959. Geologic history of the chevier plain, Southwest Louisiana. *Gulf Coast Assoc. Geol. Socs., Trans.*, **9**, 261–270.

Grant, W. D., and Madsen, O. S., 1979a. Combined wave and current interaction with a rough bottom. *J. Geophys. Res.*, **84**, 1797–1808.

Grant, W. D., and Madsen, O. S., 1979b. *Bottom Friction Under Waves in the Presence of a Weak Current*. NOAA Tech. Rept. ERL-MESA, NOAA, 150 pp. pp.

Grant, W. D., and Madsen, O. S., 1982. Movable bed roughness in unsteady oscillatory flow. *J. Geophys. Res.*, **87**, 469–481.

Greensmith, J. T., and Tucker, E. V., 1976. Major flandrian transgressive cycles, sedimentation and palaeogeography in the coastal zone of Exxex, England. *Geol. en Mijnbouw*, **55**, 131–146.

Hails, J. R., and Hoyt, J. H., 1968. Barrier development on submerged coasts: problem of sea level changes from a study of the Atlantic Coastal Plain of Georgia and part of the East Australian Coast. *Z. Geomorphol.*, **7**, 224–55.

Harms, J. C., Southard, J. B., and Walker, R. G., 1982. Structures and sequences in Clastic Rocks. *Soc. Econ. Paleon. Mineral. Short Course 9*.

Harris, R. L., 1954. *Restudy of Test Shore Nourishment by Offshore Deposition of Sand Long Branch, New Jersey*. Beach Erosion Board, Tech. Memo. 62, pp. 1–18.

Harrison, W., and Morales-Almo, R., 1964. *Dynamic Properties of Immersed Sand at Virginia Beach, Virginia*. Coastal Eng. Res. Center Tech. Memo. 9, 52 pp.

Heward, A. P., 1981. A review of wave dominated clastic shorelines. *Earth Sci. Rev.*, **17**, 223–227.

Hopkins, T. S., and Swoboda, A. L., in press. The nearshore circulation off Long Island, August 1978. *Continental Shelf Res.*

Housley, J. G., and Taylor, D. C., 1957. Application of the solitary wave theory to shoaling oscillatory waves. *Amer. Geophys. Union Trans.*, **38**, 56–61.

Howard, J. D., and Reineck, H. E., 1972. Georgia Coastal Region, Sapelo Island, USA: sedimentology and biology. IV. physical and biogenic sedimentary structures of the nearshore shelf. *Senckenberg. Marit.*, **4**, 81–123.

Howard, J. D., and Reineck, H. E., 1981. Depositional facies of high energy beach to offshore sequence: comparison with low energy sequence. *Amer. Assoc. Petrol. Geol. Bull.*, **65**, 807–830.

Howard, J. D., and Scott, R. M., 1983. Comparison of Pleistocene and Holocene Barrier Island beach to offshore sequences, Georgia and Northeast Florida Coasts, USA. *Sed. Geol.*, **34**, 167–183.

Hoyt, J. H., 1967. Barrier island formation. *Geol. Soc. Amer. Bull.*, **78**, 1125–1136.

Huff, L. C., and Friske, D. A., 1980. Development of two sediment transport instrument systems. Proc. 17th Conf. Coastal Engr., Amer. Soc. Civil Engr.

Hunt, R. E., Swift, D. J. P., and Palmer, H., 1977. Constructional Shelf topography, Diamond Shoals, North Carolina. *Geol. Soc. Amer. Bull.*, **88**, 299–311.

Hunter, R. E., 1980. Depositional environments of some Pleistocene coastal terrace deposits, Southwestern Oregon. Case history of a progradational beach and dune sequences. *Sed. Geol.*, **27**, 241–262.

Huthnance, J. M., 1982a. On one mechanism forming linear sand banks. *Est., Coastal and Shelf Sci.*, **14**, 79–99.

Huthnance, J. M., 1982b. On the formation of sand banks of finite extent. *Est., Coastal and Shelf Sci.*, **15**, 277–299.

Inman, D. L., and Bowen, A. J., 1963. Flume experiments on sand transport by waves and currents. Proc. 8th Conf. Coastal Engr., Amer. Soc. Civil Engr., pp. 137–150.

Inman, D. L., and Quinn, W. H., 1952. Currents in the surfzone. *Proc., 2nd Conf. Coast. Engr.*, 24–36.

Ippen, A. T., 1966. *Estuary and Coastline Hydrodynamics*. McGraw-Hill, New York, 774 pp.

Ippen, A. T., and Eagleson, P. S., 1955. *A Study of Sediment Sorting by Waves Shoaling on a Plane Beach*. Beach Erosion Board Tech. Memo. 63, 81 pp.

Jago, C. F., 1981. Sediment response to waves and currents, North Yorkshire Shelf, North Sea. *In*: Nio, S-D., Shuttenhelm, R. O., and Van Weering, J. C. E., (eds.), *Holocene Marine Sedimentation in the North Sea Basin*. Intl. Assoc. Sedimentologists Spec. Publ. 5, pp. 283–301.

Jago, C. F., and Borusseau, J. P., 1981. Sediment entrainment on a wave-graded shelf, Rousillon, France. *Mar. Geol.*, **42**, 279–299.

Johnson, D. W., 1919. *Shoreline Processes and Shoreline Development*, John Wiley & Sons, New York, 584 p.

Kalkanis, G., 1964. *Transport of Bed Material Due to Wave Action*. U.S. Army Corps of Engr., CERC Tech. Memo. 2, 38 pp.

Keller, J. B., 1948. The solitary wave and periodic waves in shallow water. *Comm. Appl. Math.*, **1**, 323–329.

Keulegan, G. H., and Patterson, G. W., 1940. Mathematical theory of irrotational translation waves. *J. Res. Natl. Bur. Std.*, **24** RP 1272:47.

Kim, J. H., and Gardner, W. S., 1974. *Geomarine Investigation, Final Design Study, Wastewater Ocean Outfalls for Ocean County, New Jersey Sewage Authority*. Woodward Gardner Assoc., Plymouth Meeting, PA, 66 pp.

Kinsman, B., 1965. *Wind Waves*. Prentice-Hall, Englewood Cliffs, NJ, 676 pp.

Komar, P. D., 1976. *Beach Processes and Sedimentation*. Prentice-Hall, Englewood Cliffs, NJ, 429 pp.

Korteweg, D. J., and deVries, G., 1895. On the change of form of long waves advancing in a rectangular canal, and on a new type of long stationary waves. *Phil. Mag., Ser. 5*, **39**, 422–443.

Krumbien, W. C., and Sloss, L. C., 1963. *Stratigraphy and Sedimentation*. W. H. Freeman and Company, San Francisco, 660 pp.

Langford-Smith, T., and Thom, B. G., 1969. New South Wales coastal morphology. *J. Geol. Soc. Australia*, **16**, 572–580.

Larsen, L. H., 1982. A new mechanism for seaward dispersion of midshelf sediments. *Sedimentology*, **29**, 279–284.

Leatherman, S. P., in press. Geomorphic and stratigraphic analysis of Fire Island, New York. *Mar. Geol.*

Leatherman, S. P., Williams, A. T., and Fisher, J. S., 1977. Overwash sedimentation associated with a large-scale northeaster. *Mar. Geol.*, **24**, 109–121.

Littman, W., 1957. On the existence of periodic waves near critical speed. *Comm. Pure App. Math.*, **10**, 241–269.

Longuet-Higgins, M. S., 1953. Mass transport in water waves. *Phil. Trans. Roy. Soc. London, Ser. A*, **245** (903), 535–581.

Longuet-Higgins, M. S., 1956. The refraction of sea waves in shallow water. *J. Fluid Mech.*, **1**, 163–176.

Longuet-Higgins, M. A., and Stewart, R. W., 1964. Radiation stress in water waves; a physical discussion, with applications. *Deep Sea Res.*, **11**, 529–562.

Madsen, O. S., and Grant, W. D., 1976. Sediment transport in the coastal environment. Report. No. 209, Ralph M. Parsons Laboratory, Massachusetts Inst. of Tech. Cambridge, MA, 105 pp.

Maldonado, A., Swift, D. J. P., Young, R. A., Han, G., Nittrouer, C. E., Demaster, D., Rey, J., Palomo, C., Acosta, J., Ballester, A., and Castellvi, I., in press. Sedimentation on the Valencia Continental Shelf: preliminary results. *Continental Shelf Res.*

Mann, R. G., Swift, D. J. P., and Perry, R., 1981. Size classes of flow-transverse bedforms in a subtidal environment, Nantucket shoals, North American Atlantic shelf. *Geo-Marine Letters*, **1**, 39–43.

Manohar, M., 1955. *Mechanics of Bottom Sediment Movement Due to Wave Action*. U. S. Army Corps of Engr., Beach Erosion Board, Tech. Memo. 75, 121 pp.

McCave, I. N., 1971. Sand waves in the North Sea off the coast of Holland. *Mar. Geol.*, **10**, 199–225.

McCowan, J., 1891. On the solitary wave. *Phil. Mag.*, Ser. 5, **32**, 45–58.

McGee, W. D., 1890. Encroachments of the sea. *The Forum*, **9**, 437–499.

Meade, R. H., 1969. Landward transport of bottom sediments in estuaries of the Atlantic coastal plain. *J. Sed. Petrol.*, **39**, 222–234.

Miche, R., 1944. Undulatory movements of the sea in constant and decreasing depth. *Ann. de Ponts et Chaussees*, May–June, July–August, pp. 25–78, 131–164, 270–292, 369–406.

Moody, D. W., 1964. *Coastal Morphology and Processes in Relation to the Development of Submarine Sand Ridges off Bethany Beach, Delaware*. Ph.D. Dissert., John Hopkins Univ., 167 pp.

Munch-Peterson, A., 1950. Munch-Peterson's littoral drift journals. *Beach Erosion Board Bull.*, **4**, 1–31.

Murray, S. P., 1975. Trajectories and speeds of wind-driven currents near the coast; *J. Phys. Oceanogr.*, **5**, 347–360.

Neumann, G., and Pierson, Jr., W. J., 1966. *Principles of Physical Oceanography*. Prentice-Hall, Englewood Cliffs, NJ, 545 pp.

Niedoroda, A. W., 1980. *Shoreface Surf-Zone Sediment Exchange Processes and Shoreface Dynamics*. NOAA Tech. Memo. OMPA-1, 89 pp.

Niedoroda, A. W., and Swift, D. J. P., 1981. Maintenance of the shoreface by wave orbital currents and mean flow: observations from the Long Island Coast. *Geophys. Res. Letters*, **8**, 337–348.

Niedoroda, A. W., Ma, C. M., Mangarella, P. A., Cross, R. H., Huntsman, S. R., and Treadwell, D. D., 1982. *Measured and Computed Coastal Ocean Bedload Transport*. Proc. 18th Int'l Conf. Coastal Engineering, Amer. Soc. Civil Engr., **2**, 1353–1368.

Niedoroda, A. W., Swift, D. J. P., Figueiredo, A. G., and Freeland, G. L., 1985. Barrier island evolution, Middle Atlantic Shelf, U.S.A. part II: evidence from the Shelf floor. *Mar. Geol.* (in press).

Niedoroda, A. W., Swift, D. J. P., Hopkins, T. S., and Ma, C. M., 1984. Shoreface morphodynamics on wave-dominated coasts. *Mar. Geol.*, **60**, 331–354.

Otvos, Jr., E. G., 1970. Development and migration of barrier islands, northern Gulf of Mexico. Geol. Soc. Amer. Bull., **81**, 241–246.

Otvos, Jr., E. G., and Price, W. A., 1979. Problems of chenier genesis and terminology—an overview. Mar. Geol., **31**, 251–263.

Parker, G., Lanfredi, N., and Swift, D. J. P., 1982. Substrate response flow in a southern hemisphere ridge field: Argentina Inner Shelf. *Sed. Geol.*, **33**, 195–216.

Pedlowsky, J., 1979. *Geophysical Fluid Dynamics*. Springer-Verlag Inc., New York, 624 pp.

Pingree, R. D., and Maddock, L., 1979. The tidal physics of headland flows and offshore tidal bank formation. *Mar. Geol.*, **32**, 269–289.

Price, W. A., 1955. *Correlation of Shoreline Type with Offshore Bottom Conditions*. Project 63, Dept. of Oceanography, Texas A&M Univ., 28 pp.

Putnam, J. A., Munk, W. H., and Traylor, M. A., 1949. The predication of longshore currents. *Amer. Geophys. Union Trans.*, **30**, 337–345.

Rattray, Jr., M., 1957. On the offshore distribution of tide and tidal current. *Amer. Geophys. Union Trans.*, **38**, 675–680.

Rayleigh, L., 1876. On waves. *Phil. Mag., Ser.* 5, **1**, 257–279.

Reimnitz, E., Toimil, L. J., Shepard, F. P., and Guiterrez-Estrada, M., 1976. Possible rip current origin for bottom ripple zones to 30 ft. depth. *Geology*, **4**, 395–400.

Reineck, H. E., and Singh, I. B., 1971. Der Golf Von Gaeta (Tyrrhenisches Meer). III. Die gefuge von Vorstrand und schelf sedimentatien. *Senckenberg. Marit.*, **3**, 135–183.

Rhodes, E. G., 1982. Depositional model for a chenier plain, Gulf of Carpentaria, Australia. *Sedimentology*, **29**, 201–221.

Robinson, A. H. W., 1980. Erosion and accretion along part of the Suffolk Coast of East Anglia, England. *Mar. Geol.*, **37**, 133–146.

Russell, R. C. H., and Macmillan, D. H., 1952. *Waves and Tides*. Hutchinson's Scientific and Technical Publications, London, pp. 102–104.

Russell, R. C. H., and Osorio, J. D. C., 1958. *An Experimental Investigation of Drift Profiles in a Closed Channel*. Proc. 6th Conf. Coastal Engr. Ch. 10, pp. 171–193.

Schwartz, M. L., 1973. *Barrier Islands*. Dowden, Hutchinson and Ross, Stroudsburg, PA, 451 pp.

Schwing, F. B., Kjerfve, B. J., and Sneed, J. E., 1983. Nearshore coastal currents on the South Carolina continental shelf. *J. Geophys. Res.*, **88**, 4719–4728.

Scott, J. T., and Csanady, G. T., 1976. Nearshore currents off Long Island. *J. Geophys. Res.*, **81**, (30), 5401–5409.

Seilacher, A., 1982. General remarks about event deposits. *In*: Einsele, G., and Seelucher, A., (ed.), *Cyclic and Event Stratification*. Springer-Verlag, New York, pp. 161–174.

Shepard, F. P., 1948. *Submarine Geology*. Harper and Bros., New York, 551 pp.

Shepard, F. P., Emery, K. O., and LaFond, E. C., 1941. Rip currents: a process of geological importance. *Jour. Geol.*, **49**, 337–369.

Shepard, F. P., and Inman, D. L., 1951. *Nearshore Circulation*. Proc. First Conf. Coastal Engr. Council on Wave Research, Univ. of California, pp. 50–59.

Shi, N. C., and Larsen, L. H., 1984. Reverse sediment transport induced by amplitude modulated waves. *Mar. Geol.*, **54**, 181–200.

Skjelbreia, L., 1959. *Gravity Waves, Stokes Third Order Approximations, Table of Functions*. Council on Wave Research, Engr. Foundation, Univ. of California, Berkeley.

Skjelbreia, L., and Hendrickson, J. A., 1961. *Fifth Order Gravity Wave Theory*. Proc. 7th Conf. Coast. Engr., pp. 184–196.

Stokes, G. G., 1847. On the theory of oscillatory waves. *Cambridge Phil. Soc. Trans.*, **8**:441. (Also *In*: *Mathematical and Physical Papers*, Vol. 1. Cambridge Univ. Press, London, 1880, pp. 197–229.)

Stokes, G. G., 1880. On the theory of oscillatory waves. *In*: *Mathematical and Physical Papers*, Vol. 1. Cambridge Univ. Press, London, pp. 314–326.

Sundborg, A., 1967. Some aspects of fluvial sediments and fluvial morphology: general views and graphic methods. *Geogr. Ann.*, **49**A(2-4), 333–343.

Sverdrup, H. U., 1927. Dynamics of tides on the North Siberian Shelf. *Geofysiske Publikasjoner*, **4** (5), 1–75.

Swift, D. J. P., 1975. Barrier Island genesis: evidence from the central Atlantic Shelf, eastern USA. *Sed. Geol.*, **14**, 1–43.

Swift, D. J. P., 1976. Coastal sedimentation. *In*: Stanley, D. J., and Swift, D. J. P. (eds.), *Marine Sediment Transport and Environmental Management*. John Wiley and Sons, Inc., New York, 602 pp.

Swift, D. J. P., Figueiredo, Jr., A. G., Freeland, G., and Oertel, G., 1983. Hummocky cross-stratification and megaripples, a geological double standard? *J. Sed. Petrol*, **53**, 1295–1317.

Swift, D. J. P., and Freeland, G. L., 1978. Mesoscale current lineations on the Inner Shelf Middle Atlantics Bight of North America *J. Sed. Petrol.*, **48**, 1257–1266.

Swift, D. J. P., Freeland, G. L., and Young, R. A., 1979. Time and space distributions of megaripples and associated bedforms, Middle Atlantic Bight, North American Atlantic Shelf. *Sedimentology*, **26**, 389–406.

Swift, D. J. P., Niedoroda, A. W., Vincent, C. E., and Hopkins, T. S., 1985. Barrier island evolution, Middle Atlantic Shelf, U.S.A. Part I: Shoreface dynamics. *Mar. Geol.* (in press).

Swift, D. J. P., Parker, G., Lanfredi, N. W., Perillo, G. and Figge, K., 1978. Shoreface-connected sand ridges on American and European Shelves: A comparison. *Est. Coastal Mar. Res.*, **17**, 257–273.

Swift, D. J. P., Sanford, R., Dill, Jr., C. E., and Avignone, N., 1971. Textural differentiation on the shoreface during erosional retreat of an unconsolidated coast, Cape Henry to Cape Hatteras western North Atl. Shelf. *Sedimentology*, **16**, 221–250.

Swift, D. J. P., Sears, P. C., Bohlke, B., and Hunt, R., 1978. Evolution of a shoal retreat massif, North Carolina inferences from Areal Geology. *Mar. Geol.*, **27**, 19–22.

Swift, D. J. P., Tillman, R. W., Rine, J. M., and Seimers, C. T., in press. Fluid Process and seafloor response on a modern storm-dominated shelf: Middle Atlantic shelf of North America. Part II: Response of the shelf floor. Sedimentology of shelf sands and sandstones. Canadian Society of Petroleum Geologists.

Taney, N. E., 1961. *Geomorphology of the South Shore of Long Island, New York*. Corps of Engr. Tech. Memo. 128, 91 pp.

Tanner, W. F., 1960. Florida coastal classification. *Gulf Coast Assoc. Geol. Socs. Trans.*, **10**, 259–266.

Ursell, F., 1953. The long-wave paradox in the theory of gravity waves. *Cambridge Phil. Soc. Proc.*, **49**, 685–694.

Van Straaten, L. M. J., 1965. Coastal barrier deposits in south and north Holland—in particular in the area around Scheveningen and Ijmuden. *Meded. Geol. Sticht., NS*, **17**, 41–75.

Van Veen, J., 1938. Water movements in the Straits of Dover. *Conseil Perm. Intl. Explor. de la Mer, J. Conseil*, **13**, 7–36.

Vincent, C. E., Swift, D. J. P., and Hillard, B., 1981. Sediment transport in the New York Bight, North American Atlantic Shelf. *Mar. Geol.*, **42**: 369–398.

Vincent, C. E., Young, R. A., and Swift, D. J. P., 1983. Sediment transport on the Long Island shoreface, North American Atlantic Shelf: role of waves and currents in shoreface maintenance. *Continental Shelf Res.* 2: 163–181.

von Arx, W. S., 1962. *An Introduction to Physical Oceanography*. Addison-Wesley Publ. Co., Reading, MA, 442 pp.

Wells, J. T., and Coleman, J. M., 1981. Periodic mudflat progradation, north-eastern coast of South America: a hypothesis. *J. Sed. Petrol.*, **51** (4), 1069–1075.

Winant, C. D., 1979. Coastal current observations. *Rev. Geophys. Space Phys.*, **17**, 89–98.

Wiseman, Jr., W. J., and Rouse, L. J., 1980. A coastal jet in the Chukchi Sea. *Arctic*, **33**, 21–29.

Woodruff, C. D., Curtis, R. J., and McLean, R. F., 1983. Development of a chenier plain, Firth of Thames. *New Zealand Mar. Geol.*, **53**, 1–22.

Young, R. A., Merrill, J. T., Clarke, T. L., and Proni, J. R., 1982. Acoustic profiling of suspended sediments in the marine bottom boundary layer. *Geophys. Res. Letters*, **9**, 175–178.

9

Coastal Stratigraphic Sequences

John C. Kraft and Michael J. Chrzastowski

Introduction

When James Hutton formulated the Law of Uniformitarianism, "the present is the key to the past," he based his hypothesis on the assumption that forces and processes operating on the earth's surface were the same in the past as in the present. This single hypothesis has been the key to most geologic studies that ensued during the past two centuries. Most geologists today agree that present processes and forces are very similar to those that operated throughout geologic time. However, the geologic–tectonic setting must have varied greatly, and the organisms, which potentially may have interacted with the processes and forces of action, have changed greatly over the Phanerozoic era.

Much of the stratigraphic–sedimentologic record of shoreline development through geologic time is observed from rocks that have not necessarily been formed in tectonic–geologic settings similar to those of today's coasts. For example, a great portion of the stratigraphic record may be attributed to depositional settings ranging from the margins of continents to their interiors. In addition, complexities of past plate tectonics actions preclude our fully understanding the relationship of paleo-shorelines and their true positions relative to the continents of their time from at least the end of the Mesozoic Era back toward the origin of the earth. However, shoreline sediments and their sequences of facies interrelationships must have had much in common throughout all of geologic time, given that wave energies and nearshore water depth and sediment regimes were somewhat similar to those of the present. The study of modern shorelines and their ancient equivalents or analogs then becomes an exercise in reasoning based on the partial information available to us in the stratigraphic record. It is also an exercise in ontologic methodology based on observation of present interrelationships of sedimentary depositional body shapes and structures versus the forces impinging upon, molding, and constantly reforming the present shorelines. After two centuries of development of the science of geology, the Doctrine of Uniformitarianism may well be better stated to include the variant: The present is the key to the past but the past is equally a key to the present and to the prediction of the future.

The Doctrine of Uniformitarianism is one of several conceptual tools that aid in the intrepertation of coastal stratigraphic sequences. Other aids include an understanding of the physical processes of coastal change, knowledge of the frequency and magnitude of sea level change, and the vertical relationship of coastal sedimentary facies. Using a combination of such conceptual tools allows coastal stratigraphic sequences to be "read" to better understand past and present coastal evolution, and to predict future trends in coastal change. This chapter presents a general discussion of the concepts and principles applicable to the interpretation of coastal stratigraphic sequences. It also presents several stratigraphic models and actual stratigraphic records from coastal environments where there are contrasting sequences because of different processes influencing the coastal evolution.

Processes of Coastal Change

The world's coastal areas might be depicted in terms of the morphology of sedimentary bodies and erosional features that are created by the forces and processes influencing the coastal area. Along depositional coasts the internal sedimentary parameters and the organic contributors to the sediment are dependent on these forces and processes. These sediment characteristics are significant in the interpretation of coastal stratigraphic sequences and in identifying the areal distribution of coastal sedimentary environments. Thus it is important to develop an understanding of the large number of processes that may influence the coastal area resulting in coastal change. Table 9-1 lists some of those factors that most importantly influence a shoreline configuration and the nature of the sedimentary bodies and erosional forms that may be developed.

In studying coastal sedimentary bodies or coastal stratigraphic sequences, an important perspective to maintain is that the unit being examined is the result of a depositional environment that is not fixed in time and space. Rather, this depositional environment will migrate laterally and vertically depending on the net effect of influential processes acting on the coastal area. A shoreline is a stable feature only if there is a balance of all forces and processes that are tending to move the shoreline landward or seaward. If a balance is not achieved, a marine transgression or regression may result, and

Table 9-1. Processes of coastal change.

A. Local tectonism
—possibility of short-term emergence or submergence
B. Regional tectonism
—the possibility of complete emergence of the basin or possibility of continuous or discontinuous submergence of the entire basin
C. Eustatic change
—world's sea level fluctuation which, in its simplest form, will lead to direct transgression or regression of the marine environment
D. Tectonism in the hinterland
—uplift or downwarp and its effect on increasing or decreasing amount of erosion
E. Climatic change in the hinterland
—effect on erosion and transportation of sedimentary products out of the area to the shoreline
F. Climatic change in the depositional area
—effect on the ability of deposition to take place in "normal" process forms and the potential for rapid alteration of the depositional product
—in the tropics, the possibility of carbonate depositional forms; in the Arctic, severe alteration of process forms
—the strong possibility of altering the vegetational forms that may control deposition in the shoreline area
G. The overall tectonic–depositional setting
—geosynclinal basin—length, width, and thickness
—compaction rates
—geologic history of the basin
H. Ocean currents and wave regime
—the most important single factor in the formation of a shoreline is the imposition of waves and their form as they come in contact with shallow marine areas and modify their form and energy toward the shoreline
—the nature and frequency of high-intensity events, such as hurricanes
—the possibility of directed current, such as oceanic currents, impinging on the shoreline area
—tidal effect, tidal currents, tidal range frequencies, etc.
I. Source and type of sediment
—distance from detrital sediment source
—transportation energies and types
—contribution source—is it direct from the continent or is it indirect and by reformation of other nearby shoreline elements of the same geologic age and time event?
—no terrigenous sediments available—biogenic and chemical precipitants only available source
—number of cycles of sediment particle erosion, transportation, and redeposition over geologic time (sediment maturity)
J. Human activity in the coastal zone
—human alteration of the hinterland and human-induced changes in the sedimentary patterns of erosion, transportation, and deposition

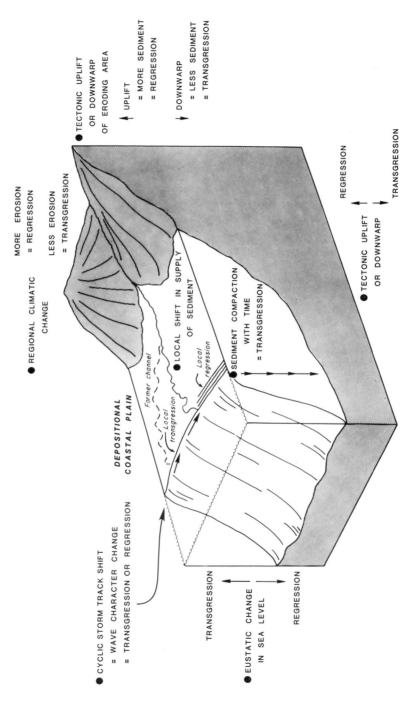

Figure 9-1. A schematic diagram of the forces and processes impinging on a coastal area and the potential response in terms of shoreline migration (after Kraft, 1972).

the stratigraphic sequences resulting from a transgression or regression are significantly different. Figure 9-1 is a schematic diagram illustrating some of the forces and processes that may contribute to a marine transgression or regression. For any coastal sedimentologic or stratigraphic analysis, there is an overwhelming need to consider the interrelationships of these forces or processes. Interpreting the form or migratory response of a coastal area should not be based on only one influential element or one set of parameters. Figure 9-1 shows that eustatic or world-wide, change in sea level is only one of several potential elements that may contribute to a marine transgression or regression. However, in studying the coastal stratigraphic sequences of modern-day coasts, eustatic change in sea level is of major significance. World-wide, the continental shelves and coastal areas have been influenced by the eustatic sea level rise since the end of the most recent continental glaciation. This eustatic sea level rise has been a major influence on the development and character of Holocene coastal sequences, and thus the factor of sea level change is a process of coastal change that deserves special consideration.

Sea Level Change

Sea level change is an important process not only influencing the relative position of shorelines but also the character of coastal stratigraphic sequences. The causes of sea level change include long-term tectonic changes, glacial isostasy, hydro-isostasy, geoidal changes, and sea level changes resulting from alternating glacial and interglacial episodes (Bowen, 1978). The geologic record suggests that sea level change is a continuing process. Shorelines have migrated back and forth across the world's continental shelves and inland seas a relatively large number of times over the past 100 million years.

The overwhelming event during the present geologic period—the Quaternary Period—has been the occurrence of major world-wide climatic fluctuations and multiple glacial and interglacial events. Accompanying these events have been major fluctuations in eustatic, or world-wide, sea level. From oxygen isotope analysis of deep sea sediments, evidence suggests that there have been as many as nine glacial and ten interglacial events in just the past 700,000 years (Shackleton and Opdyke, 1973). During this time there has been at least one interglacial event with sea level higher than present, and during the maximum of the recently past Wisconsin glaciation, sea level was approximately 100 m below its present level. If the present interglacial epoch should progress to a "full" interglacial with melting of the Antarctic and Greenland ice caps, then the world ocean could rise approximately 60 to 75 m above present sea level (Russell, 1964).

In view of these past and potential glacio-eustatic sea level fluctuations, the world shorelines in their present position can hardly be considered "typical" or in any form of stability. Accordingly, in dealing with Quaternary shorelines a major consideration must be eustatic sea level change and the immediately resultant changes in configuration and lateral migration of world shoreline positions.

Fluctuation of levels of the world's oceans must have had an overwhelming effect on positioning and configuration of shorelines that may have developed during the Quaternary Period. In addition, stillstands of sea level during these times of change would have allowed processes to concentrate on a single coastal area and thus to have an even greater effect on the shoreline morphology that might develop in space or time. Whether or not fundamental differences exist between the types of shorelines and the sedimentary sequences that may have developed by migration of shorelines in nonglacial times as compared with glacial times is not known at present. The great wealth of evidence does suggest that large numbers of studies of shoreline configurations, sediment patterns, lithosome distributions, and sedimentary environmental sequences can be effectively applied to the search for a better understanding of the history of shoreline configurations throughout the geologic record.

The change in sea level pertinent to the study of modern coastal areas is the Holocene Epoch sea level rise over the past 12,000–13,000 years. There are different interpretations as to the shape of the eustatic sea level curve for the Holocene. There are curves based on a connection of age and depth control points resulting in sea level curves with "peaks and valleys" (Figure 9-2). There are also "best-fit curves" that are smoothed versions of age and depth control points. There is continuing debate, however, as to whether or not there is, or can be, a Holocene sea level curve of world-wide application. If a Holocene sea level curve is important to any local study of coastal stratigraphic sequences, then it is necessary to refer to, or develop, a local relative sea level curve. Such curves not only represent glacio-eustatic sea level fluctuation, but they also account for local influence from such factors as tectonism, isostasy, and local sediment compaction.

Continental Margins

In discussing coastal stratigraphic sequences there is an element of scale. On a relatively small scale, coastal stratigraphic studies in the vicinity of active shorelines can be considered. On a much larger scale are the stratigraphic sequences of the seaward margin of the continents, that is, the continental shelves. Some of the continental shelves are narrow and bordered by cliffed coasts that are projections of the continental geology, an example being the continental shelf along the Pacific Coast of the United States. In contrast, a

RADIOCARBON AGE 10^3 YEARS BEFORE PRESENT

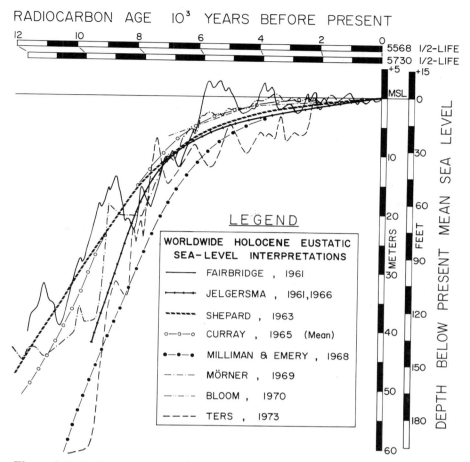

LEGEND

WORLDWIDE HOLOCENE EUSTATIC
SEA-LEVEL INTERPRETATIONS

——— FAIRBRIDGE , 1961

—+—+ JELGERSMA , 1961,1966

— — — SHEPARD , 1963

o—o—o CURRAY , 1965 (Mean)

•—•—• MILLIMAN & EMERY , 1968

—·—·— MÖRNER , 1969

—··—··— BLOOM , 1970

— — — — TERS , 1973

Figure 9-2. Variants and opinions of various geologists as to the pattern of world-wide sea level during the Holocene epoch. The dominant cause of sea level change has been an interglacial climatic warming; however, effects of tectonics and isostasy are surely also involved (curves assembled by Belknap and Kraft, 1977).

large portion of the world's continental shelves is bordered by low-lying coastal plains such as along the Atlantic Coast of the United States. It is significant to note that the world's coastal plains are in reality an emergent part of the continental shelves. The coastal plains and continental shelves are in fact a single geologic province. At any given time it is a happenstance of geologic factors such as sea level, sedimentation, and/or tectonics as to whether or not any part of the coastal plain–continental shelf is emergent or submergent.

Because of tectonics, the continental margins may be relatively unstable. Trailing-edge continental margins are frequently in a subsiding tectonic configuration resulting in continental margin sedimentary basins of accumulation, or geosynclines, of relatively large dimension. The continental margin geosynclines include some of the world's greatest thicknesses of sedimentary sequences ranging from continental to shallow marine sedimentary deposits.

Subsurface geologic studies of the world's continental shelves are beginning to provide information on the detailed history of depositional and sedimentary environmental patterns that have migrated back and forth across the continental shelves through geologic time. It is now known that a large portion of the world's continental shelves contain sediments deposited from mid-Mesozoic time to the present. A great deal of additional work remains to determine whether or not a continuum of shelf development occurs within pre-Mesozoic geosynclines or whether or not a major break in plate tectonic patterns occurred sometime in the mid-Mesozoic, shifting loci of major sedimentary deposition to the present continental margin geosynclines.

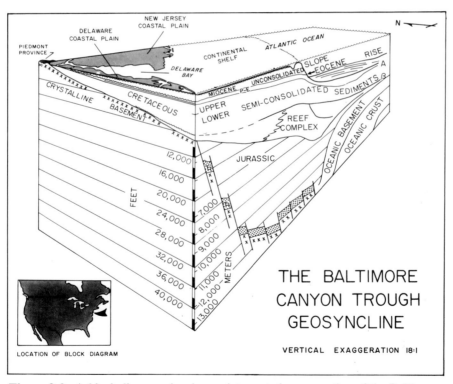

Figure 9-3. A block diagram showing an interpreted cross section of the Baltimore Canyon Trough geosyncline of the Atlantic continental margin (based on the work of Drake *et al.*, 1959; Kraft *et al.*, 1971; and Sheridan, 1974).

Figure 9-3 shows an interpreted geologic cross-section of the Baltimore Canyon Trough geosyncline, which has an axis trending generally parallel to the U.S. Atlantic coast of southern New Jersey. This is one of several major depositional basins along the Atlantic coast of North America between Florida and Newfoundland. The nature of these depositional basins was first delineated by Drake et al. in 1959. Intensive studies since that time have determined that the continental margin geosynclines of eastern North America are much deeper and the sedimentary sections are much thicker than initially estimated (Minard et al., 1974). There is as much as 12,000 m of sediment accumulation in the Baltimore Canyon Trough geosyncline (Figure 9-3). Figure 9-4 shows a similar diagram interpreting the continental shelf along the northern periphery of the Gulf of Mexico. Here, in the major axis of deposition known as the Gulf Coast geosyncline, lie the sediments that were transported by the many rivers that have drained the interior of the continent of North America over the past 100 million years. Examination of both of these cross sections of North America's continental shelves, plus the widespread extent of the continental shelves themselves, shows that shoreline configurations have been highly variable around the periphery of the world's continents of the past. For instance, major gaps in depositional record occurred during the Oligocene Epoch along the mid-Atlantic continental shelf. At other times, such as in early Late Cretaceous time along the mid-Atlantic region, shorelines extended inland across the continental shelf to its margin along the Piedmont province (Figure 9-3).

Obviously, then, the subsurface geologic records of the world's continental shelves should be an optimal place to study the history of shoreline migration over a lengthy period of geologic time. Whether or not the models and concepts thus developed can be directly applied to shorelines interpreted from still earlier sedimentary records of marine transgression or regression across the continents remains to be seen. Certainly the basic Law of Uniformitarianism must apply insofar as processes and responses are involved. However, the tectonic setting or the nature of the eustatic effect on the world's sea levels for the time of the shoreline under discussion may have had an overwhelming effect in terms of the differentiation of types of shorelines that may have been developed.

Lagoon–Barrier Systems

In discussing coastal stratigraphic sequences it is useful to examine a coastal sedimentary environment that has world-wide distribution and one that also provides an excellent opportunity for stratigraphic analysis. Coastal barriers and associated environments are such features. Barriers are long, narrow sand accumulations that are generally parallel to the mainland shore. They

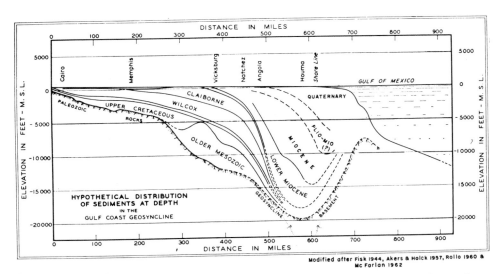

Figure 9-4. An interpreted cross section of the Gulf Coast geosyncline by Bernard and LeBlanc (1965), modified after Fisk (1944), Akers and Holck (1957), Rollo (1960), and McFarlan (1961). Copyright by Princeton University. Reprinted by permission of Princeton University Press.

are predominantly wave-built coastal features and are usually sufficiently above high tide to include dunes and vegetated zones. Barriers are typically removed from the mainland, separated by a lagoon or estuary. The landward side of the barrier may include areas of marsh or swamp. In distinguishing barrier types, baymouth barriers partly or totally enclose an inlet or embayment. Barrier islands are discrete segments often recurved at either end due to sediment transport into the coastal bay landward of the barrier. Barrier spits are formed by littoral transport of sediment eroded from older barriers or highlands at the sea shore and deposited as linear coastal beach–berm–dune features into waters of estuaries or marine embayments. The barrier combined with its associated sedimentary environments such as the lagoon, marshes, inlets, and tidal deltas may be referred to as a barrier system.

World-wide distribution of barriers is dependent on such factors as tectonic setting, coastal geomorphology, sediment supply, and tidal range. Barriers are most prevalent where a low-gradient continental shelf is adjacent to a low-relief coastal plain. These conditions typically occur along trailing edges of continents and on marginal sea coasts. Development of barriers is also dependent on an abundance of sediment supply and a moderate to low tidal range, generally less than 4 m (Glaeser, 1978; Hayes, 1979). Along the United States, these conditions are met along most of the Atlantic coast and Gulf of Mexico coast, and this coastal stretch is characterized by a rather continuous series of lagoon–barrier systems. Barriers are also common along the Arctic coast of Alaska. Along the Pacific and Hawaiian coasts, barriers do occur, but they are more limited in extent and distribution. Figure 9-5 is an oblique aerial photograph of a baymouth barrier on the Atlantic coast of Delaware. Rehoboth Bay lagoon is seen in the distance. This particular lagoon–barrier coast is transgressive, but superficially, probably with parallel beach ridges, a regressive coast would appear similar.

The origin of barriers has been a subject of considerable debate, and the various theories of formation have been discussed at length in the literature. For a review of classical and contemporary papers dealing with the origin of barriers, the reader is referred to Schwartz (1973). There are four hypotheses of barrier origin: (1) littoral transport of sand along a barrier spit tying one headland to another and then migrating landward or seaward (Gilbert, 1885); (2) the emergence of offshore bars in a stable or rising sea level condition (deBeaumont, 1845); (3) a rising sea level drowning shore-parallel features such as coastal dunes (Hoyt, 1967; McGee, 1890); and (4) a multiple causality combining elements of the other three processes (Schwartz, 1971). The question of origin remains unresolved mainly because of coastal evolution resulting in the modification or destruction of the relevant sedimentary and stratigraphic evidence. In a regressive lagoon–barrier system, such evidence concerning origin may be preserved and careful analysis may be informative on an individual case. However, many

Figure 9-5. An oblique aerial photograph of the baymouth barrier at Rehoboth Bay lagoon on the Atlantic coast of Delaware. Other than the presence of parallel beach ridges, a stratigraphic analysis would be necessary to distinguish this from a regressive lagoon–barrier system.

transgressive lagoon–barrier systems likely formed at some distance out on the continental shelf and they have migrated landward to their present position during the Holocene sea level rise. Thus, it is not possible to study the origin of these transgressive lagoon–barrier systems in their present position.

In studying coastal stratigraphic sequences, lagoon–barrier systems are a sedimentary environment complex that provide a distinct record of how coastal sedimentary environments respond to changes in the forces and processes impinging on the coast. Barrier systems also provide an example of how coastal stratigraphic sequences differ under transgressive and regressive conditions. Furthermore, the subsurface analysis of modern barrier systems is of major significance in better understanding and interpreting the ancient coastal record. This factor is important in terms of resource exploration, since many ancient lagoon–barrier systems include coal deposits and petroleum reservoirs.

Interpreting Sedimentary Sequences

As modern stratigraphic principles have been developed, applied, and modified through the mid-twentieth century, geologists have become more aware of the concept of sedimentary environmental lithosomes, or facies. These depositional bodies delineate the shape or distribution pattern of individual sedimentary environmental units in space and time. When coastal sedimentary environmental lithosomes are considered in a developmental context in terms of the lateral and vertical relationship to each other at the time they formed, then these lithosomes provide a relatively complete historic–geologic record of events of coastal change. Accordingly, the first and most important clue toward a full understanding of a coastal area, the forces and processes influencing this area, and the rates and formats of coastal change is to compare the present geographic or areal distribution of depositional environments with those observed in the immediately adjacent vertical (subsurface) sequences.

The interpretation of historic–geologic events from vertical sedimentary sequences is based on the Law of Correlation of Facies, or Walther's Law, presented by Johannes Walther in 1894. This law is the single most important conceptual tool in understanding the relationship between depositional environmental units presently observed in areal distribution, and the possibility of how similar depositional units will be related if observed in a vertical sequence. Walther's Law states:

> . . . the various deposits of the same facies areas and similarly the sum of the rocks of different facies areas are formed beside each other in space though in cross section we see them lying on top of each other. As with biotopes, it is a

basic statement of far-reaching significance that only those facies and facies areas can be superimposed primarily which can be observed beside each other at the present time (Middleton, 1973, p. 979).

Walther's Law has been the impetus for modern geologic thought and stratigraphic–sedimentologic reasoning that has enabled geologists of the twentieth century to form the present intensive study of the world's coastal depositional areas. It is important to note that Walther's Law can only be applied to vertical sequences without a break in the stratigraphic record. For example, in modern-day coastal depositional environments, dune facies are not laterally adjacent to shallow marine facies, and if these facies were superimposed in a vertical sequence, Walther's Law would predict a break in the stratigraphic record. Another important aspect in understanding Walther's Law is that it does not imply that the vertical sequence of sedimentary facies always reproduces the horizontal sequence. Certain horizontal facies may represent a transition, stage, or phase in the continuing development of a sedimentary sequence, and such facies may not have any vertical record (Middleton, 1973). In examining the areal and subsurface characteristics of barrier systems, it is quite obvious that coastal environmental variants are large in number and type. Accordingly, vertical sequences that are cored might also vary greatly, and interpretation of this record must be based on ideal or "model" sequences. Based on the pioneering work of Russell (physical geographic aspects of modern sedimentary environments: 1964 and others), Fisk (three-dimensional aspects of varied genetic depositional units: 1944 and others), Krynine (precision particle and mineralogic description and classification of sediments related to sedimentologic and climatologic environments: 1948 and others), and many other geologists, analyses of the origins and geologic settings of discrete sedimentary environmental lithosomes and sedimentary petrology advanced rapidly in the mid-twentieth century. With the impetus of new stratigraphic and sedimentologic concepts and a renewed application of Walther's Law, geologists in major petroleum companies such as Shell and Exxon made major breakthroughs in our understanding of the nature of coastal sedimentary environmental unit evolution and vertical sequence analysis in the 1950s (for instance, the early unpublished works of Bernard and LeBlanc along the Texas and Louisiana coastal zones). In order to utilize best the vertical section in the interpretation of the past geologic record, Visher (1965) formed several "sedimentary stratigraphic models" for the types of vertical sequences that would typically develop in different sedimentary environments. As stated by Visher:

It is the sedimentary process (for example, regression and transgression) that is both the unifying and the most readily identifiable feature in a sedimentary sequence. Therefore the sedimentary process permits not only description of the vertical profile, but also an understanding of the mechanism of its formation. Each fundamental sedimentary process produces both a specific environmental distribution and a specific vertical profile. With this concept as a starting point, it is possible to show why there are specific orderly environmental successions within the models that have been proposed (Visher, 1965, p. 42).

Visher's (1965) analysis of vertical sequence models was one example of the continuing interest in developing and refining facies models for different sedimentary environments. Vertical sequence models have been developed for transgressive and regressive shorelines, deltas, chenier plains, and tidal inlets, to name a few of the various coastal environments of deposition (Bernard *et al.*, 1959; Bernard and LeBlanc, 1965; Kraft, 1971b). Such models aid in interpreting the ancient record, understanding modern shoreline form and processes, and projecting future trends of coastal change. For a detailed discussion of coastal environment facies models the reader is referred to Elliott (1979) and Reinson (1979).

With frequent and lateral migration of coastal sedimentary environments, and assuming a high preservation potential of depositional units, then it should be the norm rather than the exception to have these coastal sedimentary environmental elements stacked in vertical sequences that reflect the nature of coastal change. Unfortunately, total preservation of coastal depositional units is not always achieved. Where uninterrupted sequences do occur, Walther's Law applies to the logical association of the modern areal distribution of sedimentary environmental sequences with vertical sequences that may be anticipated in the immediate coastal area. Thus, Walther's Law is of overwhelming significance in interpreting and distinguishing transgressive and regressive coastal stratigraphic sequences. If the downward progression of sedimentary facies corresponds to those facies observed in a landward horizontal succession, then the facies superposition records a transgressive sequence. If the downward progression of sedimentary facies corresponds to those facies observed in a seaward horizontal succesion, then the facies superposition records a regressive sequence. Figure 9-6 illustrates these differences between the internal geometry of sedimentary facies in models of idealized transgressive and regressive barriers. This figure clearly demonstrates the correlation of horizontal and vertical facies as stated in Walther's Law. Figure 9-6 also illustrates the internal geometry of isochronous depositional surfaces, or time lines, in transgressive and regressive sequences. Although time lines have different configurations in the two types of sequences, a common factor is that in both cases they cross stratigraphic units.

The Transgressive Coast

Many of the world's continental shelves are partly emerged as low-lying coastal plains. As sea level has continued to rise from the late Wisconsin Age and early Holocene Epoch toward the present, shorelines have tended to

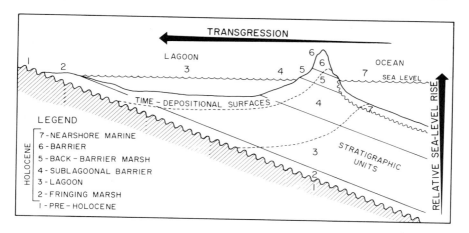

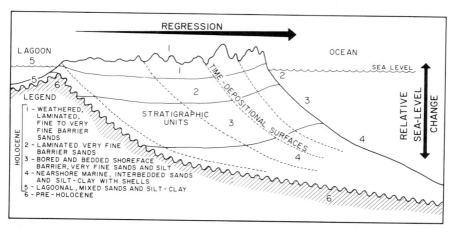

Figure 9-6. Models of transgressive and regressive barrier stratigraphic sequences illustrating the concept of Walther's Law and showing the difference in internal geometry of stratigraphic units and time lines. From Kraft and John, 1979; transgression model based on work of Kraft and John, 1979; regression model based on work of Bernard and Le Blanc (1965) and others.

migrate landward and upward in space and time. Even in those coastal areas of the world where major deltas intrude into the marine environment, it may be readily observed that simple cutoff of the source of sediment in a distributary will lead to reversal of marine regression and to the transformation to coastal erosion and transgression. Accordingly, it is obvious that the dominant Holocene Epoch shoreline movement throughout the world, given tectonic setting as neutral or negative, is in the form of a transgression.

Conceptually, a shoreline may be transgressed by the marine environment by simple erosion continuing through time and space. However, a great

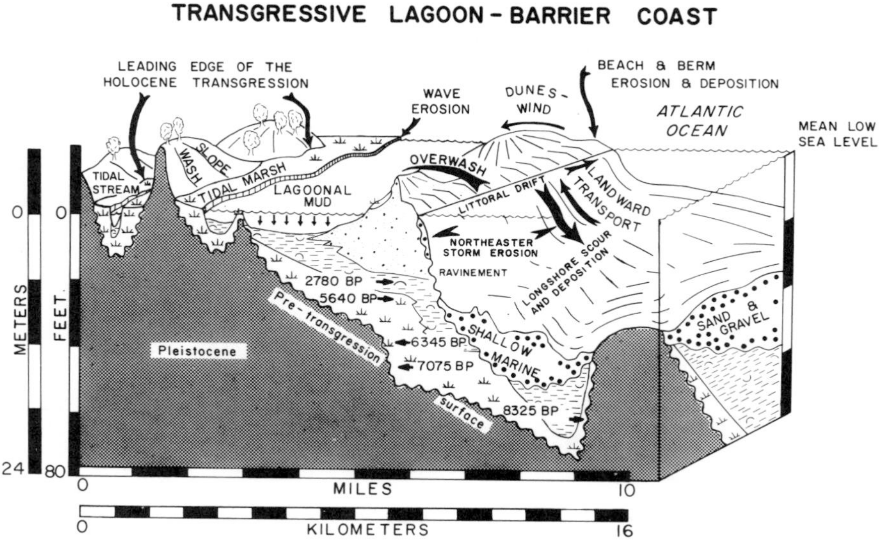

TRANSGRESSIVE LAGOON - BARRIER COAST

Figure 9-7. A block diagram cross-sectional analysis of the transgressive lagoon–barrier coast at Rehoboth Bay lagoon on the Atlantic coast of Delaware. Dates are from radiocarbon analysis of shells, wood fragments, and basal peat (modified from Kraft *et al.*, 1973).

acceleration to this rate of shoreline transgression occurs when relative sea level rise occurs at the same time. If the shoreline of transgression is in the middle of a subsiding tectonic basin or geosyncline, then the transgression rate will increase. Similarly, if eustatic sea level change is upward, the transgression rate will possibly accelerate. Minor changes in rate of sea level rise or in rate of tectonic subsidence or reversal could lead to stillstands, strandline fluctuation, or regression. The stratigraphic record abounds with examples of shorelines migrating for long distances in one direction with minor reversals or fluctuations in the opposite. It should, accordingly, be considered normal for a shoreline dominantly transgressive to have minor regressive fluctuations.

Major transgressive coasts occur throughout the world. They are perhaps best developed along the Atlantic coast of North America. Here, in an extensive coastal area of barrier–lagoons and estuaries, lie the elements of coastal transgression caused dominantly by rise in relative sea level accompanied by a relatively low contribution of sediments from the streams and rivers draining into the area.

Figure 9-7 shows a block diagram and cross-sectional analysis of part of the lagoon–barrier coast of Delaware. This area is typical of many areas along the U.S. Atlantic coast where long, linear, and relatively thick sand barriers lie between coastal lagoons and the Atlantic Ocean. Along such a lagoon–barrier coast, the lagoon and peripheral lagoon environments are

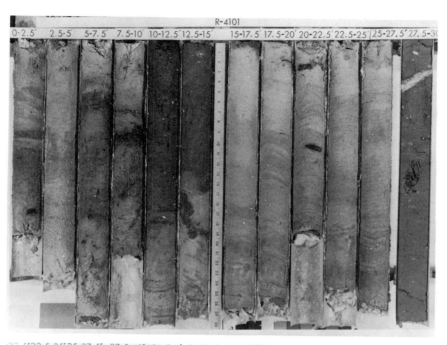

Figure 9-8. Photograph of core R-4101 through the transgressive sequence beneath the baymouth barrier at Rehoboth Bay, Delaware. A log and interpretation of the core lithology is shown in Figure 9-9 (photograph courtesy of Shell Development Company, Houston, Texas).

sheltered from the intensive wave activity of the marine environment by the barrier. However, these relatively sheltered environments are in fact marginal marine and strongly influenced by oceanic water flowing through inlets and embayments into the lagoonal areas. The significance of these marginal-marine environments in terms of the transgressive sequence is shown in Figure 9-7. The leading edge of the Holocene transgression in this lagoon–barrier environment is not along the seashore but at the landward side of the lagoon and extending up the tidal reach of streams. It is in these areas that the basal unit of a lagoon–barrier transgressive sequence is formed.

Figure 9-8 is a photograph of a core taken through the baymouth barrier diagrammed in Figure 9-7. A log of sediment characteristics and an interpretation of sedimentary environments is shown in Figure 9-9. Note that the basal unit of the transgressive sequence is marsh or marsh fringe organic muds. These are overlain by lagoonal muds with oyster and other micro- and macrofauna typical of coastal lagoons. The lagoonal material is overlain by the sand of the sublagoonal barrier. Above this unit may occur back-barrier marsh sequences that have trapped this layer of organic mud in the barrier sequence. The back-barrier marsh is overlain by barrier sand. Either the barrier sand or sublagoonal barrier sand may include high-angle cross-bedded tidal delta deposits. The uppermost units of this vertical sequence are washover beach–berm and dune sands of the surficial part of the barrier. As can be seen from Figure 9-9 and also the block diagram shown in Figure 9-7, the greatest part of the barrier sand is in fact submarine in depositional loci. Comparing these two figures provides an excellent example of the facies relationship described in Walther's Law with the vertical sedimentary sequence equating with the present areal distribution of sedimentary environments. A process that could significantly modify this "ideal" transgressive sequence is the lateral migration of tidal inlets along the barrier. Lateral migration of these inlets could destory part or all of the transgressive sequence and result in a new sequence dominated by sand facies of tidal-channel and marginal spit–beach environments (Reinson, 1979).

Research continues by geologists as to whether or not dominant sediment motion on continental shelves with transgressive shorelines is landward and upward in space and time or seaward. However, simple examination of models that may be constructed from environmental sequences along a transgressive shoreline shows the improbability of dominant sediment motion being from the shelf to the barrier. Should such an event occur, the sand–gravel sediments should accumulate in tremendous piles in the shoreline area. Accordingly, transgression would cease. This is not the case. Such mechanisms as littoral drift processes are available to remove and transport or redistribute sediment moving from the shelf to the barrier. In addition, migration of inlets along the barriers and the development of tidal deltas into lagoonal areas and into the marine area, plus frequent development of spits into deeper marine areas along shore, provide sediment sinks toward which

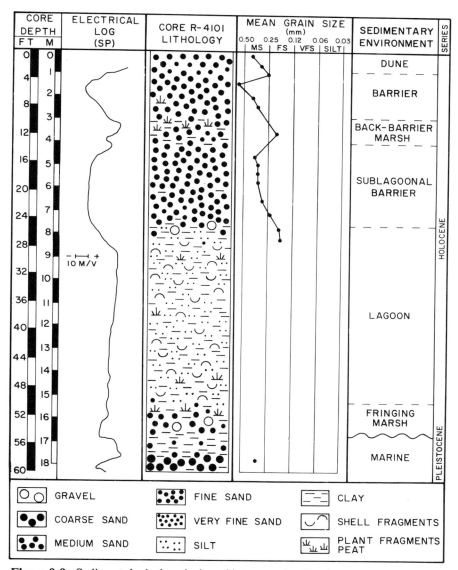

Figure 9-9. Sedimentological analysis and interpreted sedimentary environments for core R-4101. The vertical sequence is typical of a transgressive lagoon–barrier sequence with migration of sedimentary environments landward and upward in space and time (modified from Kraft and John, 1979).

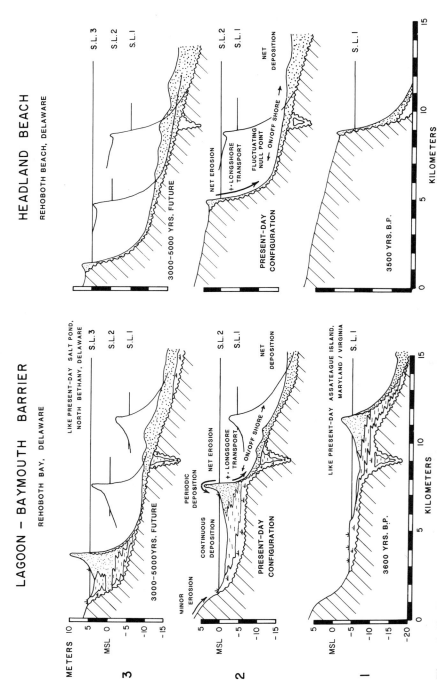

Figure 9-10. Evolution of transgressive coastal systems with conditions of net shoreface erosion and relative sea level rise (courtesy of D. F. Belknap).

sediment may flow as it tends to accumulate on the barrier. Meade (1969) has suggested that sediment moves from the shelf toward the land at the mouths of the major estuaries of the world during a transgression. It appears illogical to assume that major movement could be from the shelf to the shoreline for areas other than sediment sinks, such as the major estuarine intrusions or embayments into coastal plains.

With continued rise of local relative sea level, and with conditions of net shoreface erosion, the transgressive coastal area evolves through space and time. Figure 9-10 shows an example of the landward and upward migration of a lagoon–baymouth barrier. Examples of the three stages of development are present today along the Atlantic coast of the Delmarva Peninsula. The barrier system will continue to be a distinct coastal feature until an upland of steeper topographic relief is encountered. Then there is less potential for retaining a lagoonal area and the final expression of the barrier will have it separated from the highland by a marsh area. With continued net shoreface erosion and relative sea level rise, the back-barrier marsh will be transgressed and ultimately the barrier sand will be juxtaposed to the highland forming a headland beach at which time a barrier no longer exists as a distinct coastal feature. With net surface erosion and relative sea level rise, the headland beach also migrates landward and upward in space and time (Figure 9-10). Notice that in this evolution the sediment removed by net shoreface erosion is deposited offshore and/or also moved parallel to the coast by longshore transport. Also note in Figure 9-10 that the best potential for preservation of the transgressive units has been in pretransgressive topographic depressions such as ancestral stream valleys.

The Regressive (Progradational) Coast

Regressive coastal areas occur world-wide. Although some local variations occur, an example of a progradational coastal area is portions of the Gulf of Mexico coast between the Mississippi Delta and the United States–Mexico border. Along this coast stretch is Galveston Island, Texas, a "classic" regressive barrier. Mainly because of its close proximity to Houston, a petroleum center for North America, intensive oil company investigations have been conducted in the Galveston Island area. In addition, the work of Bernard and LeBlanc (1965) and others have been definitive in determining the sedimentological and stratigraphic characteristics of this regressive barrier system.

The distinction of a regressive coast is its seaward progradation. Such prograding coastal areas are usually characterized by parallel beach-accretion ridges behind the active shoreline with each beach ridge marking a former position of the beach face. These former beach faces may be preserved as major bedding surfaces that dip gently seaward and are

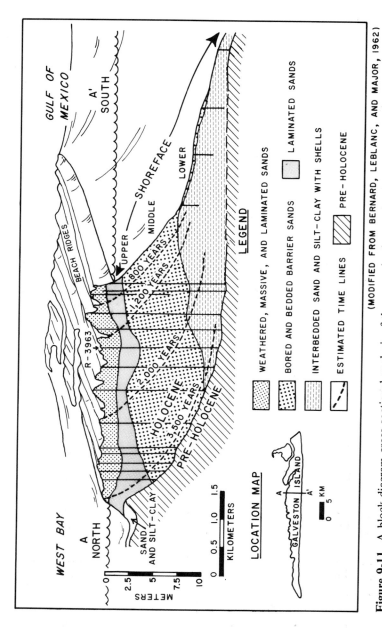

Figure 9-11. A block diagram cross-sectional analysis of the regressive lagoon–barrier coast at Galveston Island and West Bay on the Gulf coast of Texas. Cross-section dates are in radiocarbon years based on mollusks. Sand sediment grain size increases upward from the lower shoreface to upper shoreface and beach–berm.

exhibited in cores or excavations through the regressive barrier (Van Straaten, 1965). The seaward progradation of the regressive system results in beach and dune deposits developing in a seaward direction and thus overlying nearshore-marine deposits. Vertical sequences obtained from numerous borings in the Galveston Island and West Bay lagoonal area show such a superposition of sedimentary environments. These stratigraphic data, supported by radiocarbon data from shells within the barrier, clearly show that Galveston Island is an offlapping parallel beach-ridge accretion feature typical of a regressive shoreline (Figure 9-11).

Figure 9-12 is a photograph of a core taken through the Galveston Island barrier. Figure 9-13 is a log of this core's sedimentary characteristics and the interpreted sedimentary environments. The vertical sequence records an offlap relationship that is gradational at the base and coarsens upward and is an excellent example of facies relationships defined in Walther's Law. Starting from the base of the sequence, the surface of regression is overlain by interbedded sand and clay–silt of the shallow marine lower shoreface. These deposits are in turn gradational laterally into, and overlain by, the shallow marine sands of the middle and upper shoreface, which are overlain by the beach–berm and finally dune sand sequences. Thus, the vertical sequence is dominated by shoreface, foreshore, backshore, and dune facies, all of which are components of the seaward side of the barrier. Marsh and

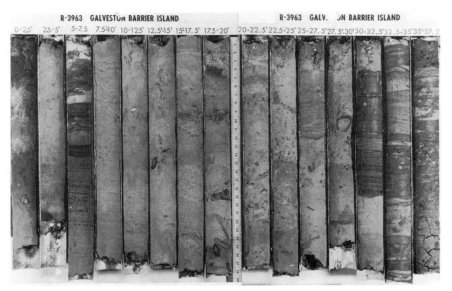

Figure 9-12. Photograph of core R-3963 through the barrier island regressive sequence at Galveston Island, Texas. A log and interpretation of the core lithology is shown in Figure 9-13 (photograph courtesy of Shell Development Company, Houston, Texas).

lagoonal facies are not major stratigraphic units as in the transgressive barrier sequence described earlier. Ultimately, if relative sea level rise is occurring with the regression or with optimal development of the landward sequences, lagoons and marsh environments may rise over regressive barriers and bury them. However, this is an unstable situation, and without proper tectonic considerations or equivalent relative sea level change, upper regressive features would normally be destroyed by the erosion accompanying the regression. It should be noted that, as with the "ideal" transgressive sequence, the "ideal" regressive sequence can be significantly modified by the lateral migration of tidal inlets.

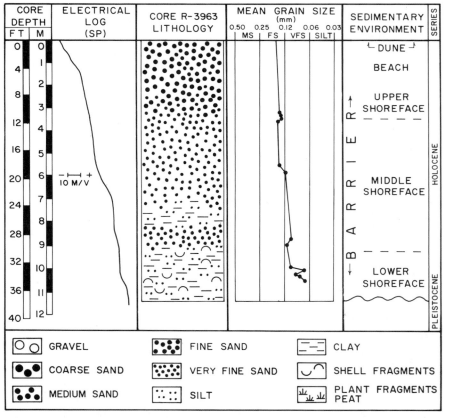

Figure 9-13. Sedimentological analysis and interpreted sedimentary environments for core R-3963. The vertical sequence is typical of a regressive coast or barrier island with seaward progradation resulting in beach and dune deposits overlying shoreface deposits (modified from Davies *et al.*, 1971, with the addition of SP log by Bernard *et al.*, 1970, from site R-3656 in the vicinity of core R-3963).

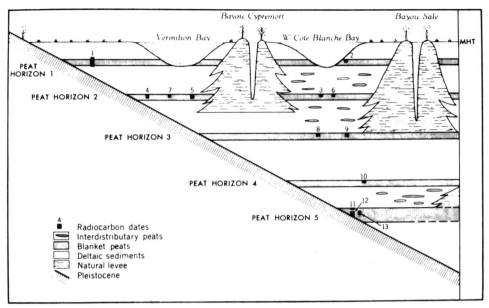

Figure 9-14. Coleman and Smith's cross section (schematic) of the relationship of delta distributaries to interdistributary muds and peats (Coleman and Smith, 1964). The succession of peat horizons is related to a local relative sea level rise (with permission reprinted from Vol. 75, *Bulletin of the Geological Society of America*).

As stated earlier, the Gulf Coast between the Mississippi Delta and the United States–Mexico border is dominantly a regressive coast at present. However, there are exceptions to this general condition and these occur in proximity to the regressive barrier of Galveston Island. Dickinson *et al.* (1972) have conducted a detailed study of the lagoon–barrier system of Padre Island and Laguna Madre, Texas. Subsurface sequences encountered under Padre Island clearly show transgressive relationships. In addition, large portions of the southern Padre Island strand are covered with a litter of shell debris that is clearly eroded from shoreface sediment sequences of the past several thousand years and then moved landward and upward to the beach face and berm area. Clearly, Padre Island is a transgressive feature in part. Wilkinson (1975) shows that nearby Matagorda Island has initial trangressive vertical sequence elements of barrier over lagoon. This is followed by parallel beach ridge regressive elements, indicating both a transgressive and a regressive phase of development of Matagorda Island.

Padre Island and Matagorda Island may be examples of temporary transgressive and regressive phases resulting in a minor variance along a dominantly regressive coast. A simple cutoff of littoral sediment supply for a period of time can change a shoreline configuration to one of wave attack, coastal erosion, and rapid shifting of vertical sedimentary environmental sequences into a transgressive sequence. Accordingly, as may be seen by these examples, major sea level fluctuations or major tectonic events do not necessarily need to occur to change coastal configurations from transgression to regression. Instead, and perhaps more frequently, the change is simply one of sediment supply and of balance with forces impinging upon the coastal area and/or a minor late Holocene sea level fluctuation.

Delta–Chenier Plains

Perhaps the most complex areal and vertical sequence models that can be developed for coastal sediment configurations are in the world's major deltas. In North America, a great deal of work has been done on the deposits of the Mississippi River Delta through the Quaternary Period. Much of the emphasis of these studies has been on Holocene Epoch sediments. It is interesting to note that so-called deltaic or regressional features, such as the huge Mississippi River Delta, occur under a dominant world transgressive situation for the world's continental shelves. Indeed, this makes the world's deltas an optimal area to study the elements that tend to cause or affect a transgression. For instance, large portions of the Mississippi River Delta are presently undergoing major transgression. It is the overwhelming supply of sediment into the deltaic area that causes the Mississippi River Delta to protrude into the marine environment on the continental shelf. A cutoff in sediment supply almost immediately leads to a local transgression caused by coastal erosion, sediment compaction, and the ultimate coastal configuration of the area, which is a slightly negative tectonic feature.

Many diagrams have been drawn showing the Mississippi Delta. It is not intended to discuss deltas as such here, for they are further developed elsewhere in this book (Chapter 1). Coleman and Smith (1964) believe that the world's eustatic sea level rose to its present point approximately 3600 to 4000 yr ago and that the sea level has remained stable since then. However, tectonic subsidence and compaction have led to a continual local relative sea level rise in the Mississippi River Delta area resulting in a vertical sequence of peat horizons as shown in Figure 9-14. Figure 9-15 shows the interrelationships between distributary channels of an older distributary system of the Mississippi River and the interdistributary "bays" and the natural levee sediments, general deltaic shallow marine and coastal marsh sediments, and organic peat layers. Compaction is an important element in development of shoreline configuration. In addition, coastal erosion tends to increase when the source of sediment supply is cut off from wave and littoral drift systems, thus redistributing sediment. Accordingly, the abandoned channels or distributary fans of the previous courses of the Mississippi River are almost immediately subjected to transgression when sediment flow ceases through them. Much of the Mississippi Delta is presently undergoing

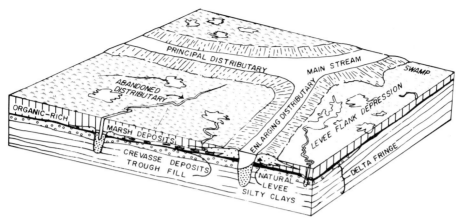

Figure 9-15. A schematic diagram of sediment patterns in the vicinity of distributary mouth bars of the Mississippi Delta showing facies patterns that would lead to a "coarsening upward" vertical sedimentary environmental sequence (from Bernard and LeBlanc, 1965, modified after Fisk, 1960).

transgression, while at the same time there is marine regression at the modern distributary systems of the bird's-foot delta south of New Orleans as well as near the mouth of the Atchafalaya River, which is presently a major element of the distributary system.

Major delta intrusions into the oceanic-marine environment provide sediment by means of wave attack and littoral drift. This sediment can potentially move out of the delta system laterally along shore and become an important source of sediment to the development of lagoon–barrier systems adjacent to the delta. In the case of the Mississippi River Delta, a major portion of the sediment is eroded and moved in a westerly fashion. Accordingly, the shoreline zone of western Louisiana is one of broad, low-lying mud plains with intermittent thin, narrow, linear sandy strand ridges known as cheniers. LeBlanc (1972), as well as many others, has interpreted these chenier plains to be formed from the massive influx of muds and sand from the Mississippi River Delta, eroded and transported west to the chenier plain area and deposited. Accordingly, chenier plain shoreline types are mainly regressive shorelines. Mud and organic sediments are the dominant components in terms of percent of mass of the sediment resulting, with thin, narrow, linear chenier ridges as outstanding geomorphic features.

Temporary short-term transgressions that may become repetitive over long periods of time erode into the mud plain, winnow and sort out the coarser clastic sands and shell debris, and pile them in linear narrow beach ridges that are noted for their great length, narrow width, and thinness. Ultimately, a greater influx of mud again isolates the sandy shoreline, and the marine environment regresses. The isolated linear sand barriers in the mud sequence remain as the chenier plain. A conceptual development diagram for chenier

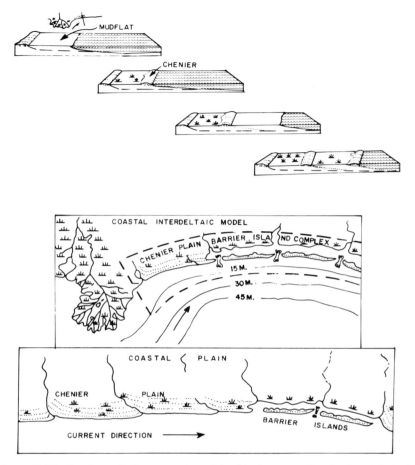

Figure 9-16. A schematic interpretation of the development of a chenier plain and its linear barrier ridges (after LeBlanc, 1972).

plain strand line was developed by LeBlanc (1972) (Figure 9-16). The vertical sequences in chenier plains would then be one of isolated, thin, relatively narrow, sand–shell barriers surrounded by organic muds both in front and in back and potentially underneath with marsh grasses in place. Should the regression continue and further sediment be supplied into the area by other means, the chenier plains might be buried and result in a stratigraphically correlatable mud unit with thin interruptions of linear barrier systems. Such correlative units would probably involve time stratigraphic units formed in centuries or millennia as opposed to long-term geologic time.

Coastal Sequence Preservation Potential

The details of the history of a transgression or regression should be reflected in the vertical sedimentary facies that result from these coastal changes. For some time, it was considered that only regressive coastal stratigraphic sequences have a good preservation potential since these are not overridden by the marine environment and thus not subjected to shoreface erosion. Fischer (1961) examined the question of preservation potential for transgressive barrier systems along the New Jersey coast. He suggested that the degree of preservation was a function of the depth of erosion across the shoreface as the transgression proceeded. In this model the important factor is incident wave energy. Low wave energy and shallow shoreface erosion results in partial retention of the transgressive sequence; high wave energy and deep shoreface erosion could result in complete loss of the transgressive sequence. It is now recognized that depth of shoreface erosion is only one of many variables related to preservation potential. Other important factors include sediment characteristics, shoreline orientation, tidal range, pre-existing topography, and the rate of relative sea level rise.

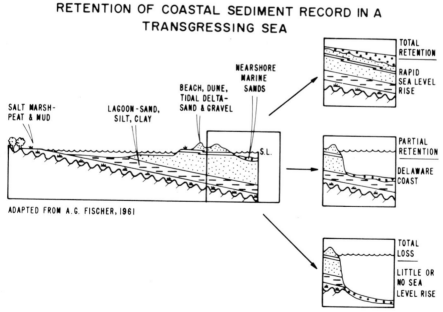

Figure 9-17. Models of potential preservation of the stratigraphic record (sedimentary environmental sequence) of a shoreline transgression (adapted from Fischer, 1961, by Kraft, 1971a).

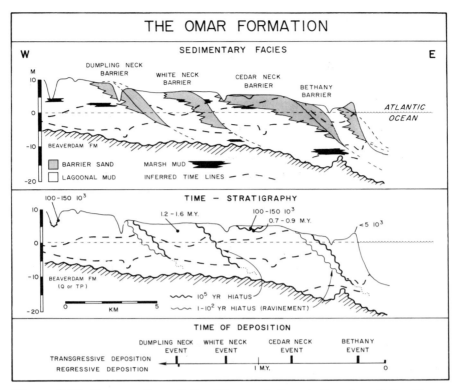

Figure 9-18. Cross section of the litho- and time-stratigraphy of the Omar Formation of southeastern Delaware. This formation represents deposition of a series of transgressive barriers each abandoned at the end of a high sea stand. Juxtaposition of these successive barriers resulted in a seaward progradation (from Demarest *et al.*, 1981).

Kraft (1971b) and Belknap and Kraft (1981) proposed a preservation model modifying Fischer's (1961) original concept to emphasize the rate of relative sea level rise as the important preservation factor. This model recognizes that although wave erosion will proceed at a rate greater than the rate of sea level rise, the more rapidly that the shallow offshore is placed below wave base, the lesser the destruction of the transgressive sequence. As shown in Figure 9-17, this model predicts that a rapid sea level rise may allow a major portion of a transgressive record to be preserved. In contrast, if relative sea level rise is slow, wave erosion may be concentrated across the shoreface for a greater period of time, and more of the formerly deposited transgressive record is removed.

Possibly the most significant factor in preservation potential of transgressive sequences is preexisting topography. Stream valleys incised in the pretransgressive surface, or other antecedent topographic lows, may provide a locale for the greatest potential accumulation and preservation of transgressive sedimentary facies.

Recent investigations along the Atlantic coast of Delaware have shown that an alternate process resulting in total retention of transgressive barrier lithosomes is the abandonment of these transgressive units at the high sea stand at the end of a transgressive event. Figure 9-18 shows the internal litho-stratigraphy and time-stratigraphy of the Omar Formation of south-eastern Delaware. This coastal stratigraphic sequence consists of a series of transgressive barriers each from a different Quaternary high sea stand depositional event, with the successive barriers juxtaposed resulting in a long-term seaward progradation. Based on sedimentological and strati-graphic characteristics, this unit satisfies the definition of a formation. However, it forces a reconsideration of the concept of a formation representing a "single" depositional event. New relative dating techniques such as amino acid racemization have shown that this is a depositional unit that represents only a minor part of the total elapsed time of deposition, and that the stratigraphic sequence represents short-term depositional events separated by major hiatuses of nondeposition and possible erosion. The Omar Formation clearly indicates that the preservation of major coastal

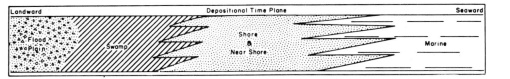

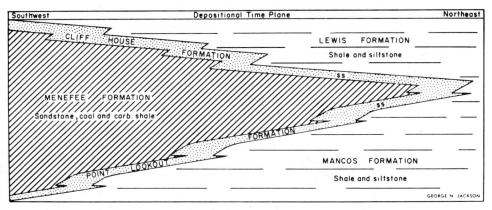

Figure 9-19. Cross-sectional analysis showing the lateral migration of the Mesa Verde littoral zone in the Coloradan-Montanan seas of the San Juan Basin (from Hollenshead and Pritchard, 1961, with permission of the American Association of Petroleum Geologists).

depositional units does not in itself imply that these are regressive sequences. Possibly many errors exist in interpreting ancient coastal stratigraphic sequences. Although the literature suggests a greater preservation potential of regressive sequences, future reexamination may show that there has been more equable preservation of transgressive and regressive sequences. In considering barrier systems, there is probably no reason for regressive barriers to be more likely preserved in the geologic record than transgressive barriers. Indeed, it is a happenstance of the history of tectonics, sedimentation, and eustacy within a basin that determines whether or not the depositional record of a barrier or lagoon–barrier sequence is preserved.

Coastal Sequences in Time and Space

It is not the purpose of this discussion to detail and examine examples of coastal sedimentary environmental sequences preserved in the ancient record. The literature abounds with such examples, sometimes interpreted correctly, sometimes not. It is the challenge of the future in the petroleum industry and in the disciplines of stratigraphy and sedimentology to attempt to sort out the maze of transgressive–regressive coastal environmental lithosome features and sequences in the large number of available subsurface and outcrop stratigraphic records. However, two prime examples should be noted to illustrate that completeness of record is available to the geologist for interpretation. Both are from the Cretaceous System of the western interior of North America, which possibly provides one of the best ancient analogs of present-day coastal environments.

The work of Hollenshead and Pritchard (1961) in the Mancos–Mesa Verde–Menefee system in the San Juan Basin of northwestern New Mexico–southwestern Colorado should be studied by all interested in transgressive–regressive shoreline systems and sequences. Here, in the Cretaceous stratigraphic record, lie the shallow, nearshore marine–littoral–coastal plain sediments of the Coloradan–Montanan seas. Coastal swamp environments with adjacent sandy barrier and shallow marine mud environments migrated back and forth as the sedimentary basin continued to subside. This sequence of events produced an outstanding example of a preserved marine regression and transgression with halts or stillstands at frequent intervals. Hollenshead and Pritchard's (1961) diagrammatic environmental interpretation of the marine, littoral, and nonmarine units of this regression–transgression and the frequent stillstands–reversals are shown in Figure 9-19. Although many other examples of migrating strandlines occur in the literature and are available for study, few have been documented in such great detail as that of the migrating Mesa Verde strandline sands.

On an even larger scale of preservation is the coastal stratigraphic sequence that includes the Fox Hills Sandstone. This uppermost Cretaceous shoreline deposit records the western shoreline and final retreat of the North American Cretaceous seaway. The Fox Hills represents depositional environments including barriers, prodelta areas, and estuary and tidal rivers (Weimer and Land, 1975). The extensive distribution of the Fox Hills Sandstone and associated formations allows a coastal paleogeographic reconstruction extending along the Rocky Mountain and High Plains region of the west-central United States (Figure 9-20).

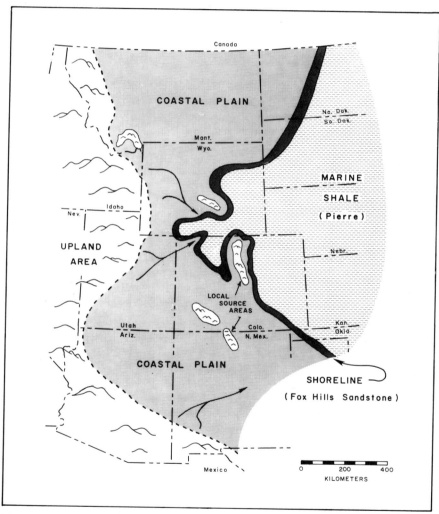

Figure 9-20. Late Cretaceous paleogeographic reconstruction across the west-central United States during late Fox Hills deposition (near the end of time of the *Baculites clinolobatus* Range Zone) (from Weimer and Land, 1975, reproduced with permission of the Geological Association of Canada).

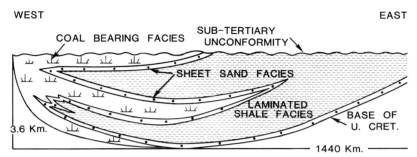

Figure 9-21. A schematic of the migrating strand line between the North American Cretaceous sea and coastal plain. Adapted from Selley (1970).

By implication, the regressions and transgressions recorded for Cretaceous time represent a relationship between intruding terrigenous clastic sediments transported into the region and regional rates of subsidence (Figure 9-21). However, it is possible that eustatic sea level changes may also have been a contributing factor.

Summary

The relationship of coastal processes and forces, or cause to effect, have been emphasized throughout this book. This chapter has demonstrated that in the study of modern-day coastal sedimentary environments not only is it important to study areal characteristics, but it is also important to study the coastal stratigraphic sequences. A true and total understanding of changes in coastal configuration, or rates of change in the position of coastal sedimentary environments, must include an understanding of the vertical sequence for the coastal area being studied. This third dimension to the coastal sedimentary environments represents the element of time, and time is of overwhelming importance in understanding coastal evolution. Without information from the third dimension one can only speculate on the nature of past and ongoing processes of coastal change. Vertical sequence analysis is equally important in determining or delineating the shape of coastal sedimentary facies through space and time. This record provides the required information for a fuller understanding of cause and effect in the landward or seaward migration of coastal sedimentary environments.

The world's literature on shoreline studies abounds with examples of misinterpretation of the sequences of coastal events based on simple areal observation. Fortunately, the past decade has shown a large number of studies that emphasize the vertical sequence in interpretation of coastal events. Nearly 90 yr have passed since Walther first published his hypothesis of the relationship between areal and vertical facies sequences. The future will likely show his fundamental concept to be the primary and guiding hypothesis for successful interpretation of ancient and modern sedimentary facies and their geographic interrelationships. Only with an examination of the sequential development of coastal sedimentary environmental units can there be a proper synthesis and understanding of coastal events and the ultimate cause of coastal configuration or coastal change.

In the future, lateral and vertical sequence analysis and an understanding of the time-depositional surfaces and hiatuses of greatly varied time duration will dominate research into utilization of modern sedimentologic analogs in the interpretation of the ancient stratigraphic record. We can no longer assume that formations in the stratigraphic record were slowly deposited over "millions of years." To the contrary, the stratigraphic units that form our geologic column may represent deposition during less than 1 to 2% of geologic time. With the possible exception of the deep ocean basins, it appears probable that nearly all environments of deposition are time transgressive. An understanding of these concepts will be critical as the sciences of stratigraphy and sedimentology are advanced through environmental interpretations, stratigraphic correlations, and reconstructions of ancient paleogeographies.

References

Akers, W. H., and Holck, A. A. J., 1957. Pleistocene beds near the edge of the continental shelf, southeastern Lousiana. *Geol. Soc. Amer. Bull.*, **68**, 983–992.

Belknap, D. F., and Kraft, J. C., 1977. Holocene relative sea-level changes and coastal stratigraphic units on the northwest flank of the Baltimore Canyon Trough geosyncline. *J. Sed. Petrol.*, **47**, 610–629.

Belknap, D. F., and Kraft, J. C., 1981. Preservation potential of transgressive coastal lithosomes on the U.S. Atlantic shelf, *Mar. Geol.*, **42**, 429–442.

Bernard, H. A., and LeBlanc, Sr., R. J., 1965. Resume of the Quaternary geology of the northwestern Gulf of Mexico Province. *In*: Wright, Jr., H. F., and Frey, D. G. (eds.), *The Quaternary of the United States*. Princeton Univ. Press, Princeton, NJ, pp. 137–185.

Bernard, H. A., Le Blanc, Jr., R. J., and Major, C. F., 1962. Recent and Pleistocene geology of southeast Texas. *In*: *Geology of the Gulf Coast and Central Texas and Guidebook of Excursions*. Houston Geological Society, pp. 175–224.

Bernard, H. A., Major, Jr., C. F., and Parrott, B. S., 1959. The Galveston barrier island and environs: a model for predicting reservoir occurrence and trend. *Gulf Coast Assoc. Geol. Socs., Trans.*, **IX**:221–224.

Bernard, H. A., Major, Jr., C. F., Parrott, B. S., and LeBlanc, Sr., R. J., 1970. Recent sediments of southeast Texas. *Guidebook No. 11*. Bureau of Economic Geology, Univ. of Texas at Austin.

Bloom, A. L., 1970. Paludal stratigraphy of Truk, Ponape, and Kusaie, Eastern Caroline Islands. *Geol. Soc. Amer. Bull.*, **81**, 1895–1904.

Bowen, D. Q., 1978. *Quaternary Geology, A Stratigraphic Framework for Multidisciplinary Work*. Pergamon Press, New York, 221 pp.

Coleman, J. M., and Smith, W. G., 1964. Late recent rise of sea level. *Geol. Soc. Amer. Bull.*, **75**, 833–840.

Curray, J. R., 1965. Late Quarternary history, continental shelves of the United States. *In*: Wright, Jr., H. E., and Frey, D. G. (eds.), *The Quaternary of the United States*. Princeton Univ. Press, Princeton, NJ, pp. 723–735.

Davies, D. K., Ethridge, F. G., and Berg, R. R., 1971. Recognition of barrier environments. *Amer. Assoc. Petrol. Geol. Bull.*, **55**, 550–565.

deBeaumont, E., 1845. Septiéme leçon; leçons de geologique. *Levéés de sables et de galet*, Vol. 1. P. Bertrand, Paris, pp. 221–253.

Demarest, J. M., Biggs, R. B., and Kraft, J. C., 1981. Time-stratigraphic aspects of a formation: interpretation of surficial Pleistocene deposits by analogy with Holocene paralic deposits, southeastern Delaware. *Geology*, **9**, 360–365.

Dickenson, K. A., Berryhill, Jr., H. L., and Holmes, C. W., 1972. Criteria for recognizing ancient barrier coast lines. *In*: Rigby, K., and Hamblin, W. K. (eds.), *Recognition of Ancient Sedimentary Environment*. Soc. Econ. Paleont. Mineral., Tulsa, OK, pp. 192–214.

Drake, C. L., Ewing, M., and Sutton, G. H., 1959. Continental margins and geosynclines—the east coast of North America north of Cape Hatteras. *In*: *Physics and Chemistry of the Earth*, Vol. 3. Pergamon Press, London, pp. 110–198.

Elliott, T., 1979. Clastic shorelines. *In*: Reading, H. G. (ed.), *Sedimentary Environments and Facies*. Elsevier, New York, pp. 143–177.

Fairbridge, R. W., 1961. Eustatic changes in sea level. *In*: *Physics and Chemistry of the Earth*, Vol. 4. Pergamon Press, New York, pp. 99–185.

Fisher, A. G., 1961. Stratigraphic record of transgressing seas in light of sedimentation on Atlantic coast of New Jersey. *Amer. Assoc. Petrol. Geol. Bull.*, **45**, 1656–1667.

Fisk, H. N., 1944. Geologic investigation of the alluvial valley of the Lower Mississippi River. *Mississippi River Comm.*, Vicksburg, MS, 78 pp.

Fisk, H. N., 1960. Recent Mississippi River sedimentation and peat accumulation. *4th Intl. Congr. Carboniferous Stratigraphy and Geology, Compt. Rend.*, pp. 187–199.

Gilbert, G. K., 1885. The topographic features of lake shores. *In*: *Fifth Annual Report of the U. S. Geological Survey*. 1883–1884. U. S. Government Printing Office, Washington, DC, pp. 69–123.

Glaeser, J. D., 1978. Global distribution of barrier islands in terms of tectonic setting. *J. Geol.*, **86**, 283–298.

Hayes, M. O., 1979. Barrier island morphology as a function of tidal and wave regime. *In*: Leatherman, S. P. (ed.), *Barrier Islands—From the Gulf of St. Lawrence to the Gulf of Mexico*. Academic Press, New York, pp. 1–27.

Hollenshead, C. T., and Pritchard, R. L., 1961. Geometry of producing Mesa Verde sandstones, San Juan basin. *In*: Peterson, J. (ed.), *Geometry of Sandstone Bodies*. Amer. Assoc. Petrol. Geol. Symposium Volume, Tulsa, OK, pp. 98–118.

Hoyt, J. H., 1967. Barrier island formation. *Geol. Soc. Amer. Bull.*, **78**, 1125–1136.

Jelgersma, S., 1961. Holocene sea-level changes in the Netherlands. *Med. Geol. Sticht., Ser. C-VI*, 101.

Jelgersma, S., 1966. Sea-level changes during the last 10,000 years. *In: Proc. Intl. Symp. on World Climates from 8000 B.C. to 0 B.C.* Roy. Meteorol. Soc., Imperial College, London, pp. 54–71.

Kraft, J. C., 1971a. *A Guide to the Geology of Delaware's Coastal Environments*. Tech. Rept. 1, College of Marine Studies, Univ. of Delaware, Newark, 220 pp.

Kraft, J. C., 1971b. Sedimentary facies patterns and geologic history of a Holocene marine transgression. *Geol. Soc. Amer. Bull.*, **82**, 2131–2158.

Kraft, J. C., 1972. *A Reconnaissance of the Geology of the Sandy Coastal Areas of Eastern Greece and the Peloponnese*. Tech. Rept. 9, College of Marine Studies, Univ. of Delaware, Newark, 158 pp.

Kraft, J. C., Biggs, R., and Halsey, S. D., 1973. Morphology and vertical sedimentary sequence models in Holocene transgressive barrier systems. *In*: Coates, D. R. (ed.), *Coastal Geomorphology*. State Univ. of New York, Binghamton, 404 pp.

Kraft, J. C., and John, C. J., 1979. Lateral and vertical facies relations of transgressive barrier. *Amer. Assoc. Petrol. Geol. Bull.*, **63**, 2145–2163.

Kraft, J. C., Sheridan, R. E., and Maisano, M., 1971. Time-stratigraphic units and petroleum entrapment models in Baltimore Canyon basin of Atlantic continental margin geosynclines. *Amer. Assoc. Petrol. Geol. Bull.*, **55**, 658–679.

Krynine, P. D., 1948. The megascopic study and field classification of sedimentary rocks. *J. Geol.*, **56**, 130–165.

LeBlanc, R. J., 1972. Geometry of sandstone reservoir bodies. *In*: Cook, T. D. (ed.), *Underground Waste Management and Environmental Implications*. Amer. Assoc. Petrol. Geol. Memoir 18, Tulsa, OK, pp. 133–190.

McFarlan, E., 1961. Radiocarbon dating of late Quaternary deposits, South Louisiana. *Geol. Soc. Amer. Bull.*, **72**, 129–158.

McGee, W. J., 1890. Encroachment of the sea. *The Forum*, **9**, 437–449.

Meade, R. H., 1969. Landward transport of bottom sediments in estuaries of the Atlantic coastal plain. *J. Sed. Petrol.*, **39**, 222–234.

Middleton, G. V., 1973. Johannes Walther's Law of the correlation of facies. *Geol. Soc. Amer. Bull.*, **84**, 979–987.

Milliman, J. D., and Emery, K. O., 1968. Sea levels during the past 35,000 years. *Science*, **162**, 1121–1123.

Minard, J. P., Perry, W., Weed, E. G. A., Rhodehamel, E. C., Robbins, E. E., and Mixon, R. B., 1974. Preliminary report on geology along Atlantic continental margin of northeastern United States. *Amer. Assoc. Petrol. Geol. Bull.*, **58**, 1169–1178.

Mörner, N. A., 1969. The Late Quaternary history of the Kattegatt Sea and the Swedish west coast. *Sveriges Geol. Under., Ser. C, NR* **640**, 487.

Reinson, G. E., 1979. Barrier island systems. *In*: Walker, R. G. (ed.), *Facies Models*. Geol. Assoc. of Canada, pp. 57–74.

Rollo, J. R., 1960. *Ground Water in Louisiana*. Louisiana Geol. Survey Bull. 2.

Russell, R. J., 1964. Techniques of eustasy studies. *Sonderdruck aus Zeitschrift fur Geomorphologie*, Vol. 8. Verlag Gebrüder Borntraeger, Berlin-Nikolasse, pp. 25–42.

Schwartz, M. L., 1971. The multiple causality of barrier islands. *J. Geol.*, **79**, 91–94.

Schwartz, M. L., 1973. *Barrier Islands*. Dowden, Hutchinson and Ross, Stroudsburg, PA, 451 pp.

Shackleton, N. J., and Opdyke, N. D., 1973. Oxygen isotope paleomagnetic stratigraphy of Equatorial Pacific core V-28-238, oxygen-isotope temperatures and ice volumes on a 10^5 year and 10^6 year scale. *Quat. Res.*, **3**, 39–55.

Shepard, F. P., 1963. 35,000 years of sea level. *In*: Clements, T. (ed.), *Essays in Marine Geology*. Univ. of Southern California Press, Los Angeles, pp. 1–10.

Sheridan, R. E., 1974. Atlantic continental margin of North America. *In*: Burk, C. A., and Drake, C. L. (eds.), *The Geology of Continental Margins*. Springer-Verlag, New York, pp. 391–407.

Ters, M., 1973. Les variations du niveau Marin depuis 10,000 ans, le long du littoral Atlantique Francais. *Le Quaternaire: Geodynamique, Stratigraphie et Environment*. Travaux Francais Recent, 9ᵉ Congrès Intl. de l'INQUA, Christchurch, New Zealand, pp. 114–135.

Van Straaten, L. M. J. U., 1965. Coastal barrier deposits in south and north Holland—in particular in the area around Scheveningen and Umuiden. *Med. Geol. Sticht., NS*, **17**, 41–75.

Visher, G. S., 1965. Use of vertical profile in environmental reconstruction. *Amer. Assoc. Petrol. Geol. Bull.*, **49**, 41–61.

Weimer, R. J., and Land, C. B., 1975. Maestrichtian Deltaic and interdeltaic sedimentation in the Rocky Mountain region of the United States. *In*: Cadwell, W. G. E. (ed.), *The Cretaceous System in the Western Interior of North America*. Geol. Assoc. of Canada, Spec. Paper 13, pp. 631–666.

Wilkinson, B. H., 1975. Matagorda Island, Texas: the evolution of a Gulf coast barrier complex. *Geol. Soc. Amer. Bull.*, **86**, 959–967.

Selected Additional References of Interest

Ager, D. J., 1973. The nature of the stratigraphic record. Halstead Press, John Wiley and Sons, New York, 114 pp.

Bloom, A. L., 1971. Glacial eustatic and isostatic controls of sea level since the last glaciation. *In*: Turekian, K. K. (ed.), *Late Cenozoic Glacial Ages*. Yale Univ. Press, New Haven, pp. 355–379.

Davis, Jr., R. A., and Ethington, R. L., 1976. *Beach and Nearshore Sedimentation*. Soc. Econ. Paleont. Mineral. Spec. Publ. 24, Tulsa, OK, 187 pp.

Kraft, J. C., 1980. *Grove Karl Gilbert and the Origin of Barrier Shorelines*. Geol. Soc. Amer. Spec. Paper 183, pp. 105–113.

Kraft, J. C., Allen, E. A., Belknap, D. F., John, C. J., and Maurmeyer, E. M., 1976. *Delaware's Changing Shoreline*. Tech. Rept. 1, Delaware Coastal Zone Management Program, Delaware State Planning Office, Dover, 219 pp.

Kraft, J. C., John, C. J., and Marx, P. R., 1981. Clastic depositional strata in a transgressive coastal environment, Holocene Epoch. *Northeastern Geol.*, **3**, 268–277.

Kumar, N., and Sanders, J. E., 1974. Inlet sequence: a vertical succession of sedimentary structures and textures created by the lateral migration of tidal inlets. *Sedimentology*, **21**, 491–532.

Leatherman, S. P., 1979. *Barrier Islands, From the Gulf of St. Lawrence to the Gulf of Mexico*. Academic Press, New York, 325 pp.

Longwell, C. R., 1949. *Sedimentary Facies in Geologic History, A Conference Meeting of the Geological Society of America*. Geol. Soc. Amer. Memoir 39, 171 pp.

Potter, P. L., 1967. Sand bodies and sedimentary environments: A review. *Amer. Assoc. Petrol. Geol. Bull.*, **51**, 337–365.

Rigby, J. D., and Hamblin, W. K., 1972. *Recognition of Ancient Sedimentary Environments*. Soc. Econ. Paleont. Mineral. Special Publ. 16, Tulsa, OK, 340 pp.

Shelton, J. W., 1973. *Models of Sand and Sandstone Deposits: A Methodology for Determining Sand Genesis and Trend*. Bull. 118, Oklahoma Geol. Survey, Norman, 122 pp.

Swift, D. J. P., 1968. Coastal erosion and transgressive stratigraphy. *J. Geol.*, **76**, 444–456.

VitaFinzi, C., 1973. *Recent Earth History*. Halsted Press, John Wiley and Sons, New York and Toronto, 138 pp.

Weimer, R. J., 1973. *Sandstone Reservoirs and Stratigraphic Concepts*. AAPG Reprint Series 7, selected papers reprinted from the Amer. Assoc. Petrol. Geol. Bull. and Memoir 18. Tulsa, OK, 190 pp.

Weimer, R. J., 1973. *Sandstone Reservoirs and Stratigraphic concepts II*. AAPG. Reprint Series 8, selected papers reprinted from the Amer. Assoc. Petrol. Geol. Bull. and *Geometry of Sandstone Bodies*. Tulsa, OK, 216 pp.

10

Modeling Coastal Environments

William T. Fox

Introduction

Several different types of computer models have been proposed to explain the development and evolution of coastal features. In most coastal studies, a profile or map forms a geometric model that provides a framework for more complex theoretical models. Geologic processes that operate within the coastal framework can be reproduced with physical models, such as wave tanks, or with various types of statistical and mathematical models. The effects that geologic processes have on the coastal environment can be studied to form a geologic response model. By combining processes and responses into a single unified model, it is possible to construct a process–response model that can be used to predict different attributes of the coastline under various conditions (Krumbein, 1961; Whitten, 1964). When a process–response model is programmed for a computer, a simulation model is developed that can be projected forward through time with the introduction of feedback loops.

In this chapter, different types of models are considered, starting with physical models and finishing with various types of computer simulation models. The conceptual model is largely inferential and used for describing patterns or relationships in the observed data. Certain obvious relationships that have been brought out by the conceptual model can be tested using a statistical model. Based on the statistical model, empirical relationships are formalized into linear equations that can be used to predict attributes of the physical environment. Empirical models based on the statistical approach work quite well in limited geographic areas but often run into difficulty when they are applied to other areas.

Mathematical models constructed from the principles of fluid mechanics are considered deterministic models. In these models, conservation of mass and conservation of momentum are used to develop equations for predicting wave motions and current velocities. Deterministic models have had varying degrees of success in predicting coastal processes. They seem to work best in conjunction with laboratory experiments where many parameters can be held constant while one parameter is varied at a time. Under field conditions, where many parameters are varying simultaneously, probabilistic models become important. Many probabilistic models are outgrowths of earlier statistical models with a pseudo-random generator used to move the model forward through time. In practice, most computer simulation models are combinations of probabilistic and deterministic models. When rigorous differential equations are available to explain processes, they are often used for parts of the model. If the mathematical theory has not been developed, or if the processes are random in nature, a probabilistic approach is used. In the final model, the deterministic and probabilistic parts are included as computer subroutines, which are controlled by a master program.

Computer simulation models can also be divided into static and dynamic models (Harbaugh and Bonham-Carter, 1970). A static model of a coastal area portrays the processes at an instant in time. In a dynamic model, in contrast, the processes move ahead through time, with feedback loops for the response of the environment. The rates at which the processes function are an important aspect that must be considered for a dynamic model. For short-term models, it is possible to estimate fairly accurate rates to plug into the model, but for longer term models, where the rates may change with time, it is often difficult to project very far into the future or back into the past.

Hydraulic Scale Models

Coastal engineers have constructed a wide variety of fixed-bed hydraulic scale models to aid in the understanding of the actions of waves, tides, and currents, and to assist in the design of structures that must survive in hostile coastal environments. Movable-bed scale models that involve sediment in the formation of sand bars, spits, and beaches are of particular interest to coastal geologists and geomorphologists. Coastal processes that have been modeled include wind waves, standing waves or seiches, seismic sea waves or tsunamis, wind setup or storm surge, and coastal currents resulting from astronomic tides. The models are used to study the effects that coastal processes have on man-made structures and the effects that the emplacement of structures have on wave and current patterns.

To model a coastal process effectively, there must be a dynamic similarity between the prototype and the scale model. Several relationships generally known as the Laws of Hydraulic Similitude are used in the solution of hydraulic problems (Sager and Seabergh, 1977). Dynamic similarity between a model and its prototype involves both geometric and kinematic similarity and Newton's Law of Motion. If the parts of the scale model have the same shape as the corresponding parts of the prototype, the two systems are considered geometrically similar. Kinematics deal with time-space relationships. A model and prototype have kinematics similarity if homologous particles are at homologous points at homologous times (Hudson *et al.*, 1979). Two systems are considered dynamically similar if there is a similarity in masses and forces.

In coastal and estuarine models, the important forces that must be considered include gravity, viscous shear, surface tension, elastic compression, and the pressure force resulting from or connected with the motion. For most coastal problems, the force of gravity predominates with viscous forces becoming important where bottom friction must be considered. Most of the remaining forces including surface tension and elastic compression are ignored in coastal models.

Dimensional analysis is used to study equations of motion in coastal models. The theory of dimensional analysis is an algebraic theory of dimensionally homogeneous functions. An equation is considered dimensionally homogeneous if the form of the equation does not depend on the units of measurement. To be physically correct any equation of motion must also be dimensionally homogeneous. Therefore, each term of the equation must contain identical powers for each of the basic dimensions including mass (M), length (L), and time (T). Dimensional analysis has been used to derive dimensionless numbers, which provide a key to scale model theory.

The two most important dimensionless numbers in scale model theory for coastal models are the Froude number, F_r, and the Reynolds number, R_e. The Froude number,

$$F_r = \frac{V}{\sqrt{gL}} \qquad (10\text{-}1)$$

where V is the velocity; g is the gravitational acceleration; and L is some characteristic length or depth, indicates the ratio of gravitational to inertial forces. The Reynolds number, R_e,

$$R_e = \frac{VL\sigma}{\mu} \qquad (10\text{-}2)$$

where V is the velocity of the flow, L is the length, σ is the density, and μ is the dynamic viscosity, is proportional to the ratio between the inertial and viscous forces. In scale models of coastal processes, gravitational forces predominate, so it is essential to have the Froude number of the model equivalent to the Froude number of the prototype. For most coastal applications, the viscous forces are relatively small and the Reynolds number is not important. However, where bottom friction is considered important, such as shallow waves moving over a long reach, the Reynolds numbers must be similar in the model and prototype.

A physical model of Masonboro Inlet, North Carolina, was constructed to investigate the interactions of tidal flow, wave action, and inlet configuration as a guide to the improvement of inlet channels and nearby shoreline protection works (Sager and Seabergh, 1977). Masonboro Inlet is a natural inlet between Wrightsville and Masonboro Beaches about 13 km southeast of Wilmington, North Carolina. Masonboro Inlet has had an interesting history of shoaling and jetty construction that makes it suitable for modeling. In June 1966, a jetty was constructed on the north side of the inlet with a low weir and deposition basin near the landward end of the jetty. A year and half after construction of the jetty, the channel migrated northward through the deposition basin and against the weir. In addition, a large sand shoal was deposited along the south side of the inlet, which presented a hazard to navigation. At that time, the Corps of Engineers decided to build a scale model of the inlet as an aid to predicting sediment movement within the inlet. Masonboro was also selected as an experimental inlet to be used for testing the usefulness and reliability of several different physical and mathematical models in predicting hydraulic characteristics of tidal inlets.

The Masonboro Inlet model was constructed at the U. S. Army Engineer Waterways Experiment Station at Vicksburg, Mississippi. The Masonboro Inlet model is a fixed-bed model with a horizontal scale of 1:300 and a vertical scale of 1:60. The model was designed and operated in accordance with Froude's model law so that the Froude numbers in the model corresponded to the Froude numbers in the prototype. The scale relations used for the model are as follows (Sager and Seabergh, 1977):

Characteristic	Scale Relations
Horizontal	$L_H = 1:300$
Vertical	$L_v = 1:60$
Slope	$L_v/L_H = 5.0$
Volume	$L_H^2 L_v = 1:5,400,000$
Velocity	$L_v^{1/2} = 1:7.746$
Discharge	$L_v^{3/2} L_H = 1:139,427$
Time—Tides	$L_H/L_v^{1/2} = 1:38.37$
Time—Wind Waves	$L_v/L_v^{1/2} = 1:7.746$

The model was 50 feet (15.3 m) wide and 150 feet (45.9 m) long, representing approximately 14.5 square miles (36.25 km²) of the prototype. The area modeled extended from the shoreward limit of the inlet's influence to the −45 foot (−13.8 m) depth contour (Figure 10-1). Because the model was of distorted scale with different scales used for the horizontal and vertical, it was necessary to add three types of artificial roughness to simulate a greater energy loss from friction than would be available with a smooth model bed. Flat metal strips were placed in the main channels to retard flow. In shallow areas, soft concrete was raked in regions where marsh grass or

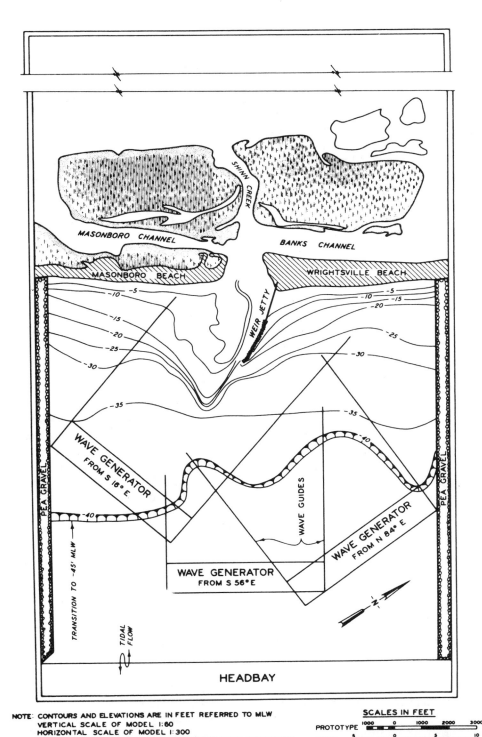

Figure 10-1. Map of model for Masonboro Inlet showing morphology, bathymetry, and location of wave generations (Sager and Seabergh, 1977).

sand shoals existed in the prototype. Stucco was placed on some shallow parts of the model to give it a rough finish.

The reproduction of tidal action was accomplished by a tide generator that maintained a differential level between pumped inflow and a gravity return flow to the model basin. Miniature tide gauges and current meters were used to monitor the tide levels and current velocities.

Three 20 foot long (6.1 m) wave generators were used in the model to reproduce short-period wave action (Figure 10-1). Both horizontal and vertical plunger-type wave generating machines were used in the model. The wave generators were set to imitate waves approaching from S16°E, S56°E, and N84°E. Waves used in the model correspond to waves with a height of about 1 m and a period of 7.4 sec in the prototype. The wave-induced surface currents varied considerably when the wave approached from the north versus the southeast.

Surface current patterns and velocities were recorded by photographs taken at hourly intervals using the time scale for the prototype (Figure 10-2). The white velocity tracks in the photograph show the direction and velocity of the current during a 4-sec time exposure. A strobe light was activated at the end of the 4 sec giving a bright spot that indicates current direction. The current pattern shown for hour 5 was taken at midflood with tidal currents moving into the inlet. Waves were approaching from the southeast with flood currents dominating over the wave-generated currents at midflood.

Based on the model tests, it was shown that the waves have a significant influence on current conditions that existed seaward of the inlet throat. Waves caused the flow along both adjacent beaches and into the sides of the throat to be strongly flood dominated. The angle of wave approach to the inlet appeared to be of relatively minor importance to the flow direction in the well-developed secondary channels. Waves from S16°E were the most significant and caused a set up in the bay and modification of the current pattern over the south shoal. Waves from both S16°E and N84°E caused a shift of the current from the main channel toward the north jetty for both ebb and flood.

Statistical Models

Statistical techniques and models have had wide application in several fields of geology and oceanography. Textbooks on the application of statistical methods to problems in geology are available for a more complete background in the field. The theoretical and mathematical development of the field is covered in depth by Miller and Kahn (1962). Statistical models and matrix techniques are introduced by Krumbein and Graybill (1965). The practical and operational aspects of data analysis as applied to the earth sciences are explained by Davis (1973). A large number of FORTRAN

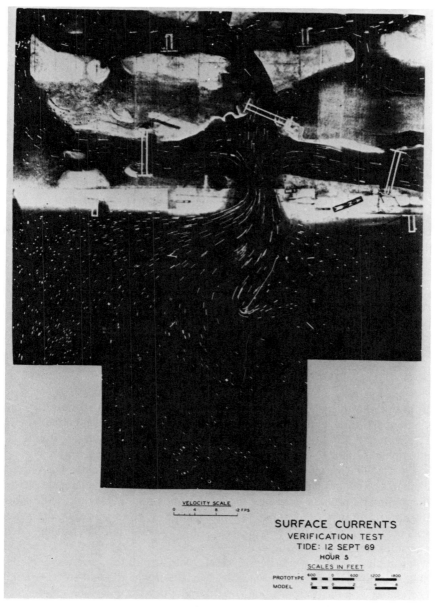

Figure 10-2. Surface currents at midflood in the Masonboro Inlet model (Sager and Seabergh, 1977).

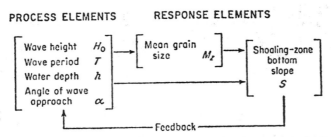

Figure 10-3. Diagrammatic representation of process–response model for shoaling-zone bottom slope (Krumbein and Graybill, 1965).

programs, from regression analysis through trend-surface analysis, are included in Davis (1973) and assume only a basic understanding of algebra. Dolan *et al.* (1977) applied eigenvector analysis to topographic profiles from the subaerial portions of barrier islands on the east and Gulf coasts.

Many of the statistical methods developed for geology as a whole can be used in the study of coastal regions. Multiple regression analysis, trend-surface analysis, and factor analysis have been applied directly to the coastal zone. A few examples that have been used in beach studies are discussed below.

Stepwise Regression Model

In many coastal problems, the objective is to predict the response of one element to other elements of the environment. In a study conducted by Harrison and Krumbein (1964) on the beach at Virginia Island, Virginia, an attempt was made to predict the beach slope based on wave energy and mean grain size. The measurements related to wave energy are included as process elements, and grain size and bottom slope as response elements in the diagrammatic process–response model (Krumbein and Graybill, 1965). For the dependent variable, nearshore bottom slope in the zone of shoaling waves was measured at low tide (Figure 10-3). At the same time, four independent variables were measured, including wave height in deep water, wave period, wave angle, and still-water depth. Mean grain size of the sand was included as a secondary independent variable, because it was thought to be strongly correlated with the bottom slope and dependent on the wave energy. Because the bottom slope has an influence on the waves, a feedback arrow is included in the process–response model (Figure 10-3).

Stepwise regression analysis provides a method for determining the contribution that each independent variable has on the prediction equation. In the beach study, the bottom slope was considered the dependent variable, Y, and the wave process variables along with the mean grain size are included as the independent variables, or X's.

Bottom slope—*S* Y
Mean grain size—*Mz* X_1
Wave period—*T* X_2
Wave height—*H*$_0$ X_3
Wave angle—*A* X_4
Water depth—*h* X_5

In stepwise regression analysis, the coefficients, β's in the regression equation, are derived from regression techniques or with multiple correlation techniques based on the general linear model (Equation 10-3):

$$Y = \beta_0 + \beta_1 X_1 + \beta_2 X_2 + \beta_3 X_3 + \beta_4 X_4 + \beta_5 X_5 + C. \quad (10\text{-}3)$$

To speed up computation on the computer, the equations are expressed in matrix terms, producing a coefficient vector. As output from the stepwise regression program, the total corrected sum of squares of Y attributable to all five X's is computed and expressed as a percentage. In the beach example, the set of five independent variables accounted for 78.7% of the total sum of squares. As shown in Table 10-1, mean grain size, X_1, taken by itself accounts for 63.1%, and wave height, X_3, and wave angle, X_4, account for 23.7% and 5.6%, respectively (Krumbein and Graybill, 1965, p. 397).

When the variables are considered two at a time, the percentage of sum of squares accounted for by the pair is greater than each individual contribution but is not equal to the sum of the individual contributions (Table 10-1). Thus, if X_1 and X_3, mean grain size and wave height, are considered together, the joint contribution is 74.1%. However, taken separately, their sum would add up to 86.7%. Therefore, if 63.1% is considered the contribution of the first independent variable, mean grain size, then an additional 11.0% is added on for the third variable, wave height. In similar fashion, when the variables are considered three at a time, the first three variables account for 75.9% with the additional 1.8% contributed by the second variable, wave period. By considering the variables four at a time and five at a time, wave angle and water depth add on 2.2% and 0.6%, respectively.

Because the linear model assumes that the variables are statistically independent, which is not the case in the beach study, this example shows how stepwise regression analysis can be used to determine meaningful relationships among a set of variables and not strictly for the production of a prediction equation. Mean grain size, which is closely correlated with bottom slope, became the most important element in the predictor equation, thereby reducing wave height to second in rank. The angle of wave approach, which should be important in wave refraction, was ranked fifth. This could be affected by position at which the wave angle was measured—deep water instead of the breaker zone.

The earlier study of beach slope by Harrison and Krumbein (1964) was expanded by Harrison (1969) using data from a 26-day time-series study. In this study, Harrison measured 17 independent variables, some at 4-hr intervals and some at high and low tide. Using a series of dimensionless

Table 10-1. Process element combinations taken one, two, three, and four at a time.[a]

X_1	X_2	X_3	X_4	X_5	Percentage of sum of squares of Y accounted for	
1					63.1	One at a time
		3			23.7	
			4		5.6	
1		3			74.1	Two at a time
1				5	66.4	
1	2				65.5	
1	2	3			75.9	Three at a time
1		3		5	74.8	
1		3	4		74.1	
1	2	3		5	78.1	Four at a time
1	2	3	4		75.9	
1	2		4	5	74.9	
1	2	3	4	5	78.7	Five at a time

Net contributions of ranked variables, %

X_1	63.1
X_2	11.0
X_3	1.8
X_4	2.2
X_5	0.6
	78.7

[a]Krumbein and Graybill (1965), p. 397.

variables, which can be expressed as ratios that have their own physical significance, Harrison developed an equation for predicting ΔQf, the quantity of sand eroded from or deposited on the foreshore in one tidal cycle. Because the data were taken at 4-hr intervals, it was possible to introduce a time lag into the predictor equations.

An expression for breaker steepness was derived from the mean breaker height, $\bar{H}_b$, and hypothetical mean wavelength at breaking, gT_b^2.

$$\text{Breaker steepness} = (\bar{H}_b/gT_b^2)^{1/2} \qquad (10\text{-}4)$$

The steepness index was originally proposed by Galvin (1968) to classify breakers on laboratory beaches. The mean wavelength was obtained by multiplying the solitary wave speed at breaking by the mean wave period.

The dimensionless ratio, h/b, where h is the hydraulic head and b is the breaker distance, would be a measure of beach saturation at different times during the tidal cycle. At low tide, the groundwater table is draining onto the foreshore and the runup traverses a saturated foreshore. The hydraulic head would be high, giving a high h/b ratio. At high water, the waves are washing across unsaturated sand and the h/b ratio would be low. According to Harrison (1969, p. 539), the interplay of breaker distance and hydraulic head on erosion and deposition of foreshore sand is complex, because the magnitude of the groundwater head may have a significant influence on the ease with which sand grains can be eroded from the foreshore surface. The breaker distance is essentially a measure of the swash energy that is available for transporting sediment up and down the slope.

The ratio $\bar{D}/\bar{z}$, where $\bar{D}$ is the mean grain size and $\bar{z}$ is the mean trough-to-bottom distance in front of a breaking wave, reflects the interplay between grain size and swash characteristics. The mean trough-to-bottom distance, $\bar{z}$, is a measure of mean swash depth, which reflects the prevailing shear stress available in the swash zone. The mean grain size, $\bar{D}$, affects bottom roughness, the size of grains in suspension transport, and the rate of groundwater flow.

The ratio of density of the liquid and sediment, ρ_L/ρ_S, provides an indication of buoyancy of the sand grains. The differences in density ratio are affected by the temperature and salinity of the water flowing out from the Chesapeake Bay at different times during the tidal cycle. The breaker angle, α_b, and beach slope, $\bar{m}$, which are already dimensionless variables, are included in the relationship. Therefore, the relationship for predicting the quantity of sand eroded or deposited on the foreshore, ΔQf, at five lag times separated by 3-hr intervals is given in Equation 10-5.

$$\Delta Qf = 5.803 - 113.151\,(\bar{H}_b/g\bar{T}_b^2)_{0.0}^{1/2} - 52.111(h/b)_{3.0}$$
$$- 7.269(\beta_b)_{6.0} + 2.724\,(\bar{D}/\bar{z})_{6.0} - 35.067\,(\bar{m})_{12.0}. \quad (10\text{-}5)$$

The subscript for each dimensionless variable indicates the time lag that has the most influence on the predictor equation.

Each term in Equation 10-5 can be interpreted in terms of its physical significance. The negative sign for breaker steepness means that as breaker steepness decreases, deposition is increased. The zero subscript indicates that breaker steepness has its greatest effect at low water. The negative sign of h/b points out that a large amount of swash water (large b) passing over an unsaturated foreshore (small h) produces erosion. The negative sign for α_b indicates that, as the breaker angle becomes larger, the longshore current would increase, resulting in foreshore erosion. With a positive correlation between ΔQf and $\bar{D}/\bar{z}$ at high water (6.0), as the size of the grains on the foreshore is increased, the percolation of swash is increased and net deposition is enhanced. For beach slope, m, an initial steep slope produces erosion, whereas an initial low slope leads to deposition.

Although statistical models are not based on the hydrodynamic laws, they are useful for studying the relationships within a set of variables. In general, the application of predictor equations developed from regression analysis is limited in geographic extent. Equations derived for one area are not usually

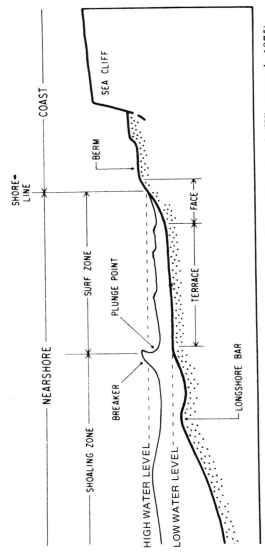

Figure 10-4. Beach profile showing characteristic features at Torrey Pines Beach (Winant *et al.*, 1975).

very successful for prediction in another area. However, if the variables are important to the process under study, the set of variables may be used in a new area to develop a regression equation with a different set of coefficients. Therefore, general statistical models are valid insofar as they point out the individual variables that are important in a general prediction equation, although the coefficients may be more restricted to a particular area or environment.

Eigenfunction Model of Beach Changes

Seasonal changes in beach profiles along the California coast have been successfully modeled using empirical eigenfunctions (Winant *et al.*, 1975; Aubrey *et al.*, 1980). Sand level changes between summer and winter beaches have been well documented since the early work of Shepard (1950). During the spring and summer, sand bars migrate toward the shore producing a steep beach face with a high berm. During the fall and winter, the berm is eroded and sediment is transported offshore forming a large bar seaward of the surf zone (Figure 10-4).

Beach and nearshore profiles were made along three lines normal to the shore at Torrey Pines Beach in San Diego County, California (Winant *et al.*, 1975). The backshore and upper foreshore elevations along each profile were measured at low tide from a bench mark seaward to wading depths. The offshore section of the profile was measured at high tide using a fathometer. Monthly surveys were made at the new moon spring tide when the maximum tidal range occurred, providing overlap between the onshore and offshore segments of the profiles. Strong winds and high waves often reduced the precision of the offshore fathometer segment of the survey. The profile program started on June 6, 1972, and extended for 2 yr.

Seasonal changes in the beach profiles were related to the onshore and offshore movement of sediment from depths of less than 10 m. During the summer months, sand bars formed in the surf zone and migrated toward the beach. Sand was added to the foreshore, building up the berm. With the onset of winter storms in November, the nearshore bars were eroded and sand was eroded from the berm. Sand was deposited on the low-tide terrace at the seaward margin of the surf zone.

Three distinctive patterns of seasonal changes in the beach, (1) construction and destruction of the subaerial berm, (2) growth and migration of nearshore bars in the surf zone, and (3) changes in the width and elevation of the low-tide terrace, can be accounted for by empirical eigenfunctions. The three eigenfunctions of the beach elevation account for a high proportion of the variation across the beach. The time variations of the three eigenfunctions also show seasonal patterns along the California coast (Winant *et al.*, 1975).

Sets of elevations along each beach profile are used to generate the empirical eigenfunctions. The elevations, h, along each profile are dependent on the distance, x, normal to the beach and the time, t, between profiles. Each data point can be represented by the symbol h_{xt} where x ranges from 1 to nx, the total number of points along the profile, and t ranges from 1 to n_t, the number of times the profiles were recorded. Then, h_{xt} can be represented in terms of a normal mode expansion of the form

$$h_{xt} = \sum_n c_{nt}\, e_{nx}. \qquad (10\text{-}6)$$

It is required that

$$\sum_x e_{nx}\, e_{mx} = \delta_{nm} \qquad (10\text{-}7)$$

where δ_{nm} is the Kronecker delta, the e_{nx} form a set of normalized modes, which are known as eigenfunctions, that are normalized to unity.

A set of empirical eigenfunctions is generated that forms the best least squares fit. A symmetric correlation matrix A is formed to generate the functions with the following elements:

$$a_{ij} = \frac{1}{n_x n_1} \sum_{t=1}^{n_t} h_{it}\, h_{jt} \qquad (10\text{-}8)$$

The diagonal elements of the correlation matrix are:

$$a_{xx} = \frac{1}{n_x n_t} \sum_{t=1}^{n_t} h_{xt}^{\,2} \qquad (10\text{-}9)$$

and represent the mean square of the data in time divided by n_x at the xt^2 point along the profiles. The sum of the diagonal elements is defined as the trace of A.

$$T_r A = \Sigma\, a_{xx} \qquad (10\text{-}10)$$

which is equal to the mean square of all the data.

The correlation matrix A contains a set of eigenvalues, λ_n, and a corresponding set of eigenfunctions, e_{nx}, which are defined by the matrix equation:

$$A e_n = \lambda_n\, e_n \qquad (10\text{-}11)$$

Therefore, the sum of all the eigenvalues is equal to the trace of the matrix, which is also equal to the mean square value of all the data. Each eigenvalue then accounts for a certain percentage of the total sum of squares for the data.

In the study conducted at Torrey Pines Beach, the first eigenfunction, which is known as the mean beach function, represents the average beach

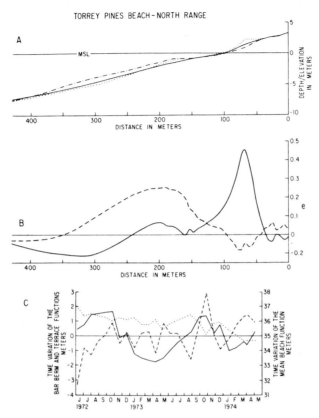

Figure 10-5. Eigenfunction of beach profile data for Torrey Pines Beach, California. (a) Solid line, mean beach function; dot-dash line, profile of April 11, 1973, and dotted line, profile of October 23, 1977. (b) Solid line, bar–berm function; dash line, terrace function. (c) Time variation of bar–berm function (solid line), terrace function (broken line), and mean beach function (dotted line) (Winant *et al.*, 1975).

profile. The beach function does not vary significantly through time and accounts for 99.75% of the total sum of squares (Figure 10-5).

The second eigenfunction, called the bar–berm function, accounts for only 0.09% of the total sum of squares, but this is equivalent to almost 40% of the variance from the mean beach function. The bar–berm function has a large maximum at the summer berm, a smaller maximum along the summer bar, and a broad minimum near the winter bar (Figure 10-5). The time dependence of the bar–berm functions shows a strong annual periodicity with positive values in the summer when sand was deposited on the summer bar and berm, and negative values during the winter when sand was eroded from the beach and stored on the offshore winter bar (Figure 10-5).

The third eigenfunction, known as the terrace function, accounts for between 30 and 45% of the variance from the mean beach function (Figure 10-5). The terrace function shows a broad high on the low-tide terrace, with a small low beneath the berm. The terrace function does not show the annual periodicity so evident on the bar–berm function. The low-tide terrace beneath the surf zone was more significantly affected by individual large storms rather than a seasonal pattern. It should also be pointed out that the bar–berm and terrace functions are linearly independent; therefore, one would not expect the annual variation to be present in both functions.

Probabilistic Model

A probabilistic model was developed to simulate the formation of Hurst Castle spit, a complex recurved spit in Hampshire, England (King and McCullagh, 1971). The different wave directions that affect the growth and development of the spit are incorporated into a series of computer subroutines. A random number generator is used to decide which subroutine is to be called and, therefore, how the spit is to grow. By varying the frequency with which different subroutines are called, it is possible to change the configuration of the spit.

The Hurst Castle spit is a shingle spit a little over 2 km long. The spit extends from the mainland coast in a southeast direction and gradually curves over to the east. At the east end of the spit, a large recurve heads back to the north (Figure 10-6). Several smaller recurves are also present on the north side of the spit.

The growth of the spit is controlled by four subroutines, WEST, STORM, SEPERP, and NOBLE. First, the predominant waves from the west (WEST) sweep down the spit, transporting sediment and elongating the spit to the southeast. As the spit extends into deeper water, the waves are refracted and the spit curves around to the east. When the spit enters deep water, the rate of growth of the spit slows down as the influence of westerly waves decreases. Second, storm waves from the west-southwest (STORM) build up the main ridge of the spit from material that has been carried along the shore by the westerly waves. Third, waves from the south or southeast (SEPERP) carry material around the tip of the spit to form the large recurve ridges. Fourth, waves coming down the Solent from the northeast (NOBLE) approach the spit from the backside. These waves tend to build up the recurve ridges at an acute angle to the main spit.

The computer model of the spit was produced by placing numbers representing different wave directions on a 50×60 matrix. Proceeding through a loop in the program, the subroutines are called one at a time in a random sequence. By controlling the distribution of random numbers assigned to each subroutine, the general shape of the spit can be determined,

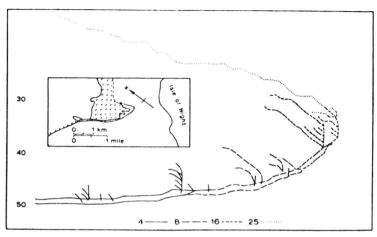

Figure 10-6. Probability simulation model of Hurst Castle spit. Stage 16 is the best fit and 25 suggests a possible future stage. The insert shows the outline of the real spit (King and McCullagh, 1971).

but the actual shape depends on the sequence of random numbers. For example, subroutine WEST could be assigned random numbers from 0 to 60; subroutine STORM from 61 to 70; subroutine SEPERP, which simulates waves from the southwest producing the vertical curves, from 70 to 80; and subroutine NOBLE, which simulates northeast waves moving down the Solent, from 81 to 100. After 50 random numbers have been called, a map is printed to show the progress of the spit. The number of maps to be printed in the sequence usually varies from 10 to 20.

Depth and refraction factors are built into subroutine WEST to influence the curvature and maximum eastward extent of the spit. For the depth factor, the water depth is specified at columns A and B where the depth becomes infinite. The refraction factor determines the curvature of the spit. When the refraction factor is low, the likelihood of jumping from one row to the next is increased, thus causing bending of the spit. The refraction numbers used in testing the model varied from 1500, which gave a highly curved spit, to 10,000, which gave a fairly straight spit.

The simulated spit plotted in Figure 10-7 shows the best fit between the observed spit and the model. The plot represents about 1000 subroutine calls, with each wave direction assigned a different number in the matrix. The 1's represent subroutine WEST and start along the lower part of the diagram in row 50. As the spit builds across the diagram, the 1's are shifted up a row at a time to form the curvature of the spit. The 2's are plotted by subroutine STORM in the row above the 1's. The storm waves build up the ridge behind the beach from the material transported down the beach by WEST. The recurves are formed by 3's, 7's, and 9's called by subroutines SEPERP and NOBLE. When subroutine SEPERP is called to initiate a recurve, a 1 is

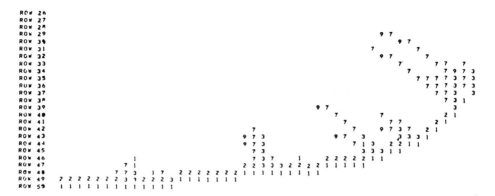

Figure 10-7. Simulated spit with different numbers for different subroutines representing waves from different directions: 1's for west; 2's for storm waves; 3's, 7's, and 9's for southwest and northeast waves resulting in recurves (King and McCullagh, 1971).

replaced by a 3. The body of the recurve is made up of 7's, which represent vertical movement, and 9's, which represent horizontal movement. It is necessary to introduce both horizontal and vertical movement so that the recurve will grow back toward the origin at an angle of 30° instead of 45°. A full listing of the program and operating instructions are given in McCullagh and King (1970).

The model was adjusted so that it closely resembled the spit at Hurst Castle by controlling the probability of calling the different subroutines. There was also a close correspondence between the frequency of winds in southeast England and the probabilities assigned to the various wave directions (King and McCullagh, 1971). The observed wind frequencies are 21% for west winds; 14% for southwest winds; of which 6% are of high enough intensity to be included as storm winds; 6.5% from the southeast; and 10% from the northeast. For the standard spit in the simulation model, frequencies of 15, 8, and 12 were assigned to subroutines STORM, SEPERP, and NOBLE, respectively. A larger set of numbers was assigned to WEST owing to the operation of both the depth and the refraction subroutines in addition to the spit elongation by subroutine WEST.

For the standard spit model in Figure 10-7, the depth factor values were set so that the depth increased linearly between columns 0 and 35 and exponentially between columns 36 and 55. The refraction factor was set up at 2300, an intermediate value that gave a reasonable curvature to the spit. The spit produced by the simulation is quite similar to the actual spit, which can be seen by comparing the inset in Figure 10-6 with the plot in Figure 10-7.

The model was tested by varying the probabilities for the different wave subroutines (Figure 10-8). By increasing the proportion of random numbers allocated to subroutine WEST, the spit is elongated and the number of recurves is reduced (Figure 10-8a). When the proportion of random numbers is decreased for WEST and increased for SEPERP, the spit produced by the simulation is shorter and has a large number of vertical recurves (Figure 10-8b). When the allocation of random numbers is increased for subroutine NOBLE, the spit produced is similar to the previous spit, but it has a greater number of lateral recurves because of the higher frequency of northeast waves (Figure 10-8c). By varying the proportions of probabilities assigned to the different subroutines, it is possible to test various hypotheses concerning the effects of waves from different directions.

The program that was developed for the Hurst Castle spit faithfully reproduced the geomorphic shape, but lacks any indication of the actual time involved in producing the spit. If the rate of formation of the spit could be

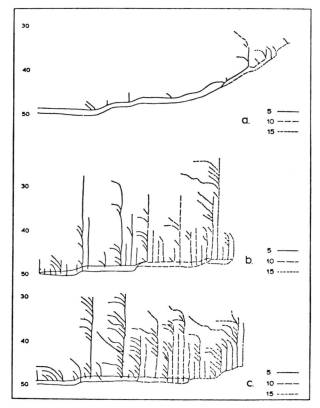

Figure 10-8. (a) Spit produced with more random numbers allocated to routine WEST. (b) Spit produced with fewer random numbers assigned to west and more to routine SEPERP. (c) Spit produced with fewer random numbers assigned to WEST and more to routine NOBLE (King and McCullagh, 1971).

determined, it might be possible to project spit development into the future. It is also possible to apply the same program to spits in other areas. The spit at Cape Lookout on the Outer Banks of North Carolina was simulated by a student as a class project at Williams College. In this simulation model, the effect of hurricane waves from the southeast was included as one of the subroutines. The broad shoal area to the south of Cape Lookout spit was also reproduced. Again, the time factor was a problem in projecting into the future.

The probabilistic spit simulation model reproduces the gross features of the spit, but it ignores the physical laws of hydrodynamics. Therefore, the probabilities of the model must be adjusted by trial and error until a close fit is achieved. Using this method, it is not possible to produce a general model that can be used for any spit based on wave height or longshore velocity.

Deterministic Model

In contrast to the probability models, Komar (1973) developed a deterministic model to simulate the effect that longshore transport has on the growth and shape of a river delta. The delta model would be considered deterministic because it does not involve an element of chance. The flow of the river, the flux of wave energy, and the longshore transport of sand are precisely determined by a series of equations based on the principles of fluid mechanics as applied to coastal hydraulics. The exact shape of the delta is predetermined once the original parameters are established for the computer run.

When a river flows into the ocean, sand is deposited to form a shoreline that takes the shape of a delta. Wave action and longshore currents tend to distribute the sand along the shore away from the mouth of the delta. If the source of supply of the sand from the river is greater than the amount of sand being transported by the longshore currents, the delta will build out into the ocean. However, if the longshore currents can transport more sand than the river supplies, the sand will be smeared out along the coast. In theory, an equilibrium is reached so that the curvature of the delta is constantly adjusted as sand is supplied by the river and redistributed by the longshore currents. The equilibrium form of a cuspate river delta was modeled by Komar (1973) using average values for sediment supplied by rivers and for wave characteristics.

In order to generalize the problem for mathematical treatment, it was necessary to make certain basic assumptions (Komar, 1973). First, it was assumed that the supply of sand from the river was constant. Second, it was assumed that the wave-energy flux was constant including a constant deep-water wave angle, height, and period. Third, it was assumed that wave refraction and diffraction can be neglected. Although these assumptions

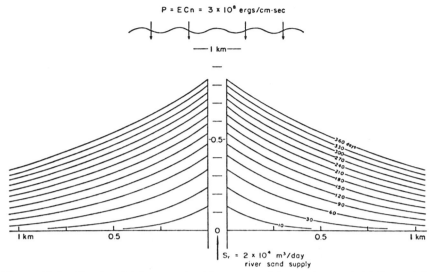

Figure 10-9. Simulated delta growth with waves arriving onshore to initially straight beach (Komar, 1973).

severely limit the application of the model to actual conditions, where rivers go through annual flood and low-water conditions; where wave angle, height, and period vary considerably during storms; and where refraction and diffraction lay an important role, they are necessary in the early stage of modeling. Once a simple model is constructed and tested, it becomes possible to introduce the complications of nature by relaxing the assumptions and adding a probabilistic component.

In the model, the shoreline configuration in the region of a river mouth was modified by waves and longshore currents transporting sediment away from the river mouth. The initial shoreline is divided into a series of cells of uniform width, ΔX, of 100 m and individual lengths, Y_i, from a baseline parallel to the shore. Changes in the shoreline are accomplished by littoral drift, S, which transfers sand from one cell to the next. Based on the continuity of sand movement, the change in volume in the ith cell is given by:

$$\Delta V_i = [(S_i - 1) - S_i]\Delta t_i \qquad (10\text{-}12)$$

where S_i is the rate of littoral transport from cell i to cell $i = 1$, and $S_i - 1$ is the littoral drift from cell i as shown in Figure 10-9 (Komar, 1973, p. 2218). The time increment is given by Δt in days.

The sediment is transported by longshore current under the action of waves arriving at an acute angle to the shoreline. The equation for longshore sand transport is based on wave-energy flux and angle of wave approach (Inman and Bagnold, 1963):

$$S = K(EC_n)_b \sin \alpha_b \cos \alpha_b \qquad (10\text{-}13)$$

where E_b is the energy of the breaking waves, Cn is the group velocity such that $(ECn)_b$ is the energy flux, and α_b is the breaker angle with the shore. The sediment transport rate, S, is related to the immersed weight sediment transport rate, 1_t, by:

$$1_t = (\rho_s - \rho)ga'S, \qquad (10\text{-}14)$$

where ρ_s and ρ are the densities of the sediment and water, respectively, and a' is correlation factor for pore space. S is also the volume discharge of sand along the beach, which is the quantity of sand used in building out the delta. If the value of the proportionality coefficient, K, is set at 0.77 and quartz sand is assumed for the beach with a density of 2.65 g/cm^3, Equations 10-13 and 10-14 can be combined to yield the transport on a volume basis:

$$S = (6.84 \times 10^{-5})(ECn)_b \sin \alpha_b \cos \alpha_b. \qquad (10\text{-}15)$$

For the first test of the model, the river supplied sediment to a straight shoreline and the wave crests moved in parallel to the coast. The river was assumed to be 200 m wide with a sediment supply of 50 tons a day per foot of stream width. For the shoreline, the cell width was set at 100 m and the nearshore water depth was constant at 4 m. A time increment of 0.1 day was used for all the tests, which were run for 360 days. The wave-energy flux was set at 3×10^8 ergs/cm/sec. The wave-energy flux used in the model corresponds to a wide range of wave height and period combinations, varying from 125 cm and 2 sec to 50 cm and 12 sec, respectively. Considering an average wave height of 63 cm for a period of 8 sec to yield the required wave-energy flux, the waves are relatively small.

The growth pattern of successive shorelines indicates a rapid initial buildup of sand followed by a steady expansion of the delta (Figure 10-9). For the first 150 to 180 days of delta growth, the shoreline angle progressively changes as the delta builds out. At that stage, the growth of the delta approaches a steady state, with increments of sediment being plastered evenly across the entire delta front. At the river mouth, the breaker angle is 38.3°, which is precisely the angle needed by the waves to transport all the sand supplied by the river. Deposition occurs equally along the entire front of the delta, so the amount of sediment in transport decreases away from the river mouth. Therefore, the breaker angle decreases from 38.3 to 7.5° at 2 km away from the river mouth.

The model was tested by varying the different input parameters. When the energy flux from the waves was varied, the shape of the delta was changed (Figure 10-10). When the wave-energy flux is increased to 5×10^8 ergs/cm/sec, or the supply of sand from the river is decreased, the equilibrium shape of the delta is flattened out. With a decrease in wave energy flux from 3×10^8 to 2×10^8 ergs/cm/sec, the delta at the river mouth built out 7.8 km from the initial straight shoreline. This strongly resembles a single distributary

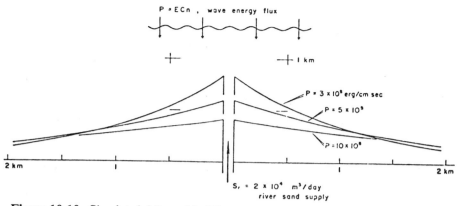

Figure 10-10. Simulated deltas with different wave-energy fluxes and constant river sediment supply (Komar, 1973).

channel in a bird's-foot delta formed under low wave conditions or high sediment discharge from the river.

Other tests were made by changing the nearshore bottom slope and angle of wave approach. When the bottom is changed from horizontal to a 0.5° slope seaward, the delta becomes more blunt near the river mouth and builds out at a slower rate. When the waves approach the shore at a small acute angle, the delta becomes slightly asymmetrical. The updrift side of the delta builds out more rapidly than the downdrift side because the prograding delta acts like a jetty and obstructs the downcoast movement of the sand. Also, the breaker angles are greater on the downdrift side, so more sediment is carried away from the lee side of the delta.

The deterministic model of delta growth under the influence of waves is useful for getting a general feeling of the processes involved. It closely resembles a laboratory experiment in which several of the variables are held constant. With the wide variety of river discharge conditions and wave energy fluctuations occurring in a normal year, holding all parameters constant for 360 days places a considerable constraint on the application of the model.

The river discharge and delta formation portion of the model could also be improved by incorporating some of the concepts from the delta model developed by Bonham-Carter and Sutherland (1968). This follows the suggestion of Bates (1953) that sediment at the mouth of a river entering a marine basin is similar to a horizontal two-dimensional plane jet. The river water is buoyed up by the salt water because of the difference in density, and vertical mixing is inhibited. Therefore, a good proportion of the sediment is carried well offshore before it is deposited. This could strongly affect the initial shape and sediment distribution pattern in the delta proposed by Komar (1973). The delta would more strongly resemble a bird's-foot delta than a cuspate delta. It would be interesting to combine the delta model proposed by Bonham-Carter and Sutherland (1968) with the model presented by Komar (1973) to stimulate the influence that waves would have on a marine delta. Wave hydrodynamics in the surf zone, which are generally represented by one wave height, one wave period, and one wave direction, have been broadened to include the statistical randomness of the sea state in a model by Collins (1976). Birkemeier and Dalrymple (1976) have developed numerical models for the prediction of wave setup and circulation in the surf zone. Wave refraction programs have been used by Goldsmith (1976) to formulate a wave climate model for the coast of Virginia. The transport of sand on beaches and the evolution of beaches has been modeled by Komar (1977). A deterministic model based on a map of barometric pressure was developed by Fox and Davis (1976a). The model provides wind speed and direction, wave height and period, and longshore current velocity as a storm passes over a coastal site.

By combining several of the above deterministic models, it would be possible to evolve a comprehensive coastal modeling system. For example, the coastal storm model (Fox and Davis, 1976a) could provide waves that would be refracted by the wave climate model (Goldsmith, 1976). As the waves approach the coast, the sea state model (Collins, 1976) and the wave setup model (Birkemeier and Dalrymple, 1976) would generate surf zone conditions that would, in turn, modify the delta constructed in the models by Komar (1973, 1977).

Fourier Model

A computer simulation model based on Fourier analysis has been developed to study relations between wind, waves, and coastal erosion on the southeastern shore of Lake Michigan (Fox and Davis, 1973). Fourier analysis was used to smooth time-series curves of the coastal processes. The daily response of the beach and nearshore bars was recorded by closely spaced profiles across the study area. These surveys were incorporated into a series of topographic maps and maps of erosion and deposition (Fox and Davis, 1976b).

Using barometric pressure, a conceptual process–response model has been constructed that shows the effects of waves and longshore currents on the morphology of the beach and nearshore environment (Davis and Fox, 1972). From the conceptual model, a mathematical model was formulated for computer simulation. In the simulation model, wave height and longshore current velocity are derived from barometric pressure. Wave and longshore current energy are used in mathematical functions to control beach and nearshore morphology through time. As the model progresses forward through time, erosion and deposition take place on the beach and the bars change position along the shore.

A diagrammatic model showing relationships among barometric pressure, wind velocity, longshore current, and breaker height is plotted in Figure 10-11. The time interval plotted across the diagram extends from 2 weeks to 2 months, depending on the frequency of storms. The curve for barometric pressure passes through two minima as frontal systems or storms move across the area. The plot of wind velocity reaches a maximum shortly after the minimum in barometric pressure. When a front passes over the shore, the winds and waves shift from southwest to northwest and longshore current reverses direction from north to south. As barometric pressure falls, breaker height increases and reaches a peak a few hours after the low has passed.

During storms, the bars oscillate back and forth, changing places with rip channels (Figure 10-12). The numbers across the bottom of Figure 10-11 correspond to maps in the sequence on Figure 10-12. In the first map (Figure 10-12-1), low-energy conditions prevail with small waves generated by southwesterly winds and weak longshore currents to the north (left). Two

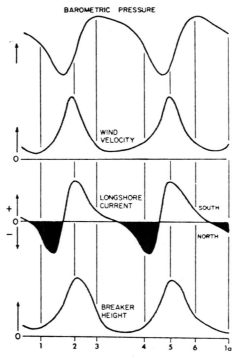

Figure 10-11. Idealized curves for barometric pressure, wind velocity, longshore current, and breaker height versus time during two storm cycles (Fox and Davis, 1973).

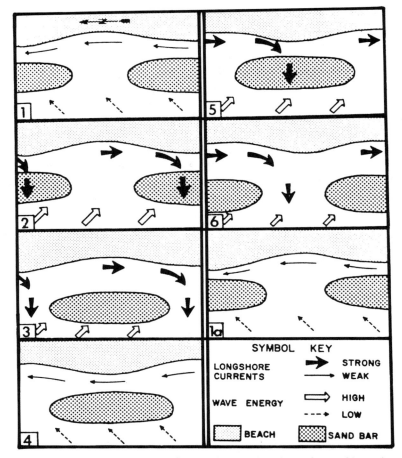

Figure 10-12. Idealized sequence of maps showing beach erosion and bar migration during two storm cycles. Map numbers relate to storm cycles shown in Figure 10-11 (Davis and Fox, 1972).

sand bars are present at either end of the study area, with a saddle that functions as a rip channel between them. The shoreline is sinuous with protuberances built out behind the bars and an embayment landward of the rip channel.

The next step in the model takes place as a storm passes over the area. Large waves approach the coast from the northwest and generate southward-flowing longshore currents that commonly reach velocities in excess of 60 cm/sec, with a maximum of 113 cm/sec. As the currents flow southward, their path is deflected seaward by the beach protuberance (cusp), thus forming a strong rip current (Figure 10-12-2). The net result of the storm is the erosion of a new rip channel through the old bar and deposition of a new bar in the position occupied by the old rip channel. The new bar and rip channel are shown at the end of the storm cycle in Figure 10-12-3. The shoreline configuration remains the same during the high-energy period, but the embayment is now located behind the bar and the protuberances are now adjacent to the rip channels.

The low-energy conditions that follow the storm show a return to low, southwest waves and slow northward-flowing longshore currents (Figure 10-12-4). The bar advances slowly shoreward as small waves transport sediment over the bar crest and down the prograding slipface. The small waves also pass through the rip channels or saddles in the bars and break directly on the beach. Erosion therefore takes place on the old protuberances at the head of the rip channels, with longshore transport and deposition in the protected areas behind the bars. Littoral drift caused by the wave swash supplements this activity by carrying the eroded sediment along the shore and depositing it behind the bar.

The return to high-energy conditions brought on by the second storm cycle (Figure 10-11) generates the same processes (Figure 10-12-5) as shown in the previous cycle (Figure 10-12-2). The beach protuberance deflects strong longshore currents to form rip currents that bisect the sand bar. The result is the formation of a new rip channel with a bar on either side (Figure 10-12-6). As barometric pressure rises and low-energy conditions return (Figure 10-11), the bars advance shoreward and the protuberance once again builds out behind the bar (Figure 10-12-1a). The last configuration in the sequence is the same as the original map (Figure 10-12-1).

The model has a definite cyclicity with three steps to each cycle. Two storm cycles are required to return the beach to its approximate original configuration. The bar form itself is not progressing in longshore direction but is being eroded and reformed as rip currents are generated and deflected by longshore currents. Because of the stronger southward component of the longshore current the net sediment transport is to the south, but the bar forms are oscillating back and forth as they are cut by rip channels and redeposited.

In the computer simulation model, wave and longshore current energies, computed from curves for breaker height and longshore current velocities, are used to modify the beach and bar topography. Several intermediate steps are necessary to accomplish the final simulation.

The first stage in the simulation model is to produce the curves for longshore current velocity and breaker height from the curve for barometric pressure. The curve for barometric pressure during July 1970 at Holland, Michigan, is plotted across the top of Figure 10-13. The observed curve for longshore current velocity closely resembles the first derivative of barometric pressure according to Equation 10-16;

$$V = \frac{dp}{dt} \qquad (10\text{-}16)$$

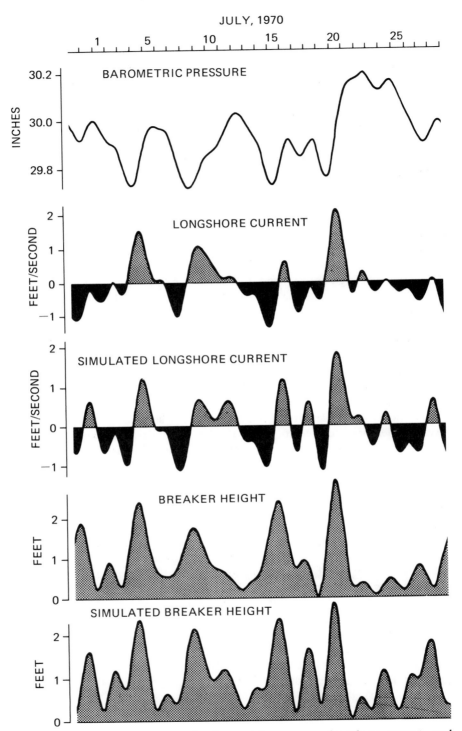

Figure 10-13. Fourier curves for barometric pressure, longshore current, and breaker height. Simulated curves for longshore current and breaker height based on first derivative and filtered version of second derivative of barometric pressure (Fox and Davis, 1973).

where V is longshore current velocity, p is the barometric pressure, and t is time.

The simulated curve for longshore current velocity based on the derivative of barometric pressure accounts for 61.2% of the total sum of squares (Figure 10-13). The change from negative to positive longshore current or from south to north takes place at the same time in the observed and simulated curves.

When a low-pressure sytem approaches, the barometric pressure is dropping and the curve has a negative slope. The steepness of slope is a function of the intensity of the low-pressure system and its rate of movement. The dropping barometric pressure is accompanied by a northward or negative longshore current. Following the passage of the low, the pressure rises and the wind and waves shift over to the northwest, generating a southward or positive longshore current. Therefore, a negative longshore current accompanies falling barometric pressure and a positive longshore current occurs with a rising pressure.

Fourier analysis is based on the summation of a series of sine and cosine curves from on a least-squares fit of the observed data. The barometric pressure and longshore current velocity curves in Figure 10-13 are the summation of the first 15 Fourier harmonics, with periods ranging from 2 to 30 days. The Fourier curves can be differentiated with ease because the derivative of a sine curve is its cosine. Therefore, to take the derivative, it is only necessary to shift the phase of the barometric pressure components by 90° (Fox and Davis, 1973).

The curve for breaker height, which is plotted as the fourth curve from the top in Figure 10-13, closely resembles the second derivative of barometric pressure as given in Equation 10-17;

$$H = \frac{d^2p}{dt^2} \tag{10-17}$$

where H is breaker height, p is barometric pressure, and t is time.

The second derivative can be obtained by taking phase shift of 180° for each harmonic times the square of the harmonic number. The observed and simulated curves for breaker height are plotted in Figure 10-13. There is a close correspondence between the major peaks in the observed and simulated curves. A better fit can be obtained by performing a filtering operation and substituting the harmonic number in place of the square of the harmonic number.

The next stage in the modeling is to reproduce the observed maps as mathematical surfaces that can be changed through time. First, the individual profiles are broken down into five component parts; beach, foreshore, plunge zone, trough, and bar. A simulated profile can be constructed by using a linear plus quadratic curve to represent the barless topography, normal curves to portray the bar, and inverted normal curves for the trough. Once an individual profile has been reconstructed, a sine curve can be used for

introducing longshore variation. For each of the profile parameters, the
mean, phase, amplitude, and wavelength of the longshore variation are given
as input parameters. The sine curve reflects the longshore rhythmic pattern of
the shoreline and nearshore bars. By varying the mean, phase, amplitude,
and wavelength for each of the nearshore parameters, it is possible to
duplicate a wide variety of map patterns.

The key to the linkage between the wave properties and map simulation
lies in the analysis of energy distribution during storm cycles and poststorm
recoveries. By integrating the area under the breaker height curve, it is
possible to compute the total wave energy per linear unit of beach during a
storm cycle. Similarly, the total longshore current energy to the north and to
the south can be obtained by integrating the areas above and below the zero
current line.

The wave and longshore current energies are used to generate map
parameters for the beach and nearshore area (Figure 10-14). Empirical
equations are used to relate total energy to bar distance, bar height, and
shoreline sinuosity. For example, the wavelength for the longshore harmonic,
L, is computed by using Equation 10-18, where Ew is the total wave
energy:

$$L = 100 \sqrt{Ew}. \qquad (10\text{-}18)$$

The bar distance, Xb, which is a function of wave energy and bottom slope,
S, is computed according to Equation 10-19:

$$Xb = \sqrt{\frac{3\,Ew}{S}} \qquad (10\text{-}19)$$

The remaining parameters, including the distance, depth, and width of the
trough and the height and width of the bar, are computed as functions of bar

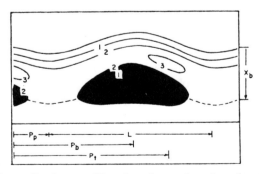

Figure 10-14. Generalized map of beach and nearshore bar showing bar distance,
X_b, plunge phase, P_p, bar phase, P_b, trough phase, P_t, and wave length, L (Fox and
Davis, 1973).

distance. The longshore position of the bar is simulated by varying the phase of the bar, *Pb*, and the phase of the trough, *Pt* (Figure 10-14), as storms move through.

The maps generated for each storm cycle and poststorm recovery closely resemble the maps that were surveyed in the field. The bars appear in the correct positions and migrate according to the conceptual model. The model apparently works quite well for the eastern shore of Lake Michigan but has a few potential weaknesses. The straightforward relationships between barometric pressure, longshore current velocity, and breaker height works only for an area where the low-pressure cells move directly onshore. The model would have to be modified for other shoreline orientations. The model also makes no provision for large swells produced by distant storms, for tides, or for storm surges, which would affect the position of the shoreline.

Coastal Storm Model

A deterministic model has been developed to hindcast or forecast changes in weather and wave conditions during the passage of a coastal storm. The "Coastal Storm Model" computes hourly predictions for barometric pressure; wind speed and direction; wave period, height, and angle of approach; and longshore current velocity at a location along a coast (Fox and Davis, 1976 a,b, 1979).

A concentric series of isobars is generated around a low pressure center to simulate the storm. The size and shape of the storm are controlled by varying the spacing of the isobars and the orientation of the major and minor axes of the storm ellipse. The drop in barometric pressure as the storm passes is portrayed by an inverted normal curve. The isobars are plotted by rotating the inverted normal curve about the center of the low pressure cell. The height of the normal curve represents the range in barometric pressure, and the width of the curve affects the gradient in pressure. By specifying the height of the normal curve and its standard deviation, along with the lengths of the major and minor axes of the storm ellipse, it is possible to simulate a wide variety of storm types.

The wind speed and direction at a coastal site can be derived from the quasi-geostrophic wind that is influenced by surface frictional effects. The quasi-geostrophic wind speed, V_g, can be calculated as follows:

$$V_g = (S/2\Omega \sin \phi) / (\Delta P/\Delta 1_h), \qquad (10\text{-}20)$$

where S is the specific volume (779 cm^3/g), Ω is the angular velocity (7.29 $\times$ 10^{-5} rad/sec), ϕ is the latitude in degrees, and ΔP is the change in barometric pressure in millibars over a horizontal distance, $\Delta 1_h$, in kilometers (Godske *et al.*, 1957, p. 370). The pressure gradient is computed along a normal curve at right angles to the isobar that passes through the shore location.

Close to the surface of the earth, the quasi-geostrophic wind speed is reduced by surface friction, and the wind direction is deflected toward the center of the low pressure system. The angle, α, between the surface wind and the quasi-geostrophic wind can be determined for a given latitude, ϕ, with the equation:

$$\cot \alpha = \tan \beta + (2 \, \Omega \, \sin \phi \, / b \, \cos \beta), \tag{10-21}$$

where b and β are empirical correction factors for wind speed and direction (Baur and Phillips, 1938). The mean values for b are $1.9 \times 10^{-4}/\text{sec}$ over land and $0.65 \times 10^{-4}/\text{sec}$ over sea. The mean angles for β are $29°$ over sea and $50°$ over land.

The ratio between the surface wind speed V_h and the quasi-geostrophic wind speed V_g are computed as follows:

$$\frac{V_h}{V_g} = 2 \, \Omega \, \sin \phi \, \sin \alpha / b \, \cos \beta. \tag{10-22}$$

The significant wave height and period are derived from the surface wind speed, wind duration, and effective fetch. The SBM method (Sverdrup, Munk, and Bretschneider) as revised by Bretschneider (1958) and plotted by the U.S. Army Coastal Engineering Research Center (1973) was used to compute wave height and period. First, it is necessary to compute the effective fetch, F_e, from the wind speed, V_h, and duration, D.

$$F_e = 10^{0.3} \, D^{1.25} \, (V_h/10)^{0.72}. \tag{10-23}$$

The effective fetch and wind speed are used to calculate the significant wave height H and period T.

$$H = 0.283 \, V_h^2 \, \tanh \, [0.125(g \, F_e/V_h^2)^{0.42}]/g \tag{10-24}$$

$$T = 3.77 \, V_h \, [0.077(g \, F_e/V_h^2)^{0.25}] \tag{10-25}$$

The breaker height H_b in the surf zone is computed according to Airy wave theory and assuming conservation of energy flux (Komar and Gaughan, 1973):

$$H_b = 0.73 + 0.388 \, g^{1/5} \, (TH^2)^{2/3} \tag{10-26}$$

When the wave enters shallow water, the angle, A_b, between the wave crest and the shoreline is determined by Snell's law:

$$\sin A_b = \sin A_w \, \tanh \left(\frac{2\pi d}{L} \right) \tag{10-27}$$

where A_w is the wave angle in deep water, d is the depth of water, and L is the wave length. The final term in the equation, $\tanh \, (2\pi d/L)$, gives the change in phase velocity as the wave moves from deep to shallow water.

The wave height in shallow water, H_r, is a function of the significant wave height, H, and the relative wave angles in deep and shallow water:

$$H_r = H \, (\cos A_w / \cos A_b)^{1/2}. \qquad (10\text{-}28)$$

The wave-generated longshore current flows parallel to the beach between the nearshore bar and the shore. The nearshore topography may vary considerably depending on the presence of nearshore bars and rip channels. As a simplifying assumption, the average longshore current velocity is computed as it flows over a planar nearshore bottom inclined slightly in a seaward direction.

Four different equations are available as options in the program for computing the velocity, V_e, of the longshore current. The first equation based on radiation stress theory was proposed by Longuet-Higgins (1970):

$$V_e = 9.0 \, m \, (g \, H_b)^{1/2} \sin A_b, \qquad (10\text{-}29)$$

where m is the gradient of the nearshore bottom slope. The second equation based on field and laboratory data was proposed by the U.S. Army Coastal Research Center (CERC)(1973):

$$V_e = 20.7 \, m \, (g \, H_b)^{1/2} \sin 2 \, A_b. \qquad (10\text{-}30)$$

When the researchers at CERC tested the Longuet-Higgins equation they found that the observed data were about 2.3 times greater than the predicted values. They also obtained a stronger correlation when $\sin 2A$ was substituted for $\sin A$. The third equation was proposed by Komar and Inman (1970), who considered longshore current velocity independent of beach slope:

$$V_e = 2.7 \, u_m \, \sin A_b \, \cos A_b, \qquad (10\text{-}31)$$

where u_m is the maximum horizontal component of the orbital velocity in the surf zone.

The fourth equation for longshore current velocity was developed by Fox and Davis (1976a) using stepwise linear regression analysis:

$$V_e = 100 \, m \, (Hb/T) \sin 4 \, A_b \qquad (10\text{-}32)$$

The regression equation accounted for almost 80% of the total sum of squares for 360 observations taken at both Sheboygan, Wisconsin, and Holland, Michigan.

A circular storm is used to demonstrate how predictions are made with the coastal storm model. The size, shape, and intensity of the model storm are based on an intense storm which passed over the Oregon coast in November 1973 (Fox and Davis, 1976a). The model storm has a radius of 450 km, a barometric pressure range of 22.9 millibars, and a velocity of 40 km/hr along a track 90° east of north. Therefore, over 30 hr, the storm would travel 1200 km along a path normal to the coast.

The shoreline configuration of the coastal site in the model is defined by the latitude, onshore azimuth, nearshore slope, and minimum fetch. The latitude of the central Oregon coast is 43°N. The shore runs north-south with the land to east. The bottom topography along the coast was assumed to be planar with a seaward-dipping slope of 0.033. The coast faces the North Pacific Ocean and the storm has a diameter of 900 km. Therefore, fetch is not a limiting factor for the model.

The output from the storm model is plotted as a series of maps that show the changes in each variable as the storm passed over the coast (Figures 10-15 and 10-16). The coastal site where predictions are being made is assumed to be at the center of the diagram. The diagrams were constructed by plotting a series of storm tracks across the map. The storm tracks were

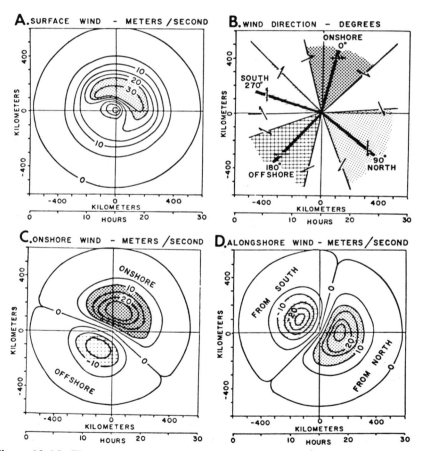

Figure 10-15. Time-distance plots of (a) and surface wind speed, (b) wind direction, (c) *onshore wind*, and (d) *alongshore wind* in a circular storm (Fox and Davis, 1979).

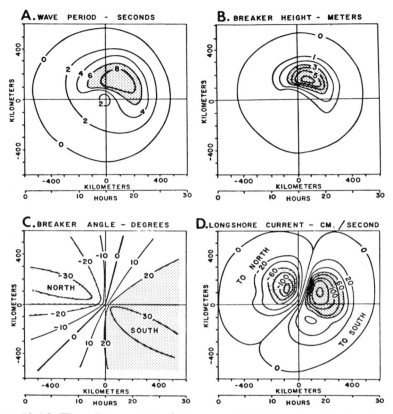

Figure 10-16. Time-distance plots of (a) wave period, (b) breaker height, (c) breaker angle, and (d) longshore current in a circular storm (Fox and Davis, 1979).

plotted at the storm location. Therefore, each map does not represent the magnitude of the variable over a given area at an instant in time, but does portray the magnitude of the variable at the coastal site when the storm center is located a certain distance and direction from the study location.

To understand the pattern of wind changes during a coastal storm, it is necessary to consider the wind speed and direction (Figure 10-15). On the contour plot of surface wind, the area of maximum wind speed occurs as a bean-shaped plot to the north of the study site. When a storm passes between 100 and 200 km north of the study site, the wind speed reaches a maximum of over 30 km/sec. When the storm center passes between 100 and 200 km south of the study site, the wind reaches a maximum between 15 and 20 m/sec. The plot of surface wind speeds is asymmetrical because of the differences in frictional effects when the wind is blowing over land and water. The contour lines indicating differences in wind direction are plotted in 30° intervals radiating outward from the center of the diagram. The arrows plotted along each contour indicate the direction that the wind would be blowing at the study site when the storm center passed over the contour (Figure 10-15b). The heavy lines indicate the major wind directions with onshore winds (0°), north winds (90°), offshore winds (180°), and south winds (270°). For example, when a storm follows a track 200 km north of the shore site, the wind direction at the study site begins from the south (274°), slowly shifts to onshore (0°), and ends up out of the northwest (52°).

The onshore and alongshore components of the wind are derived from the windspeed and direction (Figure 10-15). The onshore component is computed by taking the cosine of the wind direction times the wind speed, and the alongshore component is the sine of the wind direction times the wind speed. The line formed by the 270° and 90° contours on the wind direction plot divides the onshore and offshore components of the wind (Figure 10-15c). Storms that follow paths to the north of the study site generally have a strong onshore wind component that reaches a maximum after the storm center has passed over the coast. For the storms that pass to the south of the study site, the offshore wind component reaches a maximum before the storm reaches the coast. The line formed by the 0° and 180° contours on the wind direction plot divides the alongshore component of the wind into winds from the north and winds from the south. As the storm center approaches the coast from the west, the alongshore component of the wind is from the south. As the storm moves to the east, the wind direction shifts over to the north. If the storm passes to the north of the study site, the shift in wind direction takes place after the storm center has crossed the coast. However, if the storm passes to the south, the wind shifts before the storm center reaches the coast.

The plots for wave period and breaker height display similar patterns that closely resemble the contour plot for surface wind (Figures 10-15 and 10-16). Because it is assumed that the waves were generated by the local winds, the similarity in plots should come as no great surprise. When a storm passes 200 km north of the study site, the wave period reaches a maximum of 9.5 sec and the breaker height peaks at 5.3 m. However, when storms move onshore south of the study site the periods are generally less than 4 sec and the breaker heights are less than 2 m. The contour plots of breaker heights also closely resemble the plot for onshore winds (Figures 10-15 and 10-16).

The breaker angle is the acute angle between the wave crest and the shoreline as the wave breaks in the surf zone (Figure 10-16c). The breaker angle is influenced by the wind direction and wave refraction along the coast. Along the zero line, the wave crests are aligned parallel to the shore when the wave breaks. To the left of the zero line, the breaker angles are open to the north, and to the right of the zero line, they are open to the south (Figure 10-16c). The largest breaker angles are about 32° when they are out of the north or south parallel to the shore. In deep water, the waves would be normal to the coast, but in shallow water, the breaker angle is decreased to

32° by wave refraction. When a storm moves across the coast, the angle of wave approach shifts with the corresponding change in wind direction.

The plot for longshore current has the same general configuration as the plot for the alongshore component of the wind (Figures 10-15d and 10-16d); however, the longshore current velocities from the north are considerably stronger than those from the south. When the storm follows a path 200 km north of the study site, the longshore current velocity reaches 97.4 cm/sec to the north as the storm approaches the coast. After the storm center has passed over the coast, the current reverses direction and reaches 81.6 cm/sec to the south. When the storm comes onshore to the north of the study site, there is a lag of several hours between the times of minimum barometric pressure and current reversal. This phenomena was noted on several storms on the eastern shore of Lake Michigan where the reversal in current direction took place up to 6 hr after minimum in barometric pressure.

The comparison between the observed and hindcast values for several different variables has shown that the model generates reasonable estimates for wind speed, longshore current velocity, breaker height, and wave period during a coastal storm. The model was tested using weather and wave data from several locations including the Magdalen Islands in the Gulf of Saint Lawrence; Plum Island, Massachusetts; Cedar Island, Virginia; Sapelo Island, Georgia; Monterey, California; and Newport, Oregon. The coastal sites have different orientations relative to the paths of the storms, and they varied in the reliability and availability of observed data. In general, there was a close correspondence between observed and hindcast data when a storm moves onshore or offshore, but a poorer fit is obtained when a storm moves parallel to the coast.

Summary

In summary, four different types of models—physical, statistical, probabilistic, and deterministic—are considered in this chapter. Each model is successful in representing a different aspect of the coastal environment, but each model also has its limitations.

A hydraulic scale model of Masonboro inlet, North Carolina, was constructed by the U.S. Army Corps of Engineers to study the effects of waves and tidal currents on the shoaling of the inlet (Sager and Seabergh, 1977). The scale model was designed with a horizontal scale of 1:300 and a vertical scale of 1:60 so that the Froude numbers of the model would correspond to the Froude numbers observed in the inlet. Waves were generated by three 20-foot (6.1 m) plungers, and tides were modeled with a recirculating pump. Overhead photographs and micro-current meters were employed to measure the currents in the model. The model was successful in predicting areas of shoaling under different wave and tide conditions.

   In the statistical model proposed by Harrison and Krumbein (1964), stepwise regression analysis is employed to predict beach slope based on mean grain size, wave period, wave height, wave angle, and water depths. A linear regression equation is produced that has a set of coefficients for the independent variables. A similar regression equation was developed to predict the quantity of sand eroded or deposited on the foreshore (Harrison, 1969). Although the statistical models can be used for making predictions in a local area, they cannot be applied to other coastal regions. However, if the predictions are important for a particular area, it would be possible to undertake a short-term study to generate the necessary coefficients, then to use the regression equation for making long-term predictions.

   The eigenfunction model of beach changes is a statistical model to account for changes in beach profiles through time. The eigenfunction model was used by Winant et al. (1975) to study three profile lines at Torrey Pines Beach in California. The first eigenfunction, known as the mean-beach function, represents the mean beach profile. The second eigenfunction, called the bar–berm function, accounts for the summer bar and berm. The third function, the terrace function, is produced by the low-tide terrace beneath the surf zone. The eigenfunction model is a useful method for considering the different factors responsible for changes in beach profiles.

   King and McCullagh (1971) propose a probabilistic model to simulate the growth of the Hurst Castle spit. Four subroutines are assigned different random numbers to simulate waves approaching from the west, waves from the southwest, waves caused by storms, and waves moving down the channel between the Isle or Wight and the coast of England. The model reproduces the shape of the spit and the stages of its development, but it does not have a valid time frame. Although the model reproduces the broad features of a spit, it ignores the basic physical laws of hydrodynamics. Therefore, it has limited application as a general model for the prediction of spit growth.

   A deterministic model based on the principles of fluid mechanics was proposed by Komar (1973) to predict the growth of a delta under the influence of waves. The changes in shoreline configuration are based on littoral drift, which transfers sediment from one cell to the next along the shore. The model assumes that sediment supply from the river is constant, which causes the delta to prograde at a steady rate. Wave-energy flux is also assumed to be constant for 1 yr, which would include constant deep-water wave angle, height, and period. Because wave conditions vary considerably throughout the year, the model is limited in its application to actual delta growth. The model resembles a physical model or wave tank experiment where several parameters can be held constant, as opposed to actual field conditions where nothing is constant. If a probabilistic component were added to the model, it would more closely resemble field conditions.

   The Fourier model developed by Fox and Davis (1973) provides an empirical process-response model for beach and nearshore bar deposition on Lake Michigan. Wave height and longshore current velocity are derived from barometric pressure where the storm moves directly onshore. The shoreward migration of sand bars and the development of rip channels are related to storm conditions through a conceptual model. The conceptual model forms the basis for a computer simulation model, in which a series of maps is generated for storm cycles and poststorm recovery periods. The coastal storm model is based on empirical data and, like the statistical and probabilistic models, it is limited to the area for which it has been developed.

The coastal storm model is a deterministic model developed to predict breaker height and longshore current velocity during a coastal storm (Fox and Davis, 1979). The model computes hourly predictions for barometric pressure; wave period, height, and angle of approach; and longshore current velocity. The output from the model is plotted as a series of diagrams that show the changes in each variable as the storm passes over the coast. The model has been tested at numerous locations and gave reasonable estimates for wind speed, breaker height, and longshore current velocity.

References

Aubrey, D. G., Inman, D. L., and Winant, D., 1980. The statistical prediction of beach changes in southern California. *J. Geophys. Res.*, **86**, 3264–3276.

Bates, C. C., 1953. Rational theory of delta formation. *Amer. Assoc. Petrol. Geol. Bull.*, **37**, 2119–2162.

Baur, F., and Phillips, H., 1938. Untersuchungen der Reibung bei Luftstromungen uber dem Meer. *Ann. Hydrogr. U. Mar. Meteorol.*, **66**(6),279–296 (cited in Godske, *et al.*, 1957).

Birkemeier, W. A., and Dalrymple, R. A., 1976. *Numerical Models for the Prediction of Wave Set-Up and Nearshore Circulation*. Tech. Rept. 1, ONR Contract N00014-76-C-0342, Univ. of Delaware, 127 pp.

Bonham-Carter, G. F., and Sutherland, A. J., 1968. *Mathematical Model and FORTRAN IV Program for Computer Simulation of Deltaic Sedimentation*. Computer Contrib. 24, Univ. of Kansas, Lawrence, 56 pp.

Bretschneider, C. L., 1958. *Revisions in Wave Forecasting, Deep and Shallow Water*. Proc. 6th Conf. on Coastal Engineering, Amer. Soc. Civil Engr. Council on Wave Research.

Collins, J. I., 1976. Wave modeling and hydrodynamics. *In*: Davis, R. A., and Ethington, R. L. (eds.), *Beach and Nearshore Sedimentation*. Soc. Econ. Paleont. Mineral. Spec. Publ. 24, pp. 54–68.

Davis, J. C., 1973. *Statistics and Data Analysis in Geology*. John Wiley and Sons, New York, 550 pp.

Davis, Jr., R. A., and Fox, W. T., 1972. Coastal processes and nearshore sand bars. *J. Sed. Petrol.*, **42**, 401–412.

Dolan, R., Hayden, B. P., and Felder, W., 1977. Systematic variations in nearshore bathymetry. *J. Geol.*, **85**, 129–141.

Fox, W. T., and Davis, Jr., R. A., 1973. Simulation model for storm cycles and beach erosion on Lake Michigan. *Geol. Soc. Amer. Bull.*, **84**, 1769–1790.

Fox, W. T., and Davis, Jr., R. A., 1976a. *Coastal Storm Model*. Tech. Rept. 14, ONR Contract N00014-69-C-0151, Williams College, 122 pp.

Fox, W. T., and Davis, Jr., R. A., 1976b. Weather patterns and coastal processes. *In*: Davis, R. A., and Ethington, R. L. (eds.), *Beach and Nearshore Sedimentation*. Soc. Econ. Paleont. Mineral. Spec. Publ. 24, pp. 1–23.

Fox, W. T., and Davis, Jr., R. A., 1979. Computer model of wind, waves, and longshore currents during a coastal storm. *Math. Geol.*, 11(2),143–164.

Galvin, Jr., C. J., 1968. Breaker type classification on three laboratory beaches. *J. Geophys. Res.*, **73**, 3651–3659.

Godske, C. L., Bergeron, J., Bjeiknes, J., and Bundgaard, R. C., 1957. *Dynamic Meteorology and Weather Forecasting*. Amer. Meteorol. Soc., Boston, MA, and Carnegie Institute of Washington, 800 pp.

Goldsmith, V., 1976. Wave climate models for the continental shelf: critical links between shelf hydraulics and shoreline processes. *In*: Davis, R. A., and Ethington, R. L. (eds.), *Beach and Nearshore Sedimentation*. Soc. Econ. Paleont. Mineral. Spec. Publ. 24, pp. 24–47.

Harbaugh, J., and Bonham-Carter, G. 1970. *Computer Simulation in Geology*. John Wiley and Sons, New York, 575 pp.

Harrison, W., 1969. Empirical equations for foreshore changes over a tidal cycle. *Mar. Geol.*, **7**, 529–552.

Harrison, W., and Krumbein, W. C., 1964. *Interaction of the Beach-Ocean-Atmosphere System at Virginia Beach, Virginia*. Coastal Engineering Research Center Tech. Memo. 7.

Hine, A. C., 1977. Lily Bank, Bahamas: history of an active oolite sand shoal, *J. Sed. Petrol.*, **47**, 1554–1581.

Hudson, R. Y., Herrman, Jr., F. A., Sager, R. A., Whalin, R. W., Keulegan, G. H., Chatham, Jr., C. E., and Hales, L., 1979. *Coastal Hydraulic Models*. Spec. Rep. 5, U.S. Army, Corps of Engineers, Coastal Engineering Research Center, Fort Belvoir, VA, 531 pp.

Inman, P. L., and Bagnold, R. A., 1963. Littoral processes. *In*: Hill, M. N. (ed.), *The Sea*, Vol. 3, *The Earth Beneath the Sea*. Wiley Interscience, Inc., New York, pp. 529–553.

King, C. A. M., and McCullagh, M. H., 1971. A simulation model of a complex recurved spit. *J. Geol.*, **79**, 22–36.

Komar, P. D., 1973. Computer models of delta growth due to sediment input from rivers and longshore transport. *Geol. Soc. Amer. Bull.*, **84**, 2217–2226.

Komar, P. D., 1977. Modeling of sand transport on beaches and the resulting shoreline evolution. *In*: Goldberg, E. D., McCave, J. N., O'Brien, J. J., and Steele, H. H. (eds.), *The Sea*, Vol. 6, *Marine Modeling*. Wiley Interscience, New York, pp. 499–513.

Komar, P. D., and Gaughan, M., 1973. *Airy Wave Theory and Breaker Height Prediction*. Proc. 13th Conf. on Coastal Engineering, pp. 405–418.

Komar, P. D., and Inman, D. L., 1970. Longshore sand transport on beaches: *J. Geophys. Res.*, **75**, 5915–5927.

Krumbein, W. C., 1961. *The Analysis of Observational Data from Natural Beaches*. Beach Erosion Board Tech. Memo., 130, 59 pp.

Krumbein, W. C., and Graybill, F. A., 1965. *An Introduction to Statistical Models in Geology*. McGraw-Hill, New York, 475 pp.

Longuet-Higgins, M. S., 1970. Longshore currents generated by obliquely incident sea waves. *J. Geophys. Res.*, **75**, 6778–6789 (part 1), 6790–6801 (part 2).

McCullagh, M. J., and King, C. A. M., 1970. *Spitsym: A FORTRAN IV Computer Program for Spit Simulation*. D. F. Merriam (ed.), Computer Contrib. 50, Univ. of Kansas, Lawrence, 20 pp.

Miller, R. L., and Kahn, J. S., 1962. *Statistical Analysis in the Geological Sciences*. John Wiley and Sons, New York, 483 pp.

Sager, R. A., and Seabergh, W. C., 1977. *Physical Model Simulation of the Hydraulics of Masonboro Inlet, North Carolina*. GITI Rep. 15, U.S. Army Corps of Engineers, Coastal Engineering Research Center, Fort Belvoir, VA.

Shepard, F. P., 1950. *Beach Cycles in Southern California*. Beach Erosion Board Tech. Memo. 20, 26 pp.

U.S. Army Coastal Engineering Research Center, 1973. *Shore Protection Manual, TR4*. U.S. Army Coastal Engineering Research Center, Fort Belvoir, VA, 3 Vol.

Whitten, E. H. T., 1964. Process-response models in geology. *Geol. Soc. Amer. Bull.*, **75**, 455–463.

Winant, C. D., Inman, D. L., and Nordstorm, C. E., 1975. Description of seasonal beach changes using empirical eigenfunctions. *J. Geophys. Res.*, **80**, 1979–1986.

Index